Norwegian Petroleum Society (NPF), Special Publication No. 12

Onshore–Offshore Relationships on the North Atlantic Margin

Papers presented at the 'Onshore–Offshore Relationships on the North Atlantic Margin' Conference, 7–9 October 2002, Trondheim, Norway

Further titles in the series:

1. R.M. Larsen, H. Brekke, B.T. Larsen and E. Talleraas (Editors)
   STRUCTURAL AND TECTONIC MODELLING AND ITS APPLICATION TO PETROLEUM GEOLOGY – Proceedings of Norwegian Petroleum Society Workshop, 18–20 October 1989, Stavanger, Norway

2. T.O. Vorren, E. Bergsager, Ø.A. Dahl-Stamnes, E. Holter, B. Johansen, E. Lie and T.B. Lund (Editors)
   ARCTIC GEOLOGY AND PETROLEUM POTENTIAL – Proceedings of the Norwegian Petroleum Society Conference, 15–17 August 1990, Tromsø, Norway

3. A.G. Doré et al. (Editors)
   BASIN MODELLING: ADVANCES AND APPLICATIONS – Proceedings of the Norwegian Petroleum Society Conferernce, 13–15 March 1991, Stavanger, Norway

4. S. Hanslien (Editor)
   PETROLEUM: EXPLORATION AND EXPLOITATION IN NORWAY – Proceedings of the Norwegian Petroleum Society Conference, 9–11 December 1991, Stavanger, Norway

5. R.J. Steel, V.L. Felt, E.P. Johannesson and C. Mathieu (Editors)
   SEQUENCE STRATIGRAPHY ON THE NORTHWEST EUROPEAN MARGIN – Proceedings of the Norwegian Petroleum Society Conference, 1–3 February, 1993, Stavanger, Norway

6. A.G. Doré and R. Sinding-Larsen (Editors)
   QUANTIFICATION AND PREDICTION OF HYDROCARBON RESOURCES – Proceedings of the Norwegian Petroleum Society Conference, 6–8 December 1993, Stavanger, Norway

7. P. Møller-Pedersen and A.G. Koestler (Editors)
   HYDROCARBON SEALS – Importance for Exploration and Production

8. F.M. Gradstein, K.O. Sandvik and N.J. Milton (Editors)
   SEQUENCE STRATIGRAPHY – Concepts and Applications – Proceedings of the Norwegian Petroleum Society Conference, 6–8 September 1995, Stavanger, Norway

9. K. Ofstad, J.E. Kittilsen and P. Alexander-Marrack (Editors)
   IMPROVING THE EXPLORATION PROCESS BY LEARNING FROM THE PAST – Proceedings of the Norwegian Petroleum Society Conference, September 1998, Haugesund, Norway

10. O.J. Martinsen and T. Dreyer (Editors)
    SEDIMENTARY ENVIRONMENTS OFFSHORE NORWAY — PALAEOZOIC TO RECENT – Proceedings of the Norwegian Petroleum Society Conference, 3–5 May 1999, Bergen, Norway

11. A.G. Koestler and R. Hunsdale
    HYDROCARBON SEAL QUANTIFICATION – Papers presented at the Norwegian Petroleum Society Conference, 16–18 October, 2000, Stavanger, Norway

Norwegian Petroleum Society (NPF), Special Publication No. 12

Onshore–Offshore Relationships on the North Atlantic Margin

Proceedings of the Norwegian Petroleum Society Conference, October 2002, Trondheim, Norway

Edited by

Bjørn T.G. Wandås
Eni Norge, P.O. Box 101, Forus, 4064 Stavanger, Norway

Johan Petter Nystuen
Department of Geology, University of Oslo, P.O. Box 1047, Blindern, 0316 Oslo, Norway

Elizabeth Eide
Geological Survey of Norway, 7491 Trondheim, Norway

and

Felix Gradstein
Natural History Museum, P.O. Box 1172 Blindern, 0318 Oslo, Norway

2005
ELSEVIER

Amsterdam – Boston – Heidelberg – London – New York – Oxford – Paris – San Diego – San Francisco – Singapore – Sydney – Tokyo

ELSEVIER B.V.
Sara Burgerhartstraat 25
P.O. Box 211, 1000 AE
Amsterdam, The Netherlands

ELSEVIER Inc.
525 B Street, Suite 1900
San Diego, CA 92101-4495
USA

ELSEVIER Ltd.
The Boulevard,
Langford Lane Kidlington,
Oxford OX5 1GB UK

ELSEVIER Ltd.
84 Theobalds Road
London WC1X 8RR
UK

First edition 2005

Library of Congress Cataloging in Publication Data
A catalog record is available from the Library of Congress.

British Library Cataloguing in Publication Data
A catalogue record is available from the British Library.

ISBN: 0-444-51849-5
ISSN: 0928-8937 (Series)

♾ The paper used in this publication meets the requirements of ANSI/NISO Z39.48-1992 (Permanence of Paper).
Printed in The Netherlands.

Preface

The Conference "Onshore-Offshore Relationships on the North Atlantic Margin" was held in Trondheim 7–9th October 2002. This was a joint arrangement between the Norwegian Geological Society (NGF) and the Norwegian Petroleum Society (NPF). The aim of the conference was to create a forum for the petroleum industry and academic research to meet, present and discuss regional research from the Atlantic margin.

A paperback volume entitled "Onshore–Offshore Relationships on the Nordic Margin" was printed with 69 extended abstracts in the Hurst, A., 2002. Abstracts and Proceedings of the Norwegian Geological Society, 2: 211 pp.

The conference was divided into three different subtopics. Each subtopic had a separate day of lectures while it was possible to visit and discuss poster presentations with their authors during intermissions all three days.

The first four papers presented herein deal with "Basement control on offshore structuring" and address topics ranging from vertical movement of basement blocks and the processes responsible, to the deep structuring in the Norwegian Sea and the development of the Jan Mayen microcontinent.

The next seven papers are thematically related to "Linking uplift and erosion with subsidence and deposition" in the northern North Sea and Norwegian Sea. These papers touch upon topics related to basin infill histories from Triassic to Miocene.

The last four papers deal with "New challenges" to petroleum exploration on the Norwegian margin, and present analyses of submarine slides and the occurrences of gas hydrates and cold-water reefs.

Acknowledgements

The editors would like to thank the numerous referees who have performed the time consuming tasks of reviewing and offering constructive suggestions to the authors in order to improve the manuscripts for final publication. The editors also express their sincere thanks to the sponsors of this publication Norske Shell, Norsk Hydro, Statoil, Eni Norge and ExxonMobil.

Bjørn T.G. Wandås
Johan Petter Nystuen
Elizabeth Eide
Felix Gradstein

Stavanger, Sept. 2004

List of Contributors

T.B. ANDERSEN	*Department of Geology, University of Oslo, P.O. Box 1047, Blindern 0316 Oslo, Norway*
A. ANDRESEN	*Department of Geology, University of Oslo, P.O. Box 1047, Blindern 0316 Oslo, Norway*
K. BERG	*Norsk Hydro ASA, 0246 Oslo, Norway* *E-mail: kjell.berg@hydro.com*
C. BERNDT	*Department of Geology, University of Tromsø, 9037 Tromsø, Norway; now at: Challenger Division for Seafloor Processes, Southampton Oceanography Centre European Way, Southampton, SO14 3ZH, U.K.*
A. BRAATHEN	*Geological Survey of Norway, 7491 Trondheim, Norway now at Centre of Integrated Petroleum Research, University of Bergen, Allégt. 41, 5007 Bergen, Norway;* *E-mail: alvar.braathen@cipr.uib.no*
P. BRYN	*Norsk Hydro ASA, Vækerø, 0246 Oslo, Norway* *E-mail: petter.bryn@hydro.com*
S. BÜNZ	*Department of Geology, University of Tromsø, 9037 Tromsø, Norway*
J. CLAOUÉ-LONG	*Australian Geological Survey Organisation, G.P.O. Box 378, Canberra, ACT 2601, Australia*
F. EIDE	*Norsk Hydro ASA Exploration and Production, Sandsli, 5020 Bergen, Norway*
T. EIDVIN	*Norwegian Petroleum Directorate, P.O. Box 600, 4003 Stavanger, Norway* *E-mail: tor.eidvin@npd.no*
J.I. FALEIDE	*Department of Geosciences, University of Oslo, P.O. Box 1047, Blindern 0316 Oslo, Norway*
C.M. FANNING	*Research School of Earth Sciences, The Australian National University, Canberra, ACT 0200, Australia*
C. FICHLER	*Statoil Research Center, Rotvoll, 7005 Trondheim, Norway*
E. FJELLANGER	*Esso Norge AS, P.O. Box 60, 4064 Stavanger, Norway* *E-mail: erik.fjellanger@exxonmobil.com*
J.H. FOSSÅ	*Institute of Marine Research, 5000 Bergen, Norway* *E-mail: jhf@imr.no*
R.H. GABRIELSEN	*Institute of Geosciences, University of Bergen, Allégt. 41, 5007, Bergen Norway now at the Norwegian Research Council, Oslo, Norway* *E-mail: roy.gabrielsen@nfr.no*
A. GRØNLIE	*Statoil Exploration Department, Region North, P.O. Box 40, 9401 Harstad, Norway*
S. GUIDARD	*Department of Geology, University of Tromsø, Dramsveien 201, 9037 Tromsø, Norway*

T. HENNINGSEN — *Statoil Exploration Department, Region North, P.O. Box 40, 9401 Harstad, Norway*

S. HENRIKSEN — *Statoil Research Center, Rotvoll, 7005 Trondheim, Norway*
E-mail: shen@statoil.com

M. HOVLAND — *Statoil, 4035 Stavanger, Norway*
E-mail: mhovland@statoil.com

S.M. JONES — *Department of Geology, Trinity College, Dublin 2, Ireland*

T. KJENNERUD — *SINTEF Petroleum Research, 7465 Trondheim, Norway*

R. KYRKJEBØ — *Institute of Geosciences, University of Bergen, Allégt. 41, 5007 Bergen, Norway; now at Statoil Research Centre, Rotvoll, 7005 Trondheim, Norway*

I. LAURSEN — *Statoil Exploration Department, Region North, P.O. Box 40, 9401 Harstad, Norway*

H. LIE — *Norsk Hydro ASA, Exploration and Production, Kjørbo, 0246 Oslo, Norway*

R. LIEN — *Norsk Hydro ASA, Kjørbo, 0246 Oslo, Norway*

H. LØSETH — *Statoil Research Center, Rotvoll, 7005 Trondheim, Norway*

T. MIDTUN — *Armauer Hansens vei 13, 5081 Bergen, Norway*

J. MIENERT — *Department of Geology, University of Tromsø, Dramsveien 201, 9037 Tromsø, Norway*

A.C. MORTON — *HM Research Associates Ltd., 100 Main Street, Woodhouse Eaves, Leics LE12 8RZ, U.K. and Department of Geology and Petroleum Geology, Kings College, University of Aberdeen, Aberdeen AB24 3UE, U.K.*

J. MOSAR — *Department of Geoscience, University of Fribourg, Switzerland*

R. MÜLLER — *Department of Geology, University of Oslo, P.O.Box 1047, Blindern 0316 Oslo, Norway*
E-mail: reidar.muller@geologi.uio.no

Ø. NORDGULEN — *Geological Survey of Norway, 7491 Trondheim, Norway*

J.P. NYSTUEN — *Department of Geology, University of Oslo, P.O. Box 1047, Blindern 0316 Oslo, Norway*

B.T. OFTEDAHL — *Department of Geology, University of Oslo, P.O. Box 1047, Blindern 0316 Oslo, Norway*
E-mail: b.t.r.oftedahl@geologi.uio.no

O. OLESEN — *Geological Survey of Norway, 7491 Trondheim, Norway*

P.T. OSMUNDSEN — *Geological Survey of Norway, 7491 Trondheim, Norway*
E-mail: per.osmundsen@ngu.no

D. OTTESEN — *Geological Survey of Norway, 7491 Trondheim, Norway*
E-mail: dag.ottesen@ngu.no

C.S. PICKLES — *ChevronTexaco Upstream Europe, Seafield House, Hill of Rubislaw, Aberdeen AB15 6XL, U.K.*

I. PRINCE — *Statoil Head Office, Forus, 4035 Stavanger, Norway*

L.A. RAMSEY — *Department of Earth Sciences, University of Cambridge, Downing Street, Cambridge CB2 3EQ, U.K.*

T.F. REDFIELD — *Geological Survey of Norway, 7491 Trondheim, Norway*

D. ROBERTS — *Geological Survey of Norway, 7491 Trondheim, Norway*

Y. RUNDBERG	*Norsk Hydro ASA, Sandsli, 5020 Bergen, Norway* *E-mail: yngve.rundberg@hydro.com*
R.A. SCOTT	*CASP, Department of Earth Sciences, University of Cambridge, West Building, 181A Huntingdon Road, Cambridge CB3 0DH, U.K.* *E-mail: robert.scott@casp.cam.ac.uk*
S. SINCLAIR	*CASP, Department of Earth Sciences, University of Cambridge, West Building, 181A Huntingdon Road, Cambridge CB3 0DH, U.K.*
J.R. SKILBREI	*Geological Survey of Norway, 7491 Trondheim, Norway* *E-mail: jan.skilbrei@ngu.no*
A. SOLHEIM	*Norsk Hydro, Kjørbo, 0246 Oslo, Norway*
A. SOMMARUGA	*Department of Geoscience, University of Fribourg, Switzerland*
F. SURLYK	*Geological Institute, University of Copenhagen, Øster Voldgade 10, DK-1350 Copenhagen K, Denmark* *E-mail: Finns@geo.geol.ku.dk*
T. THORSNES	*Geological Survey of Norway, 7491 Trondheim, Norway* *E-mail: terje.thorsnes@ngu.no*
M. VANNESTE	*Department of Geology, University of Tromsø, Dramsveien 201, 9037 Tromsø, Norway* *E-mail: maarten.vanneste@ig.uit.no*
L.C. WAMSTEEKER	*ExxonMobil Canada Ltd., 237 4th Avenue S.W., P.O. Box 800, Calgary, Alberta, Canada T2P 2J7* *E-mail: Lee.Wamsteeker@exxonmobil.com*
A.G. WHITHAM	*CASP, Department of Earth Sciences, University of Cambridge, West Building, 181A Huntingdon Road, Cambridge CB3 0DH, U.K.*
S.T. WIEN	*SINTEF Petroleum Research, 7465 Trondheim, Norway* *E-mail: stein.wien@iku.sintef.no*

Contents

Vertical movements in south-western Fennoscandia: a discussion of regions and processes from the Present to the Devonian

Roy H. Gabrielsen, Alvar Braathen, Odleiv Olesen, Jan Inge Faleide, Rune Kyrkjebø and Tim F. Redfield

This review discusses regions of vertical movements of southern Norway and its continental shelf from the Present to the Devonian. The processes examined are distinguished on the basis of their effect into long-wavelength and short-wavelength.

On the mega-scale, two main configurations are identified. The older configuration relates to the late- to post-Caledonian stage, which was dominated by orogenic denudation processes that followed the Caledonian Orogeny. This probably included thermal as well as isostatic effects, which contributed to the development of an asymmetrical long-wavelength uplift area with a low-relief eastern flank towards the Baltic countries, and a western hinterland region of high relief, especially above exhumed gneissic regions. The hinterland probably had a rugged topography, similar to that of the present Himalaya.

The younger configuration is mirrored by smoother highs and lows of events that were probably, to a large extent, thermally controlled. They include Carbo-Permian, Permian and Jurassic rifts, of which the latter particularly affected the flanks of a south Norwegian high, or dome. From the earliest Tertiary, this feature seems to have been stabilised and supported by a horizontal transfer of hot material associated with the Icelandic plume. Pluming may be superimposed by more recent glacial rebound, which presently interferes with the long-term effects of the North Atlantic asthenospheric plume.

Introduction

Recent studies suggest that variations and irregularities in post-rift subsidence as well as the accumulated effects of vertical movements in passive continental margins are underestimated commonly (Cloetingh, 1986; Cloetingh et al., 1990, 1992; Cloetingh and Kooi, 1992; Sales, 1992; van der Beek et al., 1994 and Faleide et al., 2002). Such effects may be related to a variety of mechanisms, some of which are related to processes taking place at lower crustal, deep lithospheric or even asthenospheric depth (Cloetingh et al., 1994, 1995; van der Wijk and Cloetingh, 2002), whereas others are due to more localised, shallow, crustal processes. The latter, accordingly, are of more local significance. Since they occur at different depths and scales, these processes are bound to interfere with each other, implying that any subsidence- (or uplift-)curve at any point in any place reflects the sum of the effects of several processes. Subsidence/uplift analysis is further complicated by the contribution from local strain, diachronous effects due to structural and thermal heterogeneities, and the shifting rheological properties on the scale of the area under deformation.

The western Fennoscandian Shield, with its continental shelf, is one area where such complex interference of uplift/erosion and subsidence/redeposition has occurred over a time span of several hundred million years. Geologists have already acknowledged the post-glacial isostatic adjustments of Scandinavia in the 19th century, and its centre of uplift and the consequences of relative sea level change were established early (Reusch, 1901; Nansen, 1904, 1920; Ahlman, 1919). In fact, the Scandinavian Peninsula is commonly considered a classical area in this context (Lyell, 1835; Jamieson, 1882). The study of the effects of glacial and sediment loading and unloading (Bowie, 1929; Fjeldskaar, 1994; Fjeldskaar and Cathles, 1991; Fjeldskaar et al., 2000; Riis and Fjeldskaar, 1992;

Onshore–Offshore Relationships on the North Atlantic Margin
edited by B. Wandas et al. NPF Special Publication 12, pp. 1–28,
Published by Elsevier B.V., Amsterdam

Vågnes et al., 1992; Riis, 1996) directly contributes to constraining the strength of the lithosphere (Nansen, 1927; Fjeldskaar, 1994; Watts, 2001), as well as to the comprehension of the distribution of seismic activity and its relation to contemporaneous stress (Bungum et al., 1986, 1991; Fejerskov and Lindholm, 2000; Fejerskov et al., 2000).

It is, however, also acknowledged that the uplift associated with glacial effects alone cannot explain the geomorphology of Norway, including the presence of planation surfaces at high latitudes (Gjessing, 1967; Peulvast, 1985; Lidmar-Bergstrom et al., 2000). Thus, a large-scale Tertiary uplift has been suggested to have contributed to this (Holtedahl, 1953; Torske, 1972; Riis, 1996). It has also long been known that vertical movements played an important role in the post-Caledonian history and, hence, in the associated geomorphological development of the western Fennoscandia. Thus, Ramberg (1966) proposed that the lithospheric inhomogeneities inherited from Caledonian geodynamics could be the basic mechanism behind the domal features ('basement windows'), which are found along the spine of the Norwegian Caledonides. Later, the existence of a (hypothetical) Caledonian deep structure (suture) was suggested by Dewey and Bird (1970) to explain the post-Caledonian instability associated with the 'Faltungsgraben' (Goldschmidt, 1921). Although they did not associate this with the suture itself, Ramberg and Grønlie (1969) confirmed the existence of a NE–SW-trending Caledonian(?) gravity anomaly to be situated parallel to the trend of the major central south Norwegian Caledonides. More recent offshore studies have established that the Caledonian suture most likely is situated off the present coast of Norway (Hospers et al., 1989).

It is also well-established that the post-Caledonian extensional structuring and its associated uplift of the central mountain chain (Hossack, 1984; Norton, 1986, 1987; Séranne and Seguret, 1987) should be seen in the framework of late- to post-orogenic collapse (Andersen and Jamtveit, 1990) and extensional reactivation of regional-scale thrust-fault systems (Fossen, 1992, 2000). Furthermore, dating of dykes and fault rocks has documented that the extensional fault activity lasted throughout the Mesozoic and well into the Cenozoic (Færseth et al., 1976; Torsvik et al., 1992, 1997; Eide et al., 1997, 1999; Fossen and Dunlap, 1998; Andersen et al., 1999). Better examples of Mesozoic activity include the inshore Jurassic basins (Fossen et al., 1998; Sommeruga and Bøe, 2003). This implies that additional agents than those related directly or indirectly to the Caledonian Orogeny, such as the processes associated with the extension that caused the formation of the northern North Sea Basin system, have also contributed to the development of the topographic relief of the Norwegian mainland. The latest stages recorded in such chrono-tectonic studies suggest that the last reactivation of the extensional fault system may have been coeval with Tertiary uplift and tilting of the Norwegian mainland (Torske, 1972; Doré, 1992; Jensen et al., 1992; Riis and Fjeldskaar, 1992; Riis, 1996). Finally, the recent apatite fission track (AFT) studies have suggested that uplift centred in south and central Norway occurred throughout the Mesozoic and into Cenozoic times (Rohrman et al., 1995, 1996; Rohrman and van der Beek, 1996; Redfield et al., 2003, in press).

Compared to the data obtained onshore Norway, the study of the offshore regions yields more detailed and better time-constrained information, both when basin configuration (Ziegler, 1990; Odinsen et al., 2000), subsidence and uplift patterns (Cloetingh, 1986; Cloetingh et al., 1992) and patterns of erosion and deposition (Jordt et al., 1995, 2000; Faleide et al., 2002) are concerned. Thus, it is only when data from the continental shelf of Norway (e.g. Gabrielsen et al., 2001; Kjennerud et al., 2001; Kyrkjebø et al., 2001) are combined with the systematic study of onshore tectonics in Scandinavia (e.g. Lagerbäck, 1976; Olesen, 1988; Spann et al., 1991; Dehls et al., 2000; Fejerskov et al., 2000; Fejerskov and Lindholm, 2000; Anda et al., 2002) that the full complexity of the strain/stress-pattern related to the late Palaeozoic–Cenozoic of the northwestern European continental margin can be appreciated fully (e.g. Pascal and Gabrielsen, 2001).

With this background, it seems obvious that when analysing the late- to post-Caledonian to the Present pattern of vertical displacements in western Fennoscandia, one is dealing with a composite configuration resulting from the overprint of several geodynamic processes, which have been operating partly in concert, and partly in isolation. This implies that when analysing this pattern, great care must be taken in identifying, dating and estimating the effect of each of these processes.

This chapter deals with the tectono-topographical pattern of southern Norway and shelf area that resulted from complex, long-lasting movements. The progressive appearance of what appears as a high or a dome of southern Norway on a broad scale has been highlighted by, for example, Riis (1996) and Lidmar-Bergstrom et al. (2000). We apply additional components to the discussion — the importance of basin-scale processes as well as faulting at various scales. In the descriptions, we attempt to separate facts from processes; processes are highlighted in the Discussion. However, the older the event in question, the more inter-mixed these subjects become. We have therefore chosen to start with the younger and better constrained events, and then work our way back to the Devonian. At the moment, no firm answers exist to many of the questions posed in connection with the vertical movements in the western Fennoscandian Shield. Through our review and discussion, we are able to find a few answers — our aim is basically to identify the most important of the unsolved questions.

Concepts and Methods

In order to evaluate and eventually predict a subsidence or an uplift pattern, the relation between the different effects and hence, the processes responsible for each of them, must be established. This is a difficult task because some of these effects are affiliated with processes that are long extinct and from which the geologically traceable tracks are blurred and sometimes obliterated. When the more recently active processes are concerned, it is commonly possible to characterise vertical movements by their amplitude and wavelength. However, one has to keep in mind that the shape implied by 'wave-forms' in many cases can be rather imaginary, and also is more valid for large features. The analysis of wavelength and amplitude parameters may also be helpful in absolute or relative dating, thus providing a timely starting point in, for example, basin subsidence analysis. Frequently, numerical modelling provides the only means for testing which mechanism is the most relevant one. One should keep in mind, however, that the extinct, older grains sometimes exert a profound control on the younger movement patterns (e.g., Osmundsen et al., this volume).

Long wavelength effects are generally affiliated with deep crustal or lithospheric processes, or even processes within the asthenosphere (e.g. Cloetingh and Burov, 1996). The typical agents would be thermal contraction and expansion, associated with instability induced by glacial, orogenic or basin-forming processes, which induce metamorphic transformations, long-term lithostatic adjustments associated with plate-scale sedimentary and basin-wide loading or unloading processes and the thermal effects thereof (e.g. Cloetingh et al., 1994; Table 1). Because of their wide wavelengths (>100–500 km), and hence small gradients, such effects are easily overlooked in intra-basin studies, even in cases where the regional-scale vertical movements amount to several hundred metres. Typical problems occur in cases when several surfaces merge at regional unconformities (Rawson and Riley, 1982; Kyrkjebø et al., 2004). Thus, a variety of potential causes for long-wavelength effects need to be considered in all the studies of regionally significant vertical displacements.

Short-wavelength effects (<100 km) are characterised by greater horizontal gradients, and typically reflect processes that are associated with single tectonic units (fault zones or fault blocks, single faults, basin margins; Table 2). Generally speaking, these are more easily detected both in outcrop studies and in reflection seismic data. Nevertheless, they are neither always systematically evaluated nor in tune with the large-wavelength events in basin analysis.

When analysing systems where long- and short-wavelength effects interfere, it is necessary to: (i) know the deep structure of the orogen, the basin or the margin under investigation, to constrain the areas of erosion or sedimentation and to correlate these observations to deep features and the entire story of development, (ii) evaluate the potential thermal influence of earlier crustal deformation and to calculate eventual excess heat related to earlier

Table 1 Typical wavelength–amplitude dimensions for uplift related to some lithospheric-scale mechanisms.

Agent	Surface effect	Wavelength (km)	Amplitude (km)
Lithospheric thinning/thickening	Basin formation/orogenic processes	100+	1 to 10
Metamorphic transitions	Wide-angle arching/subsidence	1000	0.1
Underplating	Wide-angle arching/subsidence	100 to 1000	1
Thermal expansion/contraction	Arching or basin formation	100+	1
Sediment loading/unloading	Changing basin gradients/uplift	100	0.1 to 1
Glacial loading/unloading	Wide-angle arching/subsidence	100 to 1000	0.1 to 1+
'Intra-plate deformation'	Variable uplift/subsidence	0.1 to 1000	0.1+

Table 2 Typical wavelength–amplitude dimensions for uplift related to some intra- and upper-crustal-scale mechanisms.

Agent	Surface effect	Wavelength (km)	Amplitude (km)
Faulting	Differential uplifts/subsidence/rotation hangingwall and footwall	0.1 to 10	0.01 to 1
Elastic/isostatic response to faulting	e.g. 'cantilever effect'	0.1 to 10	0.01 to 5
Fault block sediment loading/unloading	Changing basin gradients, activation of sub-basins	1 to 10	0.1 to 1
Footwall collapse	Gravitational transport and modification of basin margins	0.1 to 2	0.01 to 0.5
Local sediment loading and compaction	Irregular subsidence	0.01 to 1	0.01 to 0.1
Gravitational effects related to surface instability	Collapse	0.1 to 100	0.01 to 0.5
Gravitational effects related to stratified instability	e.g. salt and shale-related domes and basins	0.1 to 10	0.01 to 0.5

events, (iii) evaluate potential influence from the lithospheric processes which have taken place outside the study area, but which still may have influenced its pattern of vertical movements (e.g. mantle plumes, glacial rebound), (iv) identify short-wavelength vertical movements and separate those related to shallow processes from those of more deep-seated origin and finally, (v) investigate mechanisms using analogue mechanical and numerical modelling methods.

Vertical displacements in south-western Fennoscandia

Even when dismissing the effect of the adjustments affiliated with the Caledonian Orogeny, a compilation of data from the extinct and recently to presently active regions of vertical (positive) displacement in western Scandinavia and its continental shelf reveals a complex pattern (Fig. 1 and Table 3). Several of these regions are spatially overlapping, although they may not have been simultaneously active, and some of them may be associated with major, basement involved structures (Fig. 2). Others are coeval in time and also partly overlapping in space. This implies that when analysing the upheaval/subsidence history of any given, geographically defined data point, it is likely that the total vertical displacement vector is a sum of several events of vertical displacements. In cases where these events are separated in time, methods like AFT analysis (Fitzgerald and Gleadow, 1998; Gallagher et al., 1998), may help to reveal the different components. When the events are not separate in time, a numerical modelling approach seems to be the only solution for testing the hypotheses.

Neogene — Present uplift and exhumation

At the continental shelf, sediments were supplied from the stable East Shetland Platform throughout the Miocene (Fig. 3), as extensively discussed in, for example, Faleide et al. (2002). Changes in the sedimentation patterns and the development of several unconformities of regional and sub-regional significance illustrate that the Miocene was a relatively active tectonic period. Perhaps the most conspicuous phenomenon in this context is the pattern of incised, (?)sub-aerial valleys observed in the northern North Sea, close to the eastern basin margin (Rundberg et al., 1995; Martinsen et al., 1999), suggesting considerable and fast uplift. This event was followed by subsidence in the late Miocene – Pliocene. The high of the central south Norway persisted throughout the Miocene, with a SW-directed slope in the region of the west coast (Clausen et al., 1999). Systematic intra-formational faulting in the Miocene sediments on the shelf, off southern Norway, is in harmony with the existence of a long-wavelength flank with a gradient in the order of several degrees at this time (Clausen et al., 1999).

The AFT thermal models suggesting a significant late Neogene exhumation event in western Norway are in accord with the offshore sedimentary record, which is characterized by a large Neogene clastic wedge (Faleide et al., 2002). Fault zones that exhibit non-cohesive fault rocks within their core are candidates for structures that controlled Tertiary uplift of Norway (e.g. Gabrielsen et al., 2002), albeit such fault products may also have been formed during Mesozoic activity (e.g., Fossen et al., 1998). The erosional effect on the mainland of the ice ages has removed most of the evidence of younger fault activity, with exceptions such as the neotectonic Berill fault in Romsdal (Anda et al.,

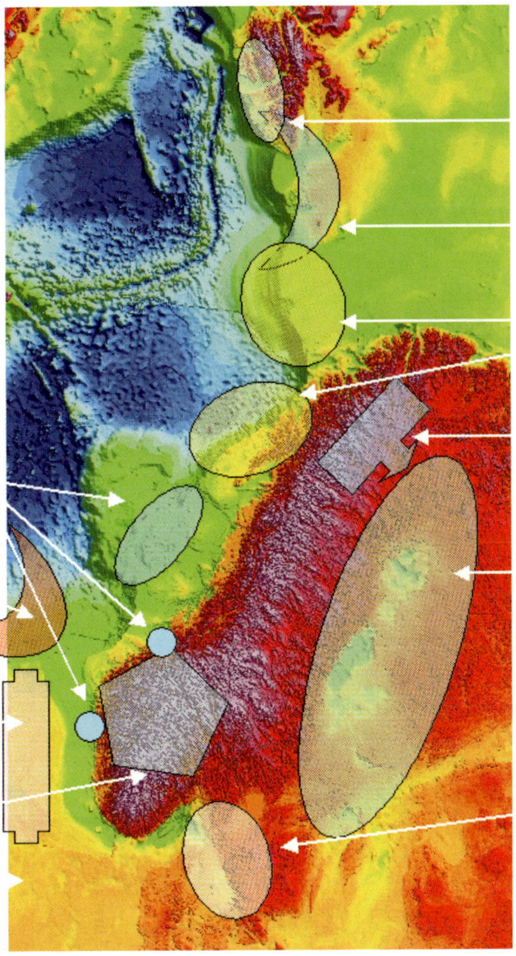

Fig. 1 Schematic Post-Caledonian centres of large-scale uplift-erosion patterns for western Scandinavia.

2002) and the Storagaurri fault in Finnmark (Olesen et al., 2004, and references therein). One can only speculate that the many faults identified as fracture lineaments of the Møre-Trøndelag region were tectonically active during the Cenozoic.

Palaeogene relief

The post-rift subsidence, affiliated with thermal cooling following the mid-Jurassic — early Cretaceous extension had ceased by the end of the Cretaceous in the northern North Sea, and the area was seen as a wide depositional low-relief, shallow basin (Gabrielsen et al., 2001), as discussed in Faleide et al. (2002). These authors depict that the Palaeocene basin was mainly sourced from the emergent East Shetland Platform and the Scottish Highlands in the west and the Norwegian mainland in the east (Fig. 3). However, the deepening of the basin took place towards the end of the Palaeocene, reaching water depths in the order of 900 m (Gradstein et al., 1994; Gilmore et al., 2001; Kyrkjebø et al., 2001). This accelerated basin subsidence was accompanied by a particularly pronounced upheaval of the area delineated by the Øygarden Fault Complex in the west, Sognefjorden to the south and the landward elongation of the Jan Mayen Lineament to the north (Faleide et al., 2002). This period is also characterised by the early Tertiary Vestbrona Volcanic Provincc (Bugge et al., 1980; Prestvik et al., 1999), which coincided in time with extensive volcanism, elsewhere in the North Atlantic.

Reduced sediment input from the easterly source in earliest Eocene times suggests that erosional levelling had taken place, and that western Scandinavia even may have been transgressed, whereas the sediment input continued from the East Shetland Platform. Still, the activation of a depocentre outside Sognefjorden probably reflects renewed tectonic uplift before the area became stabilised again and perhaps submerged during the mid- to late- Eocene (Jordt et al., 1995, 2000).

Throughout the Oligocene, the sedimentary pattern suggests an enhanced relief, probably reflecting that the basin flanks were uplifted. In late Oligocene time, even parts of the basin itself became uplifted. As described by Rohrman et al.

Table 3 Estimated vertical displacements as affiliated with thermal/metamorphic adjustments, regional tectonism, erosional/depositional loading for the Latest Caledonian, Late Phanerozoic, Late Mesozoic, Cretaceous-Tertiary and Quaternary of southern Norway. The type of tectonic environment (dominant fault type) is indicated for each case.

FACTORS EVENTS	THERMAL (T) METAMORPHIC (M)			LOAD regional tectonism			LOAD, regional erosion/deposition			EROSION removal			DETACHMENT regional fault zone		FAULT(S) local	
Late-post Caledonian extension/denudation	T, cold thrust stack	↓↑	M, eclogite transitions	Thrust stacking	↓↑	Flexural bulging	Orogen margin D	↓↑	Orogen centre E	Orogen margin	↑↓	Orogen centre	↓	Extensional detachments	↓	Supradtm. Faulting
Late Phanerozoic extension in broad province	T, slightly thinned crust	↑	M ?	Thinning	↑↕	Flexuring	Extension province D	↓↑	Shoulder E		↓	Province shoulder	↓	Faults of extensional province	↓	Shoulder faults
Late Mesozoic Viking Graben	T, around rift zone	↑		Thinning	↑↓	Flexuring	Along rift axis D	↓↑	Rift shoulder E		↓	Rift shoulder	↓	Rift margin fault zones	↓	Rift shoulder faults
Cretaceous-Tertiary North Atlantic rifting	T, near rift zone	↑↑	T, away from rift zone	Thinning, near rift	↑↓	Flexuring	Along rift axis D	↓↑	Rift shoulder E		↓	Rift shoulder	↓	Rift margin fault zones	↓	Rift shoulder faults
Quaternary glacial events				Ice sheet margin	↓↓	Ice sheet centre	Ice margin deposition D	↓↑	Sub-glacial erosion E		↓	Sub-glacial				Minor fault reactivation

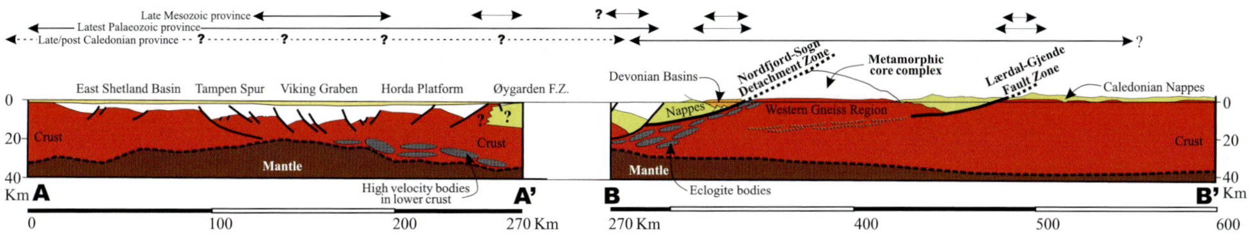

Fig. 2 Geological profile, compiled with reference in Kyrkjebøe et al. (2000) and Milnes et al. (1997). Major tectonic provinces are indicated. Important structures are shown.

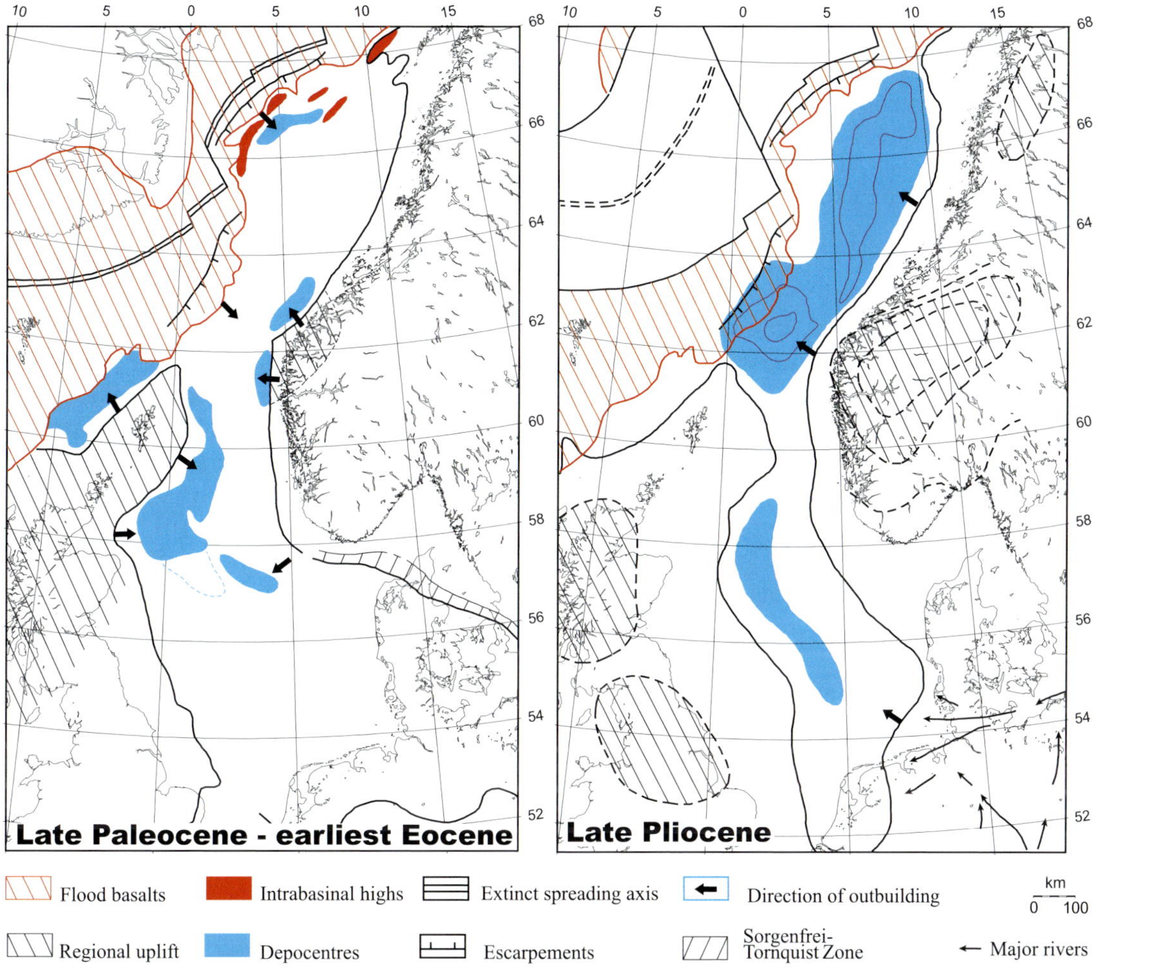

Fig. 3 Areas of regional uplift and erosion for Norway and the adjacent ocean areas for late Palaeocene—early Eocene and Late Pliocene. Modified from Faleide et al. (2002).

(1995, 1996) the topography of the southern Norwegian mainland had a dome-like shape. This rejuvenation of exhumation occurred between 40 Ma and 20 Ma, which is supported by intra-formational faults in the Miocene sequence above the Troll Field, where also the depositional pattern suggests a SW-directed slope (Clausen et al., 1999). The effect of the exhumation of southern Norway is also reflected in the sediment distribution in the Norwegian–Danish Basin (Jordt et al., 1995; Clausen et al., 2000).

A high-frequent pattern of vertical movement is superposed upon the mega-wavelength picture described earlier. Such structures are particularly evident in the Cenozoic sediment packages affiliated with the deep (Cretaceous) basins of the Mid-Norwegian continental shelf (Brekke and Riis, 1987; Doré and Lundin, 1996). However, such effects are also clearly displayed in association with master fault systems of the Barents Sea (Riis et al., 1986; Gabrielsen and Færseth, 1988; Gabrielsen et al., 1992, 1997; Faleide et al., 1993) and in the central (Fagerland, 1983; Biddle and Rudloph, 1988) and southern North Sea (e.g. Glennie and Boegner, 1981; Cartwright, 1989; Hooper et al., 1995).

There is also evidence of inversional structures in the Cenozoic, found along most of the north-western European continental margin (Roberts, 1989; Boldereel and Andersen, 1993, 1994; Doré and Lundin, 1996). These structures have been set in connection with plate tectonic reorganisation (Doré and Lundin, 1996), ridge push, intra-plate deformation associated with the Alpine Orogeny (Vågnes et al., 1998) and local isostatic adjustments (Stuevold et al., 1992). We return to the discussion of forces and processes.

Jurassic–Cretaceous relief

The Mesozoic era was characterised by major extensional events that affected the North Sea. It seems substantiated that an E–W-oriented axis of extension, parallel to the minimum horizontal principal stress axis, existed throughout the Triassic, and that this axis shifted towards NW–SE, either in the Bajocian-Bathonian (Færseth et al., 1997) or in latest Jurassic times (Doré and Gage, 1989; Ziegler, 1990; Gabrielsen et al., 1999). In the Mid-Norwegian realm, to the north of the Møre-Trøndelag Fault Complex, the NW–SE-directed extension may have been dominant, also prior to the Jurassic (Reemst and Cloetingh, 1999; Gabrielsen et al., 1999), and it seems clear that this situation prevailed into the Cretaceous (Graue, 1992).

It is well-documented that the basin-forming processes on the continental shelf were accompanied by pronounced doming centred in the central North Sea (Underhill and Partington, 1993, 1994), whereas pre-rift subsidence took place in the northern Viking Graben (Nøttvedt et al., 1995). These long-wavelength effects became overprinted by syn-rift, short wavelength structural elements, such as rotated fault blocks and development of sub-basins, some of which contributed to vertical displacement due to dynamic elastic and isostatic response (Figs. 4 and 5) (Roberts and Yielding, 1991; Yielding and Roberts, 1992). This relief soon became smoothened due to erosion and gravitational destabilisation of basin margins and fault scarps (Allihali and Damuth, 1987; Fossen and Hesthammer, 1997; Hesthammer and Fossen, 1999) and early post-rift sedimentation (Gabrielsen et al., 2001).

The pattern of erosion and non-deposition affiliated with the base Cretacaeous unconformity ('North Sea Unconformity Complex', Kyrkjebø et al., 2004) also suggests a strongly composite and polychronous nature of this surface (Rawson and Riley, 1982). This probably reflects inhomogeneous extension and also perhaps strike-parallel variance when it concerns the timing for the termination of the extension. Also, a recent detailed investigation of the post-rift geometry of the Jurassic northern North Sea Basin has revealed a more complex pattern of subsidence than perhaps expected (Fig. 5). It seems clear that this irregular subsidence pattern, which is even characterised by relative uplift in places, is partly related to the sub-basins of the basin system (Kyrkjebø et al., 2001). It may only be speculated as to whether these

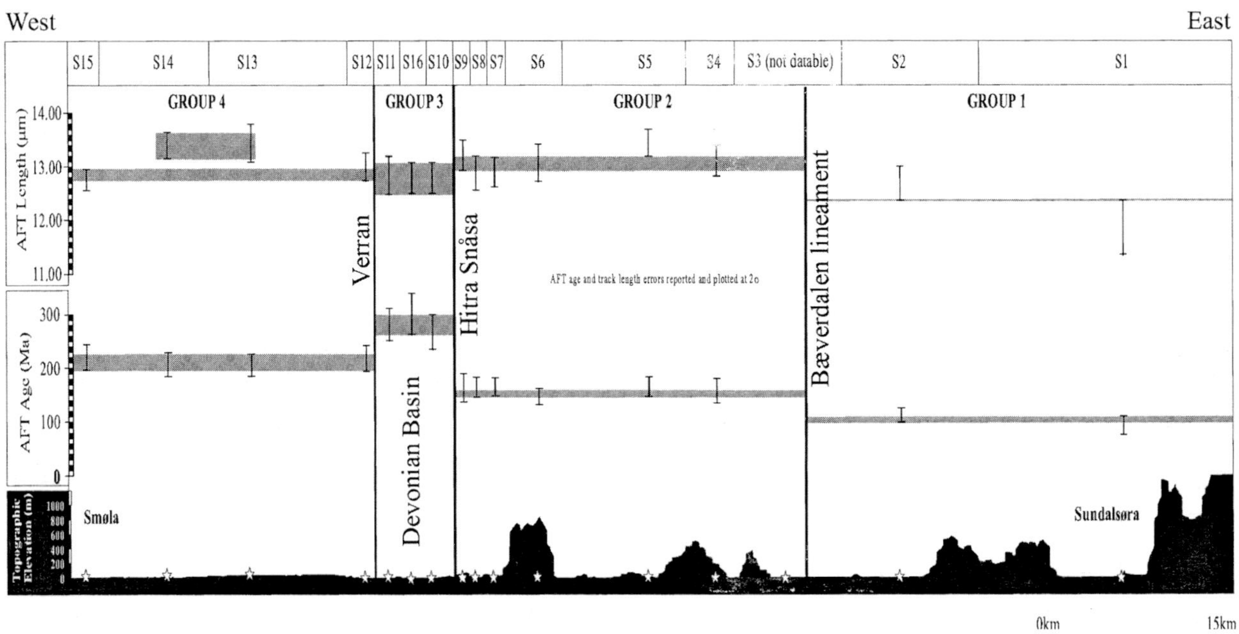

Fig. 4 Fault blocks of the Møre-Tøndelag Fault Complex, as revealed by different apatite fission track ages. Major fault strands of the fault complex are indicated. Modified form Redfield et al. (in press).

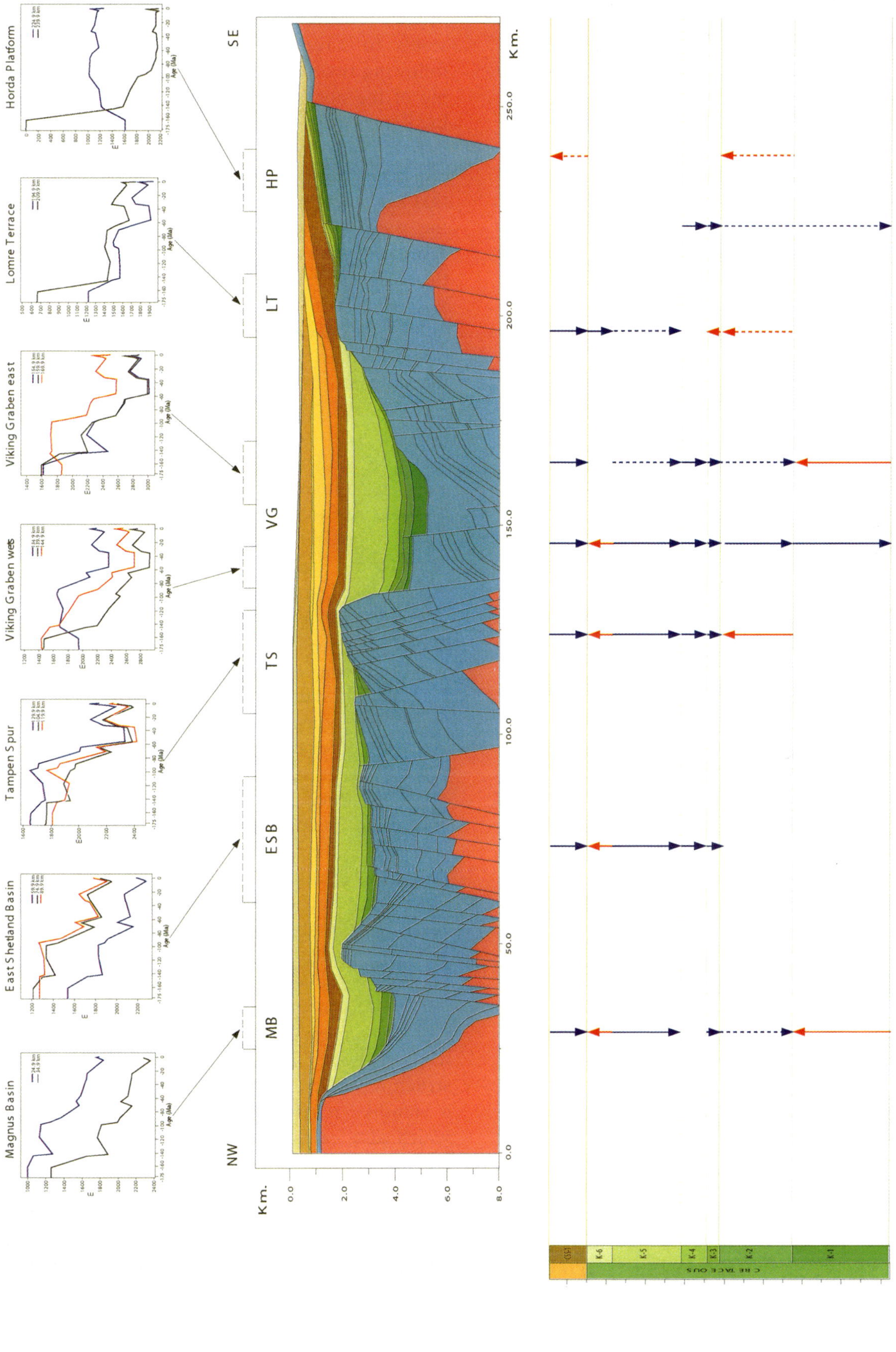

Fig. 5 Complex subsidence patterns in e.g. the post-rift development of the Viking Graben. Modified from Kyrkjebø et al. (2000).

movements are related to isostatic adjustments of intra-plate (contractional?) deformation (Grunnaleite and Gabrielsen, 1995).

The view that southern Norway was dominated throughout the Jurassic–Cretaceous by an uplifted region (Lidmar-Bergstrom, 2000), which could have been in the shape of a long-wavelength domal feature (Riis, 1996), and that it was overprinted by a short-wavelength fault-related topography along its flanks (Fig. 4; Redfield et al., in press), is supported by several types of observations. The preservation of Jurassic sediments in down-faulted segments of the coast-parallel fault system defined by the Møre-Trøndelag Fault Complex and the Hjeltefjorden Fault Complex inland of the Øygarden Fault Complex, demonstrates that the fault-related relief along the flanks of the dome was of importance (Oftedahl, 1971, 1975; Bøe, 1991, Fossen et al., 1997; Sommaruga and Bøe, 2003). Faulting along the margin of the uplifted region is also revealed in the AFT data from the Møre-Trøndelag Fault Complex. A similar faulting scenario could be argued on the basis of the AFT data for the west coast (Andriessen, pers. com.).

The near-shore Jurassic basin near Bergen shows evidence of burial to depths of 1000 m (Fossen et al., 1998), which suggests that the near-shore region and shelf subsided in the latest Mesozoic (Riis, 1996), before renewed uplift in the Tertiary. On the contrary, reports on local inversion of some fault systems in the latest Jurassic (e.g., Gabrielsen and Robinson, 1984) suggest that the stress system was not entirely homogeneous and constant throughout this period. Small-scale uplifts may have existed at several places along reactivated master faults. Furthermore, outbuilding of the Upper Jurassic sandstones in the eastern North Sea (Horda Platform) indicates that uplift of areas of an even larger extent may have occurred.

Late Palaeozoic–Triassic relief

The Late Palaeozoic–Mesozoic development of the western Fennoscandian Shield was dominated by three marked events of extension and crustal thinning, each separated from the other by a period of stress relaxation and thermal cooling. These events happened in the Latest Carboniferous – Early Permian, Permo-Triassic and Jurassic – earliest Cretaceous (Badley et al., 1984; Gabrielsen et al., 1990; Odinsen et al., 2000a, b; see Gabrielsen et al., 1999 for summary of recent literature), and terminated with the opening of the North Atlantic (Talwani and Eldholm, 1972, 1977; Skogseid and Eldholm, 1995) in early Tertiary time.

In the Late Carboniferous — Early Permian — the southern North Sea was affected by regional tectonism, related to relaxation of the Variscan orogen (Glennie, 1984; Ziegler, 1990), which also seems to have initiated the Oslo Graben. The following two events, which dominate the structuring of the northern North Sea, were characterised by contrasting basin configurations in that the Permian stretching involved a wider crustal panel than did the Jurassic crustal thinning. This indicates that the wavelength of the basin system and hence, the area affected by later thermally controlled subsidence, was wider for the Permian than for the Jurassic basin. Furthermore, a pronounced system of large half-graben units with contrasting polarity, separated by large, basement-involved transfer zones, was developed in the Permian in the northern North Sea (Gabrielsen et al., 1990; Færseth et al., 1997; Odinsen et al., 2000). The large-scale, strongly rotated fault blocks associated with this system imposed a strong, medium-scale wavelength/high-amplitude pattern on the top of these regional basin gradients (Figs. 2 and 5). The extensional stage was followed by a period of thermal relaxation and sediment loading that lasted from Scythian to Bajocian-Bathonian (Gabrielsen et al., 1990; Steel, 1993; Færseth et al., 1995).

For Norway south of the Møre-Trøndelag Fault Complex, it seems clear that the Permian–Triassic period was generally characterised by due E–W-extension (Fossen et al., 1998; Braathen, 1999; Gabrielsen et al., 1999; Valle, 2002). Although direct evidence from the central and the south Norwegian mainland is almost non-existent due to sparse locations of preserved sediments (Fossen, 1997; Sommaruga and Bøe, 2003), AFT reconnaissance study suggests that uplift of central southern Norway started in the Triassic (Rohrman et al., 1995, 1996; Rohrman and van der Beek, 1996). However, the Møre-Trøndelag Fault Complex represented a pronounced tectonic divide due to its mechanical weakness, and it is likely that it functioned as a right-lateral transtensional structure at this time. It is therefore possible that a step-like relief existed across the fault system, which also gains support from the AFT study of Redfield et al., 2004. Other onshore data from this period come from master faults of western south Norway (Nordfjord–Sogn Detachment Zone, Lærdal-Gjende Fault System). Dating shows that these structures were still active in the latest Palaeozoic (Torsvik et al., 1998, 2001; Andersen et al., 1999; Eide et al., 1999), thereby indicating that some fault-related relief could have persisted.

In its earliest stages, the late Carboniferous–Permian Oslo Graben represented negative topographic element (Ramberg and Larsen, 1978; Olaussen et al., 1997). Its development was accompanied by shifting and complex stress configurations (Heeremans 1996a, b, 1997), and thermal(?) uplift prevailed from the mid Permian times. A relatively complex thermal situation for southern Norway can be inferred. This may have been accompanied by a strong Moho relief, which may also have been essential for the development of the Oslo Graben itself (van Wijk et al., 2002).

Devonian relief and denudation of the Caledonides

A recent, intensive field analysis of the kinematics and the dynamics of the late- to post-Caledonian extension (Andersen et al., 1991, 1994; Andersen, 1993, 1998; Fossen, 1992, 2000; Chauvet and Seranne, 1994; Hartz et al., 1994; Rykkelid and Andresen, 1994; Milnes et al., 1997; Braathen et al., 2000, 2002; Mosar et al., 2002; Osmundsen et al., 2003) has demonstrated a more complex pattern of tectonic transport than that suggested in earlier studies (Hossack, 1984; Norton, 1986, 1987). Thus, it has become clear that faults of contrasting dips, various depths of detachments and changing transport directions were important factors in the mid to late Palaeozoic deformation system (Braathen et al., 2002). In this context, Fossen (1992, 1993) noted that all tectonic (extensional) transport seems to have been towards the western hinterland, thereby indicating a down-towards-the-hinterland directed tectonic system with respect to the Fennoscandian side of the orogen. The fundamental extensional detachment in southern Norway was the Nordfjord–Sogn Detachment Zone: extension-related unroofing of eclogites of the Western Gneiss Region from the footwall of this structure suggests that the footwall crustal segment initially was situated at a depth of more than 60 km before being uplifted to near-surface position (e.g. Andersen and Jamtveit, 1990; Andersen, 1998).

Recent studies (Braathen et al., 2000, 2002; Osmundsen et al., 2003) have demonstrated that transverse tectonic transport was also common during the Devonian evolution of the Caledonides in Central Norway. These authors also discuss a possible synchronous Caledonian NW–SE contraction and SW–NE orogen-parallel extension. In any case, complex tectonic transport could imply that some relief existed oblique to the strike of the orogen, adding to the expected orogen-parallel relief (Fig. 6). On the basis of modern analogues, such as the Himalayas, it is likely that the late- to post-Caledonian morphology was characterised by long-wavelengths (several hundred km) and high-amplitudes (in the order of 1000 m or more). This relief was probably further exaggerated due to thermal, elastic and isostatic response, following the extensional displacement, which took place along the large orogen-parallel fault systems. A high relief probably affected the erosional and the depositional patterns, which in turn may have influenced the climate and perhaps, even the further development of the fault system, as described from other areas with a dynamic, high relief (e.g., Franke et al., 2000). A high rate of denudation forced by rapid tectonic uplift and a high relief, would create a positive feedback loop for an additional isostatic rise of the hinterland zone.

Quantifying vertical movements of the late- to post-Caledonian, the Devonian evolution has not been straight-forward. This is because the importance of geological factors, such as thermal and metamorphic effects, thrust-stack load, erosional and depositional patterns, and regional and local faulting, still remains critical effects of unknown quantities (e.g., Dewey et al., 1993). However, a general qualitative framework can be established (Table 3). For example, it is well-established through the study of eclogites (Austrheim, 1991; Engvik et al., 2002) that the crust of western Scandinavia became significantly thickened during the Caledonian Orogeny. Burial was around 60 km in Sunnfjord (e.g., Engvik and Andersen, 2000), and more than 100 km, locally, in the Møre region (e.g., Terry et al., 2000). This likely also affected the surface altitude through the overall weight budget (Dewey et al., 1993), since eclogite has mantle density.

Several authors have compared the Scandian event of the Caledonian mountain-chain with the Present Himalayan Orogen (Andersen and Jamtveit, 1990; Andersen et al., 2002; Braathen et al., 2002). This comparison is based on similarity in the coexisting structural patterns, from thrusting near the suture zone to extension and extrusion in the uplifted hinterland. Thereby, one can argue that the surface of parts of the Caledonides was at an elevation of approximately 5000 m, or even more, above sea level. Such altitudes trigger significant erosion as well as substantial orogen-marginal deposition (e.g., Le Pichon et al., 1992). For the Caledonides, the deposition is evidenced by Devonian foreland deposits in the Baltic countries (Plink-Bjorklund et al., 2004), possibly by thrust-imbricated foreland deposits, as suggested for the Ringerike sandstone of south-east Norway (Nystuen, 1981;

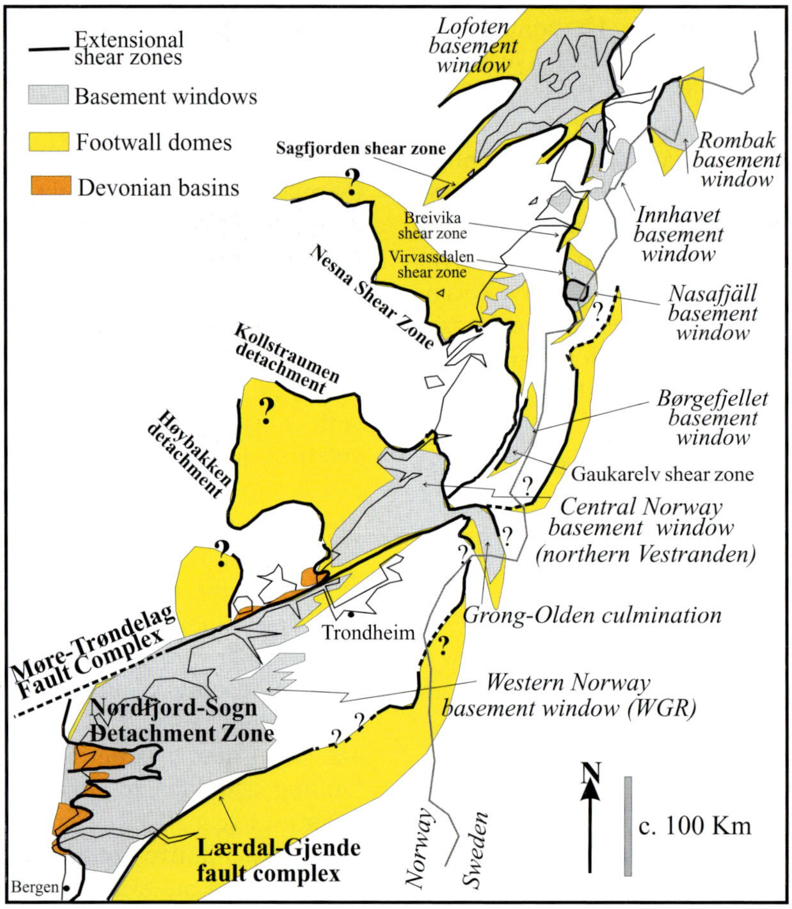

Fig. 6 Major extensional detachments active during denudation of the Scandian Caledonides. Areas of enhanced erosion in the footwall of the detachments are indicated. These areas contain high-grade metamorphic rocks of lower crustal affinity that were partly unroofed during Devonian extension of the Caledonides. The offshore extent of such areas is based on described detachments by Olesen et al. (2002).

Worsley et al., 1983; Morley, 1986), as well as by the intra-mountainous Early-Mid Devonian basins (e.g., Osmundsen et al., 1998). Deposits in the Baltic countries also reveal some information regarding the scale of the orogen, since they were affected by thrusting and therefore ended up in a piggy-back setting characterised by south-westerly nappe transport (Plink-Bjorklund et al., 2004).

The comparison with the Himalaya and other continental collision zones supports the idea of an asymmetrical topography, characterised by a rugged westerly region that, at any given time, could have had an overall shape of a positive or a negative arc. A lower relief probably characterised the foreland-dipping eastern flank (Table 3) that was made up of thrust sheets of stacked sediments, and border to the east of a prograding wedge of clastic sediments into the foreland basin. On a more local scale, faulting controlled by the underlying major extensional detachment(s) created high surface relief in more central parts of the orogen, as shown by the presence of the coarse clastic Early-Mid Devonian basins (e.g., Osmundsen et al., 1998; Osmundsen and Andersen, 2001). Other examples of supradetachment faulting include faults of Øygarden (Larsen et al., 2003) and likely at Frøya (Eliassen, 2003), Vikna (Titus et al., 2003) and within the Helgeland Nappe Complex (Braathen et al., 2002; Nordgulen et al., 2002) of northern Central Norway. In this light, the hinterland region was likely broken by steps associated with the second- and third-order faults. These features may have delineated intra-montane basins similar to the preserved Devonian basins of southern Norway.

When qualitative contributors to vertical movements are compared (Table 3), and an attempt to roughly quantify various vertical movements applied, it is clear that the displacement along the extensional detachments was of great importance. This is so, because relative uplift of detachment footwall sections in the order of 30–60 km or more occurred in a very short time span (5–10 million years), as revealed by the exhumation histories of the eclogites in the Western Gneiss Region (e.g., Andersen and Jamtveit, 1990; Engvik and Andersen, 2000; Terry et al., 2000). Recent accounts (Eide et al., 2003; Fonneland and Pedersen, 2003) present evidence that parts of the rocks of the Western Gneiss Region and Central Norway Basement Window actually surfaced during the Late Devonian(?), on the basis of the nature of deposits in the upper portions of the Devonian basins. Similar uplift can be indicated by the presence of granulites in the western part of the Central Norway Basement Window (Møller, 1988) and by eclogites in a basement window west of Nesna (e.g., Skår et al., 2002) in northern Central Norway.

In summary, it is likely that the thrust-stacking in the eastern region of the Caledonian Orogeny, and especially, the subsequent extensional event in the mountain-chain, provided a pronounced topographic relief. Maximum elevation was probably centred above the significantly exhumed basement windows, such as the Western Gneiss Region. An additional factor to the relief was that durable basement rocks progressively became present at the surface. More local, marginal highs were probably located on the shoulders of supradetachment faults, as indicated by the Devonian basins in western Norway. A similar scenario may be applied to the basement-involved Lærdal-Gjende Fault System. During the late- to post-Caledonian denudation of the orogen, there may have been an asymmetrical relief, with a rougher western flank towards the extended/thinned crust of the hinterland and a smoother eastern flank dipping towards the foreland. The eastern flank reached as far as the Baltic countries at the maximum extent of the Scandian Caledonides.

Discussion

The data presented here for the various time periods remain glimpses of information in our attempts to reconstruct the topographic evolution of the south-western Fennoscandian Shield. Numerous processes have to be evaluated; we discuss the more important ones in this chapter. Some are rather general, such as the effect of faulting and orogenic denudation, while others relate directly to Fennoscandia.

From the descriptions given earlier, it seems substantiated that the pattern of vertical uplift and subsidence of southern Norway displays great changes since the end of the Caledonian Orogeny, and two principal types of relief can be distinguished. These include (1) an asymmetrical relief, which seems to have originated in connection with the late- to post-orogenic crustal thinning that followed the Caledonian (topographic) mountain building, and (2) the Mesozoic–Cenozoic symmetrical domal pattern that developed from the start of the Mesozoic.

Uplift of southern Norway

When discussing uplift and subsidence of the past, it is crucial to do so in light of the recent and better understood processes. We therefore dwell on Neogene uplift.

Mechanisms of Neogene uplift

Gravity data provide means to study the Neogene uplift mechanisms of exhumation of the Scandinavian mountains. Assuming that the region is close to isostatic equilibrium, these uplifted mountainous areas must be supported at depth by substantial volumes of low-density material within the crust or the mantle, or at the crust/mantle or lithosphere/asthenosphere interfaces. The former models represent a Pratt-type isostasy model, while the latter two are consistent with an Airy–Heiskanen type model (Heiskanen and Moritz, 1967).

The western Scandinavian mountains (Fig. 7) are partly of the Plio-Pleistocene age (Rohrman et al., 1995; Riis, 1996; Faleide et al., 2002) and constitute a part of the circum Atlantic belt of Neogene uplifts, including the mountains in Scotland, Svalbard and East Greenland (e.g. Japsen and Chalmers, 2001). As yet, there is no generally accepted hypothesis that explains the Neogene uplift phase; however, several different models have been proposed: (i) glacial erosion, isostatic uplift (Doré, 1992; Riis and Fjeldskaar, 1992), (ii) migrating phase boundaries (Riis and Fjeldskaar, 1992), (iii) pre-subduction instability (Sales, 1992), (iv) plate reorganisation – intra-plate stress (Jensen and Schmidt, 1992), (v) mantle convection (Bannister et al., 1991; Stuevold et al., 1992; Vågnes and Amundsen, 1993), and (vi) mantle diapirism, Rayleigh–Taylor instability (Rohrman and van der Beek, 1996).

Assuming neither erosion nor sedimentation, Neogene uplift (Fig. 8a) implies that either the Moho depth has increased during this time period, or substantial volumes of low-density rocks were introduced in the crust or the mantle. Rohrman and van der Beek (1996) and Riis (1996) proposed a Neogene uplift of more than 1000 m in southern Norway from the AFT data, modelling of geomorphology and extrapolation of the offshore late Tertiary stratigraphy. Riis (1996) and Hendriks and Andriessen (2002) have also proposed a Neogene bedrock uplift of more than 1000 m in

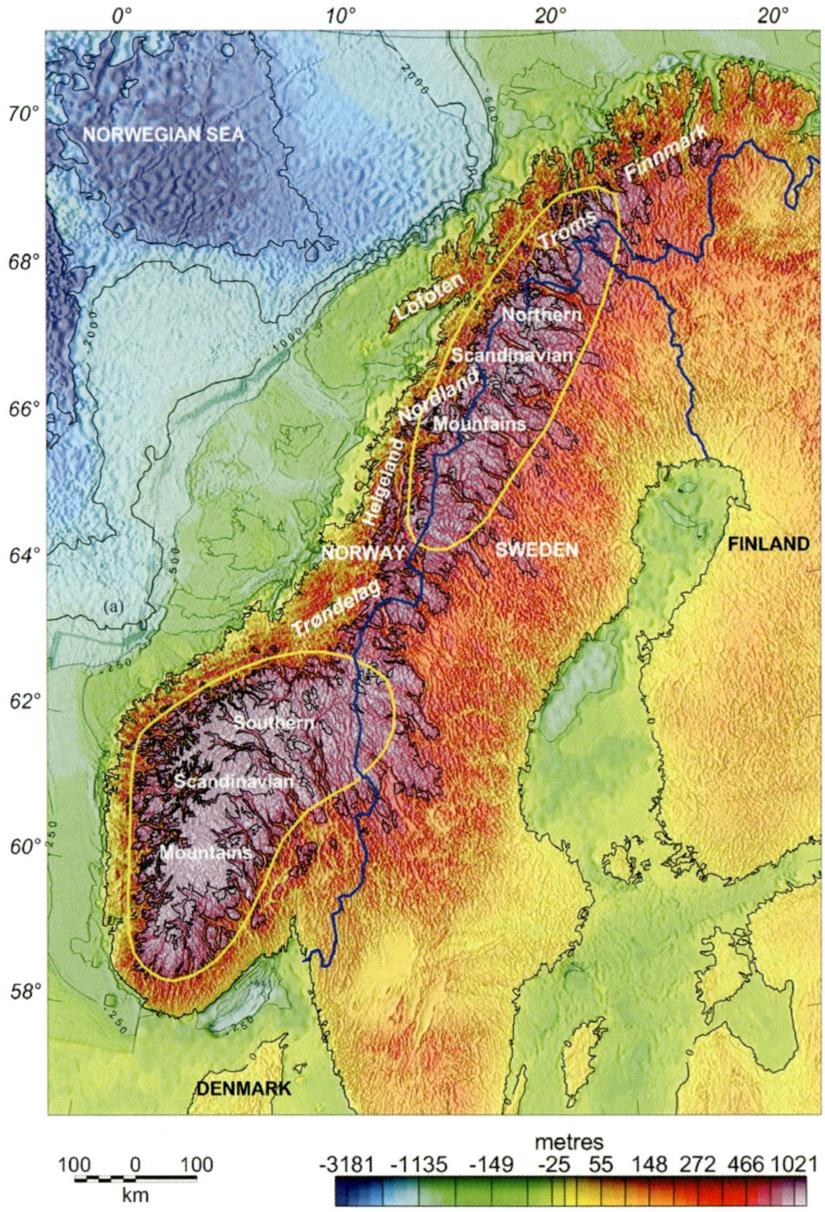

Fig. 7 Digital topography/bathymetry from Fennoscandia compiled by Dehls et al. (2000). The yellow lines depict the two mountainous areas in Scandinavia. The 500 m and 1000 m contours are outlined.

the Lofoten–Vesterålen area, and 600 m on the mainland to the east.

Modelling of isostatic Neogene uplift

An Airy–Heiskanen 'root' (Heiskanen and Moritz, 1967), shown in Fig. 8b, was calculated from the topographic and the bathymetric data set (Olesen et al., 2002). The gravitational attraction (Fig. 8c) from the 'root' was computed in the frequency domain using the AIRYROOT algorithm (Simpson et al., 1983), with a density of 2670 kg/m^3 to the mountains in Norway, and a density contrast of 330 kg/m^3 at the base of the Airy root. The calculation shows a maximum depth of the Airy root of 40 km below the mountainous part of Scandinavia. The main model is based on a 30 km Moho depth along the coast of Norway. These depths agree with Moho depths (Fig. 8d) obtained from refraction seismic studies, compiled by Kinck et al. (1991) and modified by Olesen et al. (2002). The Airy root (or 'isostatic Moho depth') below the mountainous areas is significantly different from the seismic Moho depth (Fig. 8d), indicating the occurrence of low-density rocks within the mantle or the crust.

Olesen et al. (2002) simulated an intra-mantle low-density rock body by increasing the depth to the Airy root. Furthermore, they introduced a shallower Airy root in order to test a potential

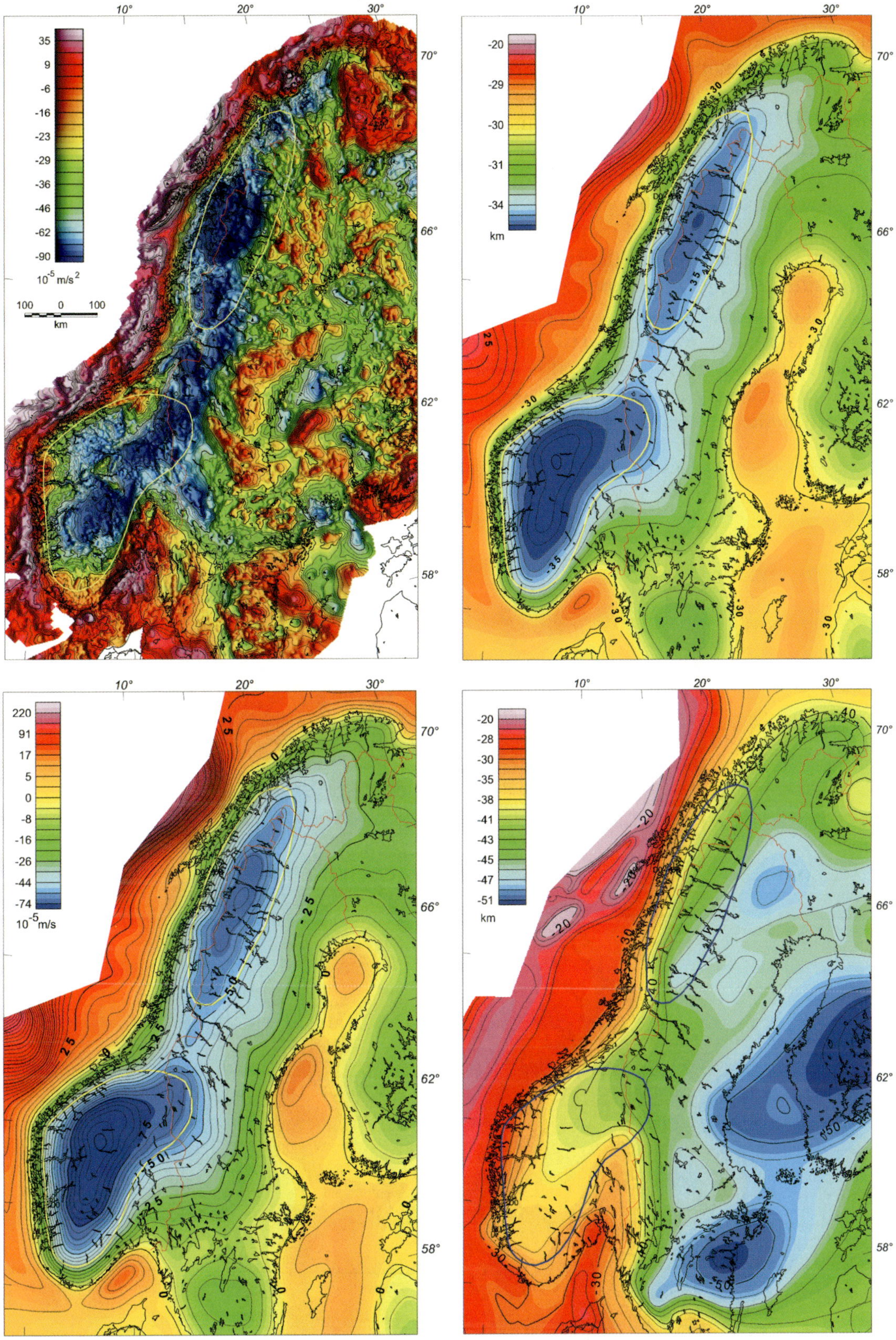

Fig. 8 (A) Bouguer gravity data from Scandinavia compiled from Korhonen et al. (1999) and Skilbrei et al. (2000). The yellow lines depict the two mountainous areas in southern and northern Scandinavia. (B) The Airy root calculated from the topography/bathymetry in Fig. 9. A Gaussian 200 km lowpass-filter has been applied to the Airy-root grid to smooth the high-frequency variation. The yellow lines depict the two mountainous areas in Scandinavia. (C) Gravity response from the Airy root in Fig. 10b. The gravitational attraction is calculated applying the AIRYROOT algorithm (Simpson et al., 1983) and a rock density of 2670 kg/m^3 on land, 2200 kg/m^3 at sea and a crust-mantle contrast of 330 kg/m^3. (D) Depth to Moho compiled from refraction seismic studies (Lund, 1979, Kinck et al., 1993, Mjelde et al., 1992, 1993, 1998 and Sellevoll, 1983) and gravity modelling (Olesen et al., 2002).

intra-crustal low-density contribution to the isostasy. The deviation of the observed Bouguer gravity data from the gravity response of Airy roots at different depths was calculated by subtracting the two data sets, as illustrated in Figs. 8a and c. Figure 9 illustrates the gravity residual for the 30 km Airy root model. The root mean squares (RMS) of the discrepancies between the two datasets within the two polygons in southern and northern Scandinavia suggest that the compensating masses are situated at different depths in the two areas (Fig. 10). Locating the isostatic compensating low-density rocks at shallow depth (10 km) in northern Norway yields a gravity field that is most similar to the observed gravity field, whereas the RMS is lowest for the 45 km Airy root model in the southern mountains. Ebbing et al. (pers. comm.) arrived upon the same conclusions applying the regional isostasy model in their study.

The compensating mass below the southern Scandinavian mountains is consequently situated at a greater depth compared with northern Norway and may be situated within the lithospheric mantle or at the lithosphere/asthenosphere interfaces (i.e. a mantle diapirism model, Rohrman and van der Beek, 1996). The results are in agreement with the conclusions of Riis (1996) and Lidmar-Bergström (1999) that the southern Norwegian plateau was partly uplifted in the Neogene, while the northern Scandinavian mountains originated mainly as a rift-shoulder in the late Cretaceous to Early Tertiary times. Hendriks and Andriessen (2002)

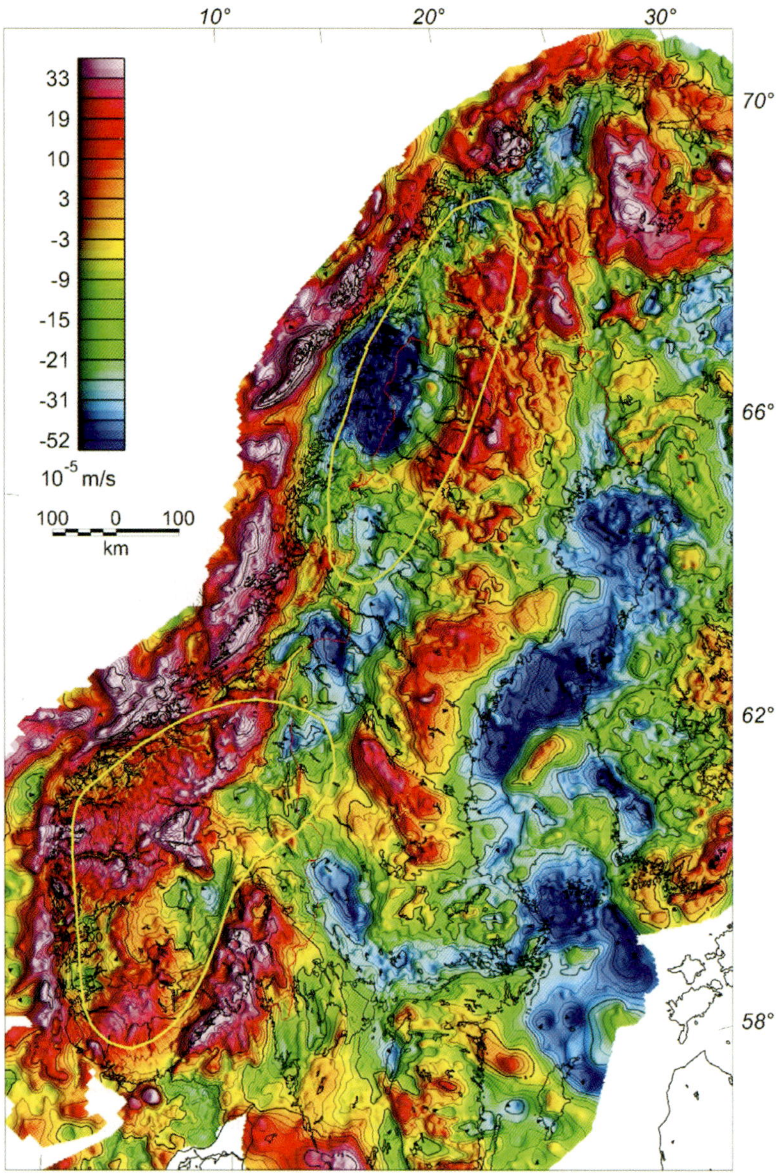

Fig. 9 Residual gravity after subtracting the gravity response of Airy roots at a depth of 30 km (Figure 8c) from the observed Bouguer gravity data (Figure 8a).

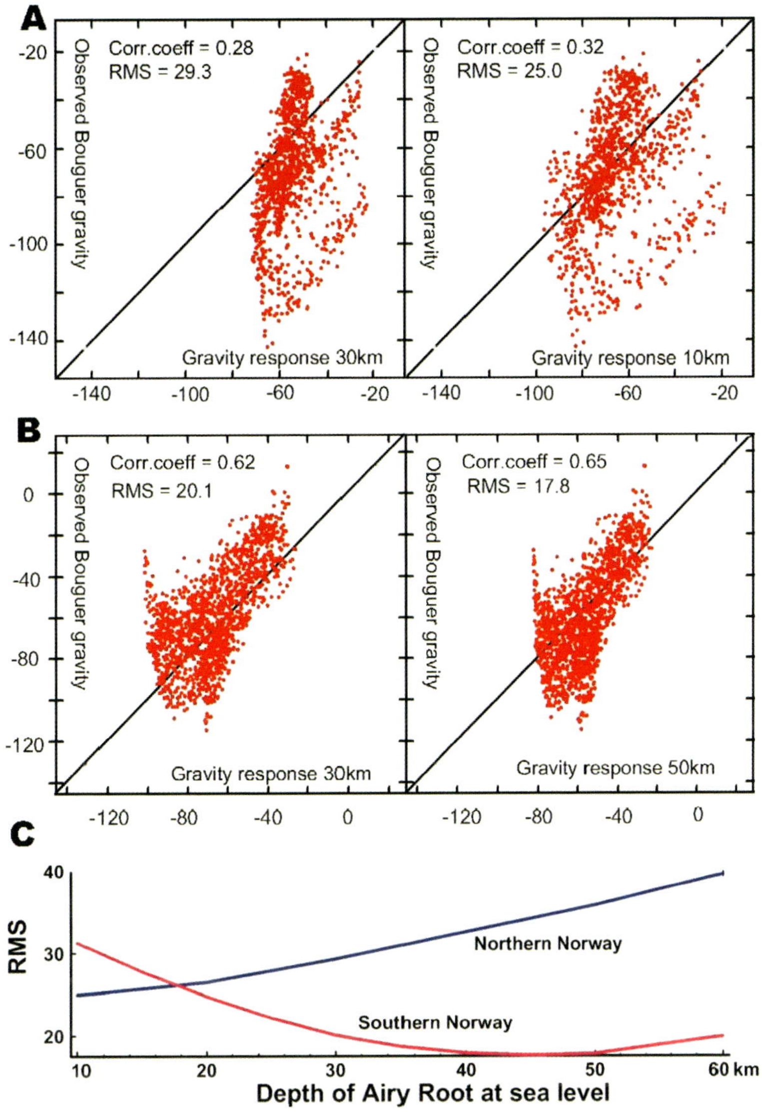

Fig. 10 Comparison of calculated gravity response of 'Airy roots' at different depths with the observed Bouguer gravity for the mountainous part (above c. 500 m asl) of northern and southern Scandinavia (Olesen et al. 2002). Root mean squares (RMS) quantify the similarity between the observed and the calculated effect of Airy roots at various depths. The best fit occurs for a shallow depth (10 km) in northern Scandinavia, while a deep source (Airy root of 40–50 km at the coast) seems to compensate for the southern Scandinavian mountains. The units of the gravity data are 10^{-5} m/s^2 (mGal).

reported that analyses of the observed AFT data, along a profile from Lofoten into Sweden fits best with those expected from a retreating scarp model.

The diapir model, based on the Rayleigh–Taylor instability (Rohrman and van der Beek, 1996), can explain equidistant (~900 km spacing) doming along the northern North Atlantic rifted margins (Figs. 7 and 8). However, the model cannot explain that the maximum uplift occurred in southern Norway. Thus it is possible that a modified mantle diapirism model linking the southern Norwegian diapir to the Iceland mantle plume can explain the observed uplift. Bijwaard and Spakman (1999) succeeded in imaging the Iceland plume to great depths in the mantle using the tomographic analysis of the teleseismic data. Lateral branches of the plume were interpreted to extend below Great Britain and southern Scandinavia. There is a need for substantial volumes of low-density mantle rocks to explain a Neogene uplift of 1000 m. An assumed density contrast of 100 kg/m^3 would imply an approximately 27 km thick slab of mantle rocks, to explain an uplift of 1000 m. Difficulties arise when finding a geological process that can emplace these extensive volumes of low-density rocks below southern Scandinavia in a relatively short time period. The initiation of a deep-seated mantle plume may have therefore started earlier, for instance in the Miocene. Alternatively, explaining the uplift by a change in Moho depth would imply an increase of the Moho depth by 8–9 km, which is geologically not very plausible. The need for large rock volumes to explain the uplift also makes the migrating phase boundary, a less viable model. The retrograded bedrock slab had to be 10–40 km thick, depending on the density reduction. The retrograde processes would need enormous quantities of water and to find such a source is difficult. The pre-subduction instability model may contribute to the uplift of the Lofoten area, but cannot explain the southern Norway uplift, which is located farther away from the oceanic crust in the Norwegian Sea.

In conclusion, therefore, our modelling shows that the main phase of the Neogene uplift of the northern Scandinavian mountains was not caused by low-density material within the mantle. We conclude from this that the uplift was neither caused by mantle convection along the boundary between a warm oceanic asthenosphere and a colder continental asthenosphere (Bannister et al., 1991; Stuevold et al., 1992; Vågnes and Amundsen, 1993), nor by Rayleigh–Taylor mantle instability (Rohrman and van der Beek, 1996). The uplift of the Lofoten area may either be explained by rift-shoulder uplift or, alternatively, by the isostatic effect following glacial erosion (Hendriks and Andriessen, 2002) similar to the Barents Sea region uplift (Doré, 1992; Riis and Fjeldskaar, 1992).

Modelling of Pleistocene to post-glacial glacio-isostatic uplift

The load of glacial ice on the crust during the last glaciation resulted in a glacio-isostatic depression and rebound of Scandinavia and a corresponding forebulge uplift, succeeded by subsidence of the shelf areas, when the ice retreated (Jamieson, 1882; Daly, 1934; Cathles, 1975). The shelf subsidence is indicated by subsided shorelines as discussed herein, but there are no precise measurements of present offshore subsidence rates. In the modelling of glacio-isostatic (and flexural) effects, it is assumed that, given a sufficiently long time after the glacial load has been applied and removed, the crust will return to the same position where it was before the load was applied.

There are, however, additional effects that are irreversible: (i) The unloading of Scandinavia by erosion and sediment loading of the depositional basins on the shelf has an isostatic effect (Riis and Fjeldskaar, 1992). (ii) Irreversible effects on the crust and mantle of cyclic glacial loading and unloading during long time-periods cannot be ruled out (Riis and Fjeldskaar, 1992). (iii) Locally, faults may be reactivated when the ice load disappears (Lagerbäck, 1979; Olesen, 1988).

There is some evidence (e.g. Mangerud et al., 1981; Sejrup, 1987) that the Norwegian coast may have been subjected to tectonic uplift in the order 0.1–0.3 mm/yr during the Quaternary, in addition to post-glacial uplift. Recent studies of the uplifted Middle and Late Weichselian marine sediments (Olsen and Grøsfjeld, 1999) do, however, show that the inland ice sheet fluctuated quite frequently during the interval 18,000–50,000 yr BP. Repeated rapid ice retreat following heavy ice loading was the most likely mechanism for depositing marine sediments of both the same and different age intervals in several uplifted positions, along the coast of Norway as well as in the inland areas of southeastern Norway. This process can also explain the elevated Weichselian marine clay on Høg-Jæren and the coastal caves above the maximum Holocene marine limit in western and northern Norway. These elevated caves have also been interpreted in terms of a Neogene tectonic uplift (Holthedahl, 1984; Sjöberg, 1988).

Geodynamic modelling of the present and the post-glacial uplift data shows that the bulk of the present uplift can be explained as a response to glacial unloading (Fjeldskaar et al., 2000). The modelled uplift in the three areas deviates, however, from the observed uplift: (1) a zone including northwestern Norway and part of eastern Norway, (2) the Lofoten-Troms area, and (3) the Bay of Bothnia area. The Bothnia area shows a negative deviation between the observed and the calculated uplift, whereas the two Norwegian areas show a positive deviation. The two areas in Norway also coincide, partly with the Neogene domes in southern Norway and Lofoten-Troms, indicating that the Neogene tectonic component is partly responsible for the present uplift.

Mechanisms of structuring in the Tertiary

Faleide et al. (2002) briefly discussed the most likely mechanisms that contributed to the Tertiary structuring of western Scandinavia. They noticed that the pattern of uplift indicates that multiple mechanisms characterised by several processes must have been involved. They particularly emphasised the uplift of the Hebrides-Shetland axis and the northwestern corner of southern Norway, uplift of the area along the incipient plate boundary, and accelerated subsidence of the North Sea, as phenomena that are crucial for understanding the driving forces. They saw the accelerated subsidence rate in the northern North Sea in the context of the arrival of the Icelandic plume and the crustal break-up of the North Atlantic, and speculated that primary or secondary thermal effects related to the plume contributed to the regional uplift of northern British Isles and the development of the Faeroe–Shetland Basin axis. In such a model, the rapid Eocene subsidence in some areas and the reduced rate of uplift in others would be related to reduction in the plume activity.

Alternatively, as discussed by Rohrman and van der Beek (1996), uplift of central southern Norway could be directly linked to the Icelandic plume. They propose a model of horizontal migration of hot material beneath southern Norway.

Mechanisms of Mesozoic uplift

In this era, the mainland became uplifted and appeared as a topographic high compared to the shelf (e.g., Riis, 1996). According to Rohrman et al. (1995, 1996), this long-lived high or domal feature was initiated, approximately 220 Ma ago, in eastern and southern south Norway, and then migrated westward. The uplifted region reached the west-coast of south Norway at approximately 160 Ma (Jurassic). This pattern is indicated by fast cooling as revealed by the AFT data.

A major uplifted region/province, or dome, represents areas of erosion and non-deposition. Recent field investigations and AFT-analysis have revealed that this apparently smooth feature probably was broken by extensional faulting in the west (Øygarden Fault Complex; Fossen et al., 1997; Redfield et al., 2003; Andriessen (pers. com.) and possibly by extensional/strike-slip faulting in the northwest (Møre-Trøndelag Fault Complex; Grønlie and Torsvik, 1989; Grønlie et al., 1990; Gabrielsen et al., 1999; Redfield et al., 2003). In this context, one has to hypothesize that the displacement related to the faulting outpaced erosional and depositional processes in order to cause relief. The AFT analysis, which records a temperature range in apatite, has to be evaluated in this light. The AFT data suggest that a total of between 1300 and 3500 m of sediments or rocks were removed over the central part of the high (Rohrman et al., 1995), which is consistent with the analyses of Riis (1996) and Lidmar-Bergström et al. (2000). Considering the diachroneity and, hence, the asymmetry of the early Mesozoic pattern of uplift in southern Norway, it is reasonable to associate this with the thermal effects and perhaps flank-uplift from the fading Permo-Triassic northern North Sea basin system and the Permo-Carboniferous Oslo Graben. Uplift of the coastal region of West Norway has also been discussed in light of Permo-Triassic magmatism (Færseth et al., 1976; Færseth, 1978; Eide et al., 1999; Fossen and Dunlap, 1999; Valle et al., 2002).

A pattern of structural juxtaposition, similar to that of the earlier periods, is associated with the Møre-Trøndelag Fault Complex, where neighbouring fault blocks record the AFT ages ranging between Permo-Triassic and possibly Palaeogene (Redfield et al. 2003, in press). These ages are consistent with the evidence that the innermost zones of the Møre-Trøndelag Fault Complex experienced cooling (exhumation) of large blocks during post-mid-Cretaceous times, quite possibly during the Teritary (Fig. 4). Although these data do not preclude the large components of strike-slip movement as postulated by other workers (Grønlie and Roberts, 1991) to have occurred across the Møre-Trøndelag Fault Complex, they suggest significant vertical block movement and partly down-westward motion in the order of 2–4 km throughout the extended periods. Movements occurred neither simultaneously nor were of equal magnitude. This opens the possibility that the now-eroded sedimentary basins had existed throughout the late Palaeozoic and Mesozoic within the realm of the Møre-Trøndelag Fault Complex, and possibly in western Norway. Inshore Jurassic basins of central (Sommeruga and Bøe, 2003) and south-western Norway (Fossen et al., 1998) are in harmony with this idea.

A major question yet, remains unsolved — Was the long-lived Mesozoic domal feature broken by faults, or is the uplift related to shoulder uplift from faulting? The width or wavelength of the uplifted region is of such a scale that shoulder uplift is less plausible, which makes it more likely that a thermal(?) process contributed to the uplift of south-central Norway. In the same period, one would expect that the thermal anomaly associated with the Oslo Graben vanished. Such possible thermal changes in south-central Norway may have stimulated continued activation along fault systems, for example the Lærdal–Gjende Fault System and its associated structures (Fig. 2).

Denudation of the Caledonides

In the analysis of the late- to post-Scandian topography of the Scandinavian Caledonides, a number of process-related parameters have to be addressed. Major contributors to surface relief within an orogenic wedge include: (i) structural framework, (ii) kinematic history including rate of deformation, (iii) spatial distribution of lithology, both with respect to mechanical behaviour as well as durability at the surface, (iv) metamorphic transitions, (v) relative strengths of basal layers and thrust-fault zones, (vi) ramp-flat geometry of thrusts and extensional faults, (vii) erosion levels and surface attitude, and (viii) sediment deposition on and in front of the orogen, at any given time (e.g., DeCelles and Mitra, 1995; Braathen et al., 1999).

Unroofing of the lower crust

The mechanism of unroofing of the lower crust is debated. It is generally accepted that unroofing of the eclogites of, for example, the Western Gneiss Region occurred during the extension of the thrust welt. Proposed processes include vertical flattening and subhorizontal stretching of the lower crust, combined with exhumation due to a footwall position of extensional shear zones, such as the Nordfjord-Sogn Detachment (Andersen and Jamtveit, 1990; Andersen et al., 1994; Krabbendam and Dewey, 1998; Engvik and Andersen, 2000). Alternatively, or additionally, as proposed by Robinson et al. (1997) and Terry et al. (2000), thrust-stacking at deep crustal levels, concurrent with transcurrent movements played a major role in addition to the extension. Other generally proposed mechanisms for exhumation of subducted crust include thrusting accompanied by erosion (Avigad, 1992) or upper plate extension (Platt, 1986; Jolivet et al., 1994), and buoyancy forces driving combined reverse and normal faulting (Chemenda et al., 1997).

Lithological and metamorphic effects

With the present knowledge of the Scandinavian Caledonides, it is impossible to establish the real effects of the various parameters. However, observations of importance for topography do exist. In the Scandinavian Caledonides, the structural framework, as revealed by the present level of erosion, is well-established, as is the basic framework of the kinematic history (e.g., Roberts and Gee, 1985; Milnes et al., 1997). The effect of lithologies is less straightforward, especially in the hinterland that was more affected by metamorphic transitions. Towards the foreland, it is well established that weak lithologies, such as the Alum Shale (Owen et al., 1990), acted as a decollement level above the Fennoscandian basement (e.g., Nystuen, 1981). Weak shales along basal detachments commonly produce overall topographic gradients in the order of 2–4° towards the foreland (Davis et al., 1983).

Another example of lithological effects relates to the aforementioned exhumation of the Western Gneiss Region (Eide et al., 2003; Fonneland and Pedersen, 2003). The abundant ortho-gneisses of this region are more durable to surface weathering and erosion than supracrustal nappe rocks. Therefore, progressive exhumation of gneisses could have caused increasing topographic elevation, which thereby triggered climate changes and related erosional process (Le Pichon et al., 1992) (Fig. 6). The total result would be higher overall altitudes and steeper topographic gradients.

In the hinterland, the effect of several metamorphic changes is important. One is the gradual enrichment of phyllosilicates in major shear zones as host-rocks were transformed into fault rocks. This commonly results in reduced shear strength (Wibberley, 1999) consistent with a strain softening-scenario (Braathen et al., in press). Weaker rocks in major shear zones weaken the orogenic wedge and will, in general, reduce the surface gradient (e.g., Smit et al., 2003). Another interesting metamorphic effect in the deeply buried parts of the Scandian orogen is the transition from granulite to eclogite (e.g., Austrheim, 1991; Engvik et al., 2000). This especially contributes to the weight budget (Dewey et al., 1993), since eclogite has a density similar to mantle rocks. Dense eclogite in the lower crust of an orogen is actually required in order to explain significant crustal thickness without unrealistic surface elevations. In the case of tectonic removal or metamorphic retrogradation of the eclogite, significant instability would result, triggering rapid surface uplift and even orogenic collapse (England and Houseman, 1988; Platt and Vissers, 1989). One can thereby speculate that the granulite to eclogite transition in the Caledonides initially cause general subsidence, unless compensated by thickening, by sedimentation or thrusting in the overlying orogenic wedge. On the contrary, tectonic removal or retrogression of the eclogites could have caused surface uplift.

Depositional provinces

The Devonian basins of western and central Norway are chiefly filled with very coarse clastic material (Steel et al., 1985; Osmundsen and Andersen, 2001), signalling a significant topographic relief in the hinterland region. This topography has been related to supradetachment faulting and continental basin formation with dominant fluvial to alluvial depositional environments. On the contrary, towards the foreland, syn-tectonic deposits are seen as finer-grained clastic rocks, reflecting lower relief and depositional environments changing between shallow marine and coastal to fluvial (e.g., Worsley et al., 1983). The weight effect of the deposits on the orogenic wedge is uncertain, since true thicknesses remain speculative. However, lower greenschist facies metamorphism in the Devonian hinterland basins signifies approximately 10 km burial (Sturt and Braathen, 2000; Svensen et al., 2001). Whether this overburden was represented by deposits or, alternatively, by fold-thrust structures or nappes, is still an open question.

Tectonic evolution

The processes affecting the surface of the Caledonides (Table 3), as highlighted earlier have to be viewed in light of the tectonic evolution in the Devonian to Early Carboniferous(?). Extensional denudation of the Scandian mountain-chain has been subdivided into two types of structural settings, both inferred to post-date orogenic contraction. They are mode I of low-angle detachment zones that reactivate the former contractional detachments and, subsequently, mode II of steeper, truncating shear zones (e.g., Nordfjord-Sogn Detachment, Lærdal-Gjende Fault System) (Fossen, 1992, 2000). New data from central Norway document contraction of the lower crust, synchronous with extension of the upper crust, in what consequently relates to a late Caledonian (latest-Scandian) phase (Terry et al., 2000; Braathen et al., 2002; Eide et al., 2002; Larsen et al., 2002; Osmundsen et al., 2003). Moreover, low-angle extensional detachments of Central and northern Central Norway reveal nearly orogen-parallel tectonic transport, in contrast to the near orogen-normal transport of western South Norway. In North Norway, orogen-normal tectonic transport is reported (e.g., Fossen and Rykkelid, 1992). On a broad scale, the complex, latest-Scandian phase was likely partly driven, both in time and space, by changes in buoyancy of the lower crust, presumably related to metamorphic transitions, as well as by orogen-parallel variations in boundary surfaces, e.g. the topography and the crust-mantel boundary. In addition, lateral plate movement during diminishing orogenic collision probably played a major role (e.g., Braathen et al., 2002; Dewey and Strachan, 2003). Interestingly, subsequent deformation in Central Norway is seen as steeper shear zones that truncated the low-angle detachments. Recent work suggests they formed in Late Devonian to Carboniferous times (Eide et al., 2002; Osmundsen et al., 2003). Thereby, there is a change from mainly detachment to shear zone/fault tectonics, which may mirror a shift in major driving forces. This shift could relate to changes in the operating processes, from those related to continental collision and orogenesis, to those related to rifting and passive margin formation (e.g., Osmundsen et al., this volume; Dewey and Strachan, 2003).

In southern Norway, a general topographic pattern could have persisted throughout the Late Palaeozoic. This is supported by recorded activity on the Hardanger Shear Zone/Lærdal-Gjende Fault System and the Nordfjord-Sogn Detachment Zone, which continued to be activated (e.g. Andersen et al., 1999; Eide et al., 1999), albeit at a much smaller scale than the orogen-related event. If relief still existed, erosion would gradually have reduced the total relief so that both the maximum elevation and the topographic gradients of Southwest and Central Norway had become drastically reduced.

Transition in to the post-orogenic period

Most of the denudation of the Scandian Caledonides was perhaps fulfilled in the Carboniferous. The following Carboniferous/Permian active crustal thinning to the west of the Norwegian mainland and in the Oslo Graben probably contributed to enhancing the overall relief towards the end of the late Palaeozoic by creating negative topographic elements to the west and the east (Fig. 4). Both these basins affected wide crustal panels. The more than 300 km wide Permian offshore basin was much wider than the later Jurassic-Cretaceous Viking Graben (Gabrielsen et al., 1990; Færseth et al., 1996; Odinsen et al., 2000). From mapping of fault systems (Ramberg and Larsen, 1978; Olaussen et al., 1994; Sundvor and Larsen, 1994) and lineament analysis (Gabrielsen et al., 2002), it is also clear that the Oslo Graben originally stretched beyond its present, preserved limits and that it originally was more than 180 km broad, at its widest. The sedimentary infilling in the Permian offshore basins as well as those in the Oslo Graben, hence, only outline parts of the extended crustal panel.

Conclusions

The history of vertical movements in south-western Fennoscandia from the Present to the Devonian can be subdivided into two main configurations and altogether into five separate stages, each characterised by its distinguishing set of mechanisms. The late- to post-Caledonian stage was dominated by orogenic denudation processes, which followed the Caledonian Orogeny. This probably included thermal as well as isostatic effects, which contributed to the development of an asymmetrical long-wavelength uplift area with a low-relief eastern flank towards the Baltic countries, and a western hinterland region of high relief, especially above exhumed gneissic regions. The hinterland probably had a rugged topography, similar to that of the present Himalaya. Highs and lows of the following events were probably, to a larger extent, thermally controlled. They include Permo-Carboniferous, Permian and Jurassic rifts, of which the latter particularly affected the flanks of a south Norwegian high, or dome. From the earliest Tertiary, this feature seems to have been established and supported by horizontal transfer of hot material, which was associated with the Icelandic plume. Pluming may be superimposed by more recent glacial rebound, which presently seems to be interfering with the long-term effects of the North Atlantic asthenospheric plume.

References

Ahlman, H.W., 1919. Geomorphological studies in Norway. Geogr. Ann. 1: 1–252.

Alhilali, K.A. and Damuth, J.E., 1987. Slide block (?) of Jurassic sandstone and submarine channels in the basal Upper Cretaceous of the Viking Graben, Norwegian North Sea. Mar. Petrol. Geol., 4: 35–48.

Anda, E., Blikra, L.H. and Braathen, A., 2002. The Berill fault—first evidence of neotectonic faulting in southern Norway. Nor. J. Geol. (NGT) 82: 175–182.

Andersen, T.B., 1993. The role of extensional tectonics in the Caledonides of south Norway: Discussion. J. Struct. Geol., 15: 1379–1380. (t)

Andersen, T.B.,1998. Extensional tectonics in the Caledonides of southern Norway, and overview. Tectonophysics, 285: 333–351.

Andersen, T.B., Torsvik, T.H., Eide, E.A., Osmundsen, P.T. and Faleide, J.I., 1999. Permian and Mesozoic extensional faulting within the Caledonides of central South Norway. J. Geol. Soc., London, 156: 1073–1080.

Andersen, T.B. and Jamtveit, B., 1990. Uplift of deep crust during orogenic extensional collapse: A model based on field studies in the Sogn-Sunnfjord Region of Western Norway. Tectonics, 9: 1097–1111.

Andersen, T., Jamtveit, B., Dewey, J.F. and Swensson, E., 1991. Subduction and eduction of continental crust: major mechanisms during continent–continent collosion and orogenic extensional collapse, a model based on the south Norwegian Caledonides. Terra Nova, 3: 303–310.

Andersen, T.B. and Osmundsen, P.T. and Jolivet, L., 1994. Deep crustal fabrics and a model for extensional collapse of the southwest Norwegian Caledonides. J. Struct. Geol., 16: 1191–1203.

Andersen, T.B., Hartz, E., Torsvik, T.H., Osmundsen, P.T., Andersen, A., Eide, E.A. and Braathen, A., 2002. The aftermath of the Caledonian collision in the North Atlantic Region: A structural template for later events? (abstract). In: A. Hurst (Editor), Onshore–Offshore Relationships on the Nordic Atlantic Margin. NGF Abstracts and proceedings 2, 2002 of the Norwegian Petroleum Society (NPF) and Norwegian Geological Society (NGF) Conference, 7–9th Oct. Trondheim, pp. 12–14.

Andersson, J., Möller, C. and Johansson, L., 2002. Zircon geochronology of migmatite gneisses along the Mylonite Zone (S Sweden): a major Sveconorwegian terrane boundary in the Baltic Shield. Precambrian Res., 114: 121–147.

Austrheim, H., 1991. Eclogite formation and dynamics of crustal roots under continental collision zones. Terra Nova, 3: 492–499.

Avigad, D., 1992. Exhumation of coesite-bearing rocks in the Dora Maira massif (western Alps, Italy). Geology, 20: 947–950.

Badley, M.E., Egeberg, T. and Nipen, O., 1984. Development of rift basins illustrated by the structural evolution of the Oseberg structure, Block 30/6, offshore Norway, J. Geol. Soc., London, 141: 639–649.

Bannister, S.C., Ruud, B.O. and Husebye, E.S., 1991. Tomographic estimates of sub-Moho seismic velocities in Fennoscandia and structural implications. Tectonophysics, 189: 37–53.

Biddle, K. and Rudolph, K.W., 1988. Early Tertiary inversion in the Stord Basin, Norwegian North Sea. J. Geol. Soc., London, 145: 603–611.

Bijwaard, H. and Spakman, W., 1999. Tomographic evidence for narrow whole mantle plume below Iceland. Earth Planet. Sci. Lett., 166: 121–126.

Boldreel, I.O. and Andersen, M.S., 1993. Late Paleocene to Miocene compression in the Faroe-Rockall area. In: J.R. Parker (Editor), Petroleum Geology of Northwest Europe: Proceedings of the 4th Conference. Geological Society, London, pp. 1925–1034.

Boldreel, I.O. and Andersen, M.S., 1994. Tertiary development of the Faroe-Rockall Plateau based on reflection seismic data. Bull. Geol. Soc. Den., 41: 162–180.

Bowie, W., 1929. Isostasy—The science of the equilibrium of the Earth's crust. E.P. Dutton, New York, 275 pp.

Braathen, A., 1999. Kinematics of post-Caledonian polyphase brittle faulting in the Sunnfjord region, western Norway. Tectonophysics 302,: 99–121.

Braathen, A., Bergh, S.G. and Maher Jr., H.D., 1999. Application of a critical wedge taper model to the Tertiary transpressional fold-thrust belt on Spitsbergen. Geol. Soc. Am. Bull., 111 (10): 1468–1485.

Braathen, A., Osmundsen, P.T., Nordgulen, ø., Roberts, D. and Meyer, G., 2002. Orogen-parallel extension of the Caledonides in northern Central Norway: an overview. Nor. J. Geol. (NGT), 82: 225–241.

Braathen, A., Osmundsen, P.T. and Gabrielsen, R.H., in press (2004). Dynamic development of fault rocks in a crustal-scale detachment; an example from western Norway. Tectonics.

Braathen, A., Nordgulen, Ø., Osmundsen, P.-T., Andersen, T.B., Solli, A. and Roberts, D., 2000. Devonian, orogen-parallel, opposed extension in the Central Norwegian Caledonides. Geology, 28 (7): 615–618.

Brekke, H. and Riis, F., 1987. Mesozoic tectonics and basin evolution of the Norwegian Shelf between 60°N and 72°N. Norsk Geol. Tidsskr., 67: 295–322.

Bugge, T., Prestvik, T. and Rokoengen, K., 1980. Lower Tertiary volcanic rocks off Kristiansund – Mid Norway, Mar. Geol., 25: 277–286.

Bungum, H., Alsaker, A., Kvamme, L.B. and Hansen, R.A., 1991. Seismicity and seismotectonics of Norway and nearby continental shelf areas. J. Geophys. Res., 24: 8249–8265.

Bungum, H., Swearingen, P.H. and Woo, G., 1986. Earthquake hazard assessment in the North Sea. Physics Earth Planet. Int., 44: 201–210.

Bøe, R., 1991. Structure and seismic stratigraphy of the innermost mid-Norwegian continental shelf: an example from the Frohavet area. Mar. Petrol. Geol., 8: 140–152.

Cartwright, J.A., 1989. The kinematics of inversion in the Danish Central Graben. In: M.A. Cooper and G.D. Williams

(Editors), Inversion Tectonics. Geol. Soc., London, Spec. Publ., 44: 153–175.
Cathles, L.M., 1975. The viscosity of the earth's mantle. Princeton Univ. Press, Princeton, N.J., 386 pp.
Chauvet, A. and Séranne, M., 1994. Extension-parallel folding in the Scandinavian Caledonides: implications for late-orogenic processes. Tectonophysics, 238: 31–54.
Chemenda, A., Matte, P. and Sokolov, V., 1997. A model of Palaeozoic obduction and exhumation of high-pressure/low-temperature rocks in the southern Urals. Tectonophysics, 276: 217–227.
Clausen, J.A., Gabrielsen, R.H., Reksnes, P.A. and Nysæther, E., 1999. Development of intraformational (Oligocene-Miocene) faults in the northern North Sea: influence from remote stresses and doming of Fennoscandia. J. Struct. Geol., 21: 1457–1475.
Clausen, O.R., Nielsen, O.B., Huuse, M. and Michelsen, O., 2000. Geological indications for Palaeogene uplift in the eastern North Sea Basin. Earth Planet. Change, 24: 175–187.
Cloetingh, S., 1986. Intraplate stresses: a new tectonic mechanism for fluctuations of reletive sea level. Geology, 14: 617–620.
Cloetingh, S. and Burov, E.B., 1996. Thermomechanical structure of European continental lithosphere: constraints from rheological profiles and EET estimates. Int. Geophys. J., 124: 695–723.
Cloetingh, S., Eldholm, O., Larsen, B.T., Gabrielsen, R.H. and Sassi, W., 1994. Dynamics of extensional basin formation and inversion: introduction. Tectonophysics, 240: 1–9.
Cloetingh, S., Gradstein, F., Kooi, H., Grant, A.C., Kaminski, M., 1990. Plate reorganization: A cause of rapid Neogene subsidence and sedimentation around the North Atlantic? J. Geol. Soc., London, 147: 495–506.
Cloetingh, S. and Kooi, H., 1992. Intraplate stresses and dynamic aspects of rifted basins. Tectonophysics, 215: 167–185.
Cloetingh, S., Reemst, P., Kooi, H., and Fanavoll, S., 1992. Intraplate stresses and the post-Cretaceous uplift and subsidence in the northern Atlantic basins. Norsk Geol. Tidsskr., 72, 229–235.
Cloetingh, S., van Vees, J.D., van der Beek, P.A. and Spadini, G., 1995. Role of pre-rift rheology in kinematics of extensional basin formation: constraints from thermomechanical models of Mediterranean and intracratonic basins. Mar. Petrol. Geol., 23(8): 793–807.
Daly, R.A., 1934. The changing world and the ice age. Yale Univ. Press., New Haven, 272 pp.
Davis, D., Suppe, J. and Dahlen, F.A., 1983. Mechanics of fold-and-thrust belts and accretionary wedges. Jour. Geophys. Res. 88 (B2): 1153–1172.
DeCelles, P.G. and Mitra, G., 1995. History of the Sevier orogenic wedge in terms of critical taper models, northeast Utah and southwest Wyoming. Geol. Soc. Am. Bull., 107: 454–462.
Dehls, J.F., Olesen, O., Olsen, L. and Blikra, L.H., 2000a. Neotectonic faulting in northern Norway; the Stuoragurra and Nordmannvikdalen postglacial faults. Quaternary Sci. Rev., 19: 1447–1460.
Dehls, J.F., Olesen, O., Bungum, H., Hicks, E., Lindholm, C.D. and Riis, F., 2000b. Neotectonic map, Norway and adjacent areas 1 : 3 mill. Norges Geologiske Undersøkelse, Trondheim, Norway.
Dewey, J.F. and Bird, J.M., 1970. Mountain belts and the new global tectonics. J. Geophys. Res., 75: 2625–2647.
Dewey, J.F., Ryan, P.D. and Andersen, T.B., 1993. Orogenic uplift and collapse, crustal thickness, fabrics and metamorphic phase changes: the role of eclogites n: H.M. Prichard, T. Alabaster, N.B.W. Harris and C.R. Neary (Editors), Magmatic Processes and Plate Tectonics. Geol. Soc., London, Spec. Publ., 76: 325–343.
Dewey, J.F. and Strachan, R.A., 2003. Changing Silurian-Devonian relative plate motion in the Caledonides: sinistral transpression to sinistral transtension. J. Geol. Soc., London, 160: 219–229.
Doré, A.G., 1992. The Base Tertiary surface of southern Norway and the northern North sea. Norsk Geol. Tidsskr., 72: 259–266.
Doré, A.G. and Gage, M.S., 1987. Crustal alignments and sedimentary domains in the evolution of the North Sea, north-east Atlantic margin and Barents shelf. In: J. Brooks and K. Glennie (Editors), Petroleum Geology of North-West Europe, Graham and Trotman, London, pp. 1131–1148.
Doré, A.G. and Lundin, E.R., 1996. Cenozoic compressional structures on the NE Atlantic margin: nature, origin and potential significance for hydrocarbon exploration. Petrol. Geosci., 2: 299–311.
Eide, E.A., Haabesland, N.E., Osmundsen, P.T., Kendrick, M.A., Andersen, T.B. and Roberts, D., 2003. 40Ar/39Ar geochronology and indirect dating of continental sedimentary rocks: Late Devonian-Early Carboniferous 'Old Red Sandstone' discovered in outer Trøndelag. In: H.A. Nakrem (Editor), Abstracts and Proceedings of Geological Society of Norway (NGF) 1, Vinterkonferansen 2003.
Eide, E.A., Torsvik, T.H. and Andersen, T.B., 1997. Absolute dating of brittle fault movements: Late Permian and Late Jurassic extensional fault breccias in western Norway. Terra Nova 9: 135–139.
Eide, E.A., Torsvik, T.H., Andersen, T.B. and Arnaud, N.O., 1999. Early Carboniferous Unroofing in Western Norway: a tale of alkali feldspar thermochronology. Journal of Geology, 107: 353–374.
Eide, E.A., Osmundsen P.T., Meyer, G.B., Kendrick, M.A. and Corfu, F. 2002. The Nesna Shear Zone, north-central Norway: An 40Ar/39Ar record of Early Devonian — Early Carboniferous ductile extension. Norsk Geol. Tidsskr., 82: 317–339.
Eliassen, H., 2003. Sen-Kaledonsk og yngre (Mesosoisk og Kenosoisk) forkastninger på Frøya, Møre-Trøndelag kysten. Unpubl. Cand. Scient. Thesis, University of Bergen.
England, P. and Houseman, G., 1988. Finite strain calculation of continental deformation 2. Comparison with the India–Asia Collision Zone. J. Geophys. Res., 91 (B3): 3664–3676.
Engvik, A.K. and Andersen, T.B., 2000. Evolution of Caledonian deformation fabrics under ecglogite and amphibolite facies at Vårdalsneset, Western Gneiss Region, Norway. J. Metamorph. Geol., 18: 241–257.
Engvik, A.K., Austrheim, H. and Andersen, T.B., 2000. Structural, mineralogical and petrophysical effects on deep crustal rocks of fluid-limited polymetamorphism, Western Gneiss Region, Norway. J. Geol. Soc., London, 157: 121–134.
Fagerland, N., 1983. Tectonic analysis of a Viking Graben border fault. AAPG Bull., 67: 2125–2136.
Faleide, J.I., Kyrkjebø, R., Kjennerud, T., Gabrielsen, R.H., Jordt, H., Fanavoll, S. and Bjerke, M.D., 2002. Tectonic impact on sedimentary processes during Cenozoic evolution of the northern North Sea and surrounding areas. In: A.G. Doré, J.A. Cartwright, M.S. Stoker, J.P. Turner and N. White (Editors), Exhumation of the North Atlantic Margin: Mechanisms and Implications for Petroleum Exploration., Geol. Soc., London, Spec. Publ., 196: 235–269.
Faleide, J.I., Vågnes, E. and Gudlaugsson, S.T., 1993. Late Mesozoic—Cenozoic evolution of the south-western Barents Sea in a regional rift-shear tectonic setting. Mar. Petrol. Geol., 10: 186–214.
Fejerskov, M. and Lindholm, C., 2000. Crustal stresses in and around Norway: An evaluation of stress generating mechanisms. In: A. Nøttvedt et al. (Editors), Geol. Soc., London, Spec. Publ., 167: 451–467. (p,b)
Fejerskov, M., Lindholm, C., Myrvang, A. and Bungum, H., 2000. Crustal stress in and around Norway: A compilation of in-situ stress observations. In: A. Nøttvedt et al. (Editors), Geol. Soc., London, Spec. Publ., 167: 441–450.
Fitzgerald, P.G. and Gleadow, A.J.W., 1988. Fission track geochronology, tectonics, and structure of the Transantarctic Mountains in Northern Victoria Land, Antarctica. Chemical. Geology, 73: 169–178.

Fjeldskaar, W., 1994. Viscosity and thickness of the asthenosphere detected from Fennoscandian uplift. Earth Planet. Sci. Lett., 126: 39–410.

Fjeldskaar, W. and Cathles, L., 1991. Rheology of mantle and lithosphere inferred from post-glacial uplift in Fennoscandia. In: R. Sabadini, K. Lambeck and E. Boschi (Editors), Glacial Isostasy, Sea Level and Mantle Rheology. Kluwer Academic Press, pp. 1–19.

Fjeldskaar, W., Lindholm, C., Dehls, J.F. and Fjeldskaar, I., 2000. Post-glacial uplift, neotectonics and seismicity in Fennoscandia. Quaternary Sci. Rev. 19: 1413–1422.

Fonneland, H.C. and Pedersen R.B., 2003. Provenance of the Hornelen Devonian basin; implications for the exhumation history of the Western Gneiss Region, Norway, Geological Society of Norway (NGF), Abstracts and Proceedings, 1-2003, pp. 28–29.

Fossen, H., 1992. The role of extensional tectonics in the Caledonides of south Norway. J. Struct. Geol., 14: 1033–1046.

Fossen, H., 1993. Linear fabrics in the Bergsdalen Nappes, southwest Norway: implications for deformation history and fold development. Norsk Geol. Tidsskr., 73: 95–108.

Fossen, H., 1998. Advances in understanding the post-Caledonian structural evolution of the Bergen area, West Norway. Norsk Geol. Tidsskr., 78: 33–46.

Fossen, H., 2000. Extensional tectonics in the Caledonides: Synorogenic or postorogenic? Tectonics, 19 (2): 213–224.

Fossen, H. and Dunlap, W.J., 1998. Timing and kinematics of Caledonian thrusting and extensional collapse, southern Norway: evidence from 40Ar/39Ar thermochronology. J. Struct. Geol., 20: 765–781.

Fossen, H. and Hesthammer, J., 1997. Structural Geology of the Gullfaks Field, northern North Sea. In: M.P. Coward, H. Johnson and T.S. Daltaban (Editors), Structural Geology in Reservoir Characterization. Geol. Soc., London, Spec. Publ., 127: 213–261.

Fossen, H., Mangerud, G., Hesthammer, J., Bugge, T., Gabrielsen, R.H., 1998. The Bjorøy Formation: a newly discovered occurrence of Jurassic sediments in the Bergen Arc System. Norsk Geol. Tidsskr., 77: 269–287.

Francke, W., Haak, V., Oncken, O. and Tanner D. (Editors), 2000. Orogenic processes; quantification and modeling in the Variscian Belt. Geol. Soc. London, Spec. Publ., 179: 456 pp.

Færseth, R.B., 1978. Mantle-derived lherzolite xonotiths and megachrysts from Permo-Triassic dykes, Sunnhordland, western Norway. Lithos 11: 23–35.

Færseth, R.B., Gabrielsen, R.H and Hurich, C.A., 1995. Influence of basement in structuring of the North Sea Basin offshore southwest Norway. Norsk Geol. Tidsskr., 75: 105–119.

Færseth, R.B., McIntyre, R.M and Naterstad, J., 1976. Mesozoic alkaine dykes in the Sunnhordland region, western Norway: ages, geochemistry and regional significance. Lithos, 9: 331–245.

Færseth, R.B., Knudsen, B.-E., Liljedahl, T., Midbøe, P.S. and Søderstrøm, B., 1997. Oblique rifting and sequential faulting in the Jurassic development of the northern North Sea. J. Struct. Geol., 19 (10): 1285–1302.

Færseth, R.B., Sjøblom, T.S., Steel, R.J., Liljedahl, T., Sauar, B.E. and Tjelland, T., 1995b. Tectonic controls on Bathonian-Volgian syn-rift successions on the Visund fault block, northern North Sea. In: R.J. Steel et al. (Editors), Sequence stratigraphy on the NW European margin. Norwegian Petroleum Society (NPF), Special Publication 5, Elsevier, Amsterdam, pp. 325–346.

Gabrielsen, R.H. and Færseth, R.B., 1988. Cretaceous and Tertiary reactivation of master fault zones of the Barents Sea (extended abstract). In: W.K. Dallmann, Y. Ohta and A. Andresen (Editors), Tertiary Tectonics of Svalbard. Extended abstracts from Symposium held in Oslo 26 and 27 April 1988. Norsk Polarinst., Rep. Ser. 46: 93–97.

Gabrielsen, R.H., Færseth, R.B., Steel, R.J., Idil, S. and Kløvjan, O.S., 1990. Architectural styles of basin fill in the northern Viking Graben. In: D.J. Blundell and A.D. Gibbs (Editors), Tectonic Evolution of the North Sea Rifts. Clarendon Press, Oxford, pp. 158–179.

Gabrielsen, R.H., Grunnaleite, I. and Ottesen, S., 1992. Reactivation of faults complexes in the Loppa High area, southwestern Barents Sea. In: T.O. Vorren, E. Bergsager, Ø.A. Dahl-Stamnes, E. Holter, B. Johansen, E. Lie and T.B. Lund (Editors), Arctic Geology and Petroleum Potential. Norwegian Petroleum Society (NPF), Special Publication 2, Elsevier, Amsterdam, pp. 631–641.

Gabrielsen, R.H., Grunnaleite, I. and Rasmussen, E., 1997. Cretaceous and Tertiary inversion in the Bjørnøyrenna Fault Complex, south-western Barents Sea. Mar. Petrol. Geol., 14 (2): 165–178.

Gabrielsen, R.H., Kyrkjebø, R., Faleide, J.I., Fjeldskaar, W. and Kjennerud, T., 2001. The Cretaceous post-rift basin configuration of the northern North Sea. Petrol. Geosci., 7: 137–154.

Gabrielsen, R.H., Odinsen, T. and Grunnaleite, I., 1999. Structuring of the Northern Viking Graben and the Møre Basin; the influence of basement structural grain, and the particular role of the Møre-Trøndelag Fault Complex. Mar. Petrol. Geol., 16: 443–465.

Gabrielsen, R.H. and Robinson, C., 1984. Tectonic inhomogeneities of the Kristiansund-Bodø Fault Complex, off-shore mid-Norway. In: A.M. Spencer et al. (Editors), Petroleum Geology of the North European Margin. Norwegian Petroleum Society (NPF), Graham and Trotman, London, pp. 397–406.

Gallagher, K., Brown, R. and Johnson, C., 1998. Fission track analysis and its applications to geological problems. Annual Review Earth Planet. Sci., 26: 519–572.

Gillmore, G.K., Kjennerud, T. and Kyrkjebø, R., 2001. The reconstruction and analysis of palaeo-water depths: a new approach and test of micropalaentological approaches in the port-rift (Cretaceous to Quaternary) interval of the Northern North Sea. In: O. Martinsen, T. Dreyer (Editors), Sedimentary Environments Offshore Norway — Palaeozoic to Recent. Norwegian Petroleum Society (NPF) Special Publication 10, Elsevier, Amsterdam, pp. 365–381.

Gjessing. J., 1967. Norway's paleic surface. Norsk Geogr. Tidsskr., 21: 69–132.

Glennie, K.W. and Boegner, P.L.E., 1981. Sole Pit inversion tectonics. In: L.W. Illing and G.D. Hobson (Editors), Petroleum Geology of the Continental Shelf of North-West Europe. (Institute of Petroleum, London), pp. 110–120.

Goldschmidt, V.M., 1921. Die kaledoische Deformation der südnorwegischen Urgebirgstafel. Skrifter, Det norske Videnskapsakademi, Oslo, Matematisk Naturvidenskabelig Klasse, 19: 1–11.

Gradstein, F.M, Kaminski, M.A, Berggren, W.A., Kristiansen, I.L. and D'Iorio, M., 1994. Cainozoic Biostratigraphy of the North Sea and Labrador Sea. Micropaleontology, 40, Suppl., 152 pp.

Graue, K., 1992. Extensional Tectonics in the Northernmost North Sea: Rifting, Uplift, Erosion and Footwall Collapse in Late Jurassic to Early Cretaceous Times. In: A.M. Spencer (Editor), Generation, Accumulation and Production of Europe's Hydrocarbons II., Special Publication of the European Association of Petroleum Geoscientists, 2, Springer Verlag, Berlin, Heidelberg, pp. 23–34.

Grønlie, A., Harder, V. and Roberts, D., 1990. Preliminary fission-track ages on fluorite mineralisation along fracture zones, inner Trondheimsfjord, Central Norway. Norsk Geol. Tidsskr., 70: 173–178.

Grønlie, A. and Roberts, D., 1989. Resurgent strike-slip duplex development along the Hitra-Snåsa and Verran faults, Møre-Trøndelag Fault Zone, Central Norway. J. Struct. Geol., 11: 295–305.

Grønlie, A. and Torsvik, T.H., 1989. On the origin and age of hydrothermal thorium-enriched carbonate veins and breccias in the Møre-Trøndelag Fault Zone, Central Norway. Norsk Geol. Tidsskr., 69: 1–19.

Grønlie, G. and Ramberg, I.B., 1973. A system of determining gravimetric terrain corrections in Norway. Norg. Geol. Unders. Bull., 294: 1–7.

Hartz, E., Andresen, A. and Andersen, T.B., 1994. Structural observations adjacent to a large-cale extensional detachment zone in the hinterland of the Norwegian Caledonides. Tectonophysics, 231:123–137.

Heeremans, M., Larsen, B.T. and Stel, H., 1996. Paleostress reconstruction from kinematic indicators in the Oslo Graben: new constraints on the mode of rifting. Tectonophysis, 266: 55–79.

Heeremans, M., Larsen, B.T. and Stel, H., 1997. Erratum to 'Paleostress reconstruction from kinematic indicators in the Oslo Graben: new constraints on the mode of rifting'. (Tectonophysis, 266, 55–79), Tectonophysics 277: 339–344.

Heiskanen, W.A. and Moritz, H., 1967. Physical Geodesy. W.H. Freeman, San Fransisco, 364 pp.

Hendriks, B.W.H. and Andriessen, P.A.M., 2002. Pattern and timing of post-Caldonian denudation of northern Scandinavia constrained by apatite fission track thermochronology. In: A.G. Doré, J.A. Cartwright, M.S. Stoker, J.P. Turner and N. White (Editors), Exhumation of the North Atlantic Margin: Mechanisms and Implications for Petroleum Exploration., Geol. Soc., London, Spec. Publ., 196: 117–137.

Hesthammer, J. and Fossen, H., 1999. Evolution and geometries of gravitational collapse structures with examples from the Statfjord Field, northern North Sea. Mar. Petrol. Geol., 16: 259–291.

Holtedahl, H., 1953. On the oblique uplift of some northern lands. Norsk Geogr. Tidsskr., 14: 132–139.

Holtedahl, H., 1984. High Pre-Late Weichselian sea-formed caves and other marine features on the Møre-Romsdal coast, West Norway. Norsk Geol. Tidsskr., 64: 75–85.

Hooper, R.J., Leng Siang Goh and Dewey, F., 1995. The inversion history of the northeastern Broad Fourteens Basin. In: J.G. Buchanan and P.G. Buchanan (Editors), Basin Inversion. Geol. Soc., London, Spec. Publ., 88: 307–318.

Hospers, J. and Ediriweera, K.K, 1988. Mapping the top of the crystalline continental crust in the Viking Graben area, North Sea. In: Norwegian research in the International Lithosphere programme. Norg. Geol. Unders., Spec. Publ., 3: 21–28.

Hossack, J.R., 1984. The geometry of listric growth faults in the Devonian basins of Sunnfjord, West Norway. J. Geol. Soc., London, 141: 629–638.

Jamieson, T.F., 1882. On the cause of the depression and re-elevation of the land during the glacial period. Geological Magazine, 2/9.

Japsen, P. and Chalmers, J.A., 2000. Neogene uplift and tectonics around the North Atlantic: overview. Global Planet. Change, 24: 165–173.

Jensen, L.N., Riis, F. and Boyd, R., (Editors) 1992. Post-Cretaceous Uplift and Sedimentation along the western Fennoscandian Shield. Norsk Geol. Tidsskr., 72 (3): 223–338.

Jensen, L.N. and Schmidt, B.J., 1993. Neogene uplift and erosion offshore South Norway: magnitude and consequences for hydrocarbon exploration in the Farsund Basin. In: A.M. Spencer (Editor), Generation, Accumulation, and Production of Europe's Hydrocarbons; III. Special Publication of the European Association of Petroleum Geoscientists, 3: 79–88.

Jolivet, L., Daniel, J.M., Truffert, C. and Goffe, B., 1994. Exhumation of deep crustal metamorphic rocks and crustal extension in arc and back-arc regions. Lithos, 33: 3–30.

Jordt, H., Faleide, J.I., Bjørlykke, K. and Ibrahim, M.T., 1995. Cenozoic sequence stratigraphy of the central northern North Sea Basin: tectonic development, sediment distribution and provenance areas. Mar. Petrol. Geol., 8: 845–879.

Jordt, H., Thyberg, B. and Nøttvedt, A., 2000. Cenozoic evolution of the central and northern North Sea with focus on differential vertical movements of the basin floor and surrounding clastic source areas. In: A. Nøttvedt (Editor), Dynamics of the Norwegian Margin. Geol. Soc., London, Spec. Publ., 167: 219–243.

Kinck, J.J., Husebye, E.S. and Larsson, F.R., 1993. The Moho depth distribution in Fennoscandia and the regional tectonic evolution from Archean to Permian times. Precambrian Res., 64: 23–51.

Kjennerud, T., Faleide, J.I., Gabrielsen, R.H., Gillmore, G.K., Kyrkjebø, R., Lippard, S.J. and Løseth, H., 2001. Structural restoration of Cretaceous-Cenozoic (post-rift) palaobathymetry. In: O.J. Marthinsen et al. (Editors), Sedimentary Environments offshore Norway – Palaeozoic to Recent. Norwegian Petroleum Society (NPF), Special Publication 9, Elsevier, Amsterdam, pp. 347–364.

Kyrkjebø, R., Gabrielsen, R.H. and Faleide, J.I., in press. Unconformities related to the Jurassic-Cretaceous syn/post-rift transition of the northern North Sea. J. Geol. Soc., London.

Kyrkjebø. R., Kjennerud, T., Gillmore, G.K., Faleide, J.I. and Gabrielsen, R.H., 2001. Cretaceous – Tertiary palaeo-bathymetry in the northern North Sea; integration of palaeo-water depth estimates obtained by structural restoration and micropalaentological analysis. In: O.J. Marthinsen et al. (Editors), Sedimentary Environments offshore Norway — Palaeozoic to Recent. Norwegian Petroleum Society (NPF), Special Publication 9, Elsevier, Amsterdam, pp. 321–345.

Lagerbäck, R., 1979. Neotectonic structures in northern Sweden. Geologiska Föreningen i Stockholm Förhandlingar, 100: 271–278.

Larsen, Ø., Skår, Ø. and Pedersen, R.B,. 2002. U-Pb zircon and titanite geochronological constraints on the late- to post-Caledonian basement evolution of the Scandinavian Caledonides in Central Nordland, northern Norway. Nor. J. Geol. (NGT), 82: 1–14.

Larsen, Ø., Fossen, H., Langeland, K. and Pedersen, R.B., 2003. Kinematics and timing of polyphase post-Caledonian deformation in the Bergen area, SW Norway. Nor. J. Geol. (NGT), 83: 149–165.

Le Pichon, X., Fournier, M., and Jolivet, L., 1992. Kinematics, topography, shortening, and extrusion in the India–Eurasia collision. Tectonics 11: 1085–1098.

Lidmar-Bergstöm, K., Ollier, C.D. and Sulebak, J.R., 2000. Landforms and uplift history of Southern Norway. Global Planet. Change, 24: 211–231.

Lyell, C., 1835. On the proofs of a gradual rising of the land in certain parts of Sweden. Phil. Trans. Royal Society London, 125: 1–38.

Manum, S., 1964. Nytt om Verran-kullene. Fossil-nytt, 3: 1–2.

Mangerud, J., Sønstegaard, E., Sejrup, H.-P. and Haldorsen, S., 1981. A continuous Eemian-Early Weichselian sequence containing pollen and marine fossils at Fjøsanger, western Norway. Boreas, 10: 137–208.

Martinsen, O.J., Bøen, F., Charnock, M.A., Mangerud, G. and Nøttvedt, A., 1999. Cenozoic development of the Norwegian margin 60–64°N: sequences and sedimentary response to variable basin physiography and tectonic setting. In: A.J. Fleet and S.A.R. Boldy (Editors), Petroleum Geology of NW Europe, Proceedings of the 5th Conference, Geological Society, London, pp. 193–304.

Milnes, A.G., Wennberg, O.P., Skår, Ø. and Koestler, A.G., 1997. Contraction, extension and timing in the south Norwegian Caledonides — The Sognefjord Transect. In: J.-P. Burg and M. Ford (Editors), Orogeny through Time. Geol. Soc., London, Spec. Publ., 121: 123–148.

Mjelde, R., Sellevoll, M.A., Shimamura, H., Iwasaki, T. and Kanazawa, T., 1992. A crustal study off Lofoten, N. Norway, by use of 3-component ocean bottom seismographs. Tectonophysics, 212: 269–288.

Mjelde, R., Sellevoll, M.A., Shimamura, H., Iwasaki, T. and Kanazawa, T., 1993. Crustal structure beneath Lofoten, N. Norway, from vertical incidence and wide-angle seismic data. Geophys. J. Int. 114: 116–126.

Mjelde, R., Digranes, P., Shimamura, H., Shiobara, H., Kodaira, S., Brekke, H., Egebjerg, T., Sørenes, N. and Thorbjørnsen, S., 1998. Crustal structure of the northern part of the Vøring Basin, mid-Norway margin, from wide-angle seismic and gravity data. Tectonophysics 293: 175–205.

Morley, C.K., 1987. Lateral and vertical changes of deformation style in the Osen-Røa thrust sheet, Oslo region. J. Struct. Geol., 9: 331–343.

Mosar, J., Eide, E.A., Osmundsen, P.T., Sommaruga, A. and Torsvik, T.H., 2002. Greenland—Norway separation: A geodynamic model for the North Atlantic. Nor. J. Geol. (NGT), 82 (4): 281–298.

Møller, C., 1988. Geology and metamorphic evolution of the Roan area, Vestranden Western Gneiss Region, Central Norwegian Caledonides. Norg. Geol. Unders. Bull., 413: 1–31.

Nansen, F., 1904. The bathymetrical features of the north polar seas, with a discussion of the continental shelves and previous oscillations of the shore-line. In: F. Nansen (Editor), The Norwegian North Polar Expedition 1893–1896. Scientific Results IV (13), Jacob Dybwad, Cristiania, 232 pp.

Nansen, F., 1920. En ferd til Spitsbergen. J. Dybwad, Oslo, 227 pp.

Nansen, F., 1927. The earth's crust, its surface forms and isostatic adjustments. Norsk Videnskapsakademi i Oslo Skrifter, Matematisk-Naturvidenskapelig Klasse, 12: 1–121.

Nordgulen, Ø., Braathen, A., Corfu, F., Osmundsen, P.T. and Husmo, T., 2002. Polyphase kinematics and geochronology of the late-Caledonian Kollstraumen detachment, north-central Norway. Nor. J. Geol. (NGT) 82: 299–316.

Norton, M.G., 1986. Late Caledonian extension in western Norway: a response to extreme crustal thickening. Tectonics, 5: 192–204.

Norton, M.G., 1987. The Nordfjord-Sogn Detachment. Norsk Geol. Tidsskr., 67: 93–106.

Nystuen, J.P., 1981. The Late Precambrian sparagmites of southern Norway: a major Caledonian allochthon—the Osen-Røa Nappe Complex. Am. J. Sci., 281: 69–94.

Nøttvedt, A., Gabrielsen, R.H. and Steel, R.J., 1995. Tectonostratigraphy and sedimentary architecture of rift basins, with reference to the northern North Sea. Mar. Petrol. Geol., 12: 881–901.

Odinsen, T., Christiansson, P., Gabrielsen, R.H., Faleide, J.I. and Berge, A.M., 2000. The geometries and deep structure of the northern North Sea rift system. In: A. Nøttvedt (Editor), Dynamics of the Norwegian Margin. Geol. Soc., London, Spec. Publ., 167: 41–57.

Odinsen, T., Reemst, P., van der Beek, P., Faleide, J.I. and Gabrielsen, R.H., 2000. Permo-Triassic and Jurassic extension in the northern North Sea: results from tectonostratigraphic forward modelling. In: A. Nøttvedt (Editor), Dynamics of the Norwegian Margin. Geol. Soc., London, Spec. Publ., 167: 83–103.

Oftedahl, C., 1972. A sideritic ironstone of Jurassic age in Beistadfjorden, Trøndelag. Norsk Geol. Tidsskr., 52: 123–134.

Oftedahl, C., 1975. Middle Jurassic graben tectonics in mid-Norway. In: Proceedings from Jurassic Northern North Sea Symposium, Norwegian Petroleum Society (NPF), 21pp.

Olafsson, I., Sundvor, E., Eldholm, O. and Grue, K.,1992. Møre Margin: Crustal structure from analysis of expanded spread profiles. Mar. Geophys. Res., 14: 137–162.

Olaussen, S., Larsen, B.T. and Steel, R., 1994. The Upper Carboniferous-Permian Oslo Rift: Basin fill in relation to tectonic development. Can. Soc. Petrol. Geol. Mem., 17: 175–197.

Olesen, O., 1988. The Stuoragurra Fault, evidence of neotectonics in the Precambrian of Finnmark, northern Norway. Norsk Geol. Tidsskr. 68: 107–118.

Olesen, O., Lundin, E., Nordgulen, Ø., Osmundsen, P.T., Skilbrei, J.R., Smethurst, M.A., Solli, A., Bugge, T. and Fichler, C., 2002. Bridging the gap between the onshore and offshore geology in Nordland, northern Norway. Norsk Geol. Tidsskr., 82: 243–262.

Olsen, L. and Grøsfjeld, K., 1999. Middle and Late Weichselian high relative sea levels in Norway: implications for glacial isostasy and ice-retreat rates. Norg. Geol. Unders. Bull., 435: 43–51.

Osmundsen, P.T. and Andersen, T.B., 2001. The middle Devonian basins of Western Norway: sedimentary response to large-scale transtensional tectonics? Tectonophysics, 332: 51–68.

Osmundsen, P.T., Braathen, A., Nordgulen, ø., Roberts, D., Meyer, G.B. and Eide, E., 2003. Late/post Caledonian extension in North-Central Norway: The Nesna shear zone and its regional implications. J. Geol. Soc., London, 160: 1–14.

Osmundsen, P.T., Braathen, A., Sommaruga, A., Skilbrei, J.R., Olesen, O., Nordgulen, Ø., Roberts, D., Andersen, T.B. and Mosar, J., 2004. Metamorphic core complexes and gneiss-cored culminations along the Mid-Norwegian margin: an overview and some current ideas. In: B.T.G. Wandås, E. A. Eide, F. Gradstein and J.P. Nystuen (Editors), Onshore-Offshore relationships on the North Atlantic Margin. Norwegian Petroleum Society (NPF), Special Publication 12. Elsevier, Amsterdam, pp. xxx-xxx.

Osmundsen, P.T., Andersen, T.B., Markussen, S. and Svendby, A.K., 1998. Tectonics and sedimentation in the hangingwall of major extensional detachment: the Devonian Kvamshesten basin, western Norway. Basin Res., 10: 213–224.

Owen, A.W., Bruton, D.L., Bockelie, J.F. and Bockelie, T.G., 1990. The Ordovician succession of the Oslo region, Norway. Norg. Geol. Unders. Spec. Publ., 4: 3–54.

Pascal, C. and Gabrielsen, R.H., 2001. Numerical modelling of Cenozoic stress patterns in the mid Norwegian Margin and the northern North Sea. Tectonics, 20 (4): 585–599.

Pascal, C., van Wijk, J.W., Cloetingh, S.A.P.L. and Davies, G.R., 2002. Effect of lithosphere thickness heterogeneities in controlling rift localization: Numerical modeling of the Oslo Graben. Geophy. Res. Lett., 29 (9): 69.1–69.4.

Peulvast, J.-P., 1985. Post-orogenic morphotectonic evolution of the Scandinavian Caledonides during the Mesozoic and Cenozoic. In: D.G. Gee and B.A Sturt (Editors), The Caledonide Orogen—Scandinavia and Related Areas. John Wiley and Sons, New York, pp. 979–995.

Platt, J.P., 1986. Dynamics of orogenic wedges and the uplift of high-pressure metamorphic rocks. Geol. Soc. Am. Bull., 97: 1037–1053.

Platt, J.P. and Vissers, R.L.M., 1989. Extensional collapse of thickened continental lithosphere: a working hypothesis for the Alboran Sea and the Gibraltar arc. Geology, 17: 540–543.

Plink-Bjorklund, P., Roberts, D. and Bjorklund, L., 2004. The Baltic Basin stratigraphic record: reflections of tectonothermal and magmatic events in the Scandinavian Caledonides and Trans-European Suture Zone. Geologiska Föreningens i Stockholm Förhandlingar, 126: 83.

Prestvik, T., Torske, T., Sundvoll, B. and Karlsson, H., 1999. Petrology of early Tertiary nephelinites off mid-Norway: additional evidence for an enriched endmember of an ancestral Iceland Plume. Lithos, 46: 317–330.

Ramberg, H., 1966. The Scandinavian Caledonides as studied by centrifuged dynamic models. Uppsala University Geological Institute Bulletin, 43, 72 pp.

Ramberg, I.B. and Grønlie, G., 1969. A crustal section across the Caledonian mountain belt (Norway) based on gravity data. Bolletino di Geofisica Teoretica ed Applicata, 11: 219–226.

Ramberg, I.B. and Larsen, B.T., 1978. Tectonomagmatic evolution. In: J.A. Dons (Editor), The Oslo Paleorift. A review and guide to excursions. Norg. Geol. Unders. Bull., 337: 55–73.

Rawson, P.F. and Riley, L.A., 1982. Latest Jurassic—early Cretaceous events and the 'Late Cimmerian Unconformity' in North Sea area. AAPG Bull., 66: 2628–2648.
Redfield, T.F., Braathen, A., Gabrielsen, R.H., Torsvik, T.H. and Andriessen, P.A.M., in press. Late Mesozoic to Early Cenozoic components of vertical separation across the Møre-Trøndelag Fault Complex, Norway. Tectonophysics.
Redfield, T.R., Torsvik, T.H., Andriessen, P.A.M. and Gabrielsen, R.H., 2004. Mesozoic and Cenozoic tectonics of the Møre-Trøndelag Fault Complex, central Norway: constraints from new apatite fission track data. Phys. Chem. Earth, 29: 673–682.
Reemst, P. and Cloetingh, S., 1999. Polyphase rift evolution of the Vøring margin (mid-Norway): constraints from forward tectonostratigraphic modeling. Tectonics, 19: 225–240.
Reusch, H. 1901. Nogle bidrag til forstaaelsen af hvorledes Norges dale og fjelde er blevne til. Norg. Geol. Unders. Bull., 32: 125–217.
Riis, F., 1996. Quantification of Cenozoic vertical movements of Scandinavia by correlation of morphological surfaces with offshore data. In: A. Solhem, F. Riis, A. Elverhøi, J.I. Faleide, L.N. Jensen and S. Cloetingh (Editors), Impact of glaciations on basin evolution: Data and models from the Norwegian margin and adjacent areas. Global Planet. Change, 12: 3311–357.
Riis, F. and Fjeldskaar, W., 1992. On the magnitude of the Late Tertiary and Quaternary erosion and its significance for the uplift of Scandinavia and Barents sea. In: R.M. Larsen, H. Brekke, B.T. Larsen and E. Talleraas (Editors), Structural and Tectonic Modelling and its Application to Petroleum Geology. Norwegian Petroleum Society (NPF), Special Publication 1, Elsevier, Amsterdam, pp. 163–188.
Riis, F., Vollset, J. and Sand, M., 1986. Tectonic development of the western margin of the Barents Sea and adjacent areas. In: M.T. Halbouty (Editor), Future petroleum provinces of the World. AAPG Mem., 40: 661–667.
Roberts, D. and Gee, D.G., 1985. An introduction to the structure of the Scandinavian Caledonides. In: D.G. Gee and B.A. Sturt (Editors), The Caledonide Orogen—Scandinavia and related area, John Wiley and Sons Ltd, pp. 55–68.
Roberts, A.M. and Yielding, G., 1991. Deformation around basin-margin faults in the North Sea/mid-Norway rift. In: A.M. Roberts, G. Yielding and B. Freeman (Editors), The Geometry of Normal faults. Geol. Soc., London, Spec. Publ., 56: 61–78.
Roberts, D.G., 1989. Basin inversion in and around the British Ilses. In: M.A. Cooper and G.D. Williams (Editors), Inversion Tectonics. Geol. Soc., London, Spec. Publ., 44: 131–150.
Robinson, P., Tucker, R.D., Solli, A., Gee, D.G., Thorsnes, T. and Terry, M.P., 1997. Re-interpretation of the Støren Nappe, outer Trondheimsfjord, Norway: bearing on relationships between thrusting and extensional exhumation in the Scandian hinterland (Abstract), Terra Nova, Abstract Supplement 1 (9): 330.
Rohrman, M., Andriessen, P. and van der Beek, 1996. The relationship between basin and margin thermal evolution assessed by fission track thermochronology: an application to offshore southern Norway. Basin Res., 8: 45–63.
Rohrman, M., van der Beek, P., Andriesen, P. and Cloetingh, S., 1995. Mesozoic-Cenozoic morphotectonic evolution of southern Norway: Neogene domal uplift infered from apatite fission track tehermochronology. Tectonics, 14 (3): 704–718.
Rohrman, M. and van der Beek, P., 1996. Cenozoic domal uplift of North Atlantic margins: An astensosheric diapirism model. Geology, 24 (10): 901–904.
Rundberg, Y. Olaussen, S. and Gradstein, F., 1995. Incision of Oligocene strata: evidence for northern North Sea Miocene uplift and key to the formation of the Utsira sands. (abstract). Geonytt, 22.
Rykkelid, E. and Andresen, A.,1994. Late Caledonian extension in the Ofoten area, northern Norway. Tectonophysics, 231: 157–169.
Sales, J.K., 1992. Uplift and subsidence of northwestern Europe: possible causes and influence on hydrocarbon productivity. Norsk Geol. Tidsskr., 72: 253–258.
Sejrup, H.-P., 1987. Molluscan and foraminiferal biostratigraphy of an Eemian-Early Weichselian section on Karmøy, southwestern Norway. Boreas, 16: 27–42.
Sellevoll, M.A., 1983. A study of the Earth in the island area of Lofoten- Vesterålen, northern Norway. Norg. Geol. Unders. Bull., 380: 235–243.
Séranne, M. and Séguret, M., 1987. The Devonian basins of western Norway: Tectonics and kinematics of extending crust. In: M.P. Coward, J.F. Dewey and P.L. Hancock (Editors), Continental extensional tectonics. Geol. Soc., London, Spec. Publ., 28: 537–548.
Simpson, R.W., Jachens, R.C. and Blakeky, R.J., 1983. AIRY-ROOT: A Fortran program for calculating the gravitational attraction of an Airy isostatic root out to 166.7 km. United States Department of the Interior, Geological Survey, Open-File Report, 83–883, 24 pp.
Sjöberg, R., 1988. Coastal caves indicating preglacial morphology in Norway. Cave Science, 15: 99–103.
Skilbrei, J.R., 1988. Geophysical interpretation of the Fosen-Namsos Western Gneiss Region and northern part of the Trondheim Region Caledonides, central Norway. Norg. Geol. Unders. Spec. Publ., 3: 70–79.
Skogseid, J. and Eldholm, O., 1995. Rifted continental margin off Mid-Norway. In: E. Banda et al. (Editors), Rifted Ocean-Continent Boundaries. Kluwer Acadeic Press, pp. 147–153.
Skår, Ø., 2002. U-Pb geochronology and geochemistry of early Proterozoic rocks of the tectonic basement windows in central Nordland, Caledonides of north-central Norway. Prec. Res., 116:265–283.
Sommaruga, A. and Bøe, R., in press. Geometry and subcrop maps of shallow Jurassic basins along the Mid-Norway coast. Mar. Petrol. Geol.
Spann, H., Brudy, M. and Fuchs, K., 1991. Stress evaluation in offshore regions of Norway. Terra Nova, 3: 148–152.
Smit, J.H.W., Brun, J.P. and Sokoutis, D., 2003. Deformation of brittle-ductile thrist wedges in experiments and nature. J. Geophys. Res., 108, (B10, ETG 9): 1–18.
Steel, R., Siedlecka, A. and Roberts, D., 1985. The Old Red Sandstone basins of Norway and their deformation: a review. In: D.G. Gee and B.A. Sturt, (Editors), The Caledonian Orogeny—Scandinavia and related areas, John Wiley and Sons.
Stuevold, L.M., Skogseid, J. and Eldholm, O., 1992. Post Cretaceous uplift events on the Vøring continental margin. Geology, 20 (10): 919–922.
Sturt, B.A. and Braathen, A., 2001. Deformation and metamorphism of Devonian rocks in the outer Solund area, western Norway—implications for models of Devonian deformation. International Journal of Earth Sciences (Geol. Rundschau) 90: 270–286.
Sundvor, E. and Larsen, B.T., 1994. Architecture and early evolution of the Oslo Rift. Tectonophysics, 240: 173–189.
Svensen, H., Jamtveit, B., Banks, D.A. and Karlsen, D., 2001. Fluids and halogens at the diagenetic-metamorphic boundary: evidence from veins in continental basins, western Norway. Geofluids, 1: 53–70.
Talwani, M. and Eldholm, O., 1972. The continental margin off Norway: A geophysical study. Geol. Soc. Am. Bull., 83: 3375–3608. (k)
Talwani, M. and Eldholm, O., 1977. Evolution of the Norwegian-Greenland Sea. Geol. Soc. Am. Bull., 88: 969–999. (k)
Terry, M.P., Robinson, P., Hamilton, M.A. and Jercinovic, M.J., 2000. Monazite geochronology of UHP and HP metamorphism, deformation and exhumation, Nordøyane, Western Gneiss Region, Norway. AAPG Mineral., 85: 1651–1664.

Torske, T.,1972. Tertiary oblique uplift of Western Fennoscandia; crustal warping in connection with rifting and break-up of the Laurasian continent. Norg. Geol. Unders. Bull., 273: 43–48.

Torsvik, T.H., Sturt, B.A., Swensson, E., Andersen, T.B. and Dewey, J.F., 1992. Palaeomagnetic dating of fault rocks: evidence for Permian and Mesozoic brittle deformation along the extensional Dalsfjord Fault, western Norway. Geophys. J. Int. 109: 565–580.

Torsvik, T.H., Andersen, T.B., Eide, E. and Walderhaug, H.J., 1997. The age and tectonic significance of dolerite dykes in western Norway. J. Geol. Soc., London 154: 961–973.

Underhill, J.R. and Partington, M.A., 1993. Jurassic thermal doming and deflation of the North Sea: implications of the sequence stratigraphic evidence. In: J.R. Parker (Editor), Petroleum Geology of Northwest Europe. Proceedings of the 4th Conference. Geological Society, London, pp. 337–346.

Underhill, J.R. and Partington, M.A., 1994. Use of maximum flooding surfaces in determining a regional control on the intra-Aalenian (Mid-Cimmerian) sequence boundary: Implications for the North Sea basin development and Exxon's sea-level chart. In: H.W. Posamentier and P.J. Wiemer (Editors), Recent Advances in Siliciclastic Sequence Stratigraphy. AAPG Mem., 58: 449–484.

Valle, P., Færseth, R.B. and Fossen, H., 2002. Devonian-Triassic brittle deformation based on dyke geometry and fault kinematics in the Sunnhordland region, SW Norway. Nor. J. Geol. (NGT), 83: 3–17.

van der Beek, P., Cloetingh, S. and Andriessen, P., 1994. Mechanisms of extensional basin formation and vertical motions at rift flanks: Constraints from tectonic modelling and fission-track thermochronology. Earth Planet. Sci. Lett., 121: 417–433.

van der Wijk, J.W. and Cloetingh, S.A.P.L., 2002. Basin migration caused by slow lithospheric extension. Earth Planet. Sci. Lett., 198: 275–288.

Vågnes, E. and Amundsen, H.E.F., 1993. Late Cenozoic uplift and volcanism on Spitsbergen: Caused by mantle convection? Geology, 21: 251–254.

Vågnes, E., Faleide, J.I. and Gudlaugsson, T., 1992. Glacial erosion and tectonic uplift in the Barents Sea. Norsk Geol. Tidsskr., 72: 333–338.

Vågnes, E., Gabrielsen, R.H. and Haremo, P., 1998. Late Cretaceous—Cenozoic intraplate contractional deformation at the Norwegian continental shelf; timing, magnitude and regional implications. Tectonophysics, 300: 29–46.

Watts, A.B., 2001. Isostasy and flexure of the lithosphere. Cambridge University Press, Cambridge, 458 pp.

Wibberley, C., 1999. Are feldspar-to-mica reactions necessarily reaction-softening processes in fault zones?. J. Struct. Geol., 21: 1219–1227.

Worsley, D., Aarhus, N., Basset, M.G., Howe, M.P.A., Mørk, A., and Olaussen, S., 1983. The Silurian succession of the Oslo region. Norg. Geol. Unders. Bull., 384, 1–57.

Yielding, G. and Roberts, A., 1992. Footwall uplift during normal faulting – implications for structural geometries in the North Sea. In: R.M. Larsen, H. Brekke, B.T. Larsen and E. Talleraas (Editors), Structural and Tectonic Modelling and its Application to Petroleum Geology. Norwegian Petroleum Societey (NPF), Special Publication 1, Elsevier, Amsterdam, pp. 289–304.

Ziegler, P.A., 1990. Geological Atlas of Western and central Europe (2nd edition), (Shell Internationale Petroleum Maatschappij B.V., The Hague), 239 pp.

Metamorphic core complexes and gneiss-cored culminations along the Mid-Norwegian margin: an overview and some current ideas

Per Terje Osmundsen, Alvar Braathen, Anna Sommaruga, Jan Reidar Skilbrei, Øystein Nordgulen, David Roberts, Torgeir B. Andersen, Odleiv Olesen and Jon Mosar

From the Palaeozoic to the Cretaceous, crustal thinning in the Mid Norway area was associated with the denudation of gneiss-cored culminations and metamorphic core complexes in the footwalls of major extensional faults. The development of the culminations led to warping and deactivation of early detachments, to the nucleation of new faults in more distal positions and to the exhumation of high-grade metamorphic rocks to more shallow levels in the crust. Some of the culminations and core complexes became part of the erosional template in Mid-Late Palaeozoic time, some were probably exhumed in the Mesozoic, whereas some may never have reached the surface. We present an overview of five types of gneiss-cored culminations and core complexes that have been identified in the field, through the interpretation of offshore, long-offset seismic reflection data. We furthermore address their mechanism(s) of formation, and their role in the progressive evolution of the Mid-Norwegian margin.

Introduction

The multi-stage development of many passive margins, the scale of differential vertical movements involved in their formation and the importance of source areas in the adjacent continent interior show that, for the most part, present-day shorelines constitute an artificial boundary in passive margin studies. Recent studies in basin dynamics emphasise the source-to-sink perspective, highlighting the importance of the processes that take place in the source areas (e.g. Leeder et al., 1998). The onshore–offshore approach to the continental margin studies provides an opportunity to address both the source and sink, and their evolution through time. In the source areas, structural studies combined with $^{40}Ar/^{39}Ar$ geochronology and apatite fission track analysis provide a means to link cooling with tectonically controlled exhumation, and to date activity on shear zones and faults. Thus, such studies allow us to assess directly the tectonothermal template that was exploited by erosion during the later rift phases.

Onshore structures are not necessarily easy to trace offshore, and, even if a successful correlation can be made, the implications and importance of this with respect to an understanding of passive margin evolution may be variable. One of the challenges in onshore–offshore studies is, thus, to define common denominators that provide links, not only between individual geological features, but more importantly, between the processes that were involved. In the present contribution, we focus on gneiss-cored culminations and metamorphic core complexes that straddle the Mid Norway passive margin. The formation of a number of these culminations is strongly linked to extensional tectonics and to the exhumation and cooling of rocks that eventually became parts of the erosional template. Most of the culminations are associated with sites of fault nucleation, re-activation and de-activation, and some appear to be controlling the location of major domain boundaries within the Mesozoic rift. The control on fault patterns by the underlying core complexes and culminations suggests that, at least locally, syntectonic sedimentation may have

Onshore–Offshore Relationships on the North Atlantic Margin
edited by B. Wandas et al. NPF Special Publication 12, pp. 29–41,

been severely affected by the denudation of the metamorphic cores.

Gneiss-cored culminations form in a variety of tectonic environments, including compressional, extensional and strike-slip settings. A common occurrence of gneiss-cored culminations is in the internal parts of orogens, where parautochthonous or autochthonous rocks are exposed in tectonic windows (e.g. Haller, 1971; Ramberg, 1980). A special type of gneiss-cored culmination is the metamorphic core complex, where rocks from the middle or lower crust have become tectonically juxtaposed with non-metamorphic sedimentary rocks. Metamorphic core complexes form in the footwalls of large-magnitude extensional detachments, and are known from a number of highly extended terranes, including the Basin and the Range Province of the western U.S. (e.g. Wernicke, 1985; Lister and Davis, 1989), the Tyrrhenian Sea area (Jolivet et al., 1991), the southwestern Norwegian Caledonides (Norton, 1986; Andersen and Jamtveit, 1990; Fossen, 1992), and from the classic rift zones, such as the Red Sea Rift (Talbot and Ghebreab, 1997). Extensional detachments have been suggested to constitute fundamental elements in the structural architecture of continental margins (e.g. Lister et al., 1991; Fossen et al., 2000; Whitmarsh et al., 2000; Manatschal et al., 2001).

The Mid Norway area (Fig. 1) experienced multi-stage crustal thinning from the Devonian to the Tertiary (e.g. Skogseid et al., 1992). The earliest phases of extension in the Devono–Carboniferous times resulted in a dramatic reduction of the Caledonian orogenic crust and in differential exhumation of the Caledonian nappe pile and basement. Structural products associated with the early phases of crustal thinning are well-preserved onshore Mid Norway (Braathen et al., 2002; Osmundsen et al., 2003), and significant advances have recently been made to understand the thermal consequences of Palaeozoic and later extension in the onshore areas (Eide et al., 2002, 2003 and 2004; Redfield, 2004). As Devono–Carboniferous, late- to post-orogenic extension gave way to successive Late Palaeozoic and Mesozoic rift phases, a complex structural hierarchy was superposed on the extended remains of the Caledonian orogen. Uplift and erosion of the Palaeozoic tectonothermal template continued into the Mesozoic and Tertiary, providing detritus that contributed to the filling of syn- and post-rift offshore basins (e.g. Sherlock, 2001). We aim at an overview of the geometry and the mechanisms of formation of gneiss-cored culminations that occur onshore and offshore Mid Norway. Whereas the exhumation history of onshore culminations is important in a provenance perspective and with respect to re-activation, the development of core complexes in Mesozoic times affected shallow-level structural development and, thus, the syndepositional rift architecture.

Differential exhumation and gneiss-cored culminations onshore Mid Norway

Late/post-Caledonian extension reworked the Caledonian nappe pile into an array of extensional allochthons bound by ductile-to-brittle, top-WSW shear zones and detachment zones (Braathen et al., 2002; Osmundsen et al., 2003). The configuration of extensional structures, strike-slip faults and gneiss-cored culminations developed progressively in Devono–Carboniferous times, commencing with deformation along low-angle, medium- to low-grade extensional shear zones. Some of these shear zones developed into large-magnitude detachment zones that eventually juxtaposed regional gneiss culminations with continental, 'Old Red' sedimentary basins. Others appear to have become deactivated as they were incised by moderate-angle, low-grade, ductile-to-brittle faults (Braathen et al., 2002; Osmundsen et al., 2003). Thus, the Palaeozoic structural framework in Mid Norway includes a large number of important shear zones and faults that cut the Caledonian nappe stack and affect the plan-view outline and the cross-sectional geometry of the onshore tectonostratigraphy.

The gneiss-cored culminations that have been identified onshore Mid Norway fall into three main types (Fig. 2; Osmundsen et al., 2002a). Unmodified, thrust-related culminations (type 1) are preserved in areas away from, and mainly east of, the major extensional shear zones that truncate the Caledonian nappe pile. These particular culminations will not be considered further here. In the areas affected by extension, two main types of culminations developed that show different characteristics with respect to strain pattern as well as to the amount of displacement on bounding shear zones. The type 2 culminations are flanked by kilometre-thick, ductile, large-magnitude (tens of kilometres) extensional detachment zones, capped by brittle detachment faults, and characterised by extension-parallel folds that developed contemporaneously with extension (Norton, 1996; Séranne, 1992; Chauvet and Séranne, 1994; Krabbendam and Dewey, 1998); i.e., the type 2 culminations are thought to have developed progressively in a

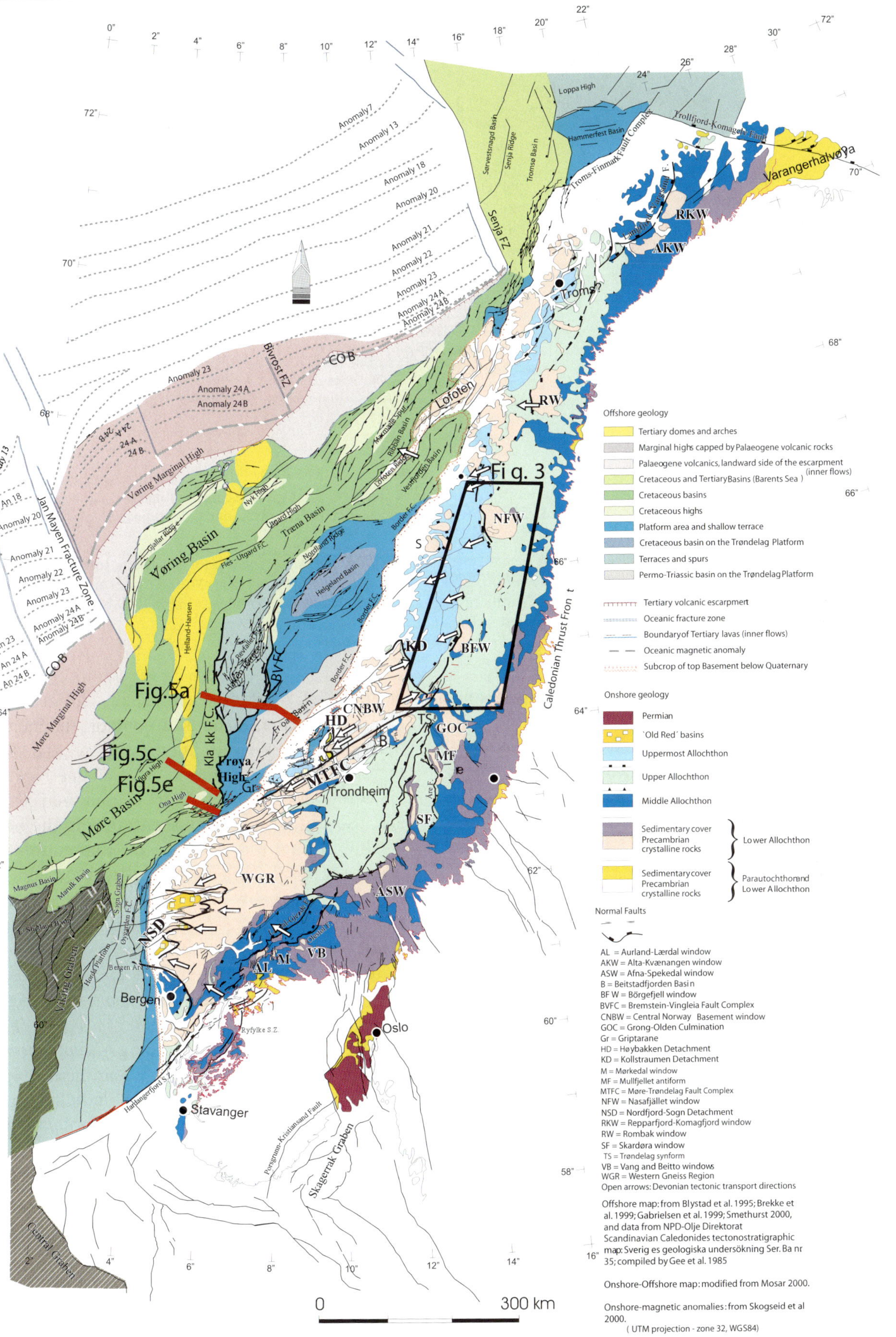

Fig. 1 Onshore-offshore map of Norway and the Norwegian margin, mainly from Mosar et al. (2002). Open arrows indicate generalised tectonic transport directions for Devono-Carboniferous, late/post-orogenic shear zones and faults.

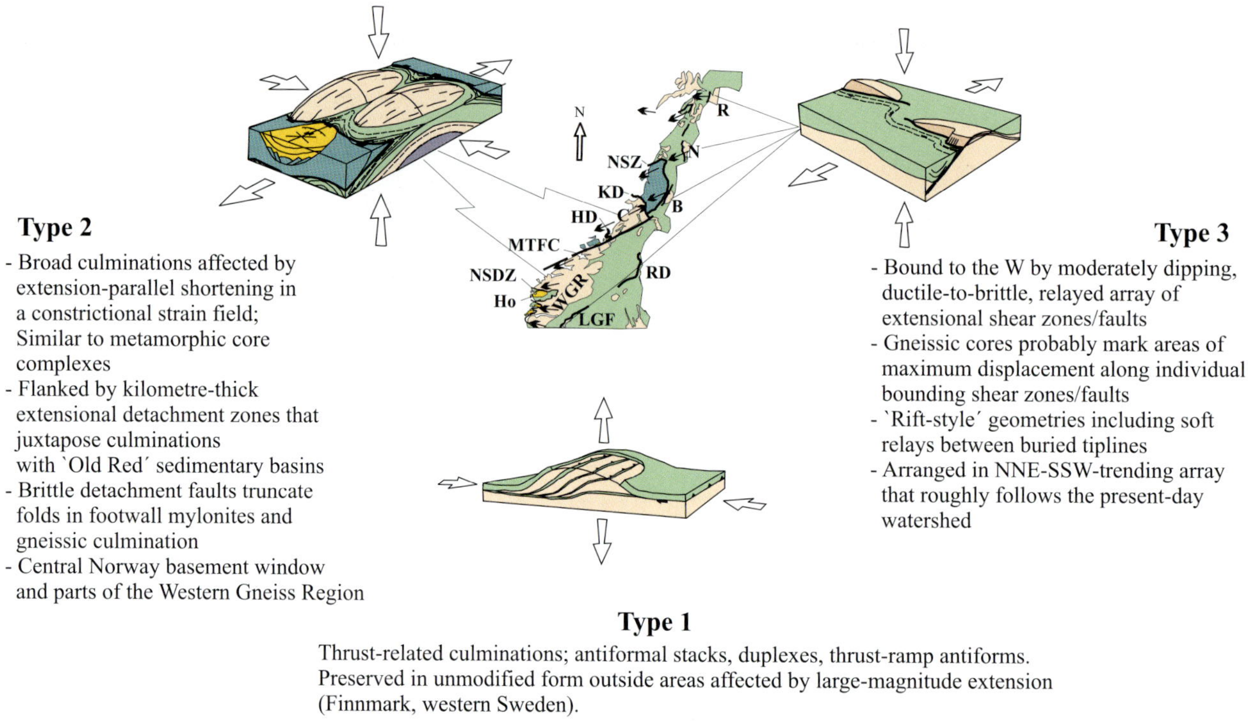

Fig. 2 Generalised types of gneiss-cored culminations found in the Scandinavian Caledonides. See text for discussion. Inset map: B–Børgefjell window; Ho – Hornelen Basin; HD – Høybakken detachment; LGF – Lærdal-Gjende Fault; MTFC – Møre-Trøndelag Fault Complex; N – Nasafjäll window; NSZ – Nesna shear zone; NSDZ – Nordfjord-Sogn detachment zone; R – Rombak window; RD – Røragen detachment; WGR – Western Gneiss Region.

constrictional strain field (e.g. Krabbendam and Dewey, 1998, Fig. 2). The type 2 culminations are juxtaposed against 'Old Red' extensional basins, believed previously to have developed from Early into Middle Devonian times (Kolderup, 1921; Jarvik, 1949; Allen, 1976). Recent results from $^{40}Ar/^{39}Ar$ geochronology indicate, however, that the upper parts of the 'Old Red' basin stratigraphy in the outer Trondheim Region are Late Devonian at the oldest, and that the material deposited in the basins was sourced in the (type 2) Central Norway Basement Window (CNBW; Fig. 1; Eide et al., 2003). Thus, some type 2 culminations were exhumed to the surface and were eroded during 'Old Red' basin deposition. The type 2 culminations in SW and Mid Norway resemble metamorphic core complexes, and have been described as such (e.g. Norton, 1986; Braathen et al., 2000).

The array of type 3 culminations (Fig. 2) generally follows the present-day watershed between Norway and Sweden (Mosar et al., 2002). Some of the type 3 culminations have been interpreted earlier as thrust-related culminations (e.g. Greiling et al., 1998), but the identification of an array of relayed, extensional, ductile-to-brittle fault zones along the western margins of the Rombak, Nasafjäll and Børgefjell windows (Fig. 1 and 3; Rykkelid and Andresen, 1994; Braathen et al., 2002; Osmundsen et al., 2003) supports extension-related structural control on gneiss-core denudation. Previous modelling of the top to magnetic basement under the Central Norwegian Caledonides (Sindre, 1998; Olesen et al., 2002) is consistent with vertical separation in the order of 3–6 km along the margins of the culminations. Thus, the displacements related to type 3 culminations were probably an order of magnitude less than the displacements associated with type 2 culminations. The type 3 culminations show little or no evidence of extension-normal shortening, and may have formed under conditions approximating plane strain or even vertical flattening.

The kinematics of bounding shear zones, the overall culmination geometry and the map-view distribution of flanking nappes indicate that the type 3 culminations resemble the footwall uplifts, commonly associated with normal faults in continental rift zones. In this scenario, the gneissic cores of the culminations mark the areas of maximum footwall uplift (and thus maximum displacement) along the ductile-to-brittle faults that bound the culminations. Correspondingly, an anomalous, ENE–WSW trend of the Caledonian nappe boundaries in the area between the Nasafjäll and Børgefjell windows has been interpreted in terms of a soft relay zone (Fig. 3a; Osmundsen et al., 2003).

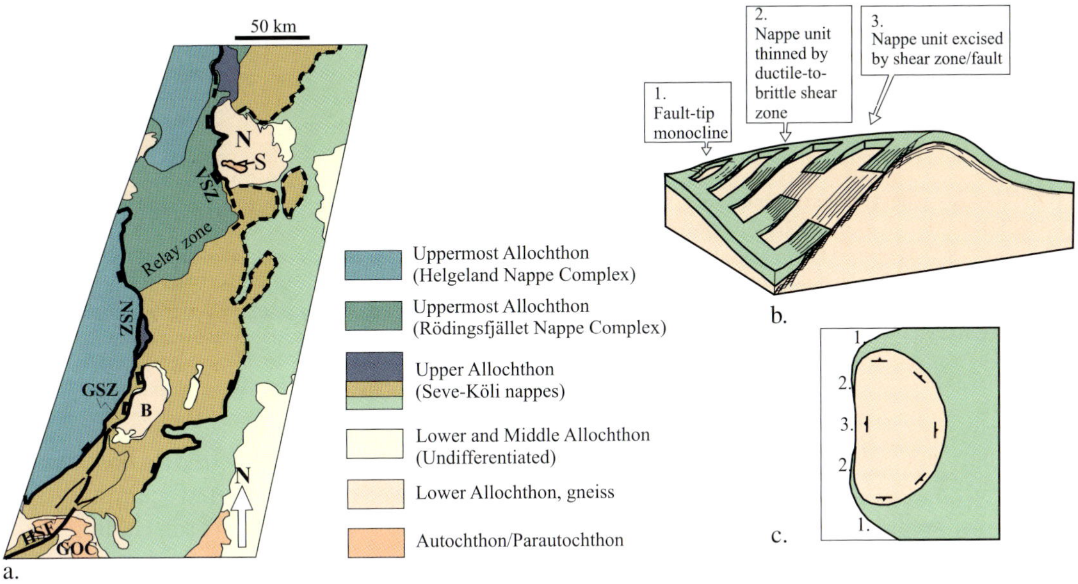

Fig. 3 a. Geological map of part of the North-central Norwegian Caledonides (simplified from Solli 1999, see Fig. 1 for location) with the Nasafjäll (N) and Børgefjell (B) culminations and adjacent nappe units. Note thinning and excision of individual units along the western boundaries of the gneiss-cored culminations. GSZ–Gaukarelv shear zone; VSZ–Virvassdalen shear zone; NSZ–Nesna shear zone (see Braathen et al., 2002 and Osmundsen et al., 2003 for descriptions of culmination-bounding shear zones); GOC–Grong-Olden Culmination; HSF–Hitra-Snåsa Fault; S–Stokkali Granite (Par-? Autochthon). b. Two-layer fault-growth model for the development of type 3 culminations such as the Nasafjäll and Børgefjell windows. c. Resultant map-view configuration. The model is consistent with: (1) the excision of nappe units along the western margins of the gneiss windows; (2) the monoclinal geometry of nappes along the northwestern and southwestern margins of the windows; and (3) the formation of soft relays with anomalous orientations of nappe boundaries between the windows, as illustrated in Fig. 3a.

An interpretation in terms of displacement gradients along the bounding faults may also explain the thinning and excision of some of the main nappe units along the western margins of the culminations (Figs. 3b–c). A northwards and southwards decrease in the amount of displacement is consistent with the tip line of individual ductile-to-brittle shear-zone segments plunging underneath the nappe pile at the northern and southern margins of the windows, leading to the deflection of nappes that is generally observed around the gneissic cores. This deflection resembles that produced by fault-tip monoclines associated with fault growth, such as described from the syn-rift stratigraphy of the Suez Rift (e.g. Gawthorpe et al., 1997). Late Devonian to Early Carboniferous ages have been assigned to top-to-the-W, low-grade ductile-to-brittle shearing along the western margin of the Rombak window (Fig. 1; Coates et al., 1999). It is likely that shearing and faulting along the Nasafjäll and Børgefjell culminations took place during the same time interval (Osmundsen et al., 2003). However, there are indications of younger rejuvenation, represented by fault slip-directions that cannot be reconciled with the Devono–Carboniferous strain pattern (Braathen et al., 2002). At present, the age of the exhumation of the gneissic cores to the surface is unknown.

The Møre–Trøndelag Fault Complex

The Møre–Trøndelag Fault Complex (MTFC) is spatially related to both the type 2 and the type 3 culminations. The Hitra-Snåsa Fault of the MTFC borders the (type 2) Central Norway basement window, and constitutes a SE transfer boundary for the extensional Høybakken detachment (Fig. 2; Séranne, 1992; Braathen et al., 2000). The array of type 3 culminations and associated, ductile-to-brittle fault zones is linked to the south with the MTFC, through a series of relayed fault strands described by Roberts (1998) that show evidence of normal/sinistral displacements (Fig. 3a). The array of fault strands can be traced towards the Grong–Olden Culmination (Fig. 3a), which is displaced 3–4 km sinistrally by the Hitra-Snåsa Fault (op. cit.). Coupled with the age of the uppermost deposits in the 'Old Red' of the outer Trondheim region, the Late Devonian–Early Carboniferous age of shearing and faulting along the gneiss culminations (Coates et al., 1999; Osmundsen et al., 2003) indicates temporal overlap between shearing and faulting, along the array of culminations and faulting/basin formation in the area of the Høybakken detachment (Fig. 2). Later activity along the MTFC occurred in multiple stages (Grønlie et al., 1991) and includes transtension of

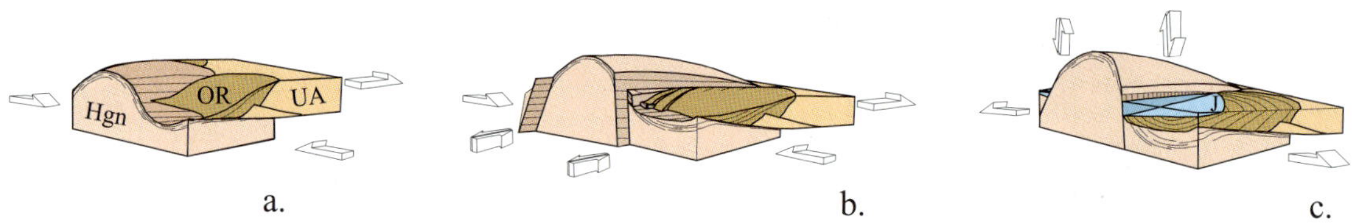

Fig. 4. Cartoon showing conceptual development of a type 2 culmination (such as the CNBW), sinistral strike-slip faults (such as the MTFC) and sedimentary basins through a) Early- to Mid Devonian, top-to the WSW extension, b) Devono-Carboniferous, continued extension and sinistral strike-slip (development of MTFC), and c) Jurassic re-activation of strands of the MTFC as normal faults. The changing roles of the MTFC have been documented by previous workers (i.e. Grønlie and Roberts, 1989; Bering, 1992; Séranne, 1992; Braathen et al., 2000). Only phases demonstrably related to basin-forming events have been included in the figure. HGn–medium- to high-grade gneisses and supracrustals, parautochthonous or in the Lower Allochthon; UA–nappe units, belonging mainly to the Upper and Uppermost Allochthon of the Caledonian nappe-stack; OR–'Old Red' sedimentary rocks; J–Jurassic sedimentary rocks.

probable Permian age (Watts, 2001), a phase of dextral re-activation (Grønlie and Roberts, 1989; Watts, 2001) as well as dip-slip re-activation, associated with sedimentation in the Jurassic Beitstadfjorden Basin (Figs. 1 and 4; Bering, 1992; Bøe, 1991; Sommaruga and Bøe, 2002). Recent apatite fission track studies show clear breaks in the derived ages across lineaments that are parallel to the trend of the MTFC; most likely, these breaks reflect Late Palaeozoic and Mesozoic fault-block rotation and differential uplift (Redfield, 2002; Redfield et al., in press). The offshore counterpart of the MTFC have been interpreted to have played an important role during Late Palaeozoic and Mesozoic rifting (Grunnaleite and Gabrielsen, 1995; Gabrielsen et al., 1999).

Gneiss-cored culminations offshore Mid Norway and their influence on rift-zone architecture

Aeromagnetic data show very strong positive signatures in the Frøya High and parts of the Trøndelag Platform area that are probably related to sources in the deep basement (Skilbrei et al., 2002). During prograde metamorphism, magnetite is commonly produced in mafic as well as intermediate rocks; only parts of the magnetite become redistributed during denudation and retrogression (e.g. Skilbrei et al., 1991). Most likely, the strong positive anomalies result from denudation of strongly magnetic, high-grade metamorphic rocks. The pattern of strong positive magnetic anomalies has been compared to the pattern of onshore type 2 and 3 culminations, to provide a basis for offshore extrapolation of detachment zones exposed in the onshore areas (Olesen et al., 2002; Skilbrei et al., 2002). In the Frøya High example, rough and shallow basement topography may add to the strong positive magnetic signature. Interpretation of long-offset seismic data has revealed two main types of deep-seated culminations. In the southern Trøndelag Platform/Halten Terrace area, a Palaeozoic detachment appears to be warped across the crest of an antiformal culmination that also controlled the ramp-flat geometry of the Mesozoic Bremstein–Vingleia Fault Complex to the east of the Njord field (Figs. 1 and 5a). Another core complex appears to be related to low-angle truncation of a strong, intra-basement reflective band west of the Frøya High by a low-angle detachment fault west of the Klakk Fault Complex (Figs. 1 and 5b). In a NW–SE oriented section, the low-angle detachment west of the Frøya High reveals horizontal separation in the order of 40 km (Osmundsen et al., 2002b). The northeastern parts of the Slørebotn Sub-basin (Fig. 5c; Blystad et al., 1995) preserve a synclinal depression with rotated, supradetachment fault blocks, a configuration commonly encountered between an exhumed core complex and the detachment breakaway (e.g. Wernicke, 1985). Thus, Mesozoic structuring in the offshore areas also involves large-magnitude normal-faulting and core-complex denudation. The pattern of offshore magnetic anomalies reflects the superposition of these processes upon the Palaeozoic structural template (Skilbrei et al., 2002).

The denudation of deep crustal rocks in the footwalls of large-magnitude Mesozoic faults can be explained by models involving excisement and/or incisement (Lister and Davis, 1989), depending on geometrical and temporal relationships between the inherited Palaeozoic template and the Mesozoic fault systems (Figs. 6a,b). Interpreted seismic lines in the southern Trøndelag Platform area reveal that a gently WNW-dipping to sub-horizontal reflection band underlies rotated, Palaeozoic-Early Mesozoic half-grabens at c. 6 s TWT (Fig. 5a). The reflection band is interpreted as a detachment zone that may represent the offshore continuation of the onshore Høybakken detachment (Osmundsen et al., 2002b; Skilbrei et al., 2002). Close to well 6407/7-1

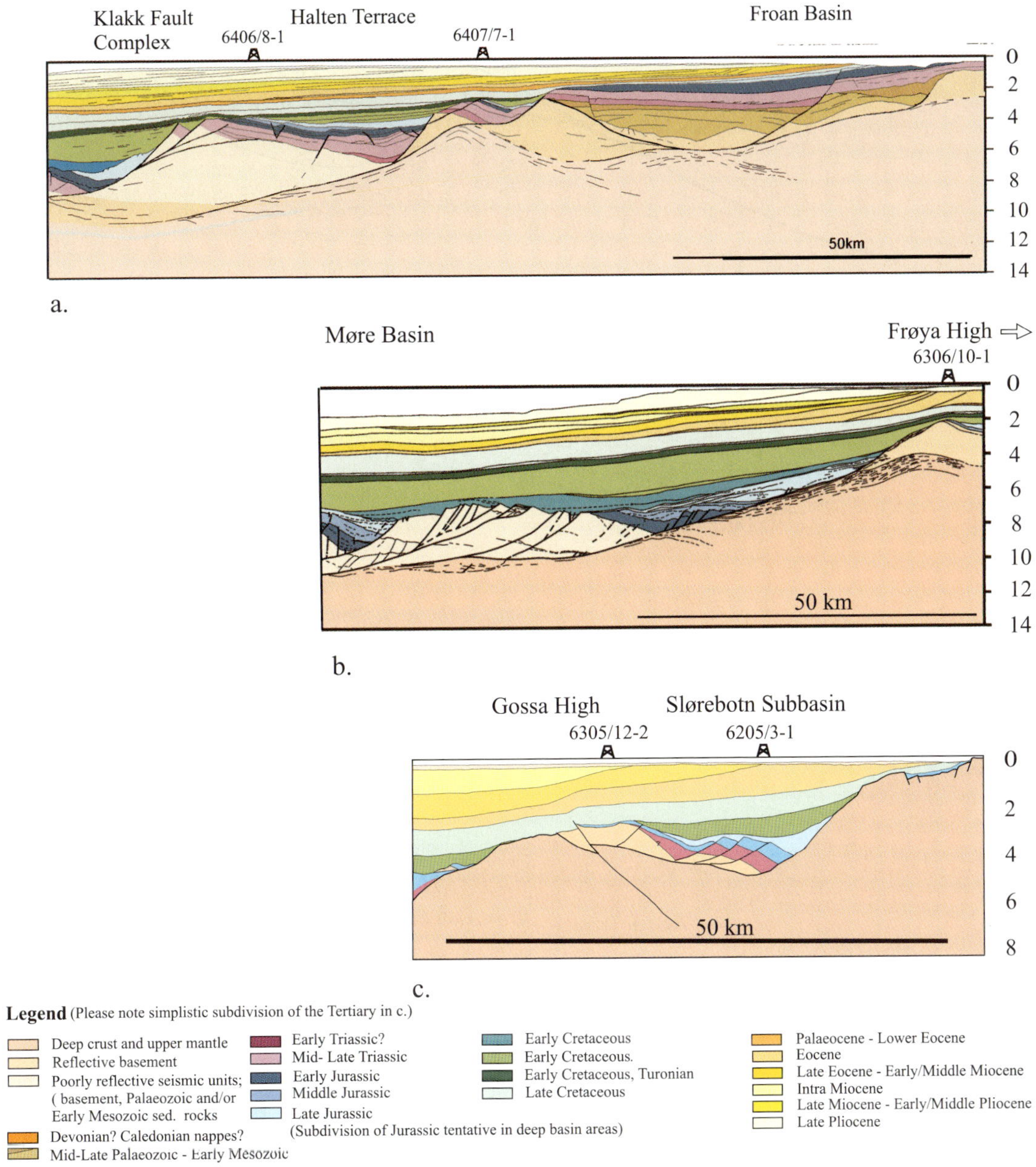

Fig. 5 Interpretations of seismic lines showing evidence for large-magnitude displacements along low-angle Mesozoic detachment faults. (a) southern Trøndelag Platform/Halten Terrace area (Osmundsen et al., 2002), (b) west of the Frøya High (op. cit.) and (c) the NE Slørebotn Sub-basin area (interpretation redrawn from Blystad et al., 1995). See fig. 1 for approximate location of seismic lines and text for further discussion.

(Fig. 5a), this reflection band appears to be warped across an antiformal culmination, with the crest positioned at c. 4.5 s TWT. Two Mesozoic fault splays, one ramp-flat and the other planar, bound the rotated fault block hosting 6407/7-1 in the southeast and in the northwest, respectively. Both faults merge at depth with the NW-dipping flank of the antiformal culmination (Fig. 5a). This fault pattern can be explained in terms of development of progressively younger detachment faults from the Palaeozoic into the Late Jurassic/Earliest Cretaceous (Figs. 6 and 7). Both in this area and in the area west of the Frøya High, the flanks of metamorphic core complexes appear to be important with respect to the location of major domain boundaries in the Mesozoic rift, such as the platform-terrace boundary in the Njord area and the boundary to the deep basin in the area of 6301/10-1 (Fig. 5b).

Most likely, the control exerted by the core complex on higher-level fault geometries, as observed in the area of 6407/7-1 (Fig. 5a), in turn affected the stratigraphic architecture in adjacent half-graben basins. The ramp-flat fault east of 6407/7-1 most likely owes its geometry to a late phase of denudation of the underlying core complex

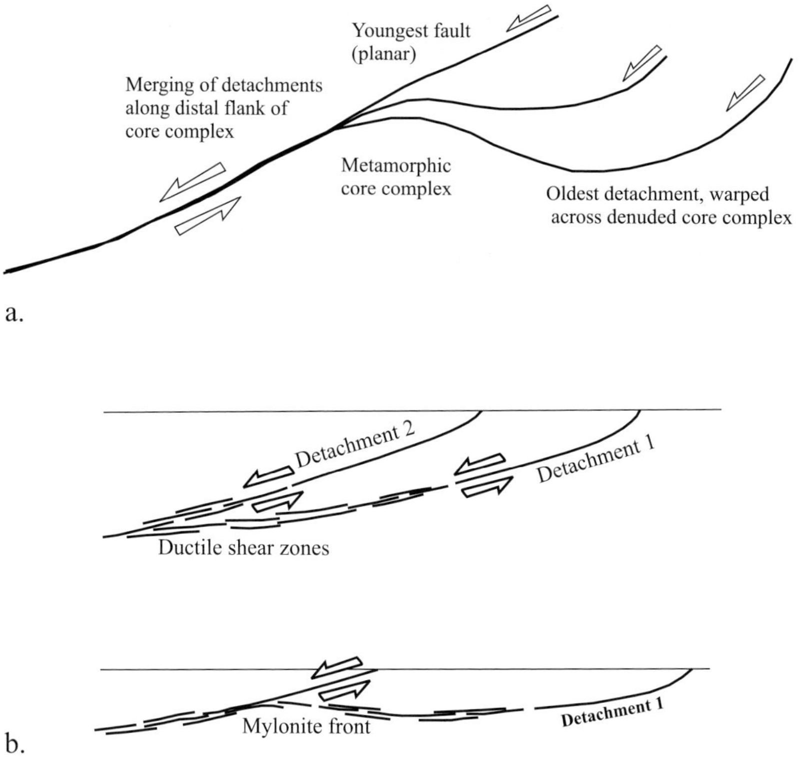

Fig. 6 Models of excision (a) and incision (b) for the formation of metamorphic core complexes (Lister and Davis, 1989). Compare with examples shown in Fig. 5. See text.

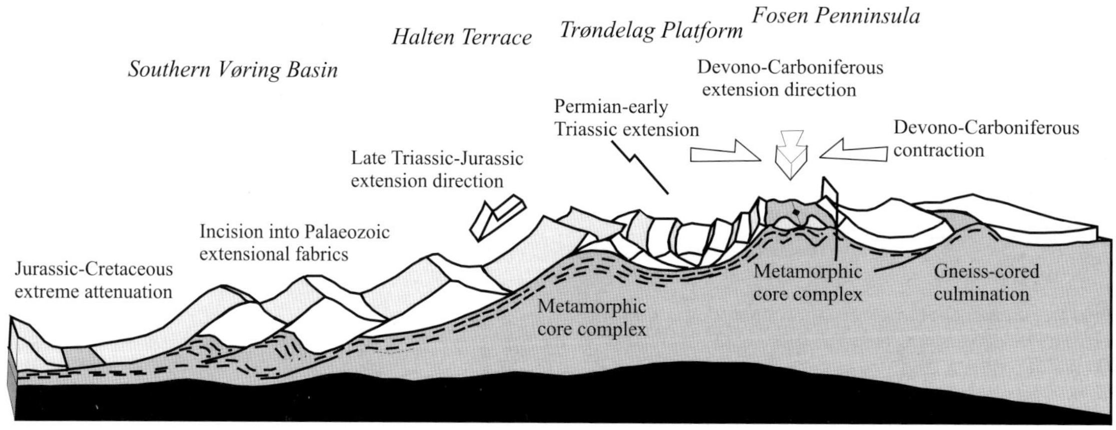

Fig. 7 Tentative model (not to scale) for the geometry of Palaeozoic and Mesozoic extensional detachments and the distribution of resultant core complexes and gneiss-cored culminations in a section that crosses the CNBW and passes offshore roughly along the trace of Fig. 5a.

(Osmundsen et al., 2002b). On the Halten Terrace, a pronounced unconformity on top of faulted Lower and lower Middle Jurassic strata indicates a main phase of faulting and rotation in Mid Jurassic time (Osmundsen et al., 2002b). The unconformity truncates a gentle syncline developed in the pre-Jurassic and Lower Jurassic strata, that may have developed during slip along the ramp-flat fault (op. cit.). To the south in the Slørebotn Sub-basin, supradetachment half-graben basins experienced phases of rotation and syntectonic sedimentation in Bathonian to Volgian times (Jongepier et al., 1996). Thus, detachment faulting and associated late exhumation of metamorphic core complexes was probably important during Mid and Late Jurassic rifting, as well as during previous Devono–Carboniferous and Permo–Triassic phases of extension. Late denudation of core complexes led to de-activation of detachment faults, such as the Palaeozoic-Early Mesozoic detachment beneath the Trøndelag Platform (Osmundsen et al., 2002b). In the Slørebotn Subbasin area (Fig. 5c), Bathonian-Volgian supra-detachment rotation had apparently ceased, prior to the onset of Aptian sedimentation (Jongepier et al., 1996). In Bathonian to Volgian times, areas east of the Frøya High underwent fault-block rotation and sedimentation (Bøe and Skilbrei, 1998), resulting in an array of Mid to Late Jurassic basins along a NE–SW trend parallel to the Møre-Trøndelag Fault Complex (Sommaruga and Bøe, 2003). The Jurassic basins are unconformable upon Devonian sedimentary rocks and upon basement rocks that belong to the 'Upper Plate' configuration (following the generalised terminology of detachment faults, e.g. Wernicke, 1995; Lister et al., 1991) of the previous Devonian structural configuration.

Discussion

Detachment faulting and subsequent de-activation through core-complex denudation or low-angle incision appears to have taken place progressively and repeatedly in the Mid Norway area since the Devonian and well into the Mesozoic. Warping, incision and de-activation of Early Devonian extensional shear zones across type 3 culminations (Fig. 2) have been interpreted from onshore structural relationships (Braathen et al., 2002; Osmundsen et al., 2003); the offshore interpretations summarised above indicate that such modes of extensional deformation are applicable to important phases of Late Palaeozoic and Mesozoic structuring. We suggest that the Late Devonian–Early Carboniferous structural template involved the onshore gneiss-cored culminations observed onshore Mid Norway and their bounding structures in a sinistral, transtensional pull-apart that probably also included a number of NE–SW-trending faults now buried beneath Middle Triassic and younger strata on the Trøndelag Platform (see also Titus et al., 2002). The rift-style geometries displayed by the type 3 culminations (Fig. 2) and associated structures contrast with the low-angle ductile shear zones that characterised the earliest phase of extension. The structures associated with the type 3 culminations, thus, appear to herald the structural styles associated with the later rift phases.

In the Mid Norway area, the maximum elongation trend changed by close to 90°, from ENE–WSW in the Devonian to NW–SE in the Late Cretaceous and Early Tertiary (Gabrielsen et al., 1999; Mosar et al., 2002). The exploitation of the Palaeozoic structural template in the Mesozoic was, thus, probably preferential and dependent on the orientation and dip direction of inherited detachments and gneiss-cored culminations. The Devono–Carboniferous structural template included extension-parallel, NE–SW-trending, megascopic fold structures, as indicated by the Trøndelag synform and the Central Norway Basement Window (CNBW, Fig. 1), as well as a large number of kilometre-scale folds. The flanks of doubly plunging, antiformal culminations wrapped by detachment zones (type 2 culminations, Fig. 2) may thus have become the preferred loci for excision or incision in the Mesozoic, even if the extension direction had changed dramatically. The Hitra-Snåsa and Verran faults of the MTFC developed along the flanks of NE–SW-trending folds (Fig. 4; Séranne, 1992; Watts, 2001). Oblique- and dip-slip re-activation of segments of the MTFC took place in the Permian and the Mesozoic, respectively (Grønlie and Roberts, 1989; Bering, 1992; Watts, 2001). The Mesozoic phase of reactivation caused the formation of the inshore Jurassic Beitstadfjorden Basin as an extensional half-graben (Bøe and Bjerkli, 1989; Bering, 1992). The array of small Jurassic half-graben basins that straddle the Norwegian coast in the Trondheimsleia area, as well as parts of the SE Møre Basin margin, may have a similar explanation (Fig. 4; Bøe and Bjerkli, 1989; Bøe, 1991; Gabrielsen et al., 1999; Sommaruga and Bøe, in press).

Along the western margin of the (type 3) Børgefjell culmination, phases of re-activation with top-to-the-SSW and top-to-the-NW polarity were superposed on the main, Devono-Carboniferous, top-WSW ductile-to brittle fault (Fig. 3). Thus, onshore, complex re-activation of the flanks of gneiss-cored culminations took place repeatedly during the formation of the passive margin. The possibility exists that individual faults in the offshore areas also experienced re-activation that involved transition from strike-slip to dip-slip or vice versa, depending on their orientations with respect to the changing stress field.

With the exception of the examples discussed above, the geographical extent of reactivation, deactivation and incision of core complexes and inherited detachments is unknown at present. We suspect, however, that variations on these themes were important during pre-Cretaceous structuring of the Mid-Norwegian margin.

Early to Mid Devonian, $^{40}Ar/^{39}Ar$ white mica cooling ages in the 395–385 Ma interval have been reported previously from gneiss-cored culminations, such as the CNBW and the Western Gneiss Region (WGR, Fig 1) (e.g. Dallmeyer et al., 1992; Berry et al., 1995; Eide et al., 2003). Thus, the Early to Middle Devonian time interval was an important one with respect to exhumation of high-grade metamorphic rocks through the 350–400°C temperature interval in the footwalls of large-magnitude extensional detachments (Chauvet et al., 1992; Andersen and Jamtveit, 1990; Eide et al., 2003; Kendrick et al., 2004). The post-Mid Devonian exhumation history of the gneiss-cored culminations is less well-known. Whereas some onshore core complexes, such as the CNBW, were exhumed to the surface in ?Late Devonian-Early Carboniferous time, yielding material to 'Old Red' sedimentary basins (Eide et al., 2003), very little is known about the late stages of exhumation of a number of other gneiss-cored culminations. Carboniferous cooling ages of around 340 Ma have been obtained through $^{40}Ar/^{39}Ar$ geochronology performed on feldspars from the Sjona window of North-Central Norway (Eide et al., 2002), indicating cooling through temperatures of around 250°C, at that time. A basement core sample from the well 6407/10-3 on the Frøya High yielded a biotite age of 395 $\pm$ 4 Ma and a K-feldspar age of 376 $\pm$ 7 Ma (Eide et al., 2003), indicating that in the Devonian, the rocks of the present-day Frøya High were being exhumed through the c. 350°C isotherm in the footwall of an extensional detachment, similar to those observed on the flanks of the type 2 culminations. Basement rocks that outcrop on the sea floor in the Griptarane area east of the Frøya High (Fig. 1), as well as on the islands in the Trondheimsleia area, belong to rock complexes that were in a high structural position in the Devonian configuration. This indicates truncation by faulting or erosion of the principal Devonian detachment between Griptarane and the Frøya High, prior to deposition of the Cretaceous strata that drape the high.

The Early to Late Devonian cooling history of the rocks positioned in the footwalls of the major, Mid Palaeozoic extensional structures provides a template for provenance studies on the Mid-Norwegian shelf. As the types 2 and 3 culminations became exposed at the surface, either through Mid Palaeozoic exhumation or by Late Palaeozoic and Mesozoic excision and incision, the culminations commenced their history as source areas for adjacent sedimentary basins. In the case of the CNBW, erosion at the surface started some time after the Mid/Late Devonian boundary, probably in the Late Devonian or Early Carboniferous, when material sourced from the CNBW was deposited in the Asenøy Basin (hangingwall of the Høybakken detachment, Fig. 1; Eide et al., 2003). Conversely, there is no evidence that coastal parts of the WGR were exhumed to the surface during deposition of the Middle Devonian basins of SW Norway (e.g. Cuthbert, 1991). Radiometric dating of mylonites related to sinistral strike-slip shear zones in the northern WGR indicate that these rocks were undergoing deformation under greenschist-facies conditions in the Mid/Late Devonian (Terry et al., 2000). Alluvial fan and fan-delta, gneiss-clast conglomerates banked against the SE margin of the Møre Basin indicate, however, that the northernmost WGR was yielding erosional products to the Slørebotn Sub-basin area in Triassic and Early Jurassic times (Smelror et al., 1994; Mørk and Stiberg, 2003).

The offshore core complexes are less well-known at present, due to their identification on a seismic grid that is not dense enough for mapping. In the area close to well 6406/7-1 (Fig. 5a), it appears that the metamorphic core resided at depth during Jurassic faulting; in the area west of the Frøya High, however, an Upper Jurassic or earliest Cretaceous unit offlaps the main detachment and onlaps the most proximal tilted fault block in the hangingwall (Fig. 5b). This indicates that the abandoned detachment fault was exposed after the main incising event and that high-grade metamorphic rocks in the footwall, as well as the tilted half-graben west of the Frøya High, may have yielded eroded debris to the adjacent Cretaceous Basin. $^{40}Ar/^{39}Ar$ geochronology performed on white micas obtained from cores through parts of the offshore Mesozoic succession reveal a spectrum of Early to Late Devonian, white mica cooling ages (Sherlock, 2001). These record Mesozoic erosion of rocks exhumed in the type 2 (and 3?) culminations. Sherlock (2001) did, however, conclude that the dated micas had experienced at least one phase of recycling since their erosion off the onshore tectonothermal template. The abandonment, partially or completely, of parts of an extensional system and its role in cannibalisation and recycling of basins is only partly understood for the Mid-Norwegian margin. An excision scenario would, however, provide an explanation for deactivation and for the commonly cited (e.g. Blystad et al., 1995) westward younging of fault activity. Both the excision and the incision models would lead to deactivation of more proximal parts of the extensional system, leaving older arrays of faults and half-grabens in a structurally high position where they could be eroded and recycled into the younger parts of the rift basin.

Conclusions

In the Mid Norway area, onshore core complexes and gneiss-cored culminations were exhumed to the surface in the footwalls of extensional shear zones and low-angle normal faults from the Late Devonian-Early Carboniferous. Since then, the culminations started to supply adjacent basins with debris that preserves the $^{40}Ar/^{39}Ar$ signature and, thus, the cooling history of the source rocks from which it was eroded. The recent AFT data show that faulting along the Møre-Trøndelag trend continued at least into the Late Mesozoic. Late Palaeozoic and Mesozoic stages of exhumation were strongly dependent on the location and geometry of new faults that formed by excisement, incisement or re-activation of the previous structures. Interpretation of long-offset seismic reflection data strongly indicates that low-angle normal-faulting and core-complex denudation continued into the Mesozoic; however, the lateral extent of the core complexes and the detachment faults interpreted in the offshore areas (Osmundsen et al., 2002b) and their influence on shallower-level extensional faulting is not known in detail. There is a considerable potential, in our view, residing in the links between deep and shallow structure and, in turn, in their influence on syn- and post-rift stratigraphic architecture. A better coverage of low-temperature thermochronological data (AFT, U-Th-He) is essential to improve our understanding of the relationship between the source area uplift and the offshore basin formation in Late Palaeozoic and Mesozoic times.

Acknowledgements

The studies summarised in this paper were conducted under the umbrella of the BAT project, hosted by the Norwegian Geological Survey (NGU). We thank the sponsors to the BAT project, including Eni Norge, BP, ChevronTexaco, ConocoPhillips, ExxonMobil, Norsk Hydro, A/S Norske Shell and Statoil. We thank Geco-Prakla for permitting an interpretation on the proprietary long-offset seismic data.

References

Allen, K.C., 1976. Devonian spores from outer Trøndelag. Norsk Geol. Tidsskr., 56, 437–448.

Andersen, T.B. and Jamtveit, B., 1990. Uplift of deep crust during orogenic extensional collapse: a model based on field studies in the Sogn-Sunnfjord area of Western Norway. Tectonics, 9: 1097–1111.

Bering, D., 1992. The orientation of minor fault plane striae and the associated deviatoric stress tensor as a key to the fault geometry in part of the Møre-Trøndelag fault zone, on-shore central Norway. In: R.M. Larsen, H. Brekke, B.T. Larsen and E. Talleraas (Editors), Structural and Tectonic Modelling and its Application to Petroleum Geology. Norwegian Petroleum Society (NPF), Special Publication 1, Elsevier, Amsterdam, pp. 83–90.

Berry, H.N., Lux, D.R., Andresen, A. and Andersen, T.B., 1995. Progressive exhumation during orogenic collapse as indicated by $^{40}Ar/^{39}Ar$ ages from different structural levels, southwest Norway. Geonytt 22 (1): 20–21.

Blystad, P., Brekke, H., Færseth, R.B., Larsen, B.T., Skogseid, J. and Tørudbakken, B., 1995. Structural elements of the Norwegian Continental Shelf. Part II: The Norwegian Sea Region. NPD, Bull., 8.

Braathen, A., Nordgulen, Ø., Osmundsen, P.T., Andersen, T.B., Solli, A. and Roberts, D., 2000. Devonian, orogen-parallel, opposed extension in the Central Norwegian Caledonides. Geology, 28: 615–618.

Braathen, A., Osmundsen, P.T., Nordgulen, Ø., Roberts, D. and Meyer, G., 2002. Orogen-parallel extension of the Caledonides in northern Central Norway: an overview. Nor. J. Geol. (NGT), 82: 225–242.

Bøe, R., 1991. Structure and seismic stratigraphy of the innermost mid-Norwegian continental shelf: an example from the Frohavet area. Mar. Petrol. Geol., 8: 140–151.

Bøe, R. and Bjerkli, K., 1989. Mesozoic sedimentary rocks in Edøyfjorden and Beitstadfjorden, Central Norway: Implications for the structural history of the Møre-Trøndelag fault zone. Mar. Geol., 87: 287–299.

Bøe, R. and Skilbrei, J.R., 1998. Structure and seismic stratigraphy of the Griptarane area, Møre Basin margin, mid-Norway continental shelf. Mar. Geol., 147: 85–107.

Chauvet, A., Kienast, J.R., Pinarden, J.L. and Brunel, M., 1992. Petrological constraints and PT path of Devonian collapse tectonics within the Scandinavian mountain belt (Western Gneiss Region, Norway). J. Geol. Soc., London, 149: 383–400.

Chauvet, A. and Séranne, M., 1994. Extension-parallel folding in the Scandinavian Caledonides: implications for late-orogenic processes. Tectonophysics, 238: 31–54.

Coates, B.H., Zeltner, D.L., Carter, B.T., Steltenpohl, M.G., Andresen, A. and Kunk, M.J., 1999. $^{40}Ar/^{39}Ar$ and structural investigations of extensional development of the North-Central Norwegian margin. Abstracts with programs, Geological Society of America, 31 (7): 118.

Cuthbert, S.J., 1991. Evolution of the Devonian Hornelen Basin, Western Norway: new constraints from petrological studies of metamorphic clasts. In: Morton, A.C., Todd, S.P. and Haughton, P.D.W. (Editors), Developments in sedimentary provenance studies. Geol. Soc., London, Spec. Publ., 57: 343–360.

Dallmeyer, R.D., Johansson, L. and Möller, C., 1992. Chronology of Caledonian high-pressure granulite-facies metamorphism, uplift and deformation within northern parts of the Western Gneiss Region. Geol. Soc. Am. Bull., 104: 444–455.

Eide, E.A., Haabesland, N.E., Osmundsen, P.T., Kendrick, M.A., Andersen, T.B. and Roberts, D., 2003. $^{40}Ar/^{39}Ar$ geochronology and indirect dating of continental sedimentary rocks: Late Devonian-Early Carboniferous 'Old Red Sandstone' discovered in outer Trøndelag. In: H.A. Nakrem (Editor), Abstracts and Proceedings of Geological Society of Norway (NGF) 1, Vinterkonferansen 2003.

Eide, E.A., Osmundsen, P.T., Meyer, G.B. and Kendrick, M.A., 2002. An $^{40}Ar/^{39}Ar$ geochronology profile through the Nesna Shear Zone, North-Central Norway: a Middle Devonian — Early

Carboniferous ductile extension and unroofing record. Nor. J. Geol. (NGT), 82: 317–339.

Fossen, H., 1992. The role of extensional tectonics in the Caledonides of southern Norway. J. Struct. Geol., 14: 1033–1046.

Fossen, H., Odinsen, T., Færseth, R.B. and Gabrielsen, R.H., 2000. Detachments and low-angle faults in the northern North Sea rift system. In: Nøttvedt et al. (Editors), Dynamics of the Norwegian margin. Geological Society, London, Special Publications, 167: 105–131.

Gabrielsen, R.H., Odinsen, T. and Grunnaleite, I., 1999. Structuring of the Northern Viking Graben and the Møre Basin; the influence of basement structural grain and the particular role of the Møre-Trøndelag Fault Complex. Mar. Petrol. Geol., 16: 443–465.

Gawthorpe, R.L., Sharp, I., Underhill, J.R. and Gupta, S., 1997. Linked sequence stratigraphic and structural evolution of propagating normal faults. Geology, 25: 795–798.

Grunnaleite, I. and Gabrielsen, R.H., 1995. The structure of the Møre Basin. Tectonophysics, 252: 221–251.

Grønlie, A. and Roberts, D., 1989. Resurgent strike-slip duplex development along the Hitra-Snåsa and Verran Faults, Møre-Trøndelag Fault Zone, Central Norway. J. Struct. Geol., 11: 295–305.

Grønlie, A., Nilsen, B. and Roberts, D., 1991. Brittle deformation history of fault rocks on the Fosen Peninsula, Trøndelag, Central Norway. Norg. Geol. Unders. Bull., 421: 39–57.

Haller, J., 1971. Geology of the East Greenland Caledonides. John Wiley and Sons. 381 pp.

Hovius, N. and Leeder, M.R., 1998. Clastic sediment supply to basins. Basin Res., 10: 1–5.

Jarvik, E., 1949. On the Middle Devonian Crossopterygians from the Hornelen field in western Norway. Universitetet i Bergen Aarbok, Naturvidenskapelig rekke, 8.

Jongepier, K., Rui, J.C. and Grue, K., 1996. Triassic to Early Cretaceous stratigraphic and structural development of the northeastern Møre Basin margin, off Mid-Norway. Norsk Geol. Tidsskr., 76: 199–214.

Kendrick, M., Eide, E.A., Roberts, D. and Osmundsen, P.T., 2004. The Middle to Late Devonian Høybakken Detachment, Central Norway: ^{40}Ar-^{39}Ar evidence for prolonged late/post-Scandian extension and uplift. Geol. Mag., 141 (3): 329–344.

Kolderup, C.F., 1921. Kvamshestens Devonfelt. Bergens Museum Aarbok 1920–21, Naturvidenskabelig Række, 4.

Krabbendam, M. and Dewey, J.F., 1998. Exhumation of UHP rocks by transtension in the Western Gneiss region, Scandinavian Caledonides. In: R.E. Holdsworth, R.A. Strachan and J.F. Dewey (Editors), Continental transpressional and transtensional tectonics. Geol. Soc., London, Spec. Publ., 135: 159–181.

Leeder, M.R., Harris, T. and Kirby, M.J., 1998. Sediment supply and climate change: implications for basin stratigraphy. Basin Res., 10: 7–18.

Lister, G.S. and Davis, G.A., 1989. The origin of metamorphic core complexes and detachment faults formed during Tertiary continental extension in the northern Colorado River region, U.S.A. J. Struct. Geol., 11: 65–94.

Lister, G.S., Etheridge, M.A. and Symonds, P.A., 1991. Detachment models for the formation of passive continental margins. Tectonics, 10: 1038–1064.

Manatschal, G., Froitzheim, N., Rubenach, M.J. and Turrin, B.D., 2001. The role of detachment faulting in the formation of an ocean-continent transition: insights from the Iberia abyssal plain. In: Wilson, R.C.L., Whitmarsh, R.B., Taylor, B. and Froitzheim, N. (Editors), Non-Volcanic Rifting of Continental Margins: A Comparison of Evidence from Land and Sea. Geological Society, London, Special Publications, 187: 161–189.

Mosar, J., Osmundsen, P.T., Sommaruga, A., Torsvik, T.H. and Eide, E.A., 2002. Greenland-Norway separation: a geodynamic model for the North Atlantic. Nor. J. Geol. (NGT), 82: 281–298.

Mørk, M.B.E. and Stiberg, J.P., 2003. Basement erosion and Mesozoic sandstone provenance in the Møre margin area. In: Norsk Petroleumsforening, extended abstracts: Petroleum Exploration and Production in Environmentally Sensitive Areas, Ålesund Norway, 5–7th May 2003, pp. 41–44.

Norton, M.G., 1986. Late Caledonian extension in western Norway: a response to extreme crustal thickening. Tectonics, 5: 195–204.

Olesen, O., Lundin, E., Nordgulen, Ø., Osmundsen, P.T., Skilbrei, J.R., Smethurst, M.A., Solli, A., Bugge, T. and Fichler, C., 2002. Bridging the gap between the onshore and offshore geology in Nordland, northern Norway. Nor. J. Geol. (NGT), 82: 243–262 .

Osmundsen, P.T. and The BAT team, 2002a. Core complexes, gneiss-cored culminations and detachments, Mid Norway. In: Eide (Editor), BATLAS — Mid Norway plate reconstruction atlas with global and Atlantic perspectives. A product of the Basin Analysis and applied thermochronology on the Mid Norwegian shelf (BAT) project, 1998–2002. ISBN: 82-7385-106-0. pp. 64–65.

Osmundsen, P.T., Braathen, A., Nordgulen, Ø., Roberts, D., Meyer, G.B. and Eide, E., 2003. The Devonian Nesna shear zone and adjacent gneiss-cored culminations, North-Central Norwegian Caledonides. J. Geol. Soc., London, 160: 137–150.

Osmundsen, P.T., Sommaruga, A., Skilbrei, J.R. and Olesen, O., 2002b. Deep structure of the Mid Norway rifted margin. Nor. J. Geol. (NGT), 82: 205–224.

Ramberg, H., 1980. Diapirism and gravity collapse in the Scandinavian Caledonides. In: Phillips, W.E.A. and Johnson, M.R.W. (Editors) Deformation and metamorphism in the Caledonian Orogen. J. Geol. Soc., London, 137: 261–270.

Redfield, T.F., 2002. Apatite fission track data from the Møre Trøndelag fault complex and the Fosen Peninsula, central Norway. In: A. Hurst (Editor), Onshore-Offshore Relationships on the Nordic Atlantic Margin. NGF Abstracts and proceedings 2, 2002 of the Norwegian Petroleum Society (NPF) and Norwegian Geological Society (NGF) Conference, 7–9th Oct. Trondheim, pp. 166–168.

Redfield, T.F., Torsvik, T.H., Andriessen, P.A.M. and Gabrielsen, R.H., 2004. Mesozoic and Cenozoic tectonics of the Møre Trøndelag Fault Complex, central Norway: constraints from new apatite fission track data. Physics and Chemistry of the Earth.

Roberts, D., 1998. High-strain zones from meso- to macro-scale at different structural levels, Central Norwegian Caledonides. J. Struct. Geol., 20: 111–119.

Rykkelid, E. and Andresen, A., 1994. Late Caledonian extension in the Ofoten area, northern Norway. Tectonophysics, 231: 157–169.

Séranne, M., 1992. Late Palaeozoic kinematics of the Møre-Trøndelag Fault Zone and adjacent areas, central Norway. Norsk Geol. Tidsskr., 72: 141–158.

Sherlock, S., 2001. Two-stage erosion and deposition in a continental margin setting; an $^{40}Ar/^{39}Ar$ laserprobe study of offshore detrital white micas in the Norwegian Sea. J. Geol. Soc., London, 158: 793–799.

Sindre, A., 1998. Tolkning av dyp til basement under de kaledonske dekkebergartene i Nordland fra gravimetriske data. Norg. Geol. Unders. Rep. 97.179, 24 pp.

Skilbrei, J.R., Olesen, O., Osmundsen, P.T., Kihle, O., Aaro, S. and Fjellanger, E., 2002. A study of basement structures and onshore-offshore correlations in Central Norway. Nor. J. Geol. (NGT), 82: 263–280.

Skilbrei, J.R., Skyseth, T. and Olesen, O., 1991. Petrophysical data and opaque mineralogy of high-grade and retrogressed lithologies: implications for the interpretation of aeromagnetic anomalies in northern Vestranden, Central Norway. Tectonophysics, 192: 21–31.

Skogseid, J., Pedersen, T. and Larsen, V.B., 1992. Vøring Basin: subsidence and tectonic evolution. In: Larsen, R.M., Brekke, H., Larsen, B.T. and Talleraas (Editors), Structural and Tectonic Modelling and its Application to Petroleum Geology. Norwegian

Petroleum Society (NPF) Special Publication 1, Elsevier, Amsterdam, pp. 55–82.
Sommaruga, A. and Bøe, R., 2002. Geometry and subcrop maps of shallow Jurassic basins along the Mid-Norway coast. Mar. Petrol. Geol., 19: 1029–1042.
Talbot, C.J. and Ghebreab, W., 1997. Red Sea detachment and basement core complexes in Eritrea. Geology, 25: 655–658.
Terry, M.P., Robinson, P., Hamilton, M.A. and Jercinovic, M.J., 2000. Monazite geochronology of UHP and HP metamorphism, deformation and exhumation, Nordøyane, Western Gneiss Region, Norway. AAPG Mineral., 85: 1651–1664.
Titus, S.J, Fossen, H., Pedersen, R.B., Vigneresse, J.L. and Tikoff, B., 2002. Pull-apart formation and strike-slip partitioning in an obliquely divergent setting, Leka Ophiolite, Norway. Tectonophysics, 354: 101–1019.
Watts, L.M., 2001. The Walls Boundary Fault Zone and the Møre-Trøndelag Fault Complex: a case study of two reactivated fault zones. Unpublished Ph.D. thesis, University of Durham, 550 pp.
Wernicke, B., 1985. Uniform-sense normal simple shear of the continental lithosphere. Can. J. Earth Sci., 22: 108–125.
Whitmarsh, R.B., Dean, S.M., Minshull, T.A. and Tompkins, M., 2000. Tectonic implicationsof exposure of lower continental crust beneath the Iberia Abyssal Plain, Northeast Atlantic Ocean: Geophysical evidence. Tectonics, 19: 919–942.

Deep structure of the Mid-Norwegian shelf and onshore–offshore correlations: Insight from potential field data

Jan Reidar Skilbrei and Odleiv Olesen

Maps of the depth to the crystalline basement and to the Base Cretaceous, and of the pre-Cretaceous sedimentary thickness have been constructed. These maps illustrate the deep geology of the study area. In addition to showing the total thickness of sedimentary basins (>12 km in the Møre and Vøring basins), the maps also show a lucid picture of structural relief, both at basement depth levels and in the pre-Cretaceous sediment thickness map. Buried rift valleys are seen in the maps. In the platform area, a picture of horsts and grabens emerges that, in most areas, is probably related to the combined late Palaeozoic to early-middle Mesozoic structuring of the area.

The geometry of structural highs and lows shown in our maps is the combined result of several rifting events that were superimposed on the Devono-Carboniferous, late- to post-orogenic extensional collapse structures. We think that the NW–SE structural grain on the shelf reflects the offshore continuations of Devono-Carboniferous shear zones and detachments that are observed onshore. Locally, these detachments were reactivated as transfer zones. Movements along these shear zones have determined the position of boundaries between the magnetic basement terranes (primarily Precambrian high-grade rocks) and the less magnetic Caledonian allochthons.

From our maps, it can be seen that 'old' sediments must occur within Permo-Triassic basins and Jurassic basins, not only on the Trøndelag Platform (previously well documented), but also farther west, beneath the Cretaceous and Cenozoic sediments of the Møre and Vøring basins. The basement topography reveals a narrow, deeply buried, rift-related relief (Jurassic rift valleys), and a wide basinal area recording prolonged subsidence in the mid and late Cretaceous, as well as the Tertiary and Quaternary subsidence. The difference between the basement map and the Base Cretaceous map supports the idea that the Cretaceous sedimentation represents a post-rift thermal and isostatic subsidence stage resulting in the infill of a pre-existing rift topography.

Comparison of the Base Cretaceous and the basement maps shows a close correspondence of trend features, demonstrating a basement influence through to the Cretaceous. The topography and the offshore basement map demonstrate similar tectonic trends, probably indicating that a basement similar to that in western Scandinavia underlies the marine areas, and that several tectonic events have affected both the land and the sea areas.

Highly magnetic, granulite-facies, felsic rocks give rise to strong magnetic anomalies along the Precambrian gneiss terranes of the coastal zone in central-northern Norway. The occurrence of these magnetic rocks is structurally related to the Devonian shear zones and the basement antiforms. We recognise these antiforms and synforms in the offshore, thereby providing a model for the distribution of low-magnetic Caledonian rocks and high-grade Precambrian intermediate rocks in the shelf areas.

The rhomboid-shaped geometry seen in the basement map is partly due to the Devonian extensional collapse structures and partly to N–S and NE–SW oriented faults active during post-Devonian rifting. Major trends responsible for the rhomboid-shaped geometry (NW–SE, N–S and ENE–WSW to NE–SW) are also found in the topography of the Precambrian and the Cambro-Silurian basement units of western Scandinavia, suggesting that Precambrian faults have been reactivated both on land and in the offshore. The Møre-Trøndelag Fault Complex (MTFC) is suggested to consist of two branches in the offshore. One branch is the extensional fault, located east of the Slørebotn Sub-basin sub-parallel to the coastline, and the other one is the fault alignment from Hitra towards Shetland.

Introduction

Information on basement topography aids in the study of rift structures at a variety of scales. On the Mid-Norwegian continental shelf, the structure of the basement surface (e.g. steep slopes) commonly shows trends that are parallel to, or coincide with, faults seen on the seismic sections. The main objectives of this study were to estimate the topography of the basement surface beneath the sedimentary rocks in the Mid-Norwegian shelf, and to outline areas where sub-volcanic intrusions occur within the sedimentary record underlying the sea areas. If the potential field data throw light on the

Onshore–Offshore Relationships on the North Atlantic Margin
edited by B. Wandas et al. NPF Special Publication 12, pp. 43–68,
Published by Elsevier B.V., Amsterdam

nature of the basement below the sea-bed and the sediments, can the Precambrian basement terranes and Caledonian nappe complexes on land be extended into the marine areas, and can different basement terranes be outlined from the available data in the sea areas? If so, do the spatial positions of these terranes relate to the Caledonian Orogeny or to Precambrian tectonic events, or to Devonian extensional events?

Many of the NW–SE to NNW–SSE- oriented tectonic trends in Scandinavia are thought to be traceable onto the shelf, suggesting a Precambrian origin of these trends in the offshore also (Henkel and Eriksson, 1987; Olesen et al., 1997a; Fichler et al., 1999). However, segments of Late Devonian structures on land show the same trend (Osmundsen et al., 2002, 2003). Therefore, one of the objectives of this work was to analyse geophysical data that cover both land and sea, in order to ascertain if structures observed on land can be projected along strike into offshore areas. By inference, the nature of the basement underlying the shelf could be appraised through a similarity with the geophysical signatures observed over the exposed basement on land. Earlier studies that used geophysical data to extend basement units into the sea areas off Mid Norway include the works of Åm (1970), Skilbrei et al. (1991a, b, 1995), Sigmond (1992), Olesen (1997a), Smethurst (2000) and Osmundsen et al. (2002).

The linking of major extensional detachments exposed in central-northern Norway with aeromagnetic signatures observed in the sea areas is speculative (Olesen et al., 2002; Skilbrei et al., 2002; Osmundsen et al., 2002). Fortunately, the deep-seismic lines existing from the Trøndelag Platform are oriented nearly perpendicular to the strike of potential field anomalies, and detachment surfaces have been recognised on these sections, although it is difficult to resolve the ages (Osmundsen et al., 2002). A combined analysis of the potential field data and the published deep-seismic sections makes it possible to address the structure of the basement, as well as the question of the offshore projection of onshore detachment zones and other major basement faults/shears in Central Norway.

Even if it is assumed that the accuracy of the applied depth methods is known, the accuracy of the magnetic depth estimates remains unknown, unless geological control exists. We have also studied the accuracy and the geological meaning of the 'magnetic basement' in the area. That is, it was expected that non-magnetic Devonian basins or Caledonian nappes occur on top of the Precambrian basement. In this case, the crystalline basement would lie above the 'magnetic basement surface'. Fortunately, some exploration wells have reached the basement. Because the study area includes the coastal zone where geological control exists, we have analysed the geological significance of the 'magnetic basement' in the near-shore areas. In this eastern part of the study area, where the basement surface is depicted from the seismic data, there is generally a good correlation between estimates made from magnetic anomalies and the depth to the Precambrian basement. The trend of magnetic anomalies observed in the shelf link up with the main tectonic trends on land. The basement rocks on the islands and in the skerries and the coastal zone also constitute the basement underlying the sediments in the sea areas. We therefore suggest that the aeromagnetic data provide a fresh insight into the concealed basement terranes within the eastern-central shelf areas. The main focus of the report is from the south-central part of the study area.

General geology of the study area

Several continental rift phases have affected the Mid-Norwegian shelf before continental separation commenced at c. 53 Ma in the Norwegian–Greenland Sea (Skogseid et al., 2000; Osmundsen et al., 2002). Major rift events occurred during Late Palaeozoic, Triassic, Jurassic and Cretaceous times, before the Tertiary break up between Greenland and North Scandinavia (e.g., Brekke, 2000). These rifts were superimposed on the crustal fabric that resulted from the late- to post-Caledonian extensional collapse of the mountains in the Devonian. In plane view, the total effect of this tectonism, that also included transpressional and transtensional events resulting from movements between Greenland/Laurentia and Europe in the Late Devonian to Early-Middle Carboniferous (Ziegler, 1988), is the rhomb-shaped geometry of the shelf platform (see Fig. 1, Doré et al., 1997; Blystad et al., 1995; Osmundsen et al., 2002).

The allochtonous sheets of the Scandinavian Caledonides were amalgamated and emplaced east- to southeastward onto the Baltican margin, and the sub-division into Lower, Middle, Upper and Uppermost Allochthons is an interpretation of emplacement order and their progressively more outboard derivations (e.g. Roberts and Gee, 1985; Eide et al., 2002). A series of gneiss-cored culminations exposes the parautochthonous or

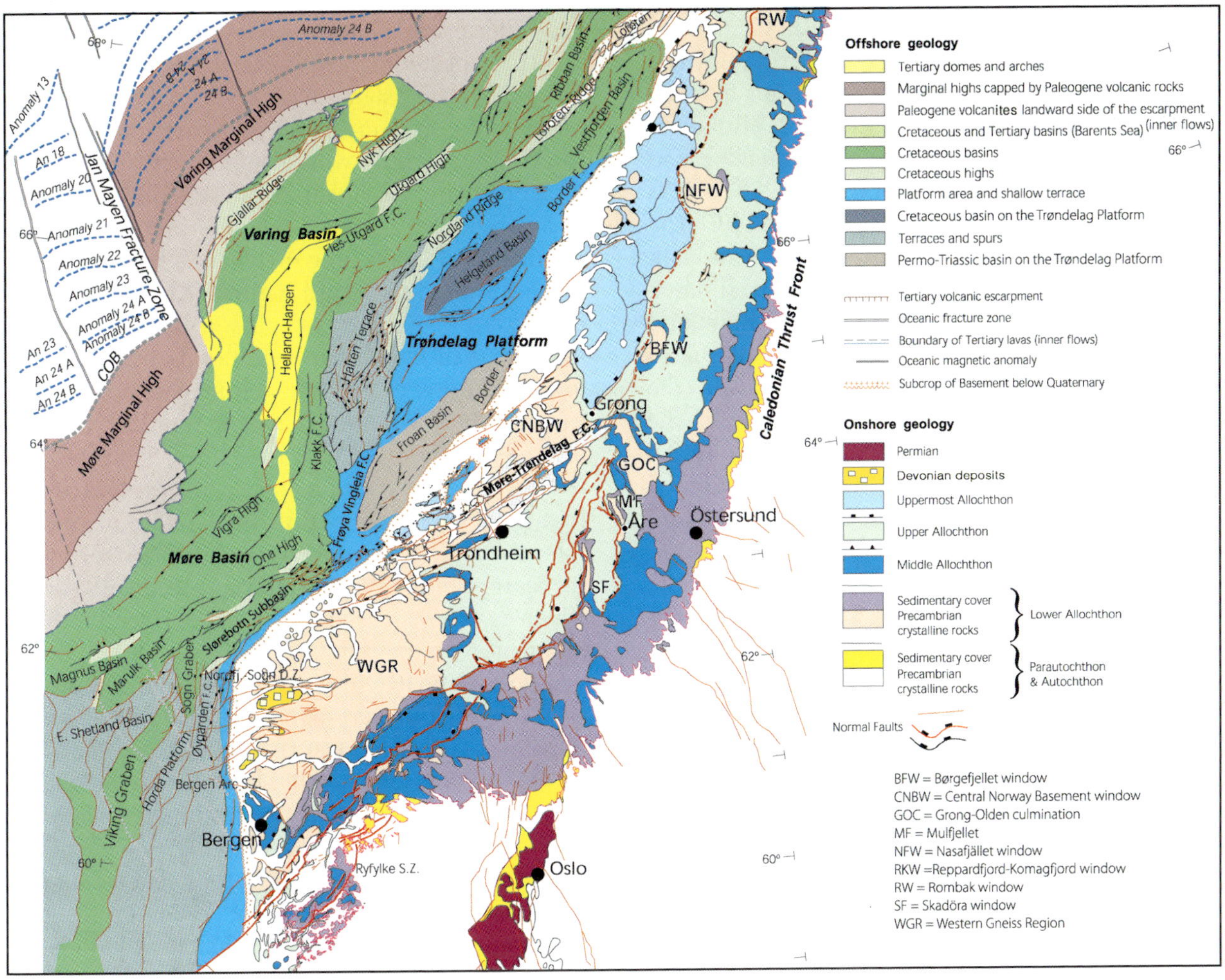

Fig. 1 Tectonostratigraphic map of the mid-Scandinavian passive margin (modified from Mosar et al., 2002).

allochthonous crystalline basement of Baltica. The structural pattern seen today in Mid Norway is the combined result of thrusting, west- to southwest extensional movements occurring during collapse of the orogen (Osmundsen et al., 2002), and faulting and rift flank uplift relating to the rifting and extensional events that affected the shelf and land areas.

The Western Gneiss Region (WGR) and the Central Norwegian Basement Window (CNBW) are mainly made up of Precambrian gneisses and granitic rocks (Sigmond et al., 1984; Sigmond, 1992) with narrow synclinal belts in which thinned remnants of the thrust sheets have been in-folded deeply. Thus, the structural fabric of most of the crustal region of interest is dominated by tight folding of basement and tectonic cover produced during Devonian phases of extensional deformation (Braathen et al., 2000; Robinson, 1995).

On the shelf, the study area covers the Trøndelag Platform, the Halten Terrace and the Frøya High (Bøen et al., 1984; Gabrielsen et al., 1984; Blystad et al., 1995; Brekke, 2000). The western portion covers the Møre Marginal High, the Møre Basin, the Klakk Fault Complex and structures associated with the tectonic transition zone, where the northern part of the Møre Basin ends at the junction between the Møre-Trøndelag Fault Complex (MTFC) and the southeastwards projection of the Jan Mayen Lineament. The MTFC constitutes the eastern margin of the Møre Basin (Gabrielsen et al., 1984; Bukovics and Ziegler, 1985). In the offshore areas, the MTFC trends ENE–WSW and is dominated by normal faults (Gabrielsen et al., 1984). A number of structural highs are situated along the MTFC. The same fault trend is also observed on land (Oftedahl, 1975; Aanstad et al., 1981) and extends into the Grong district of Central Norway (Roberts, 1998). The Klakk Fault Complex trends roughly N–S and is located on the western side of the Frøya High (Fig. 1); it meets the Bremstein Fault Complex on the north side of the Frøya High. This high comprises the northeastern margin of the Møre Basin.

On the Trøndelag Platform, the basal Cretaceous surface is underlain by a relatively uniform thickness of Jurassic sediments overlying deep basins filled by Triassic and Upper Palaeozoic sediments (Bukovics and Ziegler, 1985; Brekke, 2000; Grunnaleite and Gabrielsen, 1995; Osmundsen et al., 2002). The platform is bounded to the south by the MTFC. The Trøndelag Platform grades into the Træna and Vestfjorden basins. The Vestfjorden Basin, situated between the Lofoten Ridge, and the mainland coast of Nordland, is mainly a Cretaceous basin (Blystad et al., 1995). In the Møre Basin, the basal Cretaceous reflector reaches depths in the order of 10–12 km (Hamar and Hjelle, 1984). In the basin, the mid-Cretaceous level, which is dominated by dolerites, is nearly unstructured (Hamar and Hjelle, 1984; Skogseid et al., 2000). On reflection seismic data, the dolerites correspond to the delineation of smooth, high-amplitude reflectors that commonly terminate in a hyperbolic pattern (Gravdal, 1985). Along the eastern margin of the Møre Basin, the structural highs (Selje, Frøya, Gnausen and Gossa Highs) are overlain by a variable thickness of Late Palaeozoic to Cretaceous sedimentary rocks on top of the basement. The Vøring Basin is a large sedimentary basinal province to the north of the Jan Mayen Lineament, comprising a series of basins and highs as described in Brekke (2000).

Structurally, the shelf is dominated by Late Jurassic–Early Cretaceous extensional faults that created deep basins and intra-basinal highs within the shelf. Jurassic basins exist in Beitstadfjorden and in Frohavet as downfaulted half-grabens within the crystalline basement in Central Norway (see Fig. 1 for location).

Earlier geophysical studies

Some of the earliest published studies by the Geological Survey of Norway (NGU) demonstrated that deep sedimentary basins existed off Mid Norway (Åm, 1970). Hamar and Hjelle (1984) suggested that the Base Cretaceous was located at 10 km depth in the Møre Basin, and that Triassic evaporites were present along its eastern margin. The thickness of the pre-Cretaceous sedimentary section was not indicated. Later work has indicated pre-Cretaceous sediments on the Trøndelag Platform, based on reflection seismic data (e.g., Osmundsen et al., 2002; Brekke et al., 2000 and references therein).

A general problem with aeromagnetic data from a rifted region is that the aeromagnetic signal is often lost from the axial parts of the deepest basins or grabens, while closer to land there is sometimes a surplus of anomalies on which to perform depth analysis. In hindsight, it is easy to see that the lack of control points in the deepest parts of the basins forced the earliest workers (Åm, 1970, 1975) to underestimate basement depths from the deepest basins, in order to avoid overinterpretations. Also, the degree of smoothing involved in the contouring of the individual depth points led to an underestimation of basement depths in the deeper parts. Another well-known problem faced in making such interpretations is that the aeromagnetic anomalies can arise both from the magnetic sources within the crystalline basement (intra-basement sources), and from the intra-sedimentary dolerites/intrusions that have been observed in the Cretaceous section (Hamar and Hjelle, 1984; Gravdal, 1985), or within (beneath) the top crystalline basement surface which is the weathered erosional surface on which sediments are deposited.

Lundin and Rundhovde (1993) and Doré et al. (1997, 1999) used aeromagnetic data to illustrate the position of NW–SE transfer zones from the Møre Basin. Skilbrei et al. (1995) modelled the gravity data along the deep seismic reflection line across the Møre Margin (Olafsson et al., 1992), providing a 'minimum depth extent' of the sedimentary fill of c. 12 km in the Møre Basin. Also, the joint interpretation of deep- seismic data and the gravity modelling indicated an asymmetric Moho surface underneath the Møre Marginal High and the Møre Basin (Olafsson et al., 1992; Skilbrei et al., 1995). Magnetic modelling seems to support the existence of pronounced basement relief beneath both the western and the eastern margins of the Møre Basin, and importantly, also under the central part of the Møre Basin (Skilbrei et al., 1995; Sæterstad, 1996). The interpreted relief beneath the Møre Basin suggests the existence of pre-Cretaceous structures, since the Cretaceous sedimentary fill shows little evidence of structuring (cf. Grunnaleite and Gabrielsen, 1995; Brekke, 2000). Olesen et al. (1997a, 2002) used potential field data to interpret basement depths and offshore projection of Devonian detachments, as well as reactivated fault zones and transfer zones in the area between 65° to 71°N. Fichler et al. (1999) used image-enhanced potential field data to extend some of the onshore Precambrian shear structures into the offshore region, and presented a well-constrained gravity and magnetic model crossing the southern part of the Vøring Basin, just to the north of the Jan Mayen Lineament. These earlier works (Åm, 1970, 1975; Olesen et al., 1997a; Skilbrei et al., 1995; Fichler et al., 1999) have been incorporated into our study.

Data sets

Topography and bathymetry from the study area (Dehls et al. 2000) are shown in Fig. 2. The topography and bathymetry data are extremely valuable when studying terrain effects that are present in the aeromagnetic data sets and in the free-air gravity data (offshore).

Aeromagnetic data

NGU started a systematic surveying of the Norwegian Sea in 1965. The area from the mainland out to a water depth of approximately 1000 m was covered in the period 1965–1968, with approximately 5 km line spacing and 200 m flight altitude. In the period 1971–1973, NGU carried out more detailed measurements along the coast in order to correlate the separate measurements made over land and sea. In 1973, NGU conducted an aeromagnetic survey covering the areas that lie between water depths of approximately 1000–3000 m. The profile spacing varies between 4 km and 10 km, with a flight altitude of 500 m. This data set was digitally recorded. The navigation was based on LORAN A and DECCA for all continental shelf data.

NGU constructed aeromagnetic contour maps with a 20 nT interval. The contour lines were drawn manually after carefully adjusting for the base magnetometer data and the tie-line data. These data have been compiled into grids by digitising the contour lines, and then merging the obtained values with the data from the land and coastal areas. The resulting data set used for this study consists of a 1 km by 1 km grid. This is a grid interval that represents the offshore data adequately (c. 5 km line spacing), while depicting the high-frequency information from the land areas also at the scale of 1:500,000 and smaller. These data have been described by Åm (1970, 1975), Olesen et al. (1997a, 2002) and Skilbrei et al. (1995, 2002). The average line spacing was about 5 km for the data from the seventies, and 2 km for data acquired in the eighties and the nineties. The flight altitude was from 200 m to 500 m.

Hunting Geology and Geophysics Limited ('Hunting') carried out a high-sensitivity aeromagnetic survey of approximately 75,000 km^2, offshore central Norway in 1986. The survey was flown with a profile distance of 2 km (Skilbrei and Kihle, 1999). The Viking-93 (Smethurst, 2000) and the NAS-94 (Olesen and Smethurst, 1994) aeromagnetic surveys conducted by NGU adjoin the Hunting surveys. Two other modern data sets covering Vestfjorden and the sea areas west of Lofoten described by Olesen et al. (2002) have been analysed. We analysed both the grid images and the original profile data, in order to obtain depth estimates (read further). An aeromagnetic colour map is shown in Fig. 3.

Petrophysical data

Along the coastal zone, there is good reason to believe that the basement rocks on the continental shelf may be similar to those on the islands and skerries. Some of the magnetic and gravimetric anomalies within the project area are continuous from land onto the continental shelf. Measurements of the density and magnetic properties of the rocks on land are therefore of importance when interpreting the potential field data covering offshore areas. More than 6000 rock samples from the nearby mainland, collected during geological mapping and geophysical studies, have been measured with respect to density, susceptibility and remanence (Olesen et al., 2002; Skilbrei et al., 2002). These data, along with the magnetic property data were used in some of the forward model calculations that helped in the construction of the basement map. The petrophysical data also helped in interpreting the basement structure and the offshore projection of basement terranes. This is of particular importance where non-magnetic Caledonian or Precambrian rocks continue into the marine areas, or where low-to-intermediate density, highly magnetic basement rocks occur along the coastal zone.

Gravity data

In order to cover the coastal zone and to correlate the separate gravity measurements made over land and sea, NGU have set out gravity stations on the islands and skerries, and also on the mainland, using helicopter transportation. Measurements along roads were made using car transportation. In order to link together the land data and the gravity ship lines, new ship-borne gravity measurements were made along the coast between 61° 50' N and 65° 10' N in the coastal zone, using the research vessel 'Håkon Mosby' from the University of Bergen. A total of 1739 profile km were obtained. We have combined these gravity measurements with various data sets from the continental shelf. The data were described in Skilbrei et al. (1995, 2000).

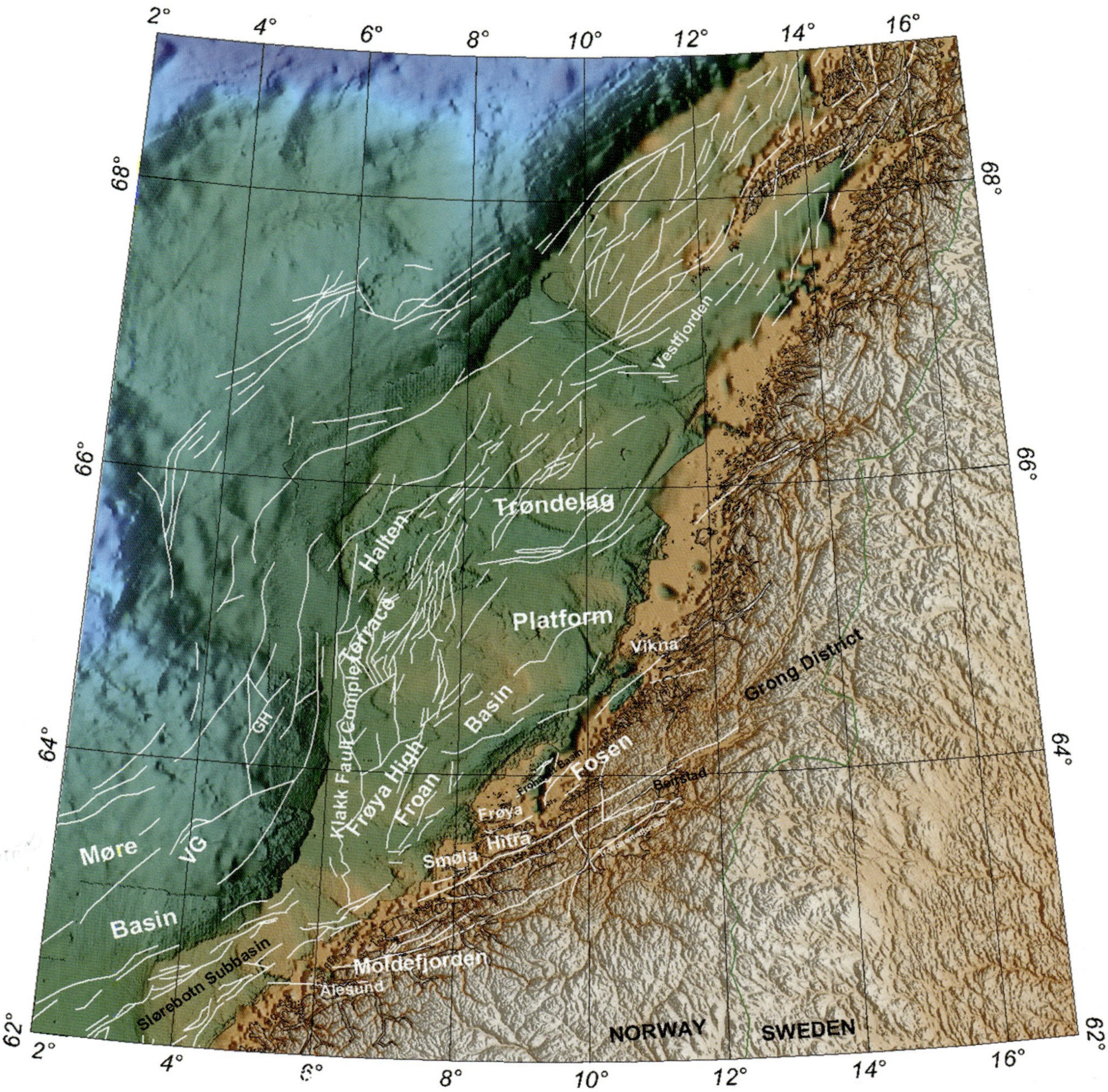

Fig. 2 Topography of the study area. Because fracture zones can be observed in the topographical data, we used this data set to represent the basement on land. A prominent ENE–WSW topographic grain that is well-expressed even at the scale of the figure characterises central Norway. Although this, to a large extent, reflects the dominant strike trend of a variety of lithological units within the Western Gneiss Region and the Central Norwegian Caledonides (Fig. 1 and Sigmond et al., 1984), several studies have indicated the existence of major regional faults of this same general trend (Oftedahl, 1975; Grønlie and Roberts, 1989). This ENE–WSW feature encompasses the Møre-Trøndelag Fault Complex that is probably also important offshore (Gabrielsen et al., 1984; Doré et al., 1997). On the topographic map, narrow linear to curvilinear features (evident in the shading and as low-amplitude steps in the colour contouring) also trend NE–SW, N–S and NW–SE to NNW–SSE. The NE–SW trend is seen primarily in the north Trøndelag area and in the southern part of the county of Nordland (northern part of the map). N–S and NW–SE to NNW–SSE features are seen throughout the coastal zone, the latter also observed as fjords. These trends are also well recorded on the shelf, as evident from the structural elements (see white lines). To the south of Hitra, ENE–WSW- trending faults dominate along the eastern margin of the Møre Basin, whereas to the west of the Trøndelag Platform itself and on the Trøndelag Platform and the Halten Terrace N–S, NE–SW and NW–SE strike trends dominate. This is well-expressed in the basement map (Fig. 5).

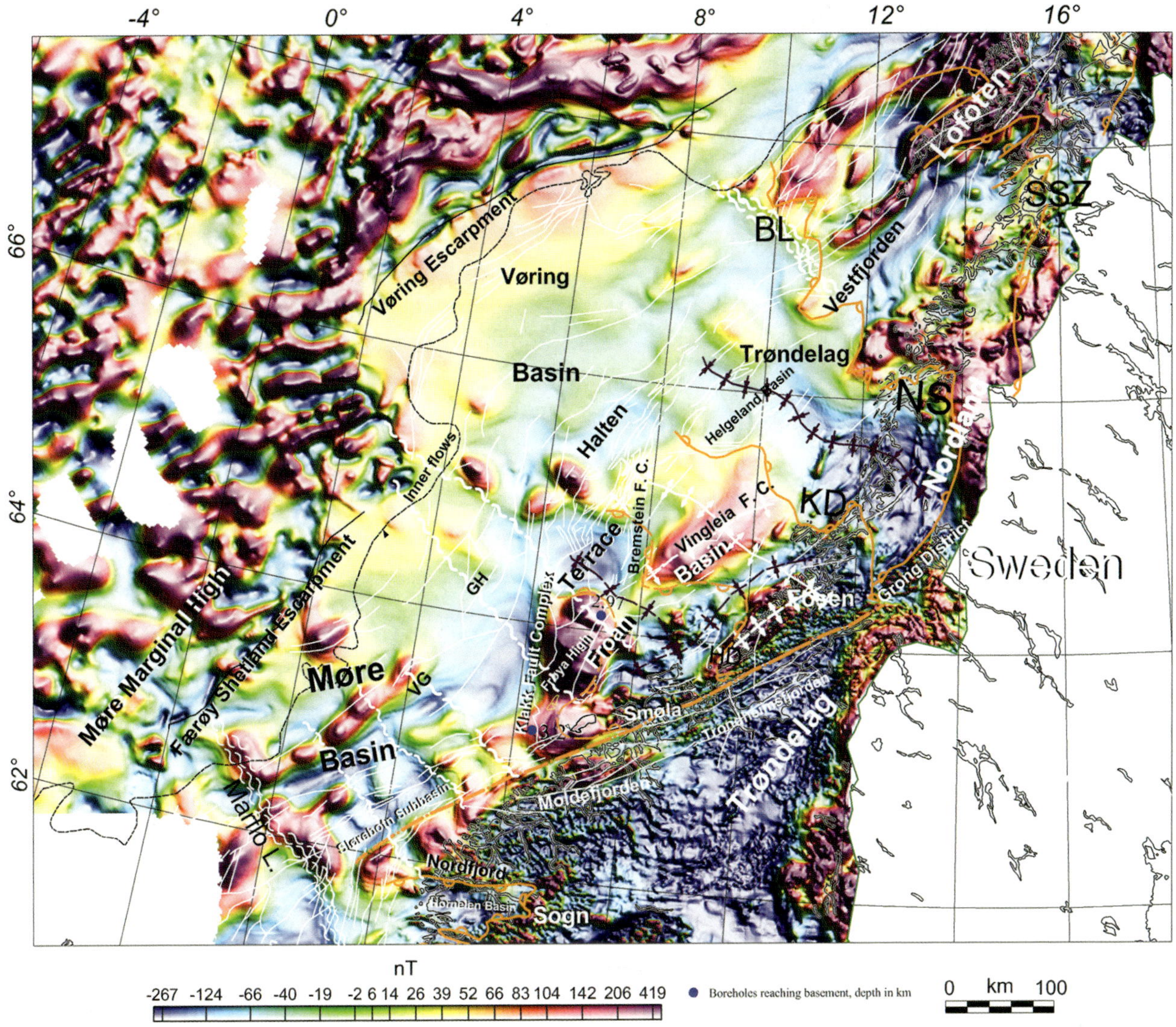

Fig. 3 Regional aeromagnetic map of the study area. Orange solid lines are the interpreted offshore projections of the Devonian detachments that have been mapped on land (Braathen et al., 2000; Nordgulen et al., 2002; Olesen et al., 2002). Note the direction of dip of detachment zones. The dips of these zones, and the magnetic data, have been used in order to infer locations of major antiforms or basement culminations (white lines with antiform symbol) and synforms in the marine areas (black line with synform symbol). We assume that magnetic and non-magnetic basement terranes on the mainland continue under the basins towards the eastern Møre and Vøring basins. Structural elements in white solid lines (from Brekke, 2000). White wiggly broken lines indicate positions of transfer zones. The Marflo Lineament is from Smethurst (2000). The transfer zones in the Møre Basin are from Doré et al. (1997) and this study. Transfer zones and detachments in the Nordland-Vestfjorden-Lofoten areas are from Olesen et al. (2002). A broad transfer zone is defined by two transfer zones at the projection of the southern and northern Jan Mayen Lineament. We suggest also that the JML on the continental margin occurs within a Devonian extensional detachment. BL = Bivrost Lineament, GH = Grip High, HD = Høybakken detachment, KD = Kollstraumen detachment, NS = Nesna Shear Zone, SSZ = Sagfjorden shear zone, VG = Vigra High.

Data enhancement

The geophysical data were processed using the Geosoft system (Geosoft, 2000). High-pass filtering, shaded-relief presentation and histogram-equalisation have been used as standard techniques in the digital image processing. In Fig. 4, a wavenumber-filtered map using a Gaussian roll-off filter shows gravity anomalies with wavelengths generally less than 200 km. Even in the interpretation maps (Figs. 5–9), the combination of colours and shading is useful for interpreting the form of the surfaces (shown in colours) and faults and steep basement slopes (depicted primarily by shading).

Methods

The report focuses on the interpretation of the depth to the magnetic sources. Magnetic models,

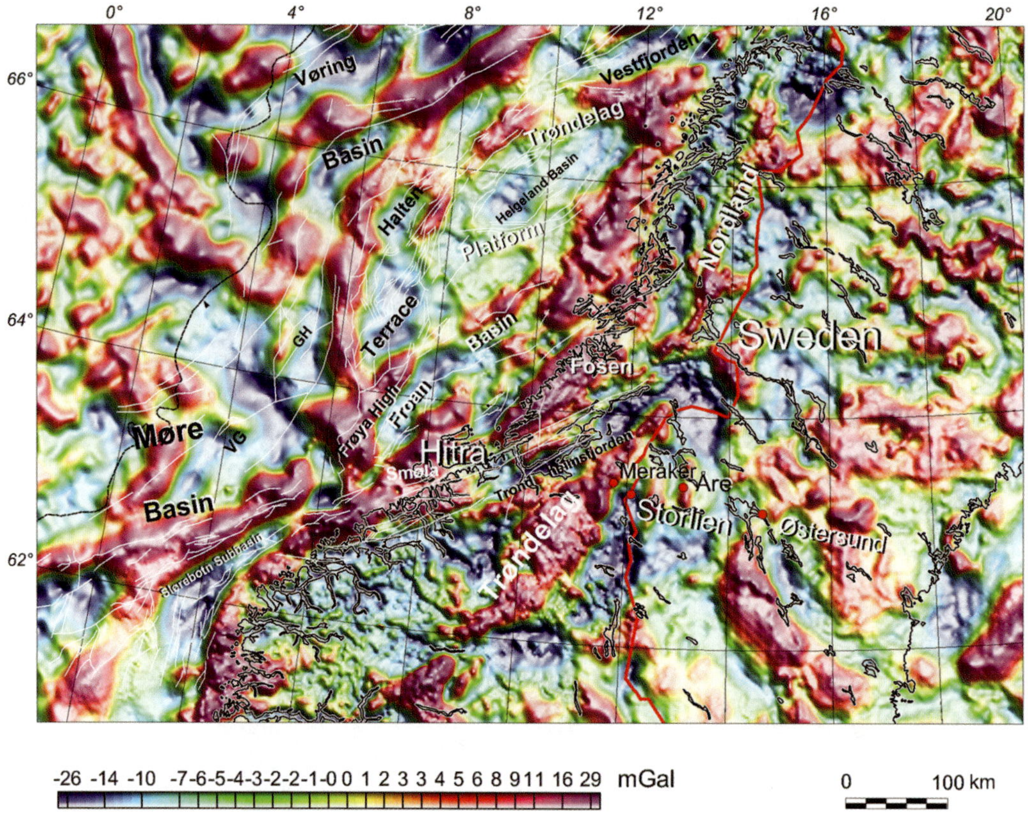

Fig. 4 Residual, high-pass filtered, gravity map (Bouguer on land, free-air at sea). VG = Vigra High, GH = Grip High. Structural elements in white solid lines. Note that many of the short-wavelength features are related to lithological units on land, as well as to the fault zones in the offshore. Note also the eyes-within-eyes pattern (see text for explanation).

gravity models and published seismic data have been used to constrain the depth solutions. Except for the short-wavelength anomalies arising from shallow sedimentary rocks and moraines, the regional magnetic anomalies may be due to intrasedimentary volcanic rocks, intra-basement sources and basement relief. An attempt has been made to differentiate between these sources.

The three-dimensional Euler deconvolution method was applied to the grid mesh. Careful analysis of the size (width) of the anomalies was performed to constrain the window size used in each sub-area in the Euler method, and all the estimated depth solutions were screened, as described by Reid et al. (1990), which included estimating and applying a so-called structural index. The structural index is a 'geological model parameter' that has to be estimated in the analysis of the Euler deconvolution. The estimates are therefore biased. Another semi-automatic technique, the autocorrelation method, was applied to the profile data. Selected anomalies were also interpreted using manual, graphical techniques, in order to estimate the depth to individual magnetic sources along profiles. This was mainly from the Frøya High areas, and from the Vigra High in the Møre Basin.

Seismic, borehole, and petrophysical data have been used to constrain the estimates, wherever possible. The different depth to magnetic source methods (Euler 3D deconvolution and the autocorrelation methods) yield errors that generally vary between 5% and 15%. Systematic errors will add to the spread shown by the different methods. These errors are reduced by constraints offered by other data. Where possible, the depth estimates were calibrated with known depths from boreholes and seismic lines (generally near the coastal zone, or where the basement is relatively shallow), from the gravity and magnetic forward models (Sæterstad, 1996; Olesen et al., 1997a; Skilbrei et al., 1995), as well as from deep seismic data (Olafsson et al., 1992; Mjelde et al., 1998; and references therein). Depths thought to represent intra-sedimentary magnetic rocks have been excluded during the contouring of the final surface shown here. As a consequence of this, the basement depths are uncertain in the far west where intra-sedimentary igneous sills and flows are known to occur.

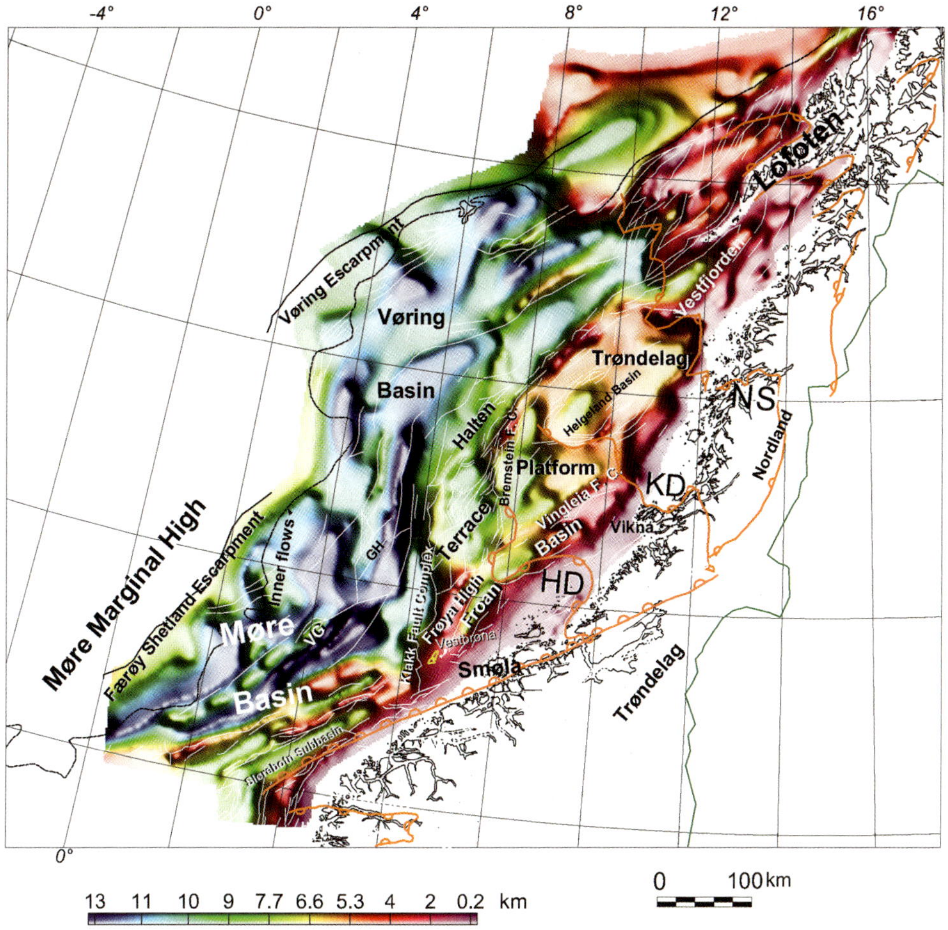

Fig. 5 The top of the crystalline basement within the central area of the Scandinavian shelf. Depths refer to the top of crystalline basement from sea level. The map represents an estimate of the total depth extent of the sedimentary basins in the area. This map is an aid in elucidating the general structure of the area and provides insights into the deep structure of this rifted region. The results have been compared with the estimates of the depth to the top of magnetic sources, enabling the accuracy as well as the geological significance of the magnetic basement in the area to be analysed (see the text for explanation). Structural elements (white solid lines) from Brekke et al. (2000). Note deep 'valleys' in the Møre Basin, and detailed topography in the northern part of the Frøya High. The latter is constrained by published seismic data (Blystad et al., 1995) and boreholes. Strong gradients in the map coincide with major basement fault zones and/or major basement flexures that define the margins of structural highs and troughs. Note that the basement and the Base Cretaceous (next figure) show similar relief (faults and slopes) orientation. HD = Høybakken detachment, KD = Kollstraumen detachment, NS = Nesna Shear Zone.

Interpretation

Topographical data

Because fracture zones can be observed in the topography, we used this data set to represent the basement on land (Fig. 2). Gabrielsen and Ramberg (1979) and Gabrielsen et al. (2002) have delineated the fundamental large-scale fracture pattern of Central Norway. A prominent ENE–WSW topographic grain, that is well-expressed in Fig. 2, characterises the area. Although to a large extent, this reflects the dominant strike trend of a variety of lithological units within the Western Gneiss Region and the Central Norwegian Caledonides (Fig. 2 and Sigmond et al., 1984), it is also the same general trend as several major regional faults (e.g. Oftedahl, 1975; Grønlie and Roberts, 1989). These ENE–WSW features include the MTFC, which is also known to be important offshore (Gabrielsen et al., 1984). On the topographic map (Fig. 2), narrow linear to curvilinear features also trend NE–SW, N–S and NW–SE to NNW–SSE. The NE–SW trend is seen primarily in northern Trøndelag and in the southern part of Nordland county (northernmost part of the map). The N–S and NW–SE to NNW–SSE features are seen throughout the coastal zone, the latter also observed as fjords. These trends are also evident on the shelf as structural elements (Fig. 1). To the south of Hitra, ENE–WSW- trending faults dominate along the eastern margin of the Møre Basin, while to the west of the Trøndelag Platform, on the Trøndelag Platform and the Halten Terrace, N–S and NE–SW strike trends dominate. In the Moldefjorden area, ENE–WSW to NE–SW trending

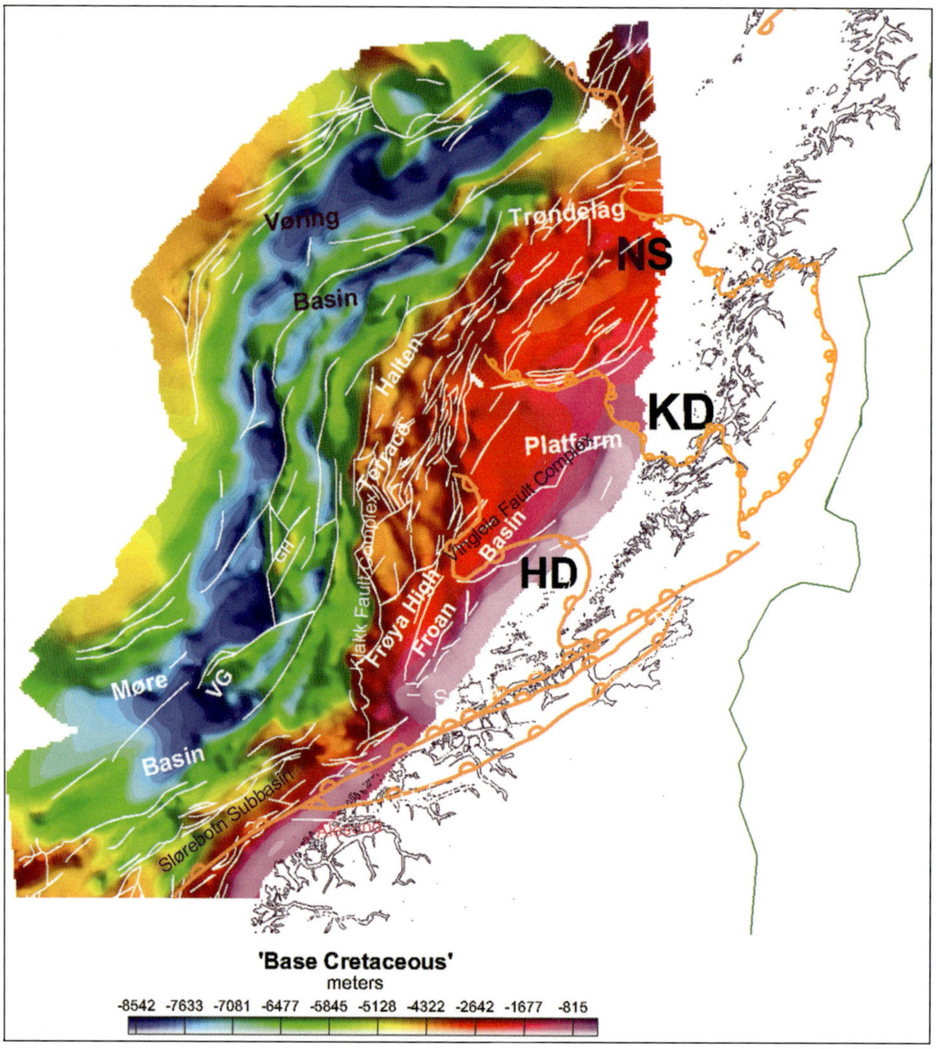

Fig. 6 Base Cretaceous unconformity map of the study area (from Brekke, 2000). White solid lines represent structural elements, from the NPD (Blystad et al., 1995). The unconformity surface is smoother than the basement map. This is probably because the top of the basement represents a steep topography (rift structures), while the Lower Cretaceous comprises post-rift sediments. This is only possible if the top basement surface was rifted and covered with syn-rift sediments. There may have been a period of tectonic quiescence following Jurassic rifting before the onset of Cretaceous sedimentation over wide areas. Alternatively, Jurassic rifting may have resulted in a less pronounced relief than the Late Palaeozoic and Early Mesozoic rift episodes. HD = Høybakken detachment, KD = Kollstraumen detachment, NS = Nesna Shear Zone.

structural trends indicate that this part of the mainland has been influenced by the stress regimes that affected the adjacent Slørebotn Sub-basin (SS). A phase of sinistral transpression occurred in the SS in the Early Cretaceous (Jongepier et al., 1996), which also influenced the Griptarane area (Bøe and Skilbrei, 1998). Some of the faults along the eastern margin of the Cretaceous SS seem to continue into the coast, probably connecting to the MTFC on land. In particular, this is the case along the E–W faults described by Blystad et al. (1995) that may run into the outlet of Moldefjorden and into the E–W fault near Ålesund (Figs. 2 and 3). A preliminary study of paleomagnetism and structural setting of fault rocks in the Moldefjord area shows fault activity from the Devonian/Carboniferous to Lower Cretaceous time (Skilbrei et al., 1995). This suggests an onshore–offshore tectonic link between faults that offset sedimentary fill within the SS and the Moldefjorden lineament.

Gravity data

While the aeromagnetic data largely have their sources within the crystalline basement, many offshore gravity anomalies reflect the density contrasts between Cenozoic and older sediments that are, in many cases, juxtaposed along fault zones. The gravity data therefore clearly illustrate the structural framework of the faults, basins and

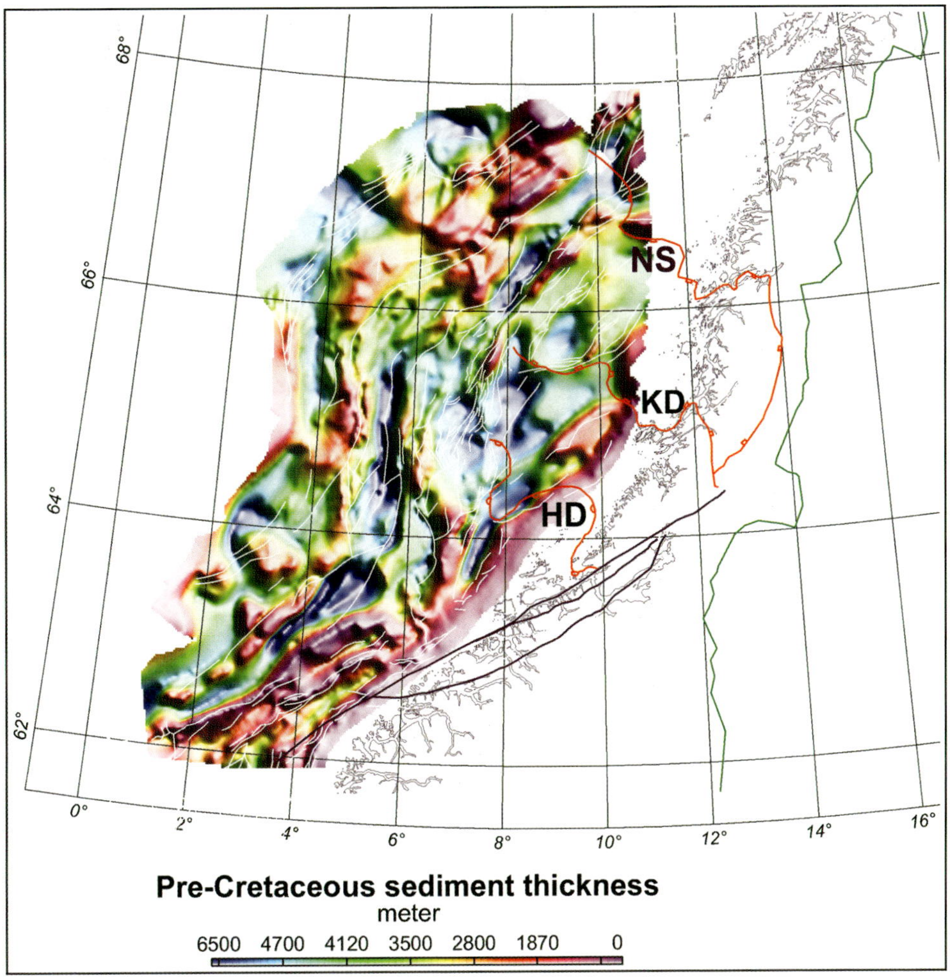

Fig. 7 Pre-Cretaceous sediment thickness map. This is the difference between the basement map and the Base Cretaceous unconformity map. These estimates are therefore to some degree, uncertain, particularly in the west. See text for explanation. HD = Høybakken detachment, KD = Kollstraumen detachment, NS = Nesna Shear Zone.

highs occurring at shallow to intermediate depths, since they pick up density contrasts from shallow features to Moho levels. The latter give rise to long-wavelength features. This is in contrasts with the magnetic anomalies that reflect intra-sedimentary volcanic and basement-related anomaly sources. The latter reflect the geometry of faulted basement rocks and of lithological contacts within the basement.

Within the MTFC, both the regional gravity gradients and the relatively short-wavelength anomalies trend mainly ENE–SSW to NE–SW, reflecting the structural grain of the area. The rather linear gradients continue from the Grong district, onto the continental shelf to the west of Molde and Kristiansund, where they interfere with a coastal edge positive anomaly that runs along, and is parallel to, the coast. It is interesting to note that the positive anomaly trend occurring between Smøla and Shetland is broken southwest of Smøla. It actually disappears on the Bouguer gravity data set (Skilbrei et al., 1995). Only local highs exist above each of the structural highs, along the southeast margin of the Møre Basin. Just to the west of the islands Smøla, Hitra and Frøya and the Froan archipelago (northeast of Frøya), a series of positive anomalies occur that trend N–S and NE–SW, resulting in the above-mentioned 'dog-leg' pattern. Gravity model calculations across this coastal edge anomaly demonstrate that it marks the contact between the basement and the southeastern side of the less dense sedimentary rocks to the northwest. This contact creates a short-wavelength anomaly, which is superimposed on the regional long-wavelength anomaly that is caused by the decrease in the depth to the Moho surface, when approaching the shelf. Sindre (1977) noted that a gravity anomaly extends onto Smøla where Caledonian mafic intrusions are exposed. The density of samples of the Caledonian plutons in the Frøya–Froan area (see Nordgulen et al., 1995 for a brief outline of the geology), lies in the range

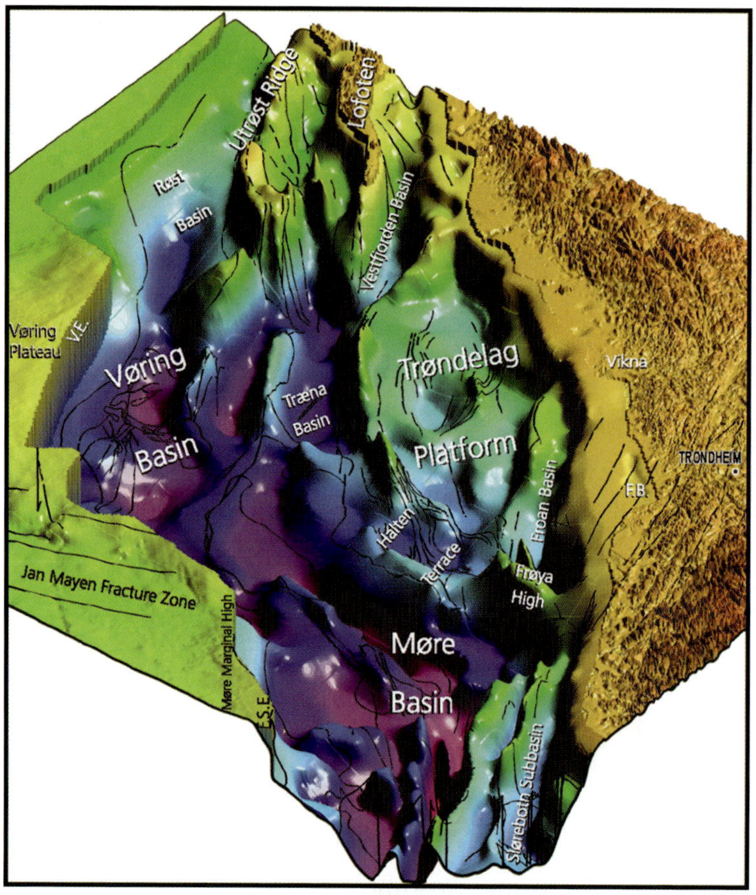

Fig. 8 Perspective view of depth to basement map of the study area. This map is an aid in elucidating the general structure of the area and provides an insight into the deep structure of this rifted region. Strong gradients in the map coincide with major basement fault zones and/or major basement flexures that define the margins of structural highs and troughs. Note deep rift valleys in the Møre and Vøring Basins, and important structural relief and subbasins on the Trøndelag Platform. The map illustrates that the continental crust has been deformed along fault zones that trend in different directions. For instance, the eastern-central area shows NNW–SSE to NW–SE, N–S, and ENE–WSW to NE–SW trending slopes. These are observed both onshore and offshore, probably indicating that the influence of an inherited, pre-existing basement structural grain is profound both offshore (assumed) and on land (e.g. Braathen et al., 2000). The figure also illustrates the rhomboidal geometry at a variety of scales. This geometry results from the effect of profound zones of weakness in the basement along N–S, NE–SW and NW–SE trends. The latter is the interpreted strike of the offshore projection of extensional detachment zones. F.B. = Frohavet Basin. V.E. = Vøring Escarpment. F.S.E. = Færøy Shetland Escarpment. The areas to the west of the V.E. and the F.S.E. represent bathymetry. On land, the figure shows topography. Solid lines represent structural elements from Blystad et al. (1995).

2620–2710 kg/m^3, with a mean that is below the average density of the nearby basement, e.g. in the Western Gneiss Region (Skilbrei, 1988a, 1989). This explains the gravity low situated in the Frøya–Froan area to the east of the Froan Basin (Fig. 11).

Aeromagnetic data

On land, much experience has been gained in our understanding of the sources of magnetic anomalies through both local studies and regional investigations. These studies have involved both model calculations of the magnetic profiles and the sampling of rock specimens for rock property studies, as well as in-situ measurements of magnetic susceptibilities of both low- and highly magnetic rocks from anomaly zones seen on the 1: 50,000 scale maps and in original profile data (Skilbrei et al., 1988b, 1991a, 2002; Olesen et al., 1997a, 2002). Both regional and more detailed interpretations from the Trøndelag area, based on ground verification work (magnetic susceptibility measurements), can be found in Skilbrei and Sindre (1991), Skilbrei et al. (1991a), Fasteland and Skilbrei (1989) and Skilbrei and Kihle (1999). These studies have shown that most of the Caledonian metasedimentary rocks show very low magnetisation values (cf. Wolff, 1984; Dyrelius, 1985; Elming, 1980). The main magnetic rocks of the Caledonian nappe pile are some of the mafic greenstone units in Trøndelag, i.e. mainly metabasalt with minor gabbro, and small pod forms of serpentinite as well as relatively small gabbroic bodies. Petrophysical data show that the

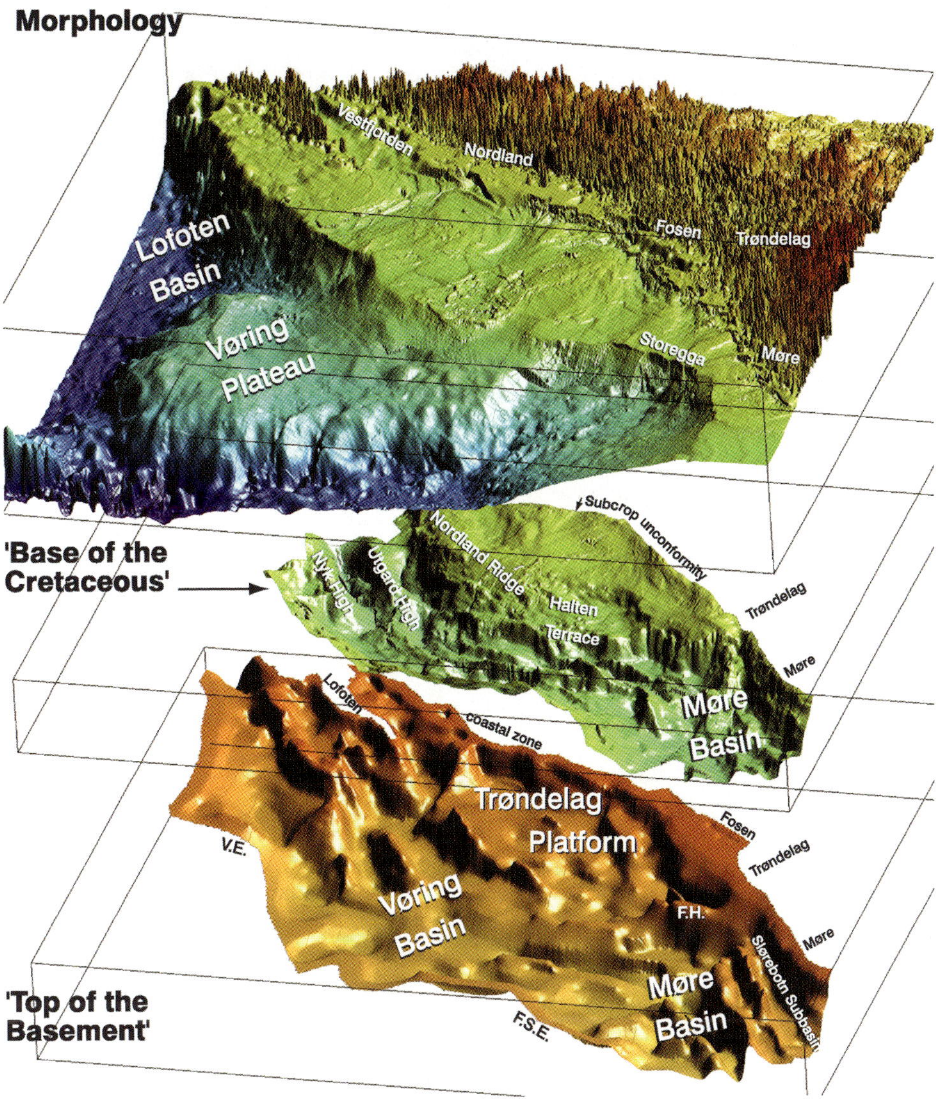

Fig. 9 Stacked perspective views of topography, the Base Cretaceous unconformity and the basement surface. The basement surface is the estimated top of the crystalline crust. The figure illustrates the effect of the net subsidence and uplift over the area, as well as the structure of the top of the basement and of the Base Cretaceous surface. The pronounced relief of the top basement surface may indicate that the structuring of some of the rifted regions largely happened before the Cretaceous, while the great depths observed today are the result also of Late Mesozoic and Cenozoic subsidence. On the Trøndelag Platform, the Cretaceous surface is much smoother than the basement surface. This difference indicates phases of pre-Cretaceous stretching and subsidence that have resulted in the formation of NE–SW oriented basins (see also Doré et al., 1999; Osmundsen et al., 2002). Deep reflection seismic data suggest that pre-Late Jurassic sedimentary rocks locally constitute more that half of the total sedimentary column on the Trøndelag Platform (Osmundsen et al., 2002). V.E. = Vøring Escarpment. F.S.E. = Færøy Shetland Escarpment.

high-grade Precambrian rocks, of intermediate to granitic composition, are responsible for high-amplitude anomalies of regional significance (Olesen et al., 1991; Skilbrei et al., 1991a,b).

From the offshore survey data, it is easy to recognise that many anomalies have their origin within the crystalline basement. Basement anomalies exist over the Frøya High, Gossa High, Manet Ridge, Selje High, Måløy Fault Block, Gnausen High, within the Trøndelag Platform and along the coastline of mainland Norway. Also noteworthy, there is a striking coincidence between magnetic anomalies and some of the structural highs, interpreted from the seismic data (Blystad et al., 1995). The depth estimates over the highs (Fig. 5) generally vary between 3 km and 7 km. Many of the depth estimates along the Frøya High, Selje High, Måløy Fault Block and the Manet Ridge are consistent with what is known from reflection seismic data and exploration drilling.

The regional magnetic anomalies in the southern half of the study area show roughly NE–SW to NNE–SSW trends. These are parallel to the Caledonian trend on land as well as to the

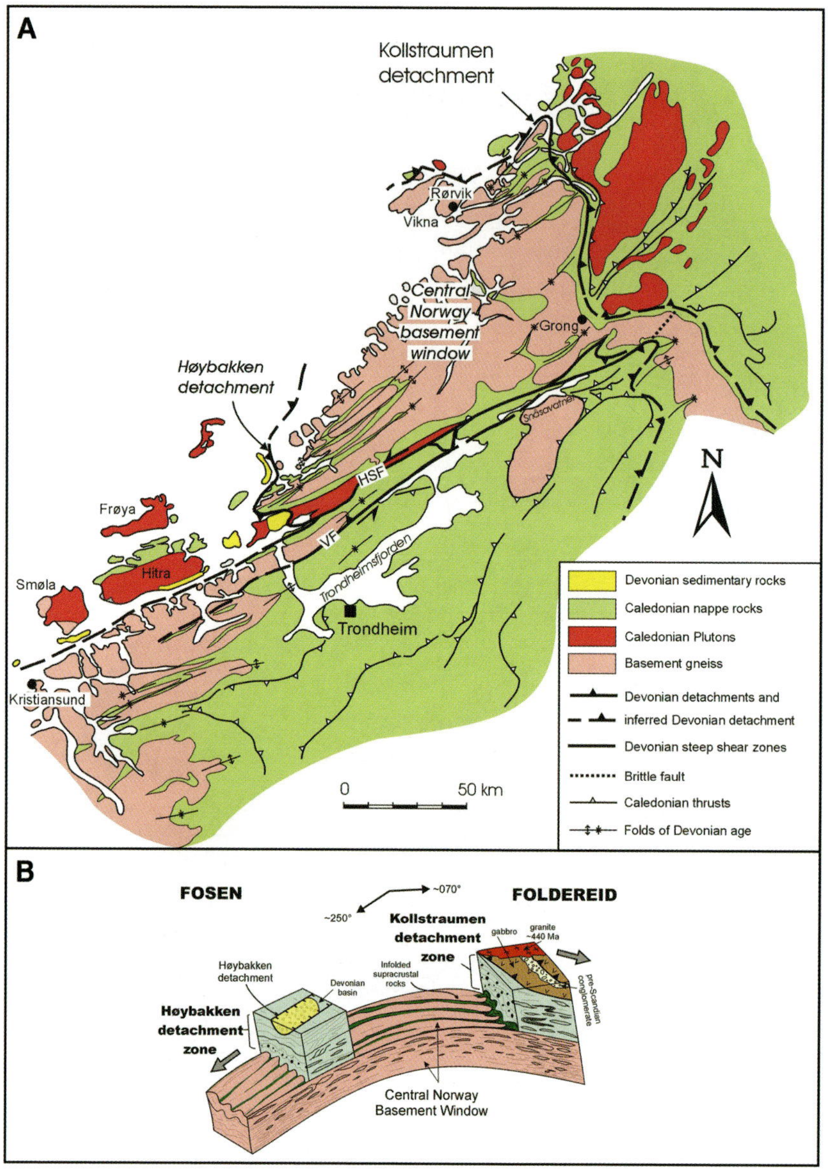

Fig. 10 (A) Simplified tectonic map of western Trøndelag and the Fosen Peninsula locating the Møre-Trøndelag Fault Complex and the Høybakken and Kollstraumen detachment zones (from Braathen et al., 2000). (B) A model for the formation of the Devonian antiform in the basement culmination. The block diagram illustrates the position and geometry of detachment surfaces and detachment zones around the Central Norway Basement Window (CNBW). Compare with Fig. 11.

principal direction of the offshore structural elements. Also, as noted earlier, the strike of the MTFC on land is parallel to the strike of the Caledonian rocks. Some of the anomalies link up with the MTFC on land and with the structural highs over the continental shelf. On the aeromagnetic maps, the anomalies to the south of the Frøya High generally trend NE–SW to ENE–WSW, while the northern half of the survey area shows N–S and NNE–SSW regional trends. There is a gradual shift in the main aeromagnetic trend from south to the north of Trondheimsfjorden (Skilbrei, 1988b). This shift in trend is related to the basement grain, which also reflects a shift in the trend of the main extensional direction. The trend of isoclinal folds and lineations within the CNBW on Fosen Peninsula (Fig. 10) is parallel to the main extensional direction (Braathen et al., 2000). From the above, we suggest that in the offshore areas from the central-east Møre Basin to within Central Norway there is parallelism between the old basement grain (Devonian?), as reflected in the potential field data, and the rift basins and their associated faults. This suggests a profound basement control on rifting and faulting (see also Aanstad et al., 1981).

A long-wavelength anomaly trends NE–SW along the coast to the west of Vikna, centered at 9°E and 64° 40' N. This anomaly may be caused by high-grade magnetic basement that occurs in an

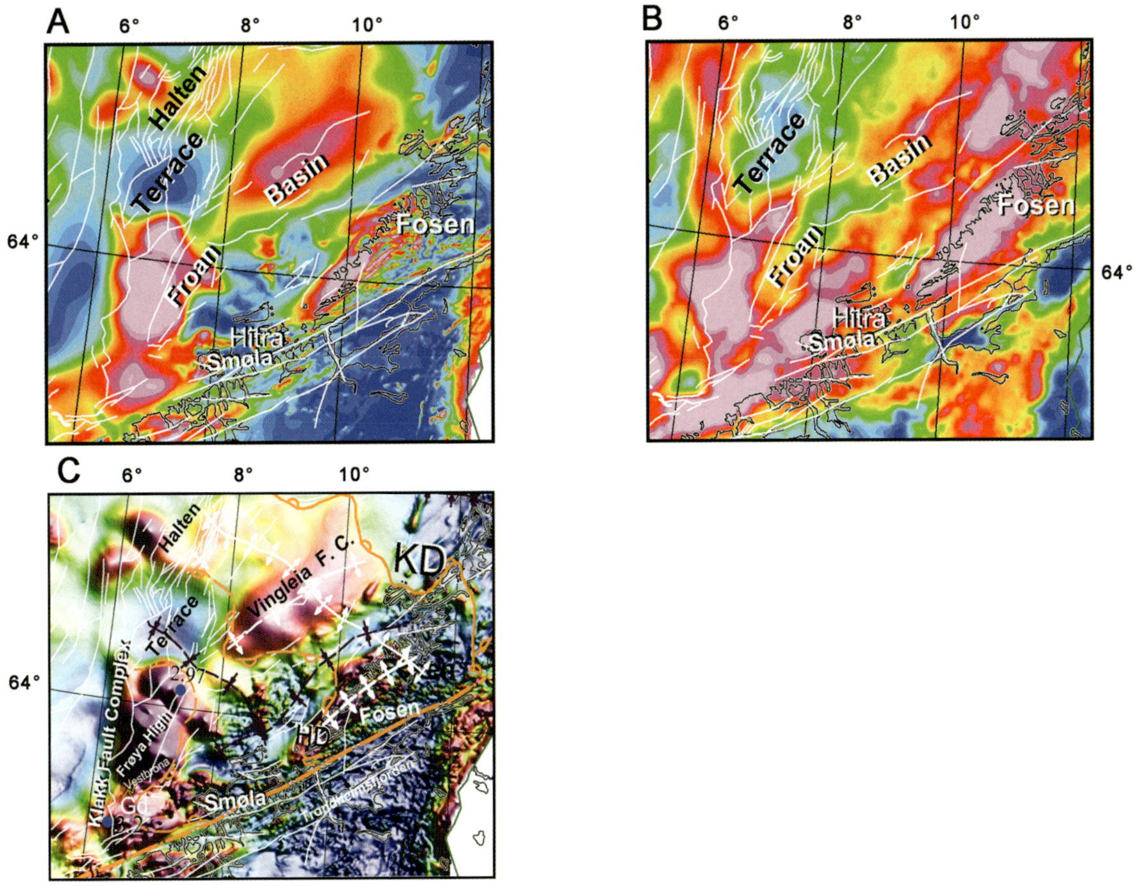

Fig. 11 Residual magnetic data A compared with residual gravity B and a relief version of magnetic data C from the Fosen area of Central Norway. Note NW–SE oriented antiforms and synforms defined by the dips of the inferred detachment faults. Note also another set of NE–SW antiforms and synforms (parallel to the stretching axis, see Fig. 10). Gd = Grip detachment (inferred). HD = Høybakken detachment, KD = Kollstraumen detachment (see Fig. 10 for geology). Violet circles show location of exploration wells and numbers represent depth to basement.

upwarp or anticlinal structure (Olesen and Smethurst, 1994; Olesen et al., 2002; Skilbrei et al., 2002; Osmundsen et al., 2002), similar to the Roan area on Fosen, where an antiform structure composed partly of highly magnetic, high-grade rocks of primarily intermediate to granitic composition, has been modelled using petrophysical data (Skyseth, 1986; Skilbrei et al., 2002). There is an antiform/basement upwarp depicted in a seismic profile north of the Frøya High, along the Bremstein–Vingleia Fault Zone (Osmundsen et al., 2002). We suggest that folds created during extension are reflected in the aeromagnetic map because of the relief of the top of the highly magnetised high-grade basement (Precambrian rocks). These rocks represent the true crystalline basement where Caledonian rocks are absent. The large amplitude magnetic anomalies are located above the crest of the antiforms, as well as above some of the flanks of horsts/structural highs.

The eastern edge of deep dolerite intrusions that exist in the Møre Basin is close to the line of the magnetic anomalies. This is also close to the eastern part of the deep-seated high-velocity material present below the western-central part of the Møre Basin and the Møre Marginal High, as interpreted by Olafsson et al. (1992). This suggests that these dolerites were melts ascending from the area where the crust is thinnest under the Møre Basin. The dolerites probably intruded in the Palaeocene (Brekke, 2000).

Basement map and basement relief

We have estimated the depth to the top of the crystalline basement of the Mid-Norwegian continental shelf using aeromagnetic and gravity data (Fig. 5), and present this surface also in different perspectives (Figs. 8 and 9). Perspective views allow visual evaluation and characterisation of many features not otherwise obvious in the 2D, shaded-relief presentations. We comment upon a few of these features here and also in the figure captions.

The greatest depth estimates of the central area of the Froan Basin are around 8–9 km and the depths are significantly shallower at the basin flanks. From the basement map, a pattern of basement highs/horsts and basins emerges that is in accordance with reflection seismic data (e.g., Brekke, 2000; Osmundsen et al., 2002). The interpreted depths to the top of the crystalline crust locally exceed 12 km in the Møre and the Vøring Basins, while the basement depths vary from 4 to 7 km on the structural highs along the east Møre Basin Margin. Depths on the Trøndelag Platform vary rather evenly, although some relief is recognised also here. This relief may represent fault zones or flexures associated with rifting (see figure texts). The most pronounced feature of the basement map is perhaps the very deep structure beneath the central parts of the Møre Basin and within the Vøring Basin. In the Møre Basin, the deepest basins occur on the eastern side of the structural highs that are associated with a chain of aeromagnetic anomalies along the central axis of the Møre Basin. Quantitative models of the gravity and magnetic fields seem to support this view (Skilbrei et al., 1995; Sæterstad, 1996). These models were used in the construction of the deepest part of the basement map. Clearly, this is speculative and involves some circular arguments. However, deep seismic data seem to support the inference (Skogseid et al., 2000; Osmundsen et al., 2002). The deepest parts of the basement map show dimensions indicative of the presence of buried valleys similar to the modern rift valleys.

If so, this may represent the main Jurassic rift valley, linking the North Viking Graben, Halten Terrace and Jameson Land Basin in East Greenland. In this model, the MTFC represent a transcurrent fault zone, or transfer fault zone, similar to some of the fault zones in the East African Rift System that link different rift valleys (notably, Lake Tanganyika and within the Kenya Rift System). The Halten Terrace and the Jurassic Jameson Land basins thus represent the marginal basins in a rifted region that includes Late Palaeozoic and Mesozoic rifted regions off the East Greenland coast (cf. Brekke, 2000). The N–S to NNE–SSW trending negative anomalies, off the Northeast Greenland shelf may represent such Late Palaeozoic and Mesozoic basins (Larsen, 1990; Skilbrei, 1995, 2000; Beyene, 2002). In a reconstructed anomaly map (unpublished data; and Mosar et al., 2002, and Torsvik et al., 2001), these anomalies may be linked with N–S faults that bound grabens and terraces in the Mid-Norwegian shelf (cf. Skilbrei 1995, 2000). The Northeast Greenland shelf shows N–S to NNE–SSW striking anomalies (Olesen et al., 1997b) that may represent buried basement ridges (high-grade Proterozoic rocks?). The pattern of anomalies is similar to that of the ranges within the extended Basin and Range Province of the western United States (Skilbrei, 2000).

On the south side of the high-grade Central Norway Basement Window (CNBW) on Fosen, the Høybakken detachment (HD) separates Devonian sedimentary rocks and their substrate from the footwall gneisses (Figs. 10 and 11). The HD and the Devonian detachment along the MTFC meet at a branching point to the north of Trondheimsfjorden. A similar 'triple point' may exist at the south side of the Frøya High, in the vicinity of Griptarane (Figs. 11 and 12), where Caledonian plutonic rocks overlie a Precambrian high-grade basement, and where sedimentary basins occur both along the MTFC (ENE–WSW), and along the NNW–SSE faults. This is where the Klakk Fault Complex, or the deeper parts of the Bremstein fault zone meet, or occur close to, the MTFC. A fragment of a fault-controlled, possibly NW–SE to N–S trending, Devonian basin occurs within the southern part of the Frøya High. This is also where the N–S and NNE–SSW trending Frøya High terminates or flexes west of Smøla, in Griptarane (Fig. 12). We think this represents a branching point of Devonian detachment. A generally N–S to NW–SE-striking detachment surface (or younger) meets the detachment along the MTFC. Also noteworthy, in the area along the southeastern margin of Griptarane, are reflections in the basement that dip up to 34° towards the southeast. At the Jurassic/basement margin south of Griptarane, the (Precambrian?) basement shows steeply dipping structures, while the Caledonian intrusions to the west of Smøla exhibit a chaotic signature (Bøe and Skilbrei, 1998). It is thus speculated that the southern flank of the Frøya High may represent an area 'joining together' the older Devonian detachments, similar to the area north of Trondheimsfjorden (Fig. 12b). This is also an area where younger faulting related to the more planar, post-Devonian, detachment surfaces occur in a 'triple point' (Bøe and Skilbrei, 1998).

Evidence for pre-Jurassic basins at intermediate depths (< 7–8 km)

The Froan Basin, the Brønnøysund Basin and the Inner Vestfjorden Basin trend NE–SW and represent half-grabens with a left-stepping, en

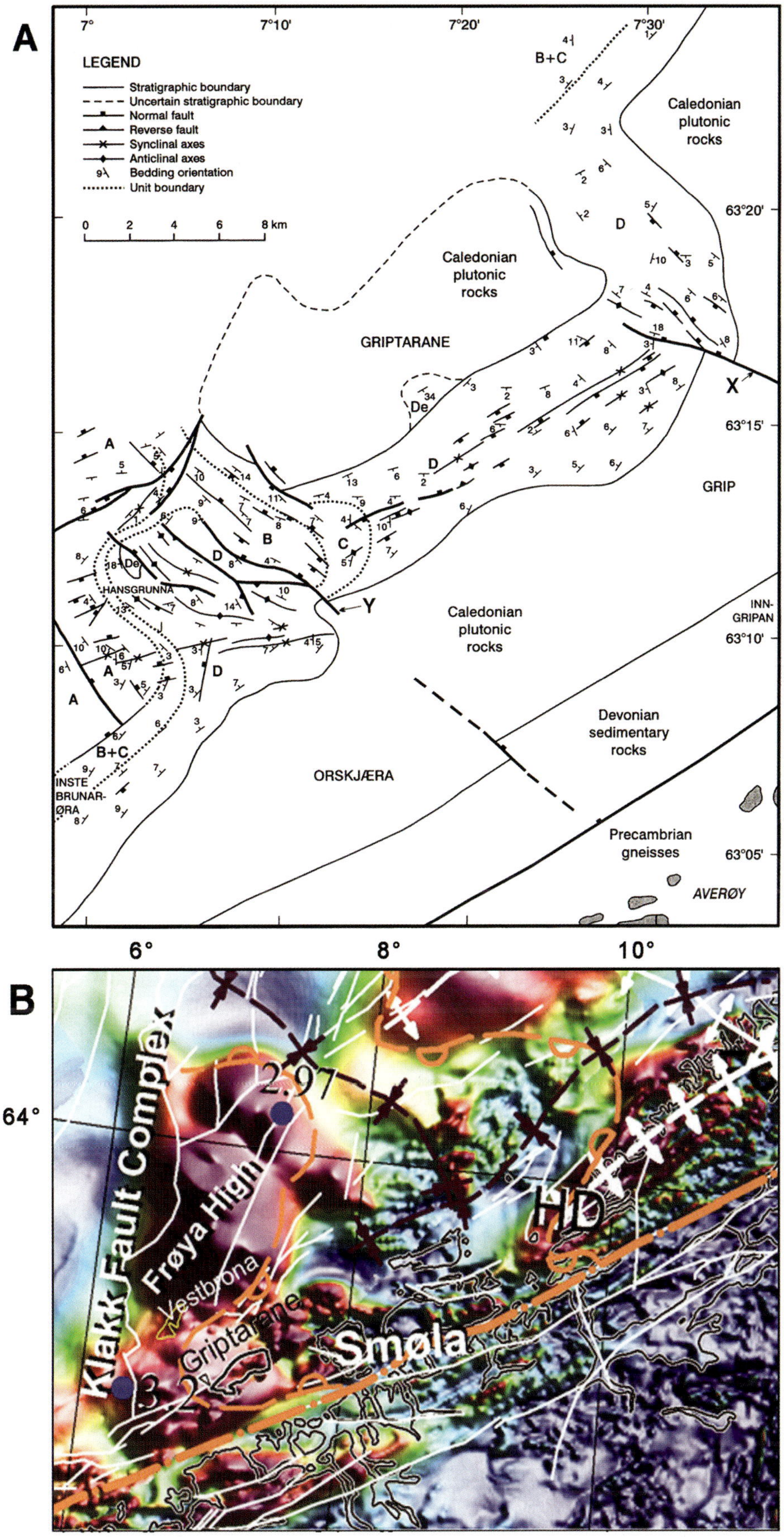

Fig. 12 (A) Structural elements and general geology of the Griptarane area, southeast of the Frøya High (from Bøe and Skilbrei, 1998). (B) Magnetic data from the Griptarane-Smøla-Frøya High region, with interpreted detachment zone. See text for explanation. Griptarane in (A) is outlined in (B) with a black solid line (note different scales). Violet circles show location of exploration wells and numbers represent depth to basement.

échelon pattern (Doré et al., 1999). Basement maps (Olesen et al., 2002; Skilbrei et al., 2002) and the geoseismic sections in Blystad et al. (1995), Brekke (2000), Doré et al. (1999) and Osmundsen (2002) indicate thick Carboniferous to Triassic deposits within the Trøndelag Platform area, underneath the Jurassic and Cretaceous sections (cf. Bugge et al., 2002). The pre-Cretaceous thickness map (Fig. 7) indicates that the pre-Cretaceous basins locally are several km thick. The structure (relief) seen in the pre-Cretaceous difference map, which can be seen as a map 'uplifted' with the thickness of the Cretaceous and younger sediments, suggests a series of structural highs and lows in the Trøndelag Platform, with a NE–SW grain separated by N–S structures. In this 'uplifted' map, in addition to the Froan Basin, six sub-basins can be distinguished within the Trøndelag Platform. This can be compared with the residual gravity map (see also Osmundsen et al., 2002). Together, these structures make up eye-within-eye patterns within the Trøndelag Platform, and one major rhomboid feature from the MTFC in the south to the Vestfjorden Basin in the north (Fig. 8). This may be the result of major faults being superimposed on Devonian extensional collapse structures. The principal trend is NE–SW in the Møre Basin and the Vøring Basin, except for the area between 64° and 65° 30' N, where the grain is N–S. The Halten Terrace trends N–S, but the internal grain of the terrace is also dominated by NE–SW trending sub-basins. In this sense, the Halten Terrace shares similarities with the Trøndelag Platform, and has probably been affected by the same tectonic events as that platform.

Offshore continuation of basement terranes on land

The CNBW occurs between the KD in the north, and the HD in the south. The CNBW is highly magnetic along the coast southwest of Namsenfjorden. We think that this magnetic terrane can be extended beneath the offshore basins, at least to the Bremstein Fault Complex and even beyond. The magnetic signature is complicated by downfaulting at the margins of the younger basins, along the structural elements seen in Fig. 1. The HD is bounded on its southern side by Caledonian intrusive rocks and by Devonian rocks on Ørlandet, as well as by the MTFC (Braathen et al., 2000). In the magnetic map, we have drawn the possible extensions of the above-mentioned detachment zones in Mid Norway. In general, 'synforms' or downfaulted units containing Caledonian nappe sequences show low-to-medium magnetisation values (except for some magnetic greenstones), while the basement in the western areas shows also very high magnetisation values in areas of granulite- and upper amphibolite- facies felsic rocks (Olesen et al., 2002; Skilbrei et al., 2002). This is reflected in the anomaly pattern both onshore and offshore. If our west-to-northwestward projection of these zones is correct, then a significant portion of the basement in the Helgeland Basin and along the Nordland Ridge constitutes low-magnetic Caledonian nappes. Conversely, a significant part of the southern Trøndelag Platform, and the Frøya High, is magnetic. This is most likely caused by downfaulting of the Helgeland Nappe Complex along the offshore extensions of the NW–SE trending, Devonian, Nesna Shear Zone and Kollstraumen detachment and by magnetic rocks being close to the top of the crystalline basement over large parts of the southern Trøndelag Platform. However, some of the low-magnetic parts of the offshore map may have also been caused by thick Devonian deposits, or low-magnetic Caledonian and Precambrian basement.

In the analysis of the region encompassing the Frøya High, Smøla, and the southwest projection of the MTFC, we have recognised that Caledonian plutons in the Smøla area give rise to magnetic anomalies (Sindre, 1977; Bøe and Skilbrei, 1999). However, considering the aerial extent and the large amplitude of the major regional offshore anomalies, we think that only the high-grade Precambrian basement rocks provide the volumes needed to explain the high-amplitude anomalies, given the depth of burial seen from the magnetic map. Of course, large volumes of basic or mafic rocks are potential sources, but we do not see any intra-basement contrasts in density to warrant such an interpretation (Skilbrei et al., 1995).

The NW–SE-trending Late Devonian KD and the Nesna shear zone, extend from the mainland northwestwards below the Helgeland, Vestfjorden and Ribban basins. The Bivrost Lineament most likely represents a detachment dipping 5–15° to the southwest and may constitute the offshore extension of the Nesna shear zone. The 4-5 km thick low-magnetic Helgeland Nappe Complex is downfaulted along these structures, causing the low magnetic field above the Helgeland Basin and the Nordland Ridge (Olesen et al., 2002).

We suggest that there is a fold pattern present that actually links the onshore coastal zone anomalies and the offshore aeromagnetic trends

(Fig. 3). High-grade rocks representing lower crustal rocks give rise to pronounced aeromagnetic anomalies, as well as to moderate-amplitude residual gravity anomalies, along segments of the Norwegian coast (e.g. Skilbrei, 1988a; Olesen et al., 1991). At least on Fosen and to the north of Moldefjorden, these rocks occur in basement antiforms or culminations (Möller, 1986; Skilbrei 1988a), probably representing metamorphic core complexes (Braathen et al., 2000; Osmundsen et al., 2002, 2003). We see a repetition of the geophysical signals in the offshore where magnetic anomalies occur over basement highs, probably representing high-grade magnetic rocks within the Frøya High (Skilbrei et al., 1995) and along the Vingleia Fault Zone (Olesen and Smethurst, 1994). The latter anomaly was named the Vingleia–Vikna magnetic anomaly by Skilbrei and Olesen (2001). Seismic data show that some of these highs represent rift flanks (e.g., Brekke 2000), possibly associated with basement culminations or metamorphic core complexes at depth as seen in a seismic profile that is located between the Frøya High anomaly and the Vingleia-Vikna magnetic anomaly (Osmundsen et al., 2002). Magnetic model calculations indicate that intra-basement anomaly sources exist, and that the basement is layered into low-magnetic Caledonian nappes that occur on top of Precambrian rocks.

In order to explain the continuation of the aeromagnetic picture into the offshore, two orthogonal sets of folds that were formed during the extensional events have been recognised. The first set is oriented NW–SE, orthogonal to the Caledonian trend, but parallel to the nappe transport direction and orthogonal to the later extension trend (Osmundsen et al., 2002; Eide et al., 2002). This first set is defined by the dip of the detachments on land. Over the sea area, it is based on the projection of the HD and the KD into the offshore. The other set of synforms and antiforms is parallel to the extension direction. It is observed on land (Nordgulen et al., 2002) and helps to explain both the onshore and the offshore pattern of aeromagnetic anomalies. The offshore extension of an orthogonal set of folds may provide a model for the distribution of weakly magnetic (primarily Caledonian rocks) and high-grade Precambrian magnetic rocks.

Within the coastal zone, the continuation of basement units into the offshore has been noted by Skilbrei (1988a), Olesen et al. (1991, 1997a, 2002), Olesen and Smethurst (1995), Sigmond (1992), Skilbrei et al. (1991a, 2002). The Caledonian rocks of Hitra, Smøla, Frøya and the Froan archipelago share similarities with rocks of the Helgeland Nappe Complex (Nordgulen et al., 1995). An explanation for this can be given from the suggested fold pattern that emerges from the combined analysis of residual gravity and magnetic maps. The negative gravity anomaly from the Halten Terrace can be explained partly by a granitoid basement, perhaps a continuation of the basement in the Froan area. This is also suggested by the fold pattern. The inferred fold pattern is speculative. However, there are other areas along the Norwegian coast where magnetic rocks occur in basement antiforms. The best example of this is from the Øygarden Complex in the Bergen Arcs. We have included a geological map of the Øygarden basement area where there is an E–W regional fold (Fig. 13a,b). A magnetic anomaly (Fig. 13b) occurs above the basement antiform. This regional fold was probably formed during Devonian extension (Larsen et al., 2003). This figure suggests that the presence of such folds can be interpreted from aeromagnetic data. Also, the residual gravity data (not shown) from Øygarden do not show positive anomalies associated with the E–W trending antiform. This suggests that the rocks involved have intermediate compositions.

Reactivation of basement structural grain

The correlation of onshore basement and the larger lineaments mapped offshore has been reviewed by Doré et al. (1997), Olesen et al. (1997a) and Fichler et al. (1999). Doré et al. (1997, 1999) found three principal fault trends, NE–SW, N–S and NW–SE. These sets have also been reported on land by Gabrielsen and Ramberg (1979) and Gabrielsen et al. (2002). We recognise these three principal trends in the basement map and in the topographical map, which testifies that these are old, deep-seated structures. Doré et al. (1997) suggested a system of NW–SE transfer zones on the Mid-Norwegian shelf, manifested as lineament terminations and offsets, and strongly segmenting the basin chain. The Jan Mayen Lineament (JML) appears to be contiguous with the oceanic fracture zone (JMFZ). Henkel and Eriksson (1987), Olesen et al. (1997a) and Fichler et al. (1999) also suggested a connection between fracture zones in the Precambrian basement and some of the offshore lineaments, based on potential field data. Steeply dipping fault zones, such as the MTFC, have been suggested to be the most probable examples of reactivation (Grønlie and Roberts, 1989). The reactivation of the low-angle Devonian shears occurred where these shears acted

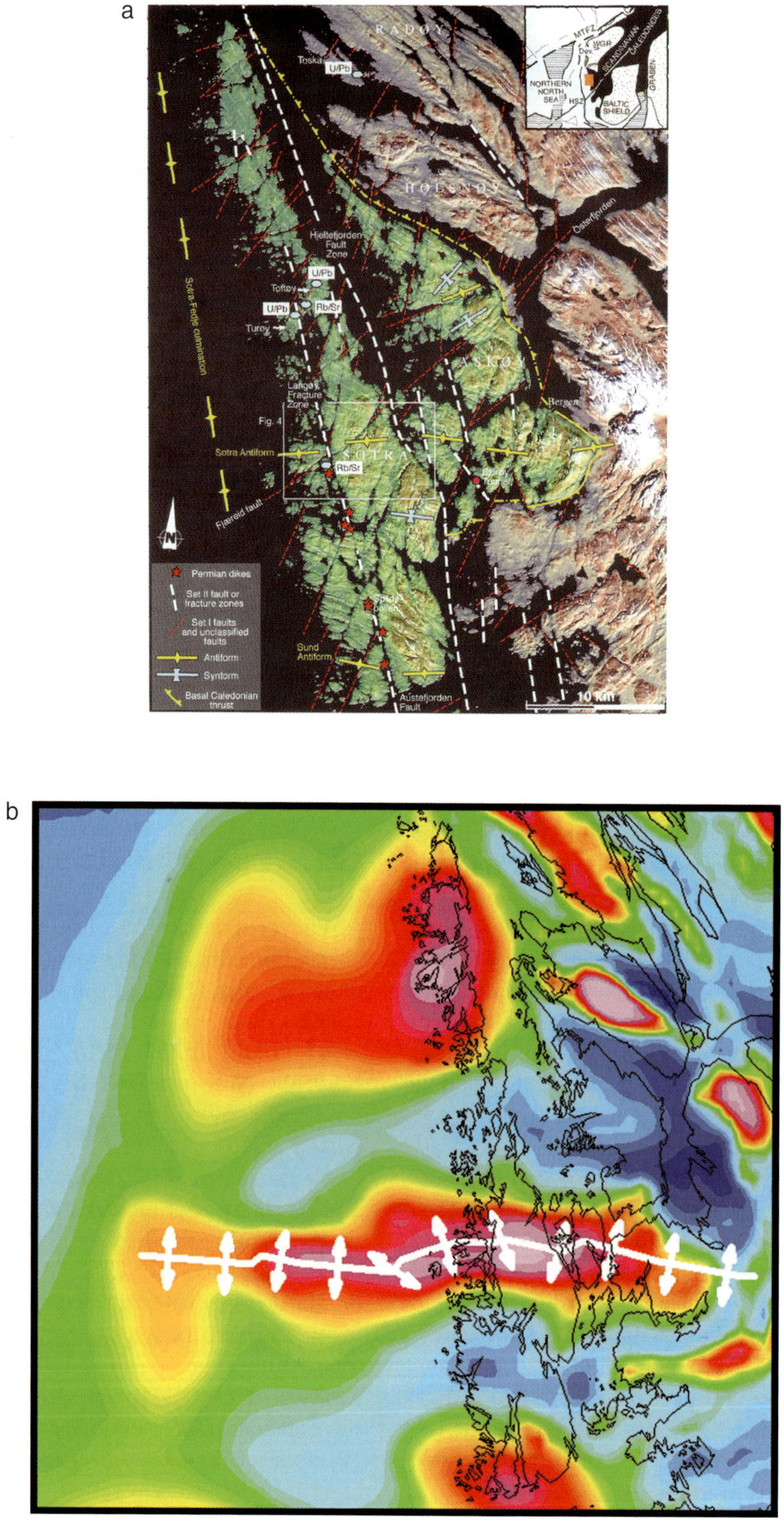

Fig. 13 (a) Western Bergen Arc System with main structural elements indicated. The basement area in the west and the Caledonian nappe area to the east are given different colours (from Larsen et al., 2003). The Sotra Antiform is an E–W- trending macro-scale fold formed after or simultaneously with W-directed Early Devonian shearing. Compare with aeromagnetic data in Fig. 13b. (b) Aeromagnetic map corresponding to the map of the upper panel, with the Sotra Antiform shown as a white solid line. Note that the scale is the same in the upper and the lower figures, but the aeromagnetic map extends farther to the west in order to show the westward continuation of the Sotra Antiform, as indicated by the anomaly. The gravity data (not shown) do not show a similar correspondence with the antiform, suggesting that the magnetic rocks along the axial plane are generally not of much higher densities than on limbs of the fold. This figure is evidence that post Caledonian folds have generated pronounced magnetic patterns along the coastal zone.

as middle–lower crustal detachments to Late Palaeozoic and Mesozoic extension (Osmundsen et al., 2002). It is also possible that some of these shears were reactivated as transfer zones, when conveniently oriented relative to stress orientation (Olesen et al., 2002). In particular, the JML represents a lineament or a transfer zone (Mosar et al., 2002), occurring along or close to a Devonian shear zone/detachment zone. On land, such Devonian shears have been mapped along the MTFC (e.g. Braathen et al., 2000). In the offshore, this is highly speculative. The evidence is that the suggested detachment and the JML both trend NW–SE, and that the JML separates high-magnetic crust from low-magnetic crust (south side of JML). This geometrical similarity and the close spatial coincidence suggest such an old and deep-seated origin of these lineaments. However, the presence of the offshore basins suggests that main faults could simply be developed as extensional structures during rifting episodes, although Osmundsen et al. (2002), in their model, suggest that the Mesozoic extensional detachment faults developed above the older (Late Devonian–Early Carboniferous) detachment zones. We suggest that the combined basement map (which is a topographical map onshore) in itself is evidence that the major faults in the basement existed before the rifting events in the offshore. One important point is that during long periods, e.g. through the Jurassic–Cretaceous, the Trøndelag area was a part of the shelf as suggested by the Jurassic basins that occur in Beitstadfjorden and in Frohavet. In the Trøndelag area, both the ENE–WSW to NE–SW structural grain and the NW–SE structural grain, existed before Mesozoic faulting occurred along the MTFC (Grønlie, 1991). This is an analogue for the nearby shelf.

Discussion

1. It has been suggested (Færseth and Lien, 2002) that the morphology of the Base Cretaceous reflects a pre-existing topography (Jurassic rift topography) rather than Early Cretaceous rifting. The difference between the Base Cretaceous map and the basement map supports this suggestion. The Base Cretaceous map is rather smooth and the long topographic wavelength of the basal Cretaceous surface suggests a passive infill of a wide subsiding basin, rather than localised rifting and formation of discrete graben and horst structures. However, in the Vøring Basin and perhaps in the Vestfjorden and the Ribban basinal areas, it is likely that there was a rifting event of Cretaceous age (e.g. Brekke, 2000, Olesen et al., 2002).
2. On the mainland, aeromagnetic lineaments trend N–S to NNW–SSE, NE–SW and ENE–WSW. These trends are also observed in the offshore. In general, geological field data, topography and structural elements mapped from seismic data, basement topography, and gravity and aeromagnetic data indicate a close directional, and sometimes spatial, correlation. The prevailing trend of basement anomalies suggests that later tectonic events have occurred along Precambrian or Caledonian zones of weakness present in the basement. In agreement with earlier workers, this suggests that the main fault zones present in the crystalline basement have imposed some control on the evolution of mid-central Scandinavia and the adjacent shelf.
3. The NW–SE trend in the mid-Norwegian shelf, here shown to represent moderate basement relief, has been previously attributed to Proterozoic shear zones (Strömberg, 1978; Henkel and Eriksson, 1987). We interpret this trend to represent the Devonian detachment or shear zones that can be followed approximately to the central part of the shelf, where it becomes too deep to be seen, and where the highest topography of the Caledonian mountain-chain was probably located. It has been suggested that the detachment zones are connected to detachments in East Greenland (Olesen et al., 2004).
4. The basement map represents deep geology, and can be used to deduce geometric features. The rhomboid-like plan sections seen inside the shelf area, recognised in the basement map, are partly due to the Devonian extensional collapse structures, and partly due to the N–S and NE–SW oriented faults, active during post-Caledonian extensional events. Sinistral movements between Greenland and Scandinavia may have played a role (e.g. Doré et al., 1997). Thus, several fault trends are responsible for the rhomboid geometry (NW–SE, N–S and ENE–WSW to NE–SW). These main trends are also found within Precambrian and Cambro–Silurian basement units of western Scandinavia, suggesting that Precambrian faults have been reactivated both on land and in the offshore. The rhomboid geometry of the entire Mid-Norwegian shelf is also easily recognised (e.g. Doré et al., 1997). This geometry actually repeats itself on a variety of scales. Transtensional and transpressional events of Late Devonian–Early Carboniferous ages have been suggested to account for the

rhomboid shape (Doré et al., 1997). Since we see these shapes in Sweden, such transpressional or transtensional events are not needed to explain these geometries. This argument does not preclude that major strike-slip movements between Greenland and Scandinavia can account for these features on the shelf. However, since the geometries are seen both in the offshore and on land, reactivation of the same fault sets on land and in the shelf is highly likely. Romer and Bax (1992) suggested that this shape is due to N–S and NW–SE Proterozoic faults that define basement blocks with rhomb-shaped plan sections. From the above, we suggest that the shape is the combined effect of Proterozoic shear zones in the basement and the NW–SE orientation of the Devonian shears, as well as the NE–SW folds and faults in the basement. In addition, N–S faults exist in the offshore due to Mesozoic rifting and an older shear system, a trend also observed on land (Gabrielsen et al., 1999). The latter may also reflect reactivated older zones of weakness. The significance of the NE–SW and N–S faults have been discussed by Doré et al. (1999). The convex-to-the-southeast and east fault pattern is the geometry resulting from the compressive Caledonian Orogeny itself; this pre-determined the Devonian collapse and the eastern limit of the shallowing of the Moho. This shallowing of the Moho must have happened contemporaneously with the Devonian collapse. The basement culminations along the coast are thus the result of post-Caledonian extensional events (Braathen et al., 2000, Osmundsen et al., 2002, 2003), and with additional localised rift flank uplift during formation of the Slørebotn Sub-basin, the Froan Basin and faulting along the offshore MTFC.

5. Doré et al. (1997) suggested that the principal lineament trends offshore are primarily a function of Mesozoic-Cenozoic, plate-wide, extensional stress fields, with the re-activation of certain basement faults. These authors also noted that the N–S faults in the basement were reactivated during Jurassic E–W extension. This N–S trend is seen on land and in the offshore, both in the gravity and the magnetic data. Because they represent both faults (structural map) and sharp gradients in the magnetic map, they are most likely old basement trends. However, basement-involved Mesozoic faulting may be responsible for some of the trends (e.g. along the margin of the Halten Terrace). The extensional faults, where striking at an angle to the Late Palaeozoic and/or Mesozoic extension directions probably predisposed the formation of major transfer zones (Doré et al., 1997). We suggest that a NW–SE- oriented transfer zone crosses the central part of the Froan Basin. This can explain the change in strike of the basin at this point (Fig. 1), and the change in the aeromagnetic signature on each side of the basin (Fig. 11). The change in strike of the basin may reflect an angle between older detachment faults and the younger extensional direction. The change in aeromagnetic signature is the 'old' basement signature inherited from the Devonian. Further to this, Brekke (2000) shows two profiles that cross the Froan Basin. We interpret these published sections to indicate that the major faults at the Late Palaeozoic and Mesozoic levels show polarity changes from the southern profile (westerly dips) to the northern profile (easterly dips). Such polarity changes between the two segments of the Froan Basin may have a deep-rooted origin whereby the zones of weakness strike at a low angle to the extension direction. In addition, the system of Devonian antiforms and synforms may have played a role in determining the polarity of half-grabens. This has been suggested for the Vestfjorden Basin (Olesen et al., 2002).
6. The Bouguer gravity shows pronounced southeastward directed gradients in the Moldefjorden area. This gradient strikes parallel to the MTFC and may reflect crustal thinning under the coastal area, as has been suggested earlier for the Fosen area, based on gravity modelling (Skilbrei, 1988a). This extension may be the combined result of Late Palaeozoic extensional events and Mesozoic rifting events that were accompanied by faulting along the preserved Jurassic basins (Bøe and Bjerklie, 1989; Smelror et al., 1994; Bøe and Skilbrei, 1998). Thus, the faulting along the MTFC may have been accompanied by a shallowing of the depth to the Moho surface.
7. The trend of the MTFC on land does not necessarily link directly with the MTFC in the offshore area (Skilbrei et al., 1995, p. 33) as defined by Gabrielsen et al. (1984). If the offshore trend of the MTFC is projected farther in the northeast direction, it may extend also across the southern part of the Frøya High, and into the region of the island Frøya. Farther east of this, major faults bound a Mesozoic graben in the Frohavet (Oftedahl, 1975) and two Mesozoic half-grabens occur off the coast of Fosen, in the region between Vikna and Frøya (Thorsnes,

1995; Sommaruga and Bøe, 2002; see also Fig. 2). This Jurassic fault activity along the coast of Trøndelag can be linked more or less directly with the faults along the eastern margin of the Møre Basin (the 'offshore MTFC'). It is perhaps likely that the MTFC on land, which is traced between the mainland and the islands of Smøla and Hitra, has its offshore continuation on the east side of the Slørebotn Sub-basin, which is indicated on the geological map of Sigmond (1992; cf. Seranne, 1992). Based on potential field data, Skilbrei et al. (1991b) and Smethurst (2000) extended the MTFC from the land and onto the south side of the SS, where it swings into a slightly more southerly trend (see orange line in Fig. 3). This is different from some of the earlier studies (e.g. Gabrielsen et al., 1984; Skilbrei 1988a) that extended the zone to Shetland. Clearly, the MTFC is a long-lived, broad zone within which elements have been active in different areas through separate phases (e.g. Grønlie, 1991). We think that one branch of the old, 'Devonian MTFC', follows the detachment along the MTFC (cf. Norton, 1987; Seranne, 1992; Kendrick et al., 2004) on the south side of the SS. Younger stretching episodes may have linked faults along the southeastern margin of the Møre Basin Margin with the coastal part of the MTFC. Thus, we suggest that the MTFC may consist of two branches in the offshore (that connect southwest of Hitra) representing different tectonic episodes.

Conclusions

The maps of the depth to the top of the crystalline basement, the depth to the Base Cretaceous, and the map of pre-Cretaceous sedimentary thickness can be used to analyse the deep geology of the study area. In addition to showing total thickness of sedimentary basins (>12 km in the Møre and Vøring Basins), the maps show a picture of pronounced structural relief, both at the basement depth level and in the pre-Cretaceous sediment thickness map. Buried, pre-Cretaceous, rift valleys can be detected in the maps. In the platform area, individual and inter-connected horsts and grabens emerge that, in most areas, are probably related to the combined result of Late Palaeozoic to Early-Middle Mesozoic structuring of the area.

The geometry of structural highs and lows seen in our maps is the combined result of several rifting events that were super-imposed on the Devonian extensional collapse structure(s). We think that the NW–SE structural grain on the shelf is caused by the offshore continuation of Devonian extensional shear zones and detachments that are observed on land. Locally, these detachments were re-activated as transfer zones (Olesen et al., 2002). Movements along these shear zones have determined the position of magnetic basement terranes (primarily Precambrian) relative to the less-magnetic Caledonian basement terranes and may explain the NW–SE- lineaments.

The top of the basement features show a picture of narrow, deeply buried, rift-related relief, and a wide basinal area recording prolonged subsidence in the Cretaceous, as well as in the Tertiary and Quaternary. The difference between the basement map and the Base Cretaceous map supports the idea that Cretaceous sedimentation represents post-rift thermal and isostatic subsidence during the infill of pre-existing rift topography (cf. Færseth and Lien, 2002).

Highly magnetic, granulite-facies intermediate rocks give rise to strong magnetic anomalies along the Precambrian gneiss terranes of the coastal zone in central-northern and western Norway. The occurrence of these magnetic rocks is structurally related to Devonian extensional shear zones and basement antiforms (Osmundsen et al., 2003). We recognise these antiforms and synforms in the offshore, thereby providing a model for the inferred distribution of low-magnetic Caledonian rocks and high-grade Precambrian intermediate rocks in the shelf areas.

The MTFC is suggested to consist of two branches in the offshore. One branch is the extensional fault that is located east of the Slørebotn Sub-basin sub-parallel to the coastline (Séranne, 1992), and the other one is the fault zone extending from the Møre-Trøndelag coast towards Shetland.

Acknowledgements

We wish to thank Per Terje Osmundsen, Alvar Braathen, Øystein Nordgulen, Peter Robinson, Elizabeth A. Eide, Erik Lundin, Mark Smethurst, Tim Redfield, Trond H. Torsvik, Jon Mosar and David Roberts for many interesting and stimulating discussions. We would also like to thank Christine Fichler and David Worsley for their constructive criticisms that helped improve the paper. David Roberts corrected the English text.

References

Aanstad, K.M., Gabrielsen, R.H., Hagevang, T., Ramberg, I.B. and Torvanger, O., 1981. Correlation of offshore and onshore structural features between 62°N and 68°N, Norway. In: Proceedings, Norwegian Symposium on Exploration, Bergen 1981, Norwegian Petroleum Society (NPF), NSE/11, 1–25.

Beyene, B.W., 2002. Basement structures in the North-East Greenland Continental Shelf interpreted from aeomagnetic, gravity and seismic data. Unpublished Master Thesis. Norwegian University for Science and Technology. 68 pp.

Blystad, P., Brekke, H., Færseth, R.B., Larsen, B.T., Skogseid, J. and Tørudbakken, B., 1995. Structural elements of the Norwegian continental shelf. Part II. The Norwegian Sea Region. NPD, Bull., 8, 45 pp.

Braathen, A., Nordgulen, Ø., Osmundsen, P.T., Andersen, T.B., Solli, A. and Roberts, D., 2000. Devonian, orogen-parallel, opposed extension in the Central Norwegian. Geology, 28: 615–618.

Brekke, H., 2000. The tectonic evolution of the Norwegian Sea Continental Margin with emphasis on the Vøring and Møre Basins. In: A. Nøttvedt (Editor), Dynamics of the Norwegian Margin. Geol. Soc., London, Spec. Publ., 167, pp. 327–378.

Bugge, T., Ringås, J.E., Leith, D.A., Mangerud, G., Weiss, H.M. and Leith, T.L., 2002. Upper Permian as a new play model on the mid-Norwegian continental shelf: Investigated by shallow stratigraphic drilling. AAPG Bull., 86: 107–127.

Bukovics, C. and Ziegler, P.A., 1985. Tectonic development of the Mid-Norway continental margin. Mar. Petrol. Geol., 2: 2–22.

Bøe, R. and Bjerklie, R., 1989. Mesozoic sedimentary rocks in Edoyfjorden and Beitstadfjorden, central Norway:implications for the structural history of the Møre-Trøndelag Fault Zone. Marine Geology, 87 (2–4); 287–299.

Bøe, R. and Skilbrei, J.R., 1998. Structure and seismic stratigraphy of the Griptarane area, Møre Basin margin, mid-Norway continental shelf. Marine Geology, 147: 85–107.

Bøen, F., Eggen, S. and Vollset, J., 1984. Structures and basins of the margin from 62° to 69° N and their development. In: A.M. Spencer et al. (Editors), Petroleum geology of the North European Margin, Norwegian Petroleum Society (NPF), Graham and Trotman, London, pp. 253–270.

Dehls, J.F., Olesen, O., Bungum, H., Hicks, E., Lindholm, C.D. and Riis, F., 2000. Neotectonic map, Norway and adjacent areas 1:3 mill. Norges Geologiske Undersøkelse, Trondheim, Norway.

Doré, A.G., Lundin, E.R., Fichler, C. and Olesen, O., 1997. Patterns of basement structure and reactivation along the NE Atlantic margin. J. Geol. Soc., London, 154: 85–92.

Doré, A.G., Lundin, E.R., Jensen, L.N., Birkeland, Ø., Eliassen, P.E. and Fichler, C., 1999. Principal tectonic events in the evolution of the northwest European Atlantic margin. In: A.J. Fleet and S.A.R. Boldy (Editors), Petroleum Geology of Northwest Europe: Proceedings of the 5th Conference. Geological Society, London, pp. 41–61.

Dyrelius, D., 1985. A geophysical perspective of the Scandinavian Caledonides. In: D.G. Gee, and B. Sturt (Editors), The Caledonide Orogen — Scandinavia and related areas, John Wiley and Sons, Chichester, pp. 185–194.

Eide, E., Osmundsen, P.T., Meyer, G.B., Kendrick, M.A. and Corfu, F., 2002. The Nesna shear Zone, north-central Norway: an 40Ar/39Ar record of Early Devonian-Early Carboniferous ductile extension and unroofing. Norsk Geol. Tidsskr., 82: 317–339.

Elming, S.Å., 1980. Density and magnetic properties of rocks in the Caledonides of Jämtland, Sweden. Geologiska Föreningens i Stockholm Förhandlingar, 102: 439–453.

Fasteland, F. and Skilbrei, J.R., 1989. Tolkning av helikoptergeofysikk, gravimetri og petrofysikk innenfor kartblad 1723 III Steinkjer, Nord-Trøndelag. Norg. Geol. Unders. Rep. 89.158, 31 pp.

Fichler, C., Rundhovde, E., Olesen, O., Sæther, B.M., Rueslåtten, H., Lundin, E. and Doré, A.G., 1999. Regional tectonic interpretation of image enhanced gravity and magnetic data covering the mid-Norwegian shelf and adjacent mainland. Tectonophysics, 306: 183–197.

Færseth, R.B. and Lien, T., 2002. Cretaceous evolution in the Norwegian Sea-a period characterized by tectonic quiescence. Mar. Petrol. Geol., 19: 1005–1027.

Gabrielsen, R.H. and Ramberg I.B., 1979. Fracture patterns in Norway from LANDSAT imagery: results and potential use. In: Proceedings of the Norwegian Sea Symposium, NSS/20, Tromsø. Norwegian Petroleum Society (NPF), pp. 1–10.

Gabrielsen, R.H., Færseth, R.B., Hamar, G. and Rønnevik, H., 1984. Nomenclature of the main structural features on the Norwegian continental shelf north of the 62nd parallel. In: A.M. Spencer et al. (Editors), Petroleum geology of the North European Margin, Norwegian Petroleum Society (NPF), Graham and Trotman, London, pp. 41–60.

Gabrielsen, R.H., Odinsen, T. and Grunnaleite, I., 1999. Structuring of the Northern Viking Graben and the Møre Basin; the influence of basement structural grain, and the particular role of the Møre Trøndelag Fault Complex. Mar. Petrol. Geol., 16: 443–465.

Gabrielsen, R.H., Braathen, A., Dehls, J. and Roberts, D., 2002. Tectonic lineaments of Norway. Norsk Geol. Tidsskr., 82: 153–174.

Geosoft, 2000. MAGMAP (2D FFT), 2-D frequency domain processing of potential field data. Tutorial and user guide. Geosoft incorporated, 67 pp.

Gravdal, N., 1985. The Møre Basin. Unpublished Cand Scient. thesis, University of Oslo.

Grunnaleite, I. and Gabrielsen, R.H., 1995. Structure of the Møre Basin, mid-Norway continental margin. Tectonophysics, 252: 221–251.

Grønlie, A. and Roberts, D., 1989. Resurgent strike-slip duplex development along the Hitra-Snåsa and Verran Faults, Møre-Trøndelag Fault Zone, Central Norway. J. Struct. Geol., 11 (3): 295–305.

Grønlie, A., 1991. Joint, fault and breccia systems in outer parts of Trøndelag, Central Norway. Doktoringeniøravhandling 1991:68, Institutt for Geologi og Bergteknikk, Norges Tekniske Høgskole, Trondheim, 57 pp.

Hamar, G.D. and Hjelle, K., 1984. Tectonic framework of the northern sea. In: A.M. Spencer et al. (Editors), Petroleum Geology of the North European Margin. Graham and Trotman, London, pp. 349–358.

Henkel, H. and Eriksson, L., 1987. Regional aeromagnetic and gravity studies in Scandinavia. Precambrian Res., 35: 169–180.

Jongepier, K., Rui, J.C. and Grue, K., 1996. Triassic to Early Cretaceous stratigraphic and structural development of the northeastern Møre Basin margin, off Mid-Norway. Norsk Geol. Tidsskr., 76: 199–214.

Kendrick, M., Eide, E.A., Roberts, D. and Osmundsen, P.T., 2004. The Mid-Late Devonian Høybakken Detachment, Central Norway: $^{40}Ar/^{39}Ar$ evidence for prolonged late/post Scandian extension and uplift. Geol. Mag., 141: 329–344.

Larsen, H.C., 1990. The East Greenland Shelf. In: A. Grantz, L. Johnsen, and J.F. Sweeney (Editors), The Arctic Ocean Region. Geological Society of America, pp. 185–210.

Larsen, Ø., Fossen, H., Langeland, K. and Pedersen, R.-B., 2003. Kinematics and timing of polyphase post-Caledonian deformation in the Bergen area, SW Norway. Norsk Geol. Tidsskr., 83: 149–165.

Lundin, E.R. and Rundhovde, E., 1993. Structural domains in the Møre Basin, Norway-from digital images of aeromagnetic data. Extended abstract, EAPG-5th Conference June 7–11, Stavanger, Norway.

Mjelde, R., Digranes, P., Shimamura, H., Shiobara, H., Kodaira, S., Brekke, H., Egebjerg, T., Sørenes, N. and Thorbjørnsen, S., 1998. Crustal structure of the northern part of the Vøring Basin, mid-Norway margin, from wide-angle seismic and gravity data. Tectonophysics, 293: 175–205.

Mosar, J., Osmundsen, P.T., Sommaruga, A., Torsvik, T. and Eide, E.A., 2002. Greenland — Norway separation: A geodynamic model for the North Atlantic. Norsk Geol. Tidsskr., 82: 281–298.

Möller, C., 1986. The Roan granulite dome, Vestranden, Western Gneiss Region, Norway. Geologiska Föreningens i Stockholm Förhandlingar, 108: 295–297.

Nordgulen, Ø., Solli, A. and Sundvoll, B., 1995. Caledonian granitoids in the Frøya-Froan area, central Norway. Norg. Geol. Unders. Bull., 427: 48–51.

Nordgulen, Ø., Braathen, A., Corfu, F. Osmundsen, P.T. and Husmo, T., 2002. Polyphase kinematics and geochronology of the late-Caledonian Kollstraumen detachment, north-central Norway. Norsk Geol. Tidsskr., 82: 299–316.

Norton, M.G., 1987. The Nordfjord-Sogn Detachment, W. Norway. Norsk Geol. Tidsskr., 67: 93–106.

Oftedahl, C., 1975. Middle Jurassic graben tectonics in mid-Norway. Proceedings Jurassic Northern North Sea Symposium 21, pp. 1–13.

Olafsson, I., Sundvor, E.O., Eldholm, O. and Grue, K., 1992. Møre Margin: Crustal Structure from Analysis of Expanded Spread Profiles. Mar. Geophys. Res., 14: 137–162.

Olesen, O., Henkel, H., Kaada, K. and Tveten, E., 1991. Petrophysical properties of a prograde amphibolite - granulite facies transition zone at Sigerfjord, Vesterålen, northern Norway. In: Wasilewski, P. and Hood, P. (Editors) Magnetic anomalies — land and sea. Tectonophysics, 192, pp. 33–39.

Olesen, O. and Smethurst, M.A., 1994. NAS-94 interpretation report, Part I: Aeromagnetic data. Norges geologiske undersøkelse report 94.074, 44 pp.

Olesen, O., Torsvik, T.H., Tveten, E., Zwaan, K.B., Løseth H. and Henningsen, T., 1997a. Basement structure of the continental margin in the Lofoten-Lopphavet area, northern Norway: constraints from potential field data, on-land structural mapping and palaeomagnetic data. Norsk Geol. Tidsskr., 77: 15–33.

Olesen, O., Gellein J., Håbrekke H., Kihle O., Skilbrei J.R. and Smethurst M., 1997b. Magnetic anomaly map Norway and adjacent ocean areas, scale 1:3 Million. Geological Survey of Norway, Trondheim.

Olesen, O., Lundin, E., Nordgulen, Ø., Osmundsen, P.T., Skilbrei, J.R., Smethurst, M., Solli, A., Bugge, T. and Fichler, C., 2002. Bridging the gap between the onshore and offshore geology in northern Norway utilizing potential field data. Norsk Geol. Tidsskr., 82: 243–262.

Olesen, O., Skilbrei, J.R., Osmundsen, P.T., and Hartz, R.T., 2004. Is the segmentation of the rifting in the Norwegian Sea — Greenland Sea controlled by Late-Caledonian detachment structures? The 26th Nordic Geological Winter meeting, January 6–9, Uppsala, Sweden. Geologiska Föreningens i Stockholm Förhandlingar, 126, Abstract Volume, pp. 82–83 (abstract).

Osmundsen, P.T., Sommaruga, A., Skilbrei, J.R. and Olesen, O., 2002. Deep structure of the Mid Norway passive margin. Norsk Geol. Tidsskr., 82: 205–224.

Osmundsen, P.T., Braathen, A., Nordgulen, Ø., Roberts, D., Meyer, G.B. and Eide, E., 2003. The Devonian Nesna shear zone and adjacent gneiss-cored culminations. North-Central Norwegian Caledonides. J. Geol. Soc., London, 160: 329–344.

Osmundsen, P.T., Braathen, A., Sommaruga, A., Skilbrei, J.R., Nordgulen, Ø., Roberts, D., Olesen, O. and Mosar, J., 2005. Metamorphic core complexes and gneiss-cored culminations along the Mid-Norwegian margin: an overview and some current ideas. In: B.T.G. Wandås, E.A. Eide, F. Gradstein and J.P. Nystuen (Editors), Onshore-Offshore relationships on the North Atlantic Margin. Norwegian Petroleum Society (NPF), Special Publication 12. Elsevier, Amsterdam, pp. 29–41 (this volume).

Reid, A.B., Allsop, J.M., Granser, H., Millett, A.J. and Sommerton, I.W., 1990. Magnetic interpretation in three dimensions using Euler deconvolution. Geophysics, 55: 80–91.

Roberts, D., 1998. High-strain zones from meso- to macro-scale at different structural levels, Central Norwegian Caledonides. J. Struct. Geol., 20 (No. 2/3): 111–119.

Roberts, D. and Gee, D.G., 1985. An introduction to the structure of the Scandinavian Caledonides. In: D.G. Gee and B. Sturt (Editors), The Caledonide Orogen-Scandinavia and related areas. Wiley, Chichester, UK, pp. 55–68.

Robinson, P., 1995. Extension of Trollheimen tectono-stratigraphic sequence in deep synclines near Molde and Brattvåg, Western Gneiss Region, southern Norway. Norsk Geol. Tidsskr., 75: 181–198.

Romer, R.L. and Bax, G., 1992. The rhombohedral framework of the Scandinavian Caledonides and their foreland. Geologische Rundschau, 81/2: 391–401.

Séranne, M., 1992. Late Palaeozoic kinematics of the Møre-Trøndelag Fault Zone and adjacent areas, central Norway. Norsk Geol. Tidsskr., 72: 141–158.

Sigmond, E.M.O., 1992. Bedrock map of Norway and adjacent ocean areas. Scale 1:3 million. Geological Survey of Norway.

Sigmond, E.M.O., Gustavsson, M. and Roberts, D., 1984. Berggrunnskart over Norge, 1:1 million. Geological Survey of Norway.

Sindre, A., 1977. Geofysiske undersøkelser innen kartblad Smøla. Norg. Geol. Unders. Skrifter, 330: 25–32.

Skilbrei, J.R., 1988a. Geophysical interpretation of the Fosen-Namsos Western Gneiss Region and northern part of the Trondheim Region Caledonides, Central Norway. Norg. Geol. Unders. Spec. Publ., 3: 59–69.

Skilbrei, J.R., 1988b. Magnetic and gravimetric interpretation of the structure of the upper crust across the Trøndelag Region of Central Norway. Abstract; 18. Nordiske Geologiske Vintermøde, København, pp. 375–376.

Skilbrei, J.R., 1989. Petrofysiske undersøkelser, Midt-Norge. Norg. Geol. Unders. Rep. 89.164, 109 pp.

Skilbrei, J.R. and Sindre, A., 1991. Tolkning av gravimetri langs ILP-profilet, Hemne-Storlien. Norges geologiske undersøkelse report 91.171, 26 pp.

Skilbrei, J.R. and Kihle, O., 1999. Display of residual profiles versus gridded image data in aeromagnetic study of sedimentary basins: A case history. Geophysics, 64: 1740–1748.

Skilbrei, J.R. and Olesen, O., 2001. Structure of the basement in the Mid-Norwegian shelf interpreted from potential field data. In: E. Eide (Editor), BAT report. Status to December 2001. Geological Survey of Norway.

Skilbrei, J.R., 1999. Interpretation of 'reconstructed' magnetic anomaly data from the Barents Sea-East Greenland region. IUGG 99 meeting, July 19–24, Birmingham, Abstract volume A, A51.

Skilbrei, J.R., 2000. Basement ridges and sedimentary basins on the Northeast Greenland shelf interpreted from 'reconstructed' aeromagnetic data, and seismic data. 24th Nordic Geological Winter meeting, January 6–9, Trondheim. Abstract, p 154.

Skilbrei, J.R., Skyseth, T. and Olesen, O., 1991a. Petrophysical data and opaque mineralogy of high grade and retrogressed lithologies: Implications for the interpretation of aeromagnetic anomalies in northern Vestranden, Western Gneiss Region, Central Norway. In: P. Wasilewski and P. Hood (Editors), Magnetic anomalies — land and sea. Tectonophysics, 192, pp. 21–31.

Skilbrei, J.R., Håbrekke, H., Olesen, O., Kihle, O. and Macnab, R., 1991b. Shaded relief aeromagnetic colour map of Norway and the Norwegian-Greenland and Barents Seas: Data compilation and examples of interpretation. Norg. Geol. Unders. Rep. 91.269, 15 pp.

Skilbrei, J.R., Sindre, A., McEnroe, S., Robinson, P. and Kihle, O., 1995. Combined interpretation of potential field-, petrophysical data and topography from Central Norway and the continental shelf between 62° N and 65° N, including a preliminary report on paleomagnetic dating of faults near Molde. Norg. Geol. Unders. Rep. 95.027, 45 pp.

Skilbrei, J.R., Kihle, O., Olesen, O., Gellein, J., Sindre, A., Solheim, D. and Nyland, B., 2000. Gravity anomaly map Norway and adjacent ocean areas, scale 1:3 Million. Geological Survey of Norway, Trondheim.

Skilbrei, J.R., Olesen, O., Osmundsen, P.T., Kihle, O., Aaro, S. and Fjellanger, E., 2002. A study of basement structures and onshore-offshore correlations in Central Norway. Norsk Geol. Tidsskr., 82: 263–279.

Skogseid, J., Planke, S., Faleide, J.I., Pedersen, T., Eldholm, O. and Neverdal, F., 2000. NE Atlantic continental rifting and volcanic margin formation. In: A. Nøttvedt (Editor), Dynamics of the Norwegian Margin. Geol. Soc., London, Spec. Publ., 167, pp. 295–327.

Skyseth, T., 1986. Geofysisk og geologisk tolkning av aeromagnetiske og gravimetriske anomalier på og rundt kartblad 1523 II Stokksund og 1623 III Roan, Sør-Trøndelag. Diploma Thesis, NTH, Trondheim, 85 pp.

Smethurst, M.A., 2000. Land-offshore tectonic links in western Norway and the northern North Sea. J. Geol. Soc., London, 157: 769–781.

Smelror, M., Jacobsen, T., Rise, L., Skarbø, O., Verdenius, J.G., and Vigran, J.O., 1994. Jurassic to Cretaceous stratigraphy of shallow cores on the Møre Basin Margin, Mid-Norway. Norsk Geol. Tidsskr., 74: 89–107.

Sola, M., 1990. Seismisk kartlegging av Frøyahøyden. Unpublished Cand. Scient. thesis, University of Bergen, 173 pp.

Sommaruga, A. and Bøe, R., 2002. Geometry and subcrop maps of shallow Jurassic basins along the Mid-Norway coast. Mar. Petrol. Geol., 19: 1029–1042.

Strömberg, A.G.B., 1978. Early tectonic zones in the Baltic Shield. In: A.F. Trendall (Editor), Evolution of the Archean terrain. Precambrian Res., 6, pp. 217–222.

Sæterstad, S., 1996. Tolkning og modellering av flymagnetiske data på midt-Norsk sokkel. Unpublished Diploma Thesis, NTNU, Trondheim, Norway, 87 pp.

Thorsnes, T., 1995. Structural setting of two Mesozoic half-grabens off the coast of Trøndelag, Mid-Norwegian shelf. Norg. Geol. Unders. Bull., 427: 68–71.

Torsvik, T.H., Van der Voo, R., Meert, J.G., Mosar, J., Walderhaug, H.J., 2001. Reconstructions of the continents around the North Atlantic at about the 60th parallel. Earth Planet. Sci. Lett., 187 (1–2): 55–69.

Wolff, F.C., 1984. Regional geophysics of the Central Norwegian Caledonides. Norg. Geol. Unders. Bull., 397: 1–27.

Ziegler, P.A., 1988. Evolution of the Arctic-North Atlantic and the western Thethys. AAPG Mem., 43: 198 pp.

Åm, K., 1970. Aeromagnetic investigations on the continental shelf of Norway, Stad-Lofoten (62–69° N). Norg. Geol. Unders., Årbok, pp. 49–61.

Åm, K., 1975. Aeromagnetic basement complex mapping north of latitude 62° N, Norway. In: A. Whiteman, D. Roberts and M.A. Selleveoll (Editors), Petroleum geology and geology of the North Sea and Northeast Atlantic continental margin. Norg. Geol. Unders. Bull., 316, pp. 351–374.

Development of the Jan Mayen microcontinent by linked propagation and retreat of spreading ridges

Robert A. Scott, Lucy A. Ramsey, Steve M. Jones, Stewart Sinclair and Caroline S. Pickles

The Jan Mayen microcontinent lies between the active Kolbeinsey Ridge spreading centre and the extinct Aegir Ridge spreading centre in post-Paleocene oceanic crust to the north of Iceland. Uncertainties concerning the age of seafloor magnetic anomalies and the precise extent of oceanic crust in this segment of the northern North Atlantic have hindered attempts to model the spreading history. Here, we propose a new, geometrically self-consistent spreading model that uses a single set of rotation poles for the entire northern North Atlantic.

In our model, the Jan Mayen microcontinent separated sequentially from the East Greenland margin during Oligocene time as a consequence of stepwise northward propagation of the Kolbeinsey Ridge and simultaneous northward retreat of the Aegir Ridge. The ridge tips were linked by a fracture zone that was periodically replaced by a new fracture zone to the north, resulting in balanced propagation/retreat of the spreading ridges and segmentation of intervening oceanic and microcontinent lithosphere. Spreading azimuths remained parallel with the West Jan Mayen Fracture Zone through the propagation/retreat phase. A number of possible fracture zones of the appropriate orientation can be identified that cut both the microcontinent and the oceanic crust to the east. Systematic sinistral offset across these fracture zones produces an apparent counterclockwise rotation of the microcontinent with respect to the adjacent continental margins, whereas structural trends within the Jan Mayen microcontinent are not rotated appreciably.

At least two factors appear to have been important in initiating the Kolbeinsey Ridge, and thus creating the Jan Mayen microcontinent: (1) the geometry of the plate boundary generated between Europe and Greenland at continental break-up (chron 24R), with the Aegir Ridge significantly offset to the east with respect to the Mohns and Reykjanes ridges; (2) a change of spreading azimuth, which acted to lock the transform system that had previously connected the southern tip of the Aegir Ridge with the northern end of the Reykjanes Ridge. The thermal effect of the Iceland plume on the overlying plates probably played little part in microcontinent generation, although the gravitational effect of the plume may have been significant.

Introduction

On the basis of spreading history, Cenozoic oceanic crust between Greenland and NW Europe can be divided into three segments, separated by two major transform fault systems (Scott, 2000; Fig. 1). The Northern Segment (north of the Jan Mayen Fracture Zone) and the Southern Segment (south of Iceland) share a relatively simple spreading history. In contrast, the Central Segment between Iceland and the Jan Mayen Fracture Zone, exhibits a complex spreading history involving an extinct spreading centre with pronounced apparent curvature in the east (the Aegir Ridge), a currently active spreading centre in the west (the Kolbeinsey Ridge), and an intervening continental fragment (the Jan Mayen microcontinent).

It is generally acknowledged that from the continental break-up at chron 24R time to at least chron 20 time, seafloor spreading in the Central Segment occurred solely on the Aegir Ridge. It is also widely accepted that since at least from chron 6 to the present day, spreading has occurred exclusively along the Kolbeinsey Ridge. However, the timing and the mechanism by which spreading was transferred from one ridge to the other between chrons 20 and 6 remains enigmatic (e.g. Talwani and Eldholm, 1977; Vogt et al., 1980; Nunns, 1983; Bott, 1985, 1987; Skogseid and Eldholm, 1987; Kuvaas and Kodaira, 1997; Skogseid et al., 2000;

Onshore–Offshore Relationships on the North Atlantic Margin
edited by B. Wandas et al. NPF Special Publication 12, pp. 69–82,

Müller et al., 2001; Lundin and Doré, 2002; Mosar et al., 2002). The uncertainty arises from problems in identifying seafloor magnetic anomalies, combined with uncertain continent–ocean boundary (COB) locations on all continental margins, especially around the Jan Mayen microcontinent. This uncertainty is compounded by a dearth of reliably dated igneous rocks and the sparse knowledge of sedimentary successions.

These uncertainties cannot be resolved with the data sets currently available, and so an alternative strategy is therefore required to constrain reconstructions. Here, we seek to demonstrate that boundary conditions imposed by the evolution of adjacent regions can be used to diminish the degrees of freedom. We derive a geometrically self-consistent kinematic model that explains seafloor spreading for the Northern, Central and Southern segments, using a single set of rotation poles. The principles, methodology and results of the modelling will be presented in detail elsewhere; here we focus on the implications of this model for the development of the Jan Mayen microcontinent. Information from this microcontinent provides a key test of the model's credibility.

Understanding the spreading history of the Central Segment is of direct relevance to hydrocarbon exploration on the adjacent continental shelves. Cenozoic inversion structures on the NW European margin are spatially associated with the projected trends of the major oceanic fracture zones that define the boundaries of the Central Segment (e.g. Doré and Lundin, 1996; Doré et al., 1997, 1999; Vågnes et al., 1998; Roberts et al., 1999; Lundin and Doré, 2002; Mosar et al., 2002). There is an implied relationship between changes of spreading geometry and periods of structural inversion because many inversion structures show important phases of growth during the Oligocene, the interval during which spreading transferred from the Aegir Ridge to the Kolbeinsey Ridge.

The Jan Mayen microcontinent occupied a central position in the Greenland–Norway rift system prior to the onset of spreading but its dimensions are a significant uncertainty in reconstructions. If the microcontinent had a topographic expression in the rift system prior to spreading, it may have played an important role in channelling, diverting or preventing clastic sediment flux from Greenland to the NW European margin. It may also have been a significant sediment source in its own right.

In the following section, we review the geology and extent of the Jan Mayen microcontinent. We then consider the earlier rotation models, and some of their inherent problems. Our new model for the tectonic evolution of the Central Segment is then outlined. Finally, we discuss the geodynamic processes that may have played a role in separating the microcontinent from the East Greenland margin.

Boundaries of the Jan Mayen microcontinent

Named after a 17th Century Dutch whaling captain, the 380 km^2 island of Jan Mayen lies close to the intersection of the Mohns spreading ridge with the West Jan Mayen Fracture Zone (Fig. 1). The NE end of the island is dominated by the active volcano, Beerensburg (2277 m). Despite its relatively high elevation, the island of Jan Mayen is probably not part of the Jan Mayen microcontinent, which begins further south in the morphologically connected, but geologically distinct, Jan Mayen Ridge. Seismic reflection and refraction profiles, potential field data and plate tectonic arguments, all indicate that the Jan Mayen Ridge contains continental crust (Fig. 2; Talwani and Eldholm, 1977; Gairaud et al., 1978; Myhre et al., 1984; Skogseid and Eldholm, 1987; Kuvaas and Kodaira, 1997; Kodaira et al., 1998). This flat-topped submarine bathymetric high trends southwards to approximately 69° N, where it is cut obliquely by the NE–SW trending Jan Mayen Trough and breaks up into a series of smaller highs, (Skogseid and Eldholm, 1987). The area of bathymetric highs south of the Jan Mayen Trough has been referred to as the Southern Ridge Complex (Pelton, 1985; Fig. 2).

The tectonic history of the Jan Mayen microcontinent is complex, owing to its involvement in two separate Cenozoic break-up events. Kuvaas and Kodaira (1997) distinguished three tectonic phases on their reflection seismic profiles: two corresponding to the break-up events and the other, to an earlier phase related to Mesozoic extension (Fig. 2).

The eastern margin of the Jan Mayen microcontinent was created during initiation of spreading in the Norwegian-Greenland Sea around the Paleocene–Eocene boundary (chron 24R). The conjugate margin lies at the western edge of the Møre Basin. The eastern margin of the Jan Mayen Ridge is characterised by seaward-dipping reflectors (SDRs), identified in multi-channel seismic reflection profiles (Skogseid and Eldholm, 1987; Gudlaugsson et al., 1988). By comparison with similar sequences on volcanic margins nearby, it is

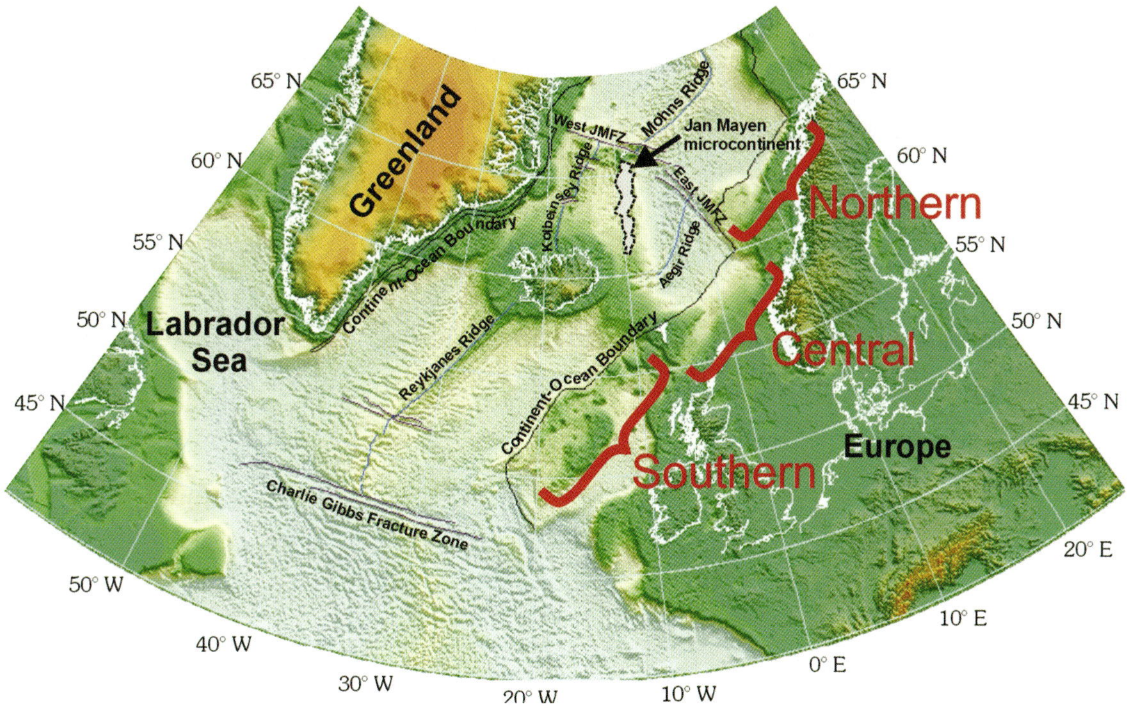

Fig. 1 Principal spreading centres and fracture zones of the northern North Atlantic on a bathymetric basemap, illustrating the division of oceanic crust into Northern, Central and Southern segments. JMFZ = Jan Mayen Fracture Zone.

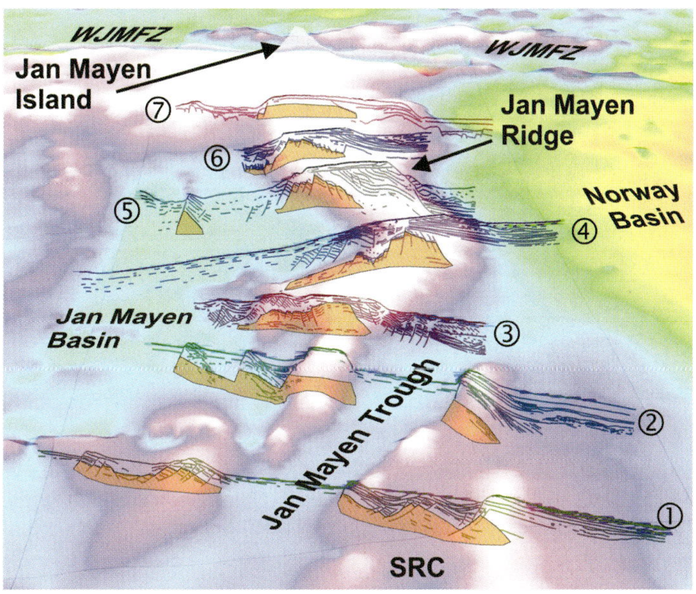

Fig. 2 Oblique view of the Jan Mayen microcontinent (looking north) showing the principal structural elements. The image covers approximately the same area as the inset map on Fig. 6. WJMFZ = West Jan Mayen Fracture Zone; SRC = Southern Ridge Complex. Shaded surface is bathymetry: the crest of the Jan Mayen Ridge is at water depths of 500–1000 m, the Jan Mayen Trough at 2000 m, and the deepest part of Norway Basin visible in this image at > 3000 m. Line drawings of seismic reflection profiles obtained from the following sources: (1), (3), (4), (5), (6) from Kuvaas and Kodaira (1997); (2) from Gudlaugsson et al. (1988); (7) from Myhre et al. (1984). The extent of continental crust is unclear owing to imaging problems and lack of well control; however, the shaded areas beneath each line represent a conservative estimate of assumed Mesozoic fault blocks, based on reflector geometry.

likely that the COB lies closer to the landward side of the SDR sequence than the seaward side (Mutter et al., 1982; Scott, 2000). However, significant uncertainties still remain because the SDR sequence is not well-imaged, particularly in the south, and because erosion on the crest of the Jan Mayen Ridge may give an erroneous impression of the original extent of SDR sequences. The eastern

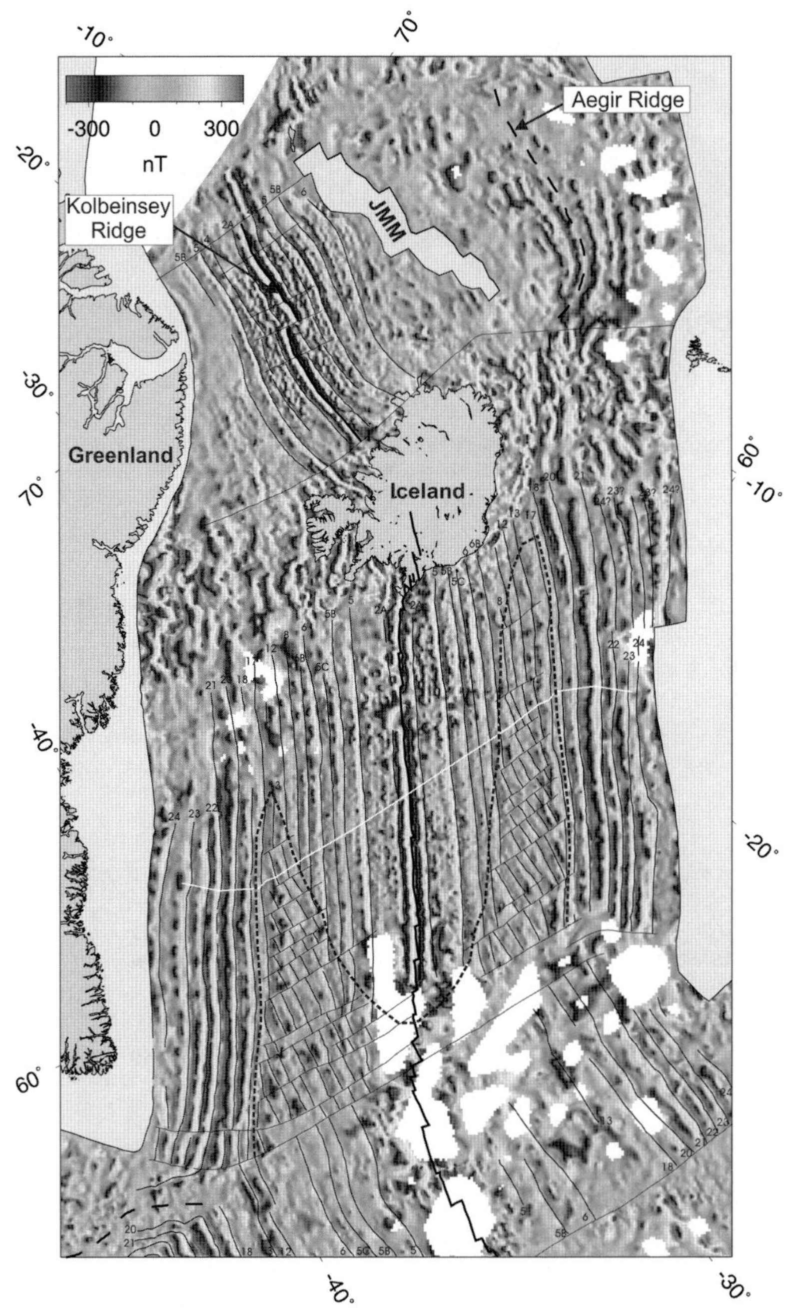

Fig. 3 Magnetic anomaly picks south of Iceland (adapted from Jones et al., 2002). Magnetic data from Verhoef et al. (1996). Dashed line marks the boundary between segmented and unsegmented crust south of Iceland. White line marks plate-spreading flowline calculated from the poles of Smallwood and White (2002). Potential picks for magnetic anomalies older than chron 6 to the north of Iceland are illustrated in Fig. 4.

margin of the microcontinent also contains a number of offsets, which have a consistent sinistral sense. These offsets are observed in the bathymetry of the Jan Mayen Ridge (and less clearly in the Southern Ridge Complex), and are also linked to offsets in magnetic lineations in the adjacent Norway Basin (Fig. 3).

The western margin of the microcontinent was formed by separation from the East Greenland margin during Oligocene time, has no clear evidence of SDRs or any underplating and is dominated by tilted fault blocks (Kuvaas and Kodaira, 1997; Fig. 2). However, despite the apparent absence of SDRs, it is on the west and south side of the microcontinent, where the COB is most difficult to pick, owing to an extensive flat-lying volcanic reflector (reflector F) of assumed earliest Miocene age, that masks the underlying structures (Gudlaugsson et al., 1988). This reflector also occurs extensively on the Iceland Plateau, where it was named the 'opaque horizon' by Eldholm and Windisch (1974).

Immediately west of the Jan Mayen Ridge is the Jan Mayen Basin, which is slightly deeper than the rest of the Iceland Plateau and is separated from it by an indistinct ridge parallel to the Jan Mayen Ridge (Fig. 2). This indistinct ridge was interpreted to be a short-lived spreading axis by Talwani and Eldholm (1977). However, evidence from the internal structure of emergent ridges (Fig. 2), from seismic refraction data and the absence of magnetic lineations suggest that the Jan Mayen Basin may also be underlain by continental crust, perhaps highly attenuated (Kuvaas and Kodaira, 1997; Kodaira et al., 1998). Magnetic anomaly patterns suggest that the western margin of the microcontinent also contains a number of sinistral offsets (Fig. 3) but, given the problems with identifying the COB, these offsets are less easy to corroborate than on the eastern margin.

The elevation of the Jan Mayen Ridge and Southern Ridge Complex decreases southward, with the bathymetric expression of the Southern Ridge Complex being lost south of 67.7° N. It is not possible to define the southern extent of the microcontinent with any degree of certainty, although the recovery of Lewisian and Jurassic zircon xenocrysts from eastern Iceland basalts (Schaltegger et al., 2002) may indicate a continuation close to the boundary between the Central and Southern segments.

Previous rotation models

A viable spreading model for the Central Segment must be able to explain the distinct curvature of the Aegir Ridge and the apparent northward divergence of magnetic anomaly stripes in adjacent oceanic crust (Figs. 3 and 4). Most authors have used these features to support models involving varying degrees of fan-shaped spreading on the Aegir Ridge, with the rotation pole situated close to the southern end (Talwani and Eldholm, 1977; Nunns, 1983; Bott, 1985, 1987; Skogseid and Eldholm, 1987; Kuvaas and Kodaira, 1997; Skogseid et al., 2000; Lundin and Doré, 2002; Mosar et al., 2002). A plausible spreading model must also provide a mechanism by which the Jan Mayen microcontinent is created. The implication of all published mechanisms invoking fan-shaped spreading is that the Jan Mayen microcontinent separated from Greenland by counterclockwise rotation (Fig. 4).

To preserve area compared with the adjacent Southern and Northern segments, a corollary to fan-shaped spreading on the Aegir Ridge is that an opposing fan-shaped geometry must develop simultaneously to the west to compensate the variable spreading rate along the Aegir Ridge. This compensating geometry may be generated by two end-member mechanisms. Nunns (1983) proposed that compensation occurred directly by oceanic spreading along the Kolbeinsey Ridge, with ~18 Ma of concurrent fan-shaped spreading, along the Aegir Ridge and Kolbeinsey Ridges, between chrons 20 and 7 (43.9–25.1 Ma) (Fig. 4). In contrast, Kuvaas and Kodaira (1997) suggested that the fan-shaped spreading on the Aegir Ridge was balanced by simultaneous extension in the East Greenland margin, with spreading on the Kolbeinsey Ridge not starting until chron 7 time. Given the uncertainty in magnetic anomaly identification, spreading on the Kolbeinsey Ridge could have begun at any stage between chron 20 and chron 7, so long as the southward-widening fan of extension/spreading on the Kolbeinsey Ridge compensates the northward-widening spreading pattern at the Aegir Ridge (Müller et al., 2001).

Geometric problems

It is possible to preserve area in the Central Segment in a general way using two opposing fan-shaped wedges of oceanic crust (or equivalent continental extension). However, this mechanism does not address the significant geometric problems created at the boundaries between the Northern, Central and Southern segments.

Spreading trajectories for the Northern and Southern Segments are well-established (e.g. Talwani and Eldholm, 1977; Nunns, 1983; Srivastava and Tapscott, 1986; Skogseid and Eldholm, 1987; Skogseid et al., 2000; Smallwood and White, 2002). Observed fracture zones in the Northern and Southern segments fall into two groups: those formed during the early stages of spreading are oriented approximately NW–SE, whereas those formed later are oriented WNW–ESE (Fig. 3). These two distinct orientations indicate one, relatively abrupt, major change of spreading azimuth during oceanic opening. There is some variation in the precise timing of this change in published studies as a result of differences in anomaly identification. In this study, we use the interpretation of Smallwood and White (2002), in which the change occurs around chron 18–17, because their model gave the best match between flowlines and fracture zones, although the precise timing of the change in spreading azimuth is not important to the geometric principles involved. The critical element

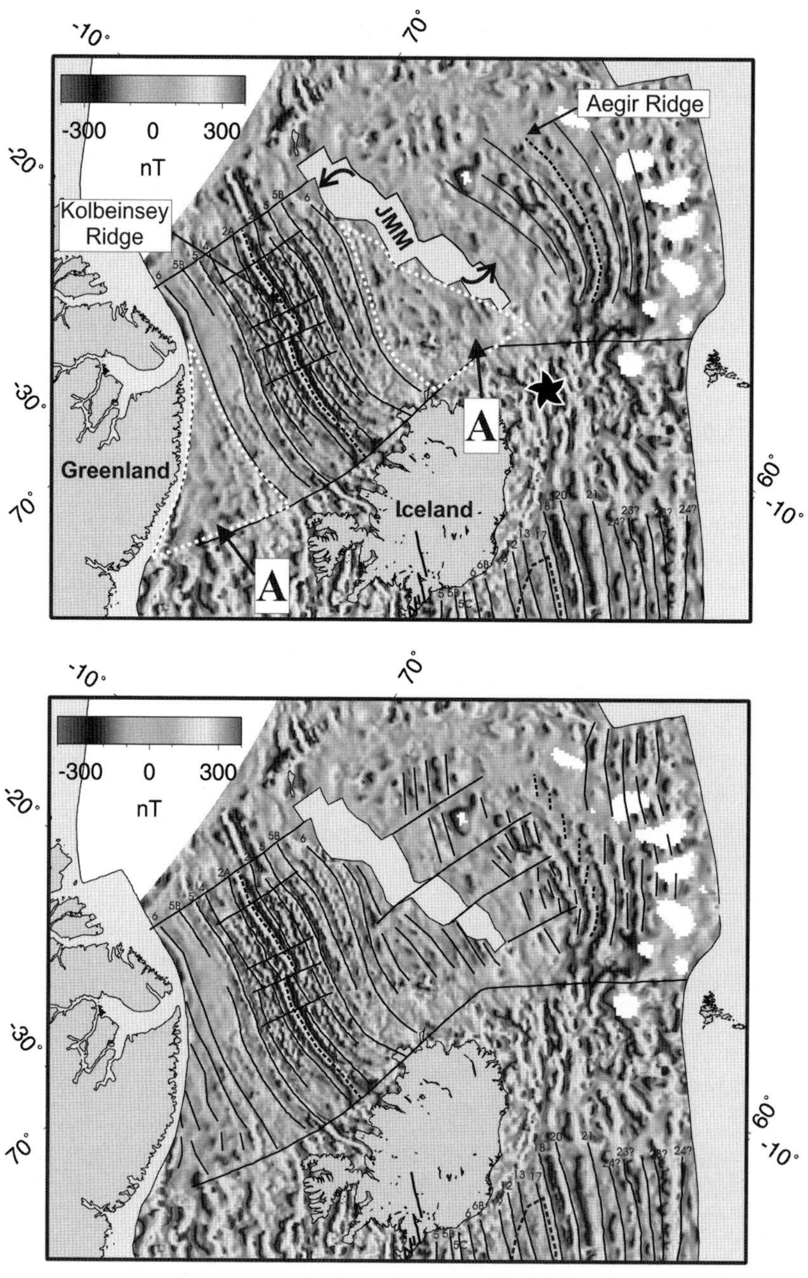

Fig. 4 Top panel shows magnetic anomalies of the Central Segment, with Nunns' (1983) fan-shaped spreading interpretation of anomalies adjacent to the Aegir Ridge. Magnetic anomaly data taken from Verhoef et al. (1996). The approximate pole of rotation for spreading on the Aegir Ridge calculated by Nunns (1983) is indicated by the star. To compensate for the fan-shaped spreading on the Aegir Ridge, the two triangular areas labelled A must contain an opposing fan with a rotation pole to the north. The Jan Mayen Microcontinent (JMM) must have rotated counterclockwise in order to accommodate the fanned spreading. Chron 6 to recent anomalies formed parallel to the Kolbeinsey Ridge. Lower panel shows the alternative interpretation favoured in this study.

is that the spreading histories of both the Northern and Southern Segments can be explained using the same stage poles, and that these poles are located at a great distance from the oceanic crust whose spreading trajectories they define. This requires that transform faults produced during spreading are virtually straight features, very slightly convex to the south.

In contrast, the putative fan-shaped spreading geometries in the Central Segment require rotation poles to be located close by (Fig. 4). The resulting fracture zones would be markedly curved — convex to the north for the oceanic crust formed at the Aegir Ridge and convex to the south for coeval oceanic crust formed at the Kolbeinsey Ridge. Juxtaposition of straight and curved transform geometries would create serious geometric problems along the boundaries between the Northern, Central and Southern Segments. Furthermore, a careful examination of the bathymetric and potential field data sets has provided no evidence to support the presence of curved fracture zones of the appropriate geometry anywhere within the Central Segment.

Non-rotation of structures in the Jan Mayen microcontinent

Two major structural trends have been identified in the microcontinent (e.g. Skogseid and Eldholm, 1987; Gudlauggsson et al., 1988) (see Fig. 6). In the northern part of the microcontinent (north of approximately 69° N) the majority of faults trend approximately N–S, which is slightly oblique to the bathymetric margin of the Jan Mayen Ridge. Further south, faults are approximately NE–SW. Fan-shaped spreading geometries in the Central Segment require an element of counterclockwise rotation as the Jan Mayen microcontinent separates from the East Greenland margin. In the case of the Nunns (1983) model, the amount of rotation is about 30° (Fig. 4).

Prior to the onset of spreading in the Eocene, the Norwegian margin and East Greenland were part of a major rift system that developed on the Caledonian Orogen. Episodes of extension with intervening periods of thermal subsidence have characterised the rift system throughout the Late Paleozoic and Mesozoic interval (Ziegler, 1988, 1989; Doré, 1991; Knott et al., 1993; Doré et al., 1999; Roberts et al., 1999). Mesozoic faults on the Norwegian and East Greenland margins have two principal trends. Figure 5 illustrates how some of the major structural elements on the Norwegian and East Greenland margins may have been related prior to seafloor spreading. Although faults of both trends have probably been active in all the major Mesozoic rift episodes, there appears to be a pronounced change at the end of the Jurassic from predominant activity on the approximately N–S trending fault set to the fault set trending more NE–SW. This change has been related to a major realignment of the rift system (Lundin and Doré, 1997; Roberts et al., 1999; Scott, 2000). The COBs formed during Cenozoic break-up predominantly follow this NE–SW Mesozoic trend (Figs. 1 and 5).

Considering its central position within the rift system, the structure of the Jan Mayen microcontinent is also likely to have been strongly influenced by Mesozoic extensional events. The approximate original dimensions of the Jan Mayen microcontinent are also illustrated in Fig. 5, and are based on underlap of the Greenland and NW European COBs. The boundaries of the microcontinent show a clear relationship with the two Mesozoic fault trends: the northern part of the microcontinent is elongated parallel to the N–S fault trend, whereas the southern tail of the microcontinent is oriented parallel to the NE–SW fault trend. These two structural trends are reflected in the change of orientation of the Greenland coast either side of Scoresby Sund, which forms a direct conjugate to the Jan Mayen microcontinent in a pre-drift reconstruction (Fig. 5).

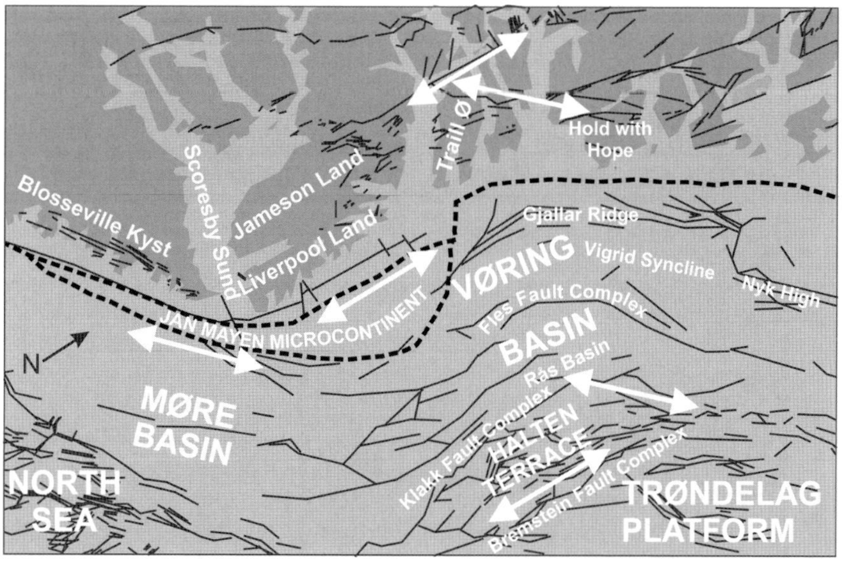

Fig. 5 Pre-drift structural correlation map between the Norwegian margin and East Greenland (adapted from Scott and Whitham, 2000). Line of eventual continental separation shown as black dashes, with the extent of the Jan Mayen microcontinent based on underlap of the conjugate margins. White double-headed arrows approximate the two Mesozoic structural trends recognised on East Greenland and on the Norwegian margin. The northern part of the Jan Mayen microcontinent is aligned with one of these trends, whereas the southern part of the microcontinent is sub-parallel to the other.

The implication of these relationships is that Mesozoic structural trends exerted a strong influence on the final shape of the microcontinent as it developed during the Cenozoic. This is consistent with the observation that assumed Mesozoic half-grabens forming the core of the microcontinent (Fig. 2) are bounded by faults that do not appear to differ significantly in orientation from the Cenozoic faults related to break-up events on the microcontinent margins (Fig. 6). It is also consistent with evidence from elsewhere that the orientation of the earliest Eocene COBs is generally parallel to the Mesozoic fault trends (Fig. 5).

If we compare the orientation of the two Mesozoic fault orientations recognised on the Norwegian margin, the East Greenland margin and the Jan Mayen microcontinent on a present-day reconstruction (i.e. after any rotations caused by seafloor spreading), it is apparent that no significant counterclockwise rotation of the Jan Mayen microcontinent has occurred with respect to either margins (Fig. 6).

Identification of magnetic anomalies

Away from the immediate vicinity of the Aegir Ridge, magnetic anomalies in the Norway Basin are generally fragmented and lack lateral continuity. Identifying specific anomalies close to the basin margins is hampered further by uncertain COB location. The Aegir Ridge itself, however, appears as a smoothly curved unbroken feature, when viewed at a regional scale in bathymetric and free-air gravity data sets (Fig. 6). This curvature has a tendency to draw the eye into interpreting magnetic lineations in the adjacent oceanic crust to have a similar, continuous curvature (Fig. 4). It is this observation that underpins models invoking fan-shaped spreading geometries. However, the aeromagnetic data also permit other interpretations (Fig. 4) and we suggest that the apparent curvature of magnetic anomalies close to the Aegir Ridge may be misleading.

Assigning ages to magnetic anomalies close to the Aegir Ridge is also subject to uncertainties.

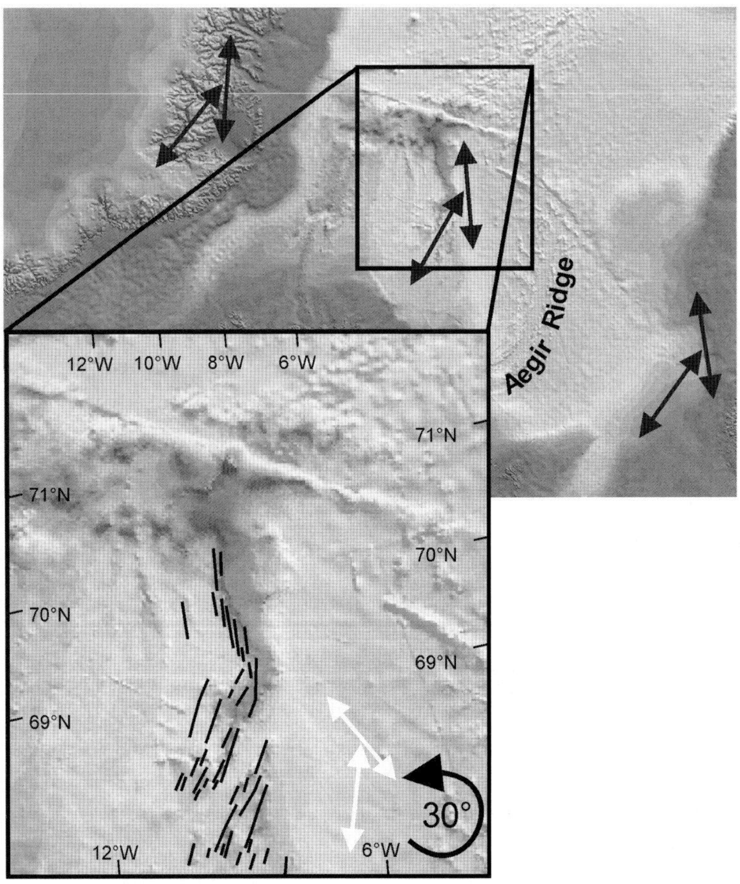

Fig. 6 Orientation of the two principal Mesozoic structural trends on East Greenland, the Norwegian margin and the Jan Mayen microcontinent. Pre-drift relationships between these structural elements are shown in Fig. 5. Inset map shows faults on the Jan Mayen microcontinent (from Gudlaugsson et al., 1988); while some of these faults clearly have a Cenozoic history, others bound assumed Mesozoic faults block and predate seaward dipping reflector (SDR) sequences of earliest Eocene age. White double-headed arrows on the inset map show the ~30° counterclockwise rotation of Norwegian margin Mesozoic structural trends implied by the Nunns (1983) model.

Deep Sea Drilling Project (DSDP) Leg 38, Site 337 was located close to the Aegir Ridge axis near its southern end. Tholeiitic basalt basement yielded K–Ar ages of 17.5–25.5 Ma (Late Oligocene to Early Miocene) which, in conjunction with stratigraphic data, were used to suggest that spreading terminated on the ridge at 25–27 Ma (chrons C7–C8) (Kharin et al., 1976; Talwani et al., 1976). Talwani and Eldholm (1977) mapped anomaly 7 as the youngest crust in the Norway Basin, and cessation of spreading along the entire length of the Aegir Ridge at this time has commonly been assumed in spreading models (e.g. Nunns, 1983; Jung and Vogt, 1997). However, the supposed Late Oligocene to Early Miocene basaltic basement at Site 337 is overlain by a Late Eocene to Early Oligocene sedimentary sequence, suggesting that the isotopic ages are unreliable. A similar problem was encountered on the Vøring Plateau, where lavas from the same DSDP leg yielded K–Ar ages of ~46 Ma, significantly younger than the oldest sediments overlying them. This observation implies that cessation of spreading at the Aegir Ridge may be much older than chron 7, at least along the southern part of the ridge.

Given the inherent uncertainties in identifying and tracing magnetic lineations in the Central Segment, we refrain from constructing a detailed anomaly map for the Norway Basin, although we sketch some possible trends on Fig. 4. However, the geometric inconsistencies implied by multiple sets of rotation poles, and the apparent lack of rotation of structures in the Jan Mayen microcontinent suggest a reappraisal of fan-shaped spreading models is necessary.

A new model

The simplest solution to avoid geometric problems along the boundaries between segments is to develop a model that uses a single set of rotation poles for the entire northern North Atlantic. The poles that define spreading in the Northern and Southern Segments are based on a well-constrained database of magnetic anomaly picks and fracture zone orientations (Fig. 3). Our modelling strategy was therefore to develop a model for the evolution of the Central Segment using the same poles, given the geological constraints. The advantage of this strategy is that identification of magnetic stripes and structural elements in the Central Segment is not a prerequisite of modelling. The model has been developed using GIS-based rotation software and the implications tested using 3-D GIS techniques for visualisation. We limit ourselves here to a brief summary of the main evolutionary stages, which are described below and presented graphically in Fig. 7. Although this is not the only possible interpretation using the same rotation parameters, we believe that this model is the most consistent with available geological data.

Stage 1. Spreading between chrons 24R to 18 had a constant NW–SE azimuth. During this stage, the Aegir Ridge was straight and parallel to the Mohns and Reykjanes ridges. In the Central Segment, this lineation trend is largely preserved east of the Aegir Ridge.

Stage 2. At chron 18–17 time, a change of spreading direction to a WNW–ESE azimuth occurred (Smallwood and White, 2002). A stepwise northward propagation of the Kolbeinsey Ridge then occurred between chron 17 to chron 6. At each

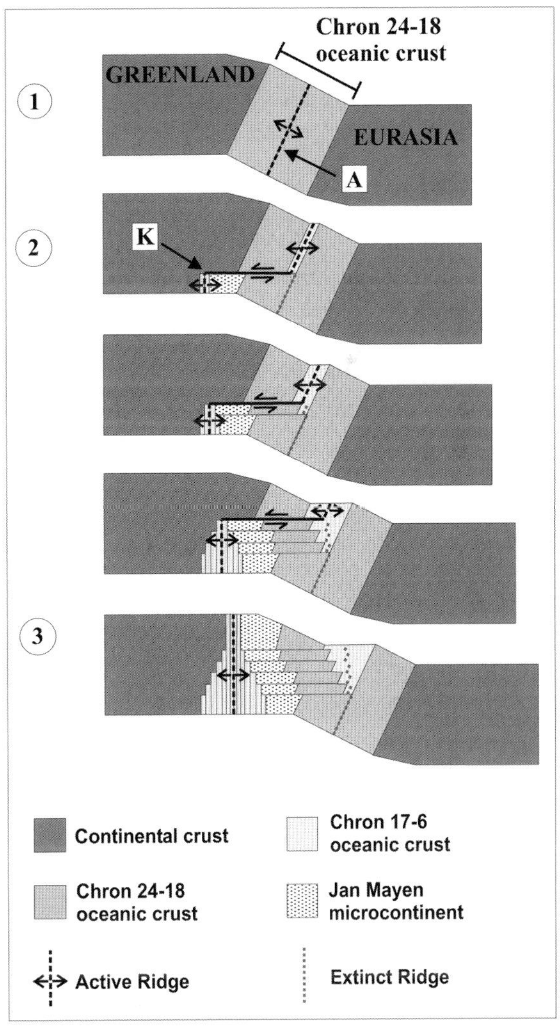

Fig. 7 Schematic sequence, showing the three principal stages of evolution of the Central Segment in the new model (see text for details). K = Kolbeinsey Ridge, A = Aegir Ridge

stage of this propagation, the northern tip of the Kolbeinsey Ridge was linked by a fracture zone to the northward-retreating southern tip of active spreading on the Aegir Ridge, so that the spreading centres never overlapped. The Jan Mayen microcontinent was separated progressively from the East Greenland margin as the linking fracture zone jumped northwards periodically, leaving behind a zone of segmented oceanic and microcontinent lithosphere. A similar configuration was sketched by Larsen (1988) in an extended abstract, but was not quantified, and entailed a soft linkage between the retreating Aegir and the advancing Kolbeinsey ridges.

Stage 3. By chron 6 time, the propagating tip of the Kolbeinsey Ridge had reached the West Jan Mayen Fracture Zone and spreading ceased on the Aegir Ridge. Spreading continued at the Kolbeinsey Ridge alone to the present day.

Supporting observations

Internal structure of the Jan Mayen microcontinent

Evidence from the Jan Mayen microcontinent and the immediately adjacent oceanic crust is critical for testing the validity of the proposed model. The mechanism of microcontinent separation implied in Stage 2 of the model provides several opportunities to test the hypothesis.

According to the new model proposed here, the Jan Mayen microcontinent was separated sequentially from the East Greenland margin by a process that does not require any significant rotation of crustal blocks. This mechanism could therefore explain the non-rotation of structures in the Jan Mayen microcontinent compared with adjacent conjugate margins (Fig. 6), whereas significant rotation of structures is a requirement of the fan-shaped spreading model (Nunns, 1983).

The new model also predicts that the northern tip of the propagating Kolbeinsey Ridge was linked by a succession of fracture zones to the northward-retreating southern tip of active spreading on the Aegir Ridge during chron 18–17 to chron 6 time. These WNW–ESE trending fracture zones should cut both the Jan Mayen microcontinent and the chron 24R to 18 oceanic crust immediately to the east, with a consistent sinistral sense of displacement (Fig. 7). They should cut both Eocene oceanic crust and at least some of the overlying sediment; the age of the fracture zones should decrease northwards.

Examination of the magnetic anomaly pattern in oceanic crust immediately east of the Jan Mayen microcontinent provides some support for the predictions outlined above: there are discontinuities in the magnetic anomaly pattern that have both the appropriate orientation and coincide with the systematic bathymetric offsets in the microcontinent margins (Figs. 4 and 8). Furthermore, the discontinuities can be traced westward through seismic lines that cross the Jan Mayen microcontinent. There is evidence of substantial faulting along these WNW–ESE trending zones, which affects post-basaltic sedimentary sequences and can be traced right through to the western side of the microcontinent, and, in some cases, out into the oceanic crust generated at the Kolbeinsey Ridge (Fig. 8).

All these observations suggest that the Jan Mayen microcontinent was segmented during Oligocene separation from the East Greenland margin. They are consistent with the model of linked propagation and retreat of spreading ridges, but are not a precondition of the fan-shaped spreading model. Unfortunately, the seismic lines available for this study are predominantly in an unfavourable orientation (parallel to the inferred fracture zones) in the southern part of the microcontinent, making it difficult to test the hypothesis further.

Oceanic crust generated at the Aegir Ridge

The Aegir Ridge is generally interpreted as a smoothly curved feature, bordered by magnetic lineations in the adjacent oceanic crust with a similar, continuous curvature (Fig. 4). We argue, however, that the Aegir Ridge can equally be interpreted as a series of short segments that are consistently offset in one direction (Figs. 4 and 7). If the apparent curve of the Aegir Ridge is ignored, the magnetic stripes to the east of the ridge can be interpreted as essentially straight and parallel to those on the Reykjanes and Mohns ridges (although individual segments are slightly offset by fracture zones).

Likewise, much of the apparent curvature anomaly lineations to the west of the ridge, can also be attributed to the systematic offset of short lineation segments by fracture zones (Fig. 4). On this side of the ridge, however, the offset of lineations appears to be much greater, breaking the oceanic crust into a series of westward-stepping blocks (from south to north). This leaves a triangular wedge of unclear magnetic lineations at the northern end of the Aegir Ridge, which

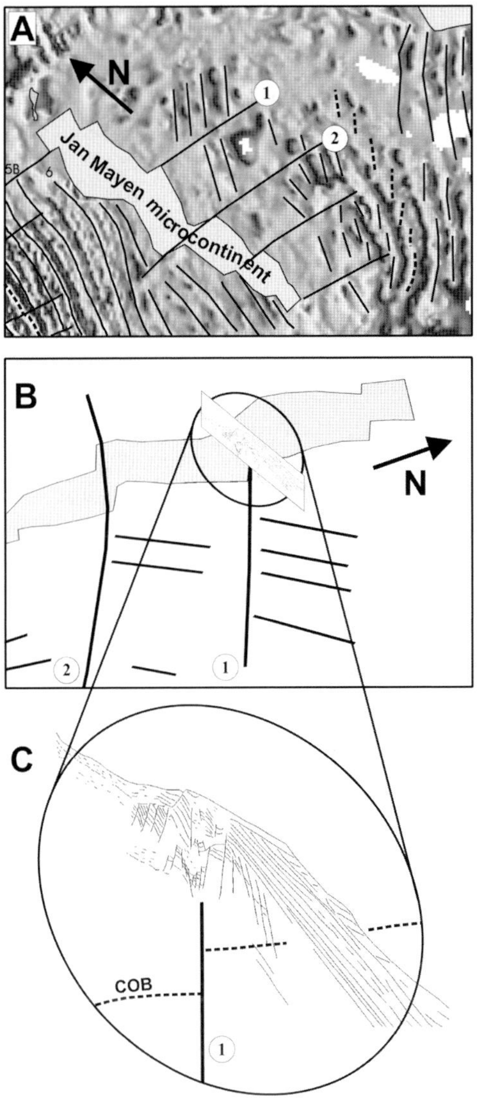

Fig. 8 Evidence for fracture zones segmenting the Jan Mayen microcontinent and adjacent oceanic crust. (A) Interpreted fracture zones and magnetic anomalies adjacent to the Jan Mayen microcontinent (see also, Fig. 4). (B) Schematic oblique sketch of the Jan Mayen microcontinent and adjacent oceanic crust of the Norway Basin viewed from the east. Post-chron 17 fracture zones labelled 1 and 2 are located on Part A. (C) Enlargement of cross-section (line drawing of seismic reflection profile 4 in Fig. 2). Faulting that postdates formation of the seaward-dipping reflector sequence (on right-hand-side of profile) is clearly visible in the central part of the profile where it intersects fracture zone 1.

presumably contains the youngest ocean floor generated by the Aegir Ridge. This observation is consistent with the northward retreat of active spreading predicted by the model proposed here.

Anomaly patterns west of the Jan Mayen microcontinent

As a corollary to the reinterpretation of the Aegir Ridge, we infer a new interpretation of the magnetic anomalies formed at the Kolbeinsey Ridge. Although magnetic lineations are difficult to pick for the early history of the Kolbeinsey Ridge, we note the lack of evidence for southward diverging anomalies and markedly curved fracture zones required by the Nunns (1983) interpretation. Instead, anomalies appear to terminate against the adjacent continental margins in a stepwise fashion consistent with the model (Fig. 4).

Processes involved in microcontinent formation

Jan Mayen is one of a number of microcontinents recognised in the present-day oceans, and the common occurrence of accreted continental fragments in convergent settings testifies their significance in the past; however, the processes that generate microcontinents are not well-understood (Müller et al., 2001). We now briefly consider the implications of the spreading model proposed here for the geodynamic processes involved in separating the Jan Mayen microcontinent from the East Greenland margin.

Two different processes have previously been considered important in the formation of the Jan Mayen microcontinent. Nunns (1983) regarded a change of spreading direction at chron 20 to be the most significant control on the initiation of the Kolbeinsey Ridge. In contrast, Müller et al. (2001) concluded that the angular difference between the two spreading azimuths was too small to be significant, and anyway, predated the onset of microcontinent separation (chron 13 according to their model). Instead, Müller et al. (2001) noted a global association between microcontinent formation and mantle plume activity, of which Jan Mayen was one of the cited examples. Rheological studies suggest that an inner continental margin is weaker than both old oceanic lithosphere and highly stretched continental lithosphere (e.g. Vink et al., 1984; Newman and White, 1997). On this basis, Müller et al. (2001) concluded that microcontinents were formed when a plume stem caused thermal weakening of <25 Ma old continental margins, thus permitting a spreading ridge to relocate closer to the plume centre. However, we doubt if thermal weakening had played a major role for three reasons: (1) active plume upwelling is unlikely at depths shallower than 100 km (Ito et al., 1999; Maclennan et al., 2001), and thus the strongest upper part of the lithosphere would not be affected; (2) conductive heating takes place on a much longer timescale, dependent on lithosphere thickness; (3) the continental margins on either side of the Kolbeinsey Ridge are characterised by modest amounts of volcanism and oceanic crustal thicknesses of 8–9 km, only slightly above the global average of 7 km (Weigel et al, 1995; Kuvaas and Kodeira, 1997; Kodeira et al., 1998).

To the south of Iceland and to the north of the Jan Mayen Fracture zone, an abrupt change of spreading azimuth from approximately NW–SE to WNW–ESE occurred during chron 18–17 time (Smallwood and White, 2002). In the model proposed here, geometrical consistency can only be maintained on the major transform systems bounding the northern and southern margins of the Central Segment if this change of spreading azimuth coincides with the initiation of the Kolbeinsey Ridge (Fig. 9). As discussed earlier by Nunns (1983), on the northern margin of the Central Segment (the Jan Mayen Fracture Zone) a change of spreading azimuth from NW–SE to WNW–ESE leads to a transtensional sense of displacement along the transform fault connecting the Aegir Ridge with the Mohns Ridge. In contrast, on the southern margin of the Central Segment, the transform system connecting the Aegir Ridge with the Reykjanes Ridge develops a transpressional relationship, and implies that restructuring must take place, if relative plate motion is to continue (Fig. 9).

There are several ways to resolve the geometrical problem on this transform system. It is significant that in the new spreading direction, the southern end of the Aegir Ridge now overlaps with the northern end of the Reykjanes Ridge, implying that the tip of one or the other ridge becomes superfluous (Fig. 9). The simplest solution is to create a

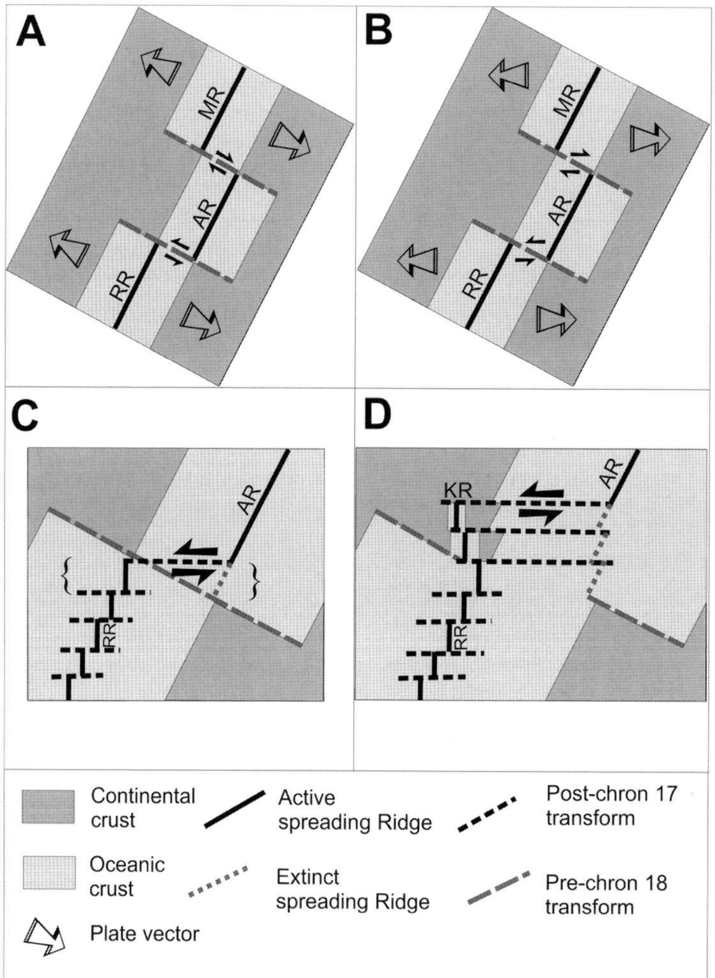

Fig. 9 The influence of changing spreading geometries on microcontinent separation. (A) Relative plate motion during pre-chron 18 spreading. (B) Change of spreading azimuth at chron 17–18 time. Note the transpressional relationship on the fracture zone connecting the Aegir Ridge to the Reykjanes Ridge. (C) Enlargement of B showing development of a new fracture zone connecting the Aegir Ridge to the Reykjanes Ridge. Change in spreading azimuth has resulted in segmentation of the Reykjanes Ridge. Brackets delimit area of ridge overlap in the new spreading direction, which results in extinction of the southern end of the Aegir Ridge. (D) Propagation of the Kolbeinsey Ridge at the expense of the Aegir Ridge during chron 6–17 time. AR = Aegir Ridge; RR = Reykjanes Ridge; MR = Mohns Ridge; KR = Kolbeinsey Ridge.

new transform fault parallel to the second spreading azimuth, somewhere within this area of ridge overlap. This new transform fault represents the first in the sequence of transforms that eventually separated the Jan Mayen microcontinent from East Greenland (Fig. 7).

The role of the Iceland plume is not easy to quantify, and may in any case ultimately be linked with the change of spreading azimuth through mantle convection processes. According to Lawver and Müller (1994), the Iceland hotspot was located beneath the Greenland margin during microcontinent initiation and separation. Although thermal weakening may have played some part in facilitating the ridge jump, we consider that gravitational forces generated by dynamic support are likely to be the most important plume influence, but only if these were sufficient to overcome the ridge push force generated at the Aegir Ridge. Dynamic elevation of the East Greenland margin relative to the Aegir Ridge would probably increase during Eocene time as Greenland moved westward relative to the hotspot, thus making it progressively more likely that ridge push forces can be overcome. However, this process is not essential for the mechanism we envisage.

Conclusions

Cenozoic ocean crust north of Iceland and south of the Jan Mayen Fracture Zone had a complex spreading history that has proved difficult to resolve using magnetic anomaly patterns generated at the two ridges involved (Aegir and Kolbeinsey). Most previous models involve rotation poles close to the southern end of the Aegir Ridge and northern end of the Kolbeinsey Ridge. Such models were invoked to explain the pronounced apparent curvature of the Aegir Ridge and the apparent northward divergence of associated magnetic anomaly lineations by fan-shaped spreading geometries. However, the curved fracture zone geometries implied by these rotation poles are incompatible with the virtually straight fracture zones of adjacent oceanic crust to the north and south. Furthermore, rotation of the Jan Mayen microcontinent during its separation from the Greenland margin, a prerequisite of the fan-shaped spreading model, is not supported by structural evidence.

We overcome these inconsistencies by proposing a new spreading model, based on one set of rotation parameters for the entire northern North Atlantic region. In this model, the curvature of the Aegir Ridge is not a primary feature but has resulted from the northward decay of active spreading in response to northward propagation of the Kolbeinsey Ridge over an interval of ~18 Ma. The propagating tip of the Kolbeinsey Ridge and the retreating tip of the Aegir Ridge were linked by a fracture zone, which stepped north through time, separating the Jan Mayen microcontinent from the East Greenland margin by a process of segmentation. There is evidence of this segmentation in the microcontinent and in adjacent oceanic crust — segmentation that is not predicted or required by the fan-shaped spreading model.

The original plate geometry at chron 24R continental break-up, and a change of spreading azimuth at chron 17–18 determined when and why the Jan Mayen microcontinent was created, whereas the presence of the Iceland plume probably smoothed the process of separation, and may have influenced how this separation occurred. This combination of processes may be significant in the formation of other microcontinents.

Acknowledgements

The CASP contribution to this research was funded by Anadarko, BP, ChevronTexaco, ExxonMobil, JNOC, Shell, Statoil and Total. Steve Jones acknowledges a NERC post-doctoral fellowship. Mark Allen, Tony Doré, Johan Petter Nystuen, Jakob Skogseid and two anonymous referees are thanked for their comments and improvements to an earlier version of this manuscript.

References

Bott, M.H.P., 1985. Plate tectonic evolution of the Icelandic transverse ridge and adjacent regions. J. Geophys. Res., 90: 9953–9960.

Bott, M.H.P., 1987. The continental margin of central East Greenland in relation to North Atlantic plate tectonic evolution. J. Geol. Soc., London, 144: 561–568.

Doré, A.G., 1991. The structural foundation and evolution of Mesozoic seaways between Europe and the Arctic. Palaeogeogr., Palaeoclim., Palaeoecol., 87: 441–492.

Doré, A.G. and Lundin, E.R., 1996. Cenozoic compressional structures on the NE Atlantic margin: nature, origin and potential significance for hydrocarbon exploration. Petrol. Geosci., 2: 299–311.

Doré, A.G., Lundin, E.R., Birkeland, ø., Eliassen, P.E. and Jensen, L.N., 1997. The NE Atlantic margin: implications of late Mesozoic and Cenozoic events for hydrocarbon prospectivity. Petrol. Geosci., 3: 117–131.

Doré, A.G., Lundin, E.R., Jensen, L.N., Birkeland, Ø., Eliassen, P.E. and Fichler, C., 1999. Principal tectonic events in the evolution of the northwest European Atlantic margin. In: A.J. Fleet and S.A.R. Boldy (Editors), Petroleum geology of

Northwest Europe: Proceedings of the 5th Conference. Geological Society, London, pp. 41–61.

Eldholm, O. and Windisch, C.C., 1974. Sediment distribution in the Norwegian-Greenland Sea. Geol. Soc. Am. Bull., 85: 1661–1676.

Gairaud, H., Jacquart, G., Aubertin, F. and Beuzart, P., 1978. The Jan Mayen Ridge: synthesis of geological knowledge and new data. Oceanologica Acta, 1: 335–358.

Gudlaugsson, S.T., Gunnarsson, K., Sand, M. and Skogseid, J., 1988. Tectonic and volcanic events at the Jan Mayen Ridge microcontinent. Geol. Soc., London, Spec. Publ., 39: 85–93.

Ito, G., Shen, Y., Hirth, G. and Wolfe, C.J., 1999. Mantle flow, melting, and dehydration of the Iceland mantle plume. Earth Planet. Sci. Lett., 165: 81–96.

Jones, S.M., White, N.J. and Maclennan, J., 2002. V-shaped ridges around Iceland: implications for spatial and temporal patterns of mantle convection. Geochemistry, Geophysics, Geosystems, 3.

Jung, W.-Y. and Vogt, P.R., 1997. A gravity and magnetic anomaly study of the extinct Aegir Ridge, Norwegian Sea. J. Geophys. Res., 102 (B3): 5065–5089.

Kharin, G.S., Udintsev, G.B., Bogatikov, O.A., Dmitriev, J.I., Raschka, H., Kreuzer, H., Mohr, M., Harre, W. and Eckhardt, F.J., 1976. K/Ar ages of the basalts of the Norwegian-Greenland Sea DSDP Leg 38. In: M. Talwani, G.B. Udintsev et al. (Editors), Initial Reports of the Deep Sea Drilling Project, Washington, DC, pp. 755–759.

Knott, S.D., Burchell, M.T., Jolley, E.J. and Fraser, A.J., 1993. Mesozoic to Cenozoic plate reconstructions of the North Atlantic and hydrocarbon plays of the Atlantic margins. In: J.R. Parker (Editor), Petroleum geology of Northwest Europe: Proceedings of the 4th Conference. Geological Society, London, pp. 953–974.

Kodaira, S., Mjelde, R., Gunnarsson, K., Shiobara, H. and Shimamura, H., 1998. Structure of the Jan Mayen microcontinent and implications for its evolution. Geophys. J. Int., 132: 383–400.

Kuvaas, B. and Kodaira, S., 1997. The formation of the Jan Mayen microcontinent: the missing piece in the continental puzzle between the Møre-Vøring Basins and East Greenland. First Break, 15 (7): 239–247.

Larsen, H.C., 1988. A multiple and propagating rift model for the NE Atlantic. Geol. Soc., London, Spec. Publ., 39: 157–158.

Lawver, L.A. and Müller, R.D., 1994. Iceland hotspot track. Geology, 22: 311–314.

Lundin, E.R. and Doré, A.G., 1997. A tectonic model for the Norwegian passive margin with implications for the NE Atlantic: Early Cretaceous to break-up. J. Geol. Soc., London, 154 (3): 545–550.

Lundin, E. and Doré, A.G., 2002. Mid-Cenozoic post-breakup deformation in the 'passive' margins bordering the Norwegian-Greenland Sea. Mar. Petrol. Geol., 19: 79–93.

Maclennan, J., McKenzie, D. and Gronvöld, K., 2001. Plume-driven upwelling under central Iceland. Earth Planet. Sci. Lett., 194: 67–82.

Mosar, J., Lewis, G. and Torsvik, T.H., 2002. North Atlantic sea-floor spreading rates: implications for the Tertiary development of inversion structures of the Norwegian-Greenland Sea. J. Geol. Soc., 159: 503–515.

Mutter, J.C., Talwani, M. and Stoffa, P.L., 1982. Origin of seaward dipping reflectors in oceanic crust off the Norwegian margin by subaerial sea floor spreading. Geology, 10: 353–357.

Müller, R.D., Gaina, C., Roest, W.R. and Hansen, D.L., 2001. A recipe for microcontinent formation. Geology, 29 (3): 203–206.

Myrhe, A.M., Eldholm, O. and Sundvor, E., 1984. The Jan Mayen Ridge: present status. Polar Res., 2: 47–59.

Newman, R. and White, N.J., 1997. Rheology of the continental lithosphere inferred from sedimentary basins. Nature, 385: 621–624.

Nunns, A.G., 1983. The structure and evolution of the Jan Mayen Ridge and surrounding regions. AAPG Mem., 34: 193–208.

Pelton, C.D., 1985. Geophysical interpretation of the structure and evolution of the Jan Mayen Ridge. Report of the Institute of Oceanographic Sciences, 205: 1–38.

Roberts, D.G., Thompson, M., Mitchener, B., Hossack, J., Carmichael, S. and Bjørnseth, H.-M., 1999. Palaeozoic to Tertiary rift and basin dynamics: mid-Norway to the Bay of Biscay — a new context for hydrocarbon prospectivity in the deep water frontier. In: A.J. Fleet and S.A.R. Boldy (Editors), Petroleum geology of Northwest Europe: Proceedings of the 5th Conference. Geological Society, London, pp. 7–40.

Schaltegger, U., Amundsen, H., Jamtveit, B., Frank, M., Griffin, W.L., Gronvold, K., Tronnes, R. and Torsvik, T., 2002. Contamination of OIB by underlying ancient continental lithosphere: U-Pb and Hf isotopes in zircons question EM1 and EM2 mantle components. Geochimica et Cosmochimica Acta, 66 (15A): A673.

Scott, R.A., 2000. Mesozoic–Cenozoic evolution of East Greenland: implications of a reinterpreted continent–ocean boundary location. Polarforschung, 68: 83–91.

Scott, R.A. and Whitham, A.G., 2000. East Greenland geology plays role in Northwest European margin. Offshore, 60 (9): 148, 150, 184.

Skogseid, J. and Eldholm, O., 1987. Early Cenozoic Crust at the Norwegian Continental Margin and the Conjugate Jan Mayen Ridge. J. Geophys. Res., 92: 11471–11491.

Skogseid, J., Planke, S., Faleide, J.I., Pedersen, T., Eldholm, O. and Neverdal, F, 2000. NE Atlantic continental rifting and volcanic margin formation. Geol. Soc., London, Spec. Publ., 167: 295–326.

Smallwood, J.R. and White, R.S., 2002. Ridge-plume interactions in the North Atlantic and its influence on continental breakup and seafloor spreading. Geol. Soc., London, Spec. Publ., 197: 15–37.

Srivastava, S.P. and Tapscott, C.R., 1986. Plate kinematics of the North Atlantic. In: P.R. Vogt and B.E. Tucholke (Editors), The Western North Atlantic Region. Geological Society of America, Boulder, CO, pp. 379–404.

Talwani, M., Udintsev, G.B. and others (Editors), 1976. Initial Reports of the Deep Sea Drilling Project, NSF SP38, Washington, DC.

Talwani, M. and Eldholm, O., 1977. Evolution of the Norwegian-Greenland Sea. Geol. Soc. Am. Bull., 88: 969–999.

Vågnes, E., Gabrielsen, R.H. and Haremo, P., 1998. Late Cretaceous-Cenozoic intraplate contractional deformation at the Norwegian continental shelf: timing, magnitude and regional implications. Tectonophysics, 300: 29–46.

Verhoef, J., Roest, W.R., Macnab, R., Arkani-Hamed, J. and Members of the Project Team, 1996. Magnetic anomalies of the Arctic and North Atlantic oceans and adjacent land areas. Open File 3125a, Geological Survey of Canada, Dartmouth, NS.

Vink, G.E., Morgan, W.J. and Zhao, W.-L., 1984. Preferential rifting of continents: a source of displaced terranes. J. Geophys. Res., 89: 10072–10076.

Vogt, P.R., Johnson, G.L. and Kristjansson, L., 1980. Morphology and magnetic anomalies north of Iceland. J. Geophys. Res., 47: 67–80.

Weigel, W. et al., 1995. Investigations of the East Greenland continental margin between 70° and 72° N by deep seismic sounding and gravity studies. Mar. Geophys. Res., 17: 167–199.

Ziegler, P.A., 1988. Evolution of the Arctic-North Atlantic and the Western Tethys. AAPG Mem., 43: 1–198 + 30 plates.

Ziegler, P.A., 1989. Evolution of the North Atlantic – an overview. AAPG Mem., 46: 111–129.

The role of East Greenland as a source of sediment to the Vøring Basin during the Late Cretaceous

Andrew C. Morton, Andrew G. Whitham, C. Mark Fanning and Jonathan Claoué-Long

Provenance-sensitive heavy mineral criteria, mineral chemistry and detrital zircon age data show that there are strong links between Cretaceous sandstones in the Vøring Basin and East Greenland areas. There are marked differences in the age spectra of detrital zircons from wells along the eastern margin of the Vøring Basin (sandstone type K1) and those in the centre and west of the basin (sandstone type K2). The K1 sandstones have relatively simple zircon age spectra with largely Mid-Late Proterozoic zircons and a number of Caledonian age zircons. By contrast, the K2 sandstones have complex zircon age spectra, with Archaean, Early Proterozoic, Permo-Triassic and mid-Cretaceous zircons that are absent in the K1 sandstones. Some sandstones of Cenomanian and younger age from East Greenland share mineralogical features with the K2 sandstone type, having overlapping ranges of critical provenance sensitive parameters, such as RuZi, MZi and CZi, and similar types of detrital tourmalines and garnets. Detrital zircon age spectra from East Greenland samples include critical Archaean, Early Proterozoic and Permo-Triassic populations found in K2 sandstones. The zircon age data, therefore, provide support for sourcing of K2 sandstones from East Greenland. However, a source for the K2 sandstones to the east of the Caledonian front in Scandinavia cannot be ruled out, neither can the recycling of older sediment previously transferred across the rift.

Introduction

Cretaceous sandstones are an important hydrocarbon exploration target in the Vøring Basin, situated on the Mid-Norwegian continental shelf between 64° and 68°N and 2° and 10°E (Fig. 1). The Nordland Ridge, the Halten Terrace and the Dønna Terrace form the eastern margin of the basin. The position and nature of the western margin is a matter of conjecture since Tertiary basalts obscure the deep Cretaceous structure. Some workers place a continental fragment nearly 200 km wide between the Vøring Basin and NE Greenland (Brekke et al., 1999). However, analysis of recent magnetic data and new picks on the continent ocean boundaries of NW Europe and NE Greenland (Scott, 2000) indicate that there is insufficient space for a microcontinent of such a size and that NE Greenland lay immediately to the west of the Vøring Basin (Fig. 2).

One of the main areas of uncertainty concerning hydrocarbon exploration in the Vøring Basin is the question of reservoir presence. Since reservoir presence is partly governed by location of the clastic source regions, provenance studies play a key role in the exploration of the region. Potential sources of coarse clastic sediment lie both to the east, in Mid Norway, and to the west, in East Greenland. This chapter presents the results of an integrated heavy mineral and detrital zircon age study aimed at identifying the provenance of deep marine Late Cretaceous sandstones in the Vøring Basin, through characterisation of sandstones in the basin and comparing with onshore equivalents in East Greenland.

Geological background

Development of the Vøring Basin

The Vøring Basin was formed in the Late Jurassic–Early Cretaceous and is filled by 9–13 km of sediment (Brekke et al., 1999). The structural and sedimentological evolution of the Cretaceous fill of the basin is still poorly understood, owing to the limited number of wells drilled and the depth to the Base Cretaceous unconformity. Brekke (2000)

Onshore–Offshore Relationships on the North Atlantic Margin
edited by B. Wandas et al. NPF Special Publication 12, pp. 83–110,

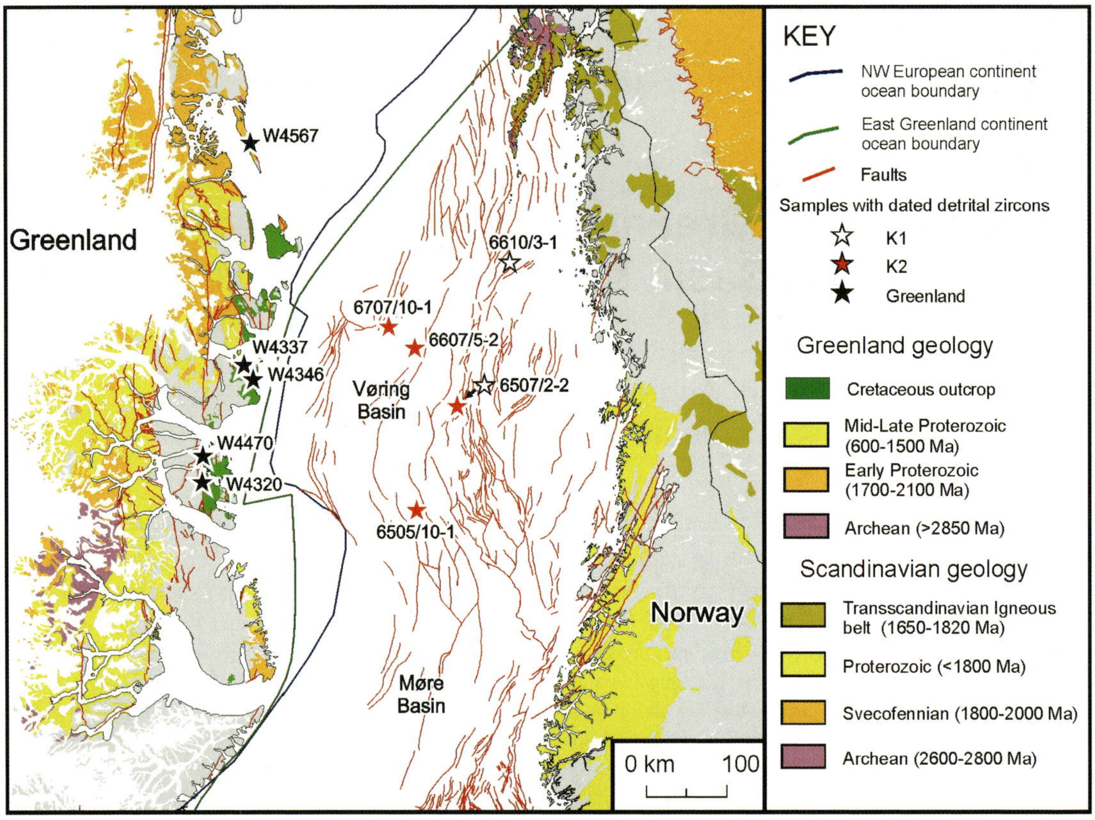

Fig. 1 Pre-rift reconstruction of the Norwegian Sea (pre-magnetic anomaly 24), showing distribution of basement terrains in Scandinavia and East Greenland and structural features on the Mid Norway continental margin.

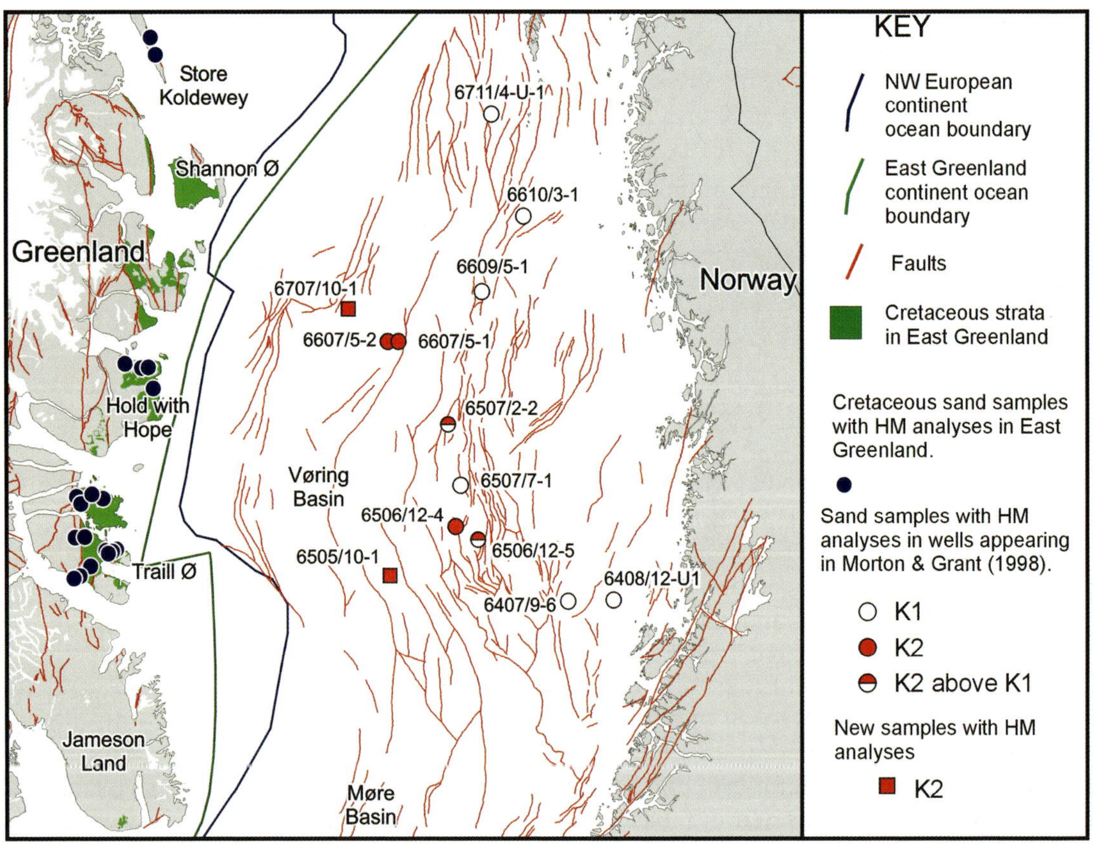

Fig. 2 Location of sample sites discussed in this chapter, showing distribution of K1 and K2 sandstones in wells in the Vøring Basin as described by Morton and Grant (1998).

divides the Cretaceous development of the basin into two phases: a pre-Cenomanian phase and a post-Cenomanian phase. The basin was initiated by rifting in the Late Jurassic–Early Cretaceous and subsequently underwent thermal subsidence until the Cenomanian, with a minor period of normal faulting in the Aptian. After the Cenomanian, subsidence was tectonically driven with intermittent phases of normal faulting and compression. A period of normal faulting is thought to have occurred in the Campanian, and a period of transpression in the Maastrichtian. This was followed later in the Maastrichtian by further rifting that preceded the onset of seafloor spreading in the Tertiary. Spencer et al. (1999) presume a simpler story for the evolution of the basin, with two main phases of rifting, the first in the Early Cretaceous and the second in the latest Cretaceous–Early Tertiary.

The evolution of the East Greenland margin during the Cretaceous is different to that proposed for the Vøring Basin. Following a major period of rifting in the Late Jurassic–Early Cretaceous, when there was substantial fault block re-organisation (Vischer, 1943; Maync, 1949; Surlyk, 1978), there were smaller rift events in the Late Barremian–Early Aptian and Mid-Albian (Whitham et al., 1999). These rift events created and maintained a sea floor dominated by tilted fault blocks. Subsequent to rifting in the Mid-Albian and prior to rifting in the latest Cretaceous–Early Tertiary, the margin underwent thermal subsidence. During this phase, the fault block topography was rapidly infilled, creating a continental margin with a shelf-break morphology. There is no evidence of rifting in this interval, as suggested by Brekke (2000), although it should be emphasised that the exposure of Santonian–Early Tertiary strata in NE Greenland is poor.

Late Cretaceous sand supply patterns

A heavy mineral study of Late Cretaceous sandstones from the Trøndelag Platform, Nordland Ridge and the deeper water Vøring Basin (Morton and Grant, 1998) showed the presence of two distinct mineralogical types (termed K1 and K2), indicating the involvement of at least two sediment source regions. A variety of mineralogical parameters were used by Morton and Grant (1998) to distinguish K1 and K2 sandstones, including provenance-sensitive ratios, tourmaline geochemistry and garnet geochemistry. The critical ratio parameters are rutile : zircon, monazite : zircon and chrome spinel : zircon. These ratios are expressed as index values (RuZi, MZi, CZi), following the notation of Morton and Hallsworth (1994, 1999). Binary plots comparing K1 and K2 sandstones are shown in Fig. 3. Tourmaline populations are characterised using the Al–Al50Fe(tot)50–Al50Mg50 ternary diagram (Fig. 4) devised by Henry and Guidotti (1985), which enables compositions to be tied back to specific source rock types. Garnet populations are compared using Fe + Mn–Mg–Ca ternary diagrams (Fig. 5).

As shown in Figs 3, 4 and 5, K1 sandstones have higher RuZi than K2, K1 having RuZi > 40 and K2 having RuZi < 40. K1 has generally higher MZi values (3–10) than K2 (mostly < 5), and has lower CZi values (< 1, compared with 2–10). The tourmalines in K1 sandstones were predominantly derived from Al-poor metasedimentary rocks, with

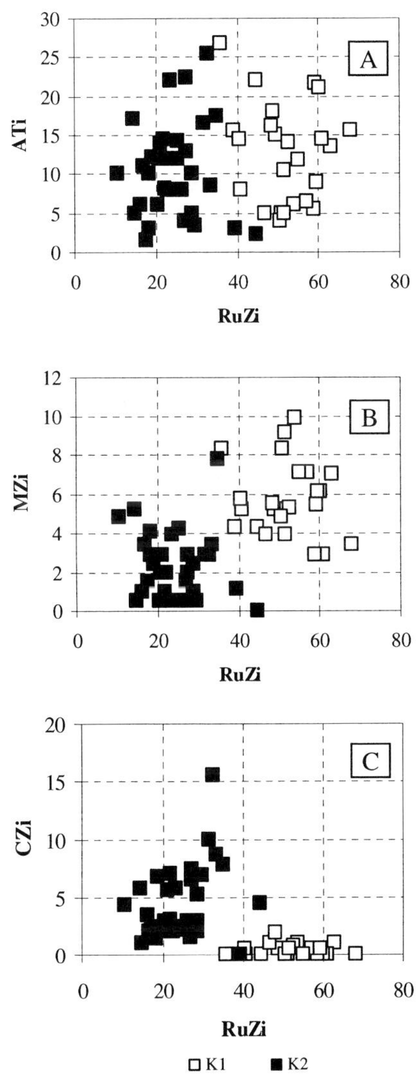

Fig. 3 Vøring Basin Cretaceous sandstones plotted on (A) ATi-RuZi, (B) MZi-RuZi and (C) CZi-RuZi crossplots, demonstrating key mineralogical contrasts between K1 and K2 samples. Data from Morton and Grant (1998).

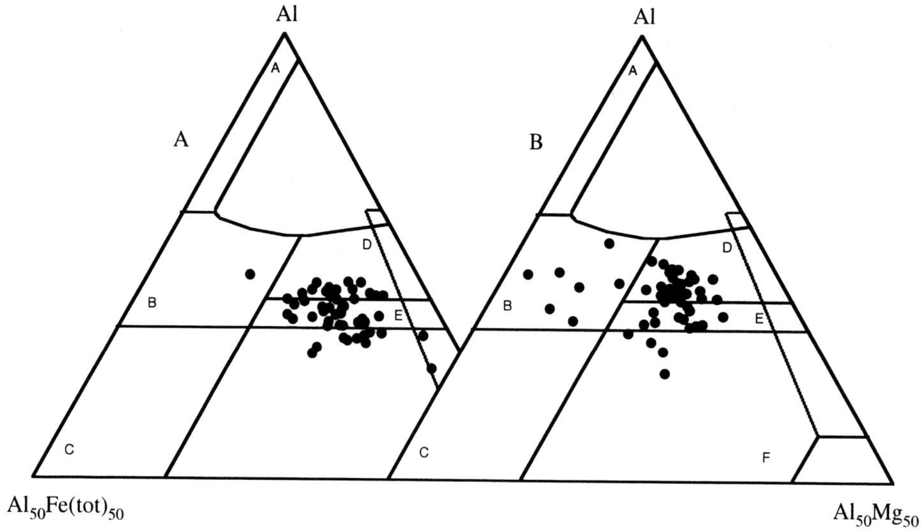

Fig. 4 Typical K1 and K2 tourmaline assemblages plotted on the provenance-discriminant ternary diagram devised by Henry and Guidotti (1985). A = 6507/2-2, 3338.5 m (K1). B = 6507/2-2, 2830.1 m (K2). Field A – Li-rich granitoids, pegmatites and aplites. Field B – Li-poor granitoids, pegmatites and aplites. Field C – hydrothermally-altered granitic rocks. Field D – Aluminous metapelites and metapsammites. Field E – Al-poor metapelites and metapsammites. Field F – Fe^{3+}-rich quartz-tourmaline rocks, calc-silicates and metapelites. Field G – Low-Ca ultramafics. Field H – metacarbonates and metapyroxenites.

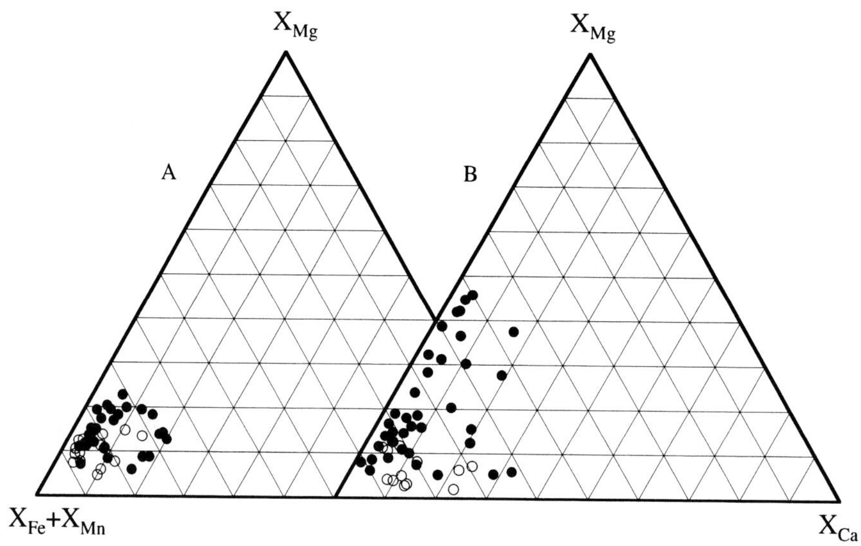

Fig. 5 Ternary plots of typical K1 and K2 garnet assemblages. A = 6610/3-1, 2523.0 m (K1). B = 6707/10-1, 3134.4 m (K2) X_{Fe}, X_{Mg}, X_{Ca}, X_{Mn} = molecular values of Fe, Mg, Ca and Mn respectively, calculated on the basis of 24 oxygens, and normalised to total Fe + Mg + Ca + Mn, as recommended by Droop and Harte (1995). All Fe calculated as Fe^{2+}. ● – garnets with $X_{Mn} < 5\%$, ○ – garnets with $X_{Mn} < 5\%$.

a subsidiary number from Al-rich metasediments and a very small number from granites. By contrast, tourmalines in K2 sandstones were derived mainly from Al-rich metasedimentary rocks, with subsidiary contributions from Al-poor metasediments and granites. Garnet assemblages in both K1 and K2 sandstones are dominated by low-Mg types with variable Ca contents, typical of amphibolite-facies metasediments. However, K2 sandstones contain a small additional group of garnets with high Mg and low Ca, typical of high-grade (granulite-facies) metasediments or charnockites.

Sandstone type K1 has been identified in the Cenomanian to Campanian of wells located on the Trøndelag Platform and Nordland Ridge, but has not been found further offshore in the Vøring Basin. Preliminary U–Pb dating of detrital zircons from one sample with K1 mineralogy indicated that the source could be readily tied back to the adjacent parts of Scandinavia (Morton and Grant, 1998). This conclusion is consistent with the sedimentological evidence, which indicates deposition on an unstable slope by debris flows and slumps, with minor reworking by bottom currents (Shanmugam et al., 1994; Morton and Grant, 1998). By contrast, sandstone type K2 has been identified in the Turonian to Maastrichtian from wells in more basinal locations in the Norwegian Sea, and occurs mainly as basin floor lobe deposits (Morton and Grant, 1998; Kittilsen et al., 1999). Detrital zircon age data from a K2 sandstone sample showed the involvement of crustal components that are not present in the adjacent parts of Scandinavia, but are present in East Greenland, which forms the western margin of the basin (Morton and Grant, 1998). This observation provided support for possible sourcing of sediment from the conjugate East Greenland margin of the Vøring Basin, a concept also supported by seismic facies mapping (Vergara et al., 2001). Such a provenance model would predict increasing net:gross ratios and increasing thickness of sandstone units towards the west, and would explain the c. 1 km thickness of Late Cretaceous sandstones with K2 mineralogical affinities and good porosity and permeability characteristics in well 6707/10-1 on the Nyk High (Kittilsen et al., 1999).

This chapter presents U–Pb detrital zircon age data from sandstones with K1 and K2 mineralogy in the Vøring Basin, and from Cretaceous sandstones from the sector of NE Greenland that formed the western margin of the Vøring Basin. The main aims of the chapter are to confirm whether the zircon age data support the mineralogical discrimination of the two sand types (as suggested by Morton and Grant, 1998), to identify the geochronological characteristics of the source areas for the two sand types, and to place constraints on the entry points of sediment to the Norwegian Sea during the Cretaceous.

Zircon age dating

Detrital zircon age data have been acquired from 9 new samples, 4 from offshore Mid Norway and 5 from East Greenland, supplementing data from the 2 samples described in brief by Morton and Grant (1998). U–Pb dating of detrital zircon was undertaken, using the sensitive high-resolution ion microprobe (SHRIMP) at the Australian National University in Canberra. The procedures employed for zircon U–Pb dating followed Claoué-Long et al. (1995). Zircons obtained from standard density and magnetic separation techniques were mounted in epoxy. Their surfaces were polished until the grains were sectioned in half, thus exposing all zones to the probe. Under vacuum, a beam of oxygen ions was focused to approximately 30 μm diameter and targeted at the desired zone of the crystal. Particles accelerated from the resulting crater were passed into a high-resolution, high-sensitivity mass spectrometer for counting of the U and Pb ions of interest, from which U–Pb ages for each probe site were calculated. The number of scans through the mass stations was reduced to four to achieve a rapid throughput of data at the expense of some counting precision per analysis. Subjectivity in zircon dating was avoided by analysing all zircons encountered during the traverse of the mount, even if the grain appeared to be partially metamict. At least 60 zircons were analysed from each sample. The zircon age spectra are shown as relative probability plots, generated by the Isoplot program (Ludwig, 1999). Zircons with isotopic compositions that deviate from concordia by an arbitrary threshold of 20% are not plotted on the relative probability plots. The full data sets are given in Tables 1–11.

Vøring Basin

K1 sandstones

The ages of detrital zircons from two sandstones with K1 mineralogy have been measured using SHRIMP, one from well 6507/2-2, 3281.0 m (Cenomanian) and one from well 6610/3-1, 2300.0 m (Campanian), data from the latter sample having been discussed in brief by Morton and Grant (1998). Most of the zircon isotopic compositions lie on or close to the concordia curve, but there is evidence of significant Pb loss in some grains whose compositions therefore only record minimum ages. In the Precambrian, the Pb loss is readily detected by departure from the concordia curve. However, leakage among Phanerozoic grains is more cryptic because the Pb loss trajectory is sub-parallel to the concordia curve. The sample from well 6610/3-1 is believed to show this phenomenon, since it contains a group of Phanerozoic zircons that lie above the concordia curve, along or close to the predicted Pb loss trajectory for zircons with 450 Ma crystallisation ages (Fig. 7). This possibility is given support by the presence of a cluster of concordant zircons at c. 450 Ma. Consequently, the apparent ages of the zircons deviating from the concordia curve are not included in the relative probability plot (Fig. 6). Two grains with apparently concordant ages younger than 450 Ma could represent a minor Palaeozoic and Mesozoic component to the detritus, but it is possible that these are also Caledonian-age grains that have leaked Pb.

Table 1 Summary of SHRIMP U–Pb zircon results for 6505/10-1, 3711.6 m. f^{206}% denotes the percentage of ^{206}Pb that is common Pb. For zircons >800 Ma, correction for common Pb made using the measured $^{204}Pb/^{206}Pb$ ratio. For zircons <800 Ma, correction for common Pb made using the measured Th/U ratio and calculated $^{208}Pb^*$ content. 100% denotes a concordant analysis.

Grain.spot	U (ppm)	Th (ppm)	f^{206}%	$^{206}Pb/^{238}U$	$\pm 1\sigma$	$^{207}Pb/^{235}U$	$\pm 1\sigma$	$^{207}Pb/^{206}Pb$	$\pm 1\sigma$	Preferred age	±	Conc.
23.1	179	81	1.09	0.0128	0.0006	0.096	0.014	0.0547	0.0072	82	4	–
47.1	100	46	0.62	0.0139	0.0005	0.106	0.013	0.0552	0.0063	89	3	–
28.1	141	78	1.39	0.0143	0.0007	0.099	0.026	0.0499	0.0128	92	4	–
60.1	102	62	5.43	0.0144	0.0009	0.122	0.059	0.0613	0.0290	92	5	–
2.1	179	105	6.04	0.0145	0.0004	0.106	0.024	0.0529	0.0115	93	2	–
21.1	82	43	1.36	0.0151	0.0007	0.107	0.035	0.0514	0.0166	96	4	–
30.1	165	244	<0.01	0.0233	0.0009	0.186	0.055	0.0579	0.0165	149	6	–
48.1	308	167	<0.01	0.0306	0.0007	0.211	0.015	0.0500	0.0032	194	4	–
46.1	295	238	<0.01	0.0318	0.0006	0.241	0.015	0.0551	0.0031	202	4	–
10.1	111	113	<0.01	0.0402	0.0013	0.345	0.065	0.0622	0.0114	254	8	–
41.1	186	123	<0.01	0.0436	0.0010	0.341	0.021	0.0568	0.0032	275	6	–
27.1	148	96	<0.01	0.0474	0.0026	0.391	0.045	0.0600	0.0058	298	16	–
19.1	2303	642	<0.01	0.0601	0.0007	0.523	0.009	0.0631	0.0008	376	4	–
12.1	211	111	0.22	0.0629	0.0012	0.467	0.026	0.0539	0.0027	393	7	–
25.1	2055	841	<0.01	0.0667	0.0010	0.509	0.012	0.0553	0.0009	417	6	–
40.1	126	133	0.27	0.0668	0.0015	0.508	0.041	0.0553	0.0041	417	9	–
15.1	824	144	0.19	0.0686	0.0009	0.567	0.015	0.0600	0.0013	427	6	–
55.1	157	110	0.66	0.0700	0.0013	0.505	0.033	0.0523	0.0031	436	8	–
32.1	139	126	0.71	0.0719	0.0019	0.469	0.056	0.0473	0.0053	447	12	–
56.1	568	392	0.13	0.0741	0.0010	0.569	0.020	0.0557	0.0018	461	6	–
57.1	683	217	<0.01	0.0757	0.0011	0.695	0.017	0.0666	0.0012	470	6	–
18.1	157	109	0.27	0.0758	0.0017	0.532	0.063	0.0509	0.0058	471	10	–
42.1	141	41	−1.16	0.0766	0.0014	0.724	0.023	0.0686	0.0017	476	8	–
36.1	140	86	<0.01	0.0780	0.0020	0.658	0.037	0.0612	0.0029	484	12	–
53.1	342	193	<0.01	0.0823	0.0011	0.961	0.028	0.0847	0.0020	510	7	39
45.1	96	85	<0.01	0.0864	0.0020	0.746	0.051	0.0626	0.0038	534	12	77
31.1	97	68	<0.01	0.1029	0.0032	0.943	0.095	0.0665	0.0061	631	19	77
20.1	240	167	<0.01	0.1188	0.0025	1.233	0.054	0.0753	0.0027	724	14	67
58.1	402	517	2.66	0.1327	0.0018	1.985	0.061	0.1085	0.0028	803	11	45
13.1	83	92	<0.01	0.1591	0.0053	1.664	0.133	0.0759	0.0052	952	30	87
37.1	210	22	<0.01	0.1613	0.0030	1.546	0.038	0.0695	0.0010	964	17	106
9.1	20	12	0.56	0.1676	0.0107	1.767	0.221	0.0765	0.0076	999	59	90
33.1	145	161	0.15	0.1986	0.0039	2.114	0.057	0.0772	0.0012	1126	32	104
16.1	582	161	<0.01	0.1970	0.0039	2.160	0.059	0.0795	0.0013	1185	33	98
39.1	249	67	0.26	0.2140	0.0029	2.389	0.047	0.0809	0.0011	1220	26	103
54.1	216	100	0.04	0.2235	0.0037	2.720	0.054	0.0883	0.0008	1389	18	94
29.1	161	177	0.19	0.2395	0.0049	3.026	0.097	0.0917	0.0020	1460	43	95
1.1	470	416	0.12	0.2322	0.0043	2.937	0.065	0.0917	0.0010	1462	20	92
11.1	204	140	0.32	0.2395	0.0044	3.118	0.081	0.0944	0.0016	1517	31	91
7.1	131	59	0.20	0.2852	0.0067	3.839	0.108	0.0976	0.0012	1580	24	102
44.1	272	148	0.46	0.2409	0.0043	3.257	0.079	0.0981	0.0014	1588	28	88
14.1	383	445	0.16	0.2681	0.0038	3.684	0.070	0.0996	0.0011	1617	21	95
26.1	183	95	0.14	0.3220	0.0077	4.647	0.133	0.1047	0.0014	1709	24	105
5.1	581	353	2.74	0.1941	0.0034	2.836	0.075	0.1060	0.0019	1731	33	66
52.1	171	55	0.15	0.3000	0.0055	4.417	0.110	0.1068	0.0016	1745	27	97
17.1	196	55	<0.01	0.3347	0.0089	5.148	0.201	0.1116	0.0029	1825	47	102
38.1	514	114	0.10	0.3173	0.0047	4.897	0.080	0.1119	0.0006	1831	10	97
34.1	511	16	<0.01	0.3507	0.0063	5.598	0.122	0.1158	0.0012	1892	18	102
3.1	511	36	<0.01	0.3539	0.0045	5.688	0.081	0.1166	0.0006	1905	9	103
6.1	212	117	0.12	0.3332	0.0053	5.363	0.103	0.1167	0.0010	1907	16	97
24.1	28	10	0.14	0.3295	0.0142	5.333	0.299	0.1174	0.0036	1917	56	96
50.1	585	58	0.01	0.3699	0.0051	5.996	0.091	0.1176	0.0006	1920	9	106
51.1	517	173	0.02	0.3779	0.0058	6.809	0.111	0.1307	0.0005	2107	7	98
59.1	39	20	<0.01	0.3532	0.0154	6.485	0.489	0.1332	0.0075	2140	102	91
35.1	370	202	0.17	0.4252	0.0124	9.775	0.306	0.1667	0.0014	2525	14	91
43.1	112	29	<0.01	0.5498	0.0136	14.030	0.393	0.1851	0.0019	2699	17	105
22.1	252	136	0.06	0.4977	0.0152	12.749	0.404	0.1858	0.0011	2705	9	96
49.1	39	35	0.04	0.5336	0.0213	13.809	0.590	0.1877	0.0021	2722	18	101
8.1	242	98	0.02	0.5179	0.0096	13.753	0.274	0.1926	0.0010	2764	8	97
4.1	85	45	0.15	0.5406	0.0111	14.544	0.332	0.1951	0.0015	2786	13	100

Table 2 Summary of SHRIMP U–Pb zircon results for 6507/2-2, 2830.1 m. f^{206}% denotes the percentage of ^{206}Pb that is common Pb. For zircons > 800 Ma, correction for common Pb made using the measured $^{204}Pb/^{206}Pb$ ratio. For zircons < 800 Ma, correction for common Pb made using the measured Th/U ratio and calculated $^{208}Pb^*$ content. 100% denotes a concordant analysis.

Grain.spot	U (ppm)	Th (ppm)	f^{206}%	$^{206}Pb/^{238}U$	±1σ	$^{207}Pb/^{235}U$	±1σ	$^{207}Pb/^{206}Pb$	±1σ	Preferred age	±	Conc.
51.1	119	53	5.82	0.0134	0.0006	0.041	0.039	0.0221	0.0209	86	4	–
58.1	147	45	2.72	0.0140	0.0004	0.100	0.019	0.0515	0.0097	90	3	–
53.1	325	167	0.86	0.0141	0.0004	0.111	0.011	0.0573	0.0051	90	3	–
7.1	300	185	0.68	0.0150	0.0005	0.094	0.011	0.0458	0.0048	96	3	–
26.1	138	69	<0.01	0.0150	0.0005	0.123	0.009	0.0593	0.0035	96	3	–
50.1	135	45	0.65	0.0156	0.0005	0.110	0.012	0.0512	0.0050	100	3	–
6.1	79	35	1.49	0.0159	0.0008	0.114	0.024	0.0522	0.0102	102	5	–
41.1	120	69	1.91	0.0160	0.0008	0.106	0.024	0.0482	0.0105	102	5	–
40.1	271	133	16.07	0.0163	0.0006	0.125	0.032	0.0558	0.0139	104	4	–
55.1	514	394	0.11	0.0390	0.0008	0.291	0.028	0.0541	0.0050	247	5	–
46.1	247	73	2.66	0.0411	0.0010	0.306	0.019	0.0541	0.0029	260	6	–
30.1	286	108	0.85	0.0453	0.0011	0.303	0.018	0.0486	0.0025	285	7	–
43.1	412	416	0.02	0.0479	0.0007	0.339	0.011	0.0514	0.0014	302	4	–
61.1	128	105	1.79	0.0499	0.0012	0.304	0.039	0.0442	0.0055	314	7	–
44.1	194	189	<0.01	0.0576	0.0014	0.446	0.029	0.0562	0.0032	361	9	–
1.1	943	4	0.04	0.0643	0.0009	0.494	0.009	0.0557	0.0006	402	5	–
5.1	169	106	<0.01	0.0706	0.0019	0.567	0.022	0.0583	0.0015	440	12	–
9.1	874	1471	4.65	0.0715	0.0010	0.563	0.034	0.0572	0.0033	445	6	–
23.1	89	57	0.55	0.0747	0.0028	0.611	0.060	0.0593	0.0052	465	17	–
24.1	151	6	<0.01	0.1016	0.0024	1.339	0.055	0.0956	0.0030	624	14	41
36.1	697	74	1.36	0.1125	0.0016	1.267	0.048	0.0817	0.0028	687	9	56
60.1	78	102	0.57	0.1024	0.0039	0.896	0.054	0.0635	0.0027	725	93	87
22.1	321	41	0.08	0.1718	0.0027	1.672	0.038	0.0706	0.0010	946	29	108
32.1	74	75	0.45	0.1826	0.0044	1.727	0.148	0.0686	0.0054	1081	24	122
15.1	100	41	<0.01	0.2387	0.0074	2.748	0.097	0.0835	0.0012	1281	28	108
8.1	273	108	0.11	0.2707	0.0046	3.653	0.078	0.0979	0.0011	1584	20	98
45.1	382	53	0.02	0.2980	0.0042	4.119	0.068	0.1003	0.0007	1629	14	103
19.1	368	120	0.12	0.2844	0.0041	4.036	0.075	0.1029	0.0011	1677	19	96
4.1	339	77	0.77	0.1951	0.0027	2.771	0.082	0.1030	0.0025	1680	46	68
21.1	341	171	0.13	0.2788	0.0037	4.013	0.063	0.1044	0.0007	1704	13	93
57.1	57	54	<0.01	0.2925	0.0105	4.216	0.220	0.1046	0.0036	1706	64	97
3.1	85	33	0.36	0.3206	0.0077	4.720	0.181	0.1068	0.0029	1745	51	103
20.1	131	31	<0.01	0.2220	0.0040	3.282	0.099	0.1072	0.0024	1753	41	74
49.1	127	13	0.11	0.2716	0.0071	4.017	0.124	0.1073	0.0015	1754	25	88
28.1	194	96	0.20	0.2782	0.0053	4.155	0.112	0.1083	0.0018	1771	31	89
52.1	656	226	0.13	0.2998	0.0044	4.542	0.080	0.1099	0.0009	1798	15	94
18.1	247	107	0.01	0.3448	0.0049	5.332	0.088	0.1122	0.0008	1835	13	104
17.1	58	37	0.30	0.3735	0.0138	5.781	0.269	0.1123	0.0027	1836	44	111
34.1	1657	102	0.05	0.2039	0.0025	3.164	0.043	0.1126	0.0005	1841	8	65
59.1	412	1	0.06	0.3070	0.0091	4.786	0.154	0.1131	0.0011	1849	17	93
10.1	884	55	0.04	0.2776	0.0038	4.399	0.074	0.1149	0.0010	1879	15	84
14.1	723	230	0.03	0.3192	0.0041	5.072	0.073	0.1153	0.0006	1884	10	95
47.1	606	41	<0.01	0.3278	0.0054	5.214	0.094	0.1154	0.0007	1886	10	97
25.1	282	60	0.10	0.2557	0.0041	4.082	0.081	0.1158	0.0011	1892	17	78
16.1	216	170	<0.01	0.3305	0.0056	5.321	0.113	0.1168	0.0013	1907	20	97
39.1	448	155	0.05	0.3235	0.0043	5.248	0.089	0.1177	0.0011	1921	17	94
42.1	106	36	<0.01	0.3392	0.0068	5.512	0.158	0.1179	0.0021	1924	33	98
11.1	49	9	0.27	0.3609	0.0103	5.909	0.208	0.1188	0.0021	1938	32	103
2.1	78	7	0.23	0.3664	0.0092	6.007	0.200	0.1189	0.0023	1940	34	104
54.1	418	19	0.12	0.3165	0.0059	5.252	0.109	0.1203	0.0009	1961	13	90
13.1	336	52	0.04	0.3518	0.0053	5.864	0.097	0.1209	0.0007	1970	10	99
29.1	345	134	0.09	0.2893	0.0056	4.833	0.102	0.1212	0.0008	1974	11	83
56.1	128	126	0.12	0.3651	0.0091	6.287	0.206	0.1249	0.0023	2027	33	99
37.1	253	136	2.64	0.2640	0.0055	4.653	0.204	0.1278	0.0046	2068	65	73
35.1	81	60	8.13	0.3921	0.0130	6.018	0.768	0.1113	0.0133	2132	61	117
38.1	388	253	0.17	0.3957	0.0059	7.393	0.136	0.1355	0.0012	2171	16	99
27.1	243	269	0.04	0.4071	0.0092	7.735	0.211	0.1378	0.0018	2200	23	100
31.1	67	135	0.01	0.4237	0.0098	8.374	0.236	0.1433	0.0019	2268	23	100
33.1	86	163	<0.01	0.5070	0.0160	11.521	0.414	0.1648	0.0023	2506	24	106
48.1	222	81	<0.01	0.4780	0.0076	10.963	0.211	0.1663	0.0015	2521	15	100
12.1	219	101	0.41	0.4980	0.0102	12.170	0.316	0.1773	0.0024	2627	23	99

Table 3 Summary of SHRIMP U–Pb zircon results for 6507/2-2, 3281.0 m. f^{206}% denotes the percentage of ^{206}Pb that is common Pb. For zircons >800 Ma, correction for common Pb made using the measured $^{204}Pb/^{206}Pb$ ratio. For zircons <800 Ma, correction for common Pb made using the measured Th/U ratio and calculated $^{208}Pb^*$ content. 100% denotes a concordant analysis.

Grain.spot	U (ppm)	Th (ppm)	f^{206}%	$^{206}Pb/^{238}U$	$\pm 1\sigma$	$^{207}Pb/^{235}U$	$\pm 1\sigma$	$^{207}Pb/^{206}Pb$	$\pm 1\sigma$	Preferred age	±	Conc.
48.1	1026	220	0.65	0.0622	0.0008	0.487	0.014	0.0569	0.0013	389	5	–
10.1	214	193	0.81	0.0698	0.0014	0.594	0.055	0.0617	0.0054	435	8	–
27.1	700	4	0.32	0.0778	0.0011	0.624	0.014	0.0582	0.0009	483	7	–
59.1	90	7	0.70	0.1563	0.0038	1.478	0.112	0.0686	0.0047	936	21	106
31.1	798	146	1.48	0.2209	0.0029	2.905	0.077	0.0954	0.0020	1535	41	84
41.1	569	165	0.19	0.1757	0.0022	2.375	0.039	0.0981	0.0009	1588	17	66
5.1	339	130	0.36	0.1624	0.0023	2.226	0.046	0.0994	0.0013	1613	25	60
37.1	812	108	0.24	0.2225	0.0032	3.051	0.054	0.0995	0.0008	1614	16	80
49.1	42	10	2.21	0.2685	0.0113	3.704	0.274	0.1001	0.0056	1625	107	94
17.1	452	105	0.01	0.2393	0.0047	3.305	0.071	0.1002	0.0007	1628	13	85
55.1	374	202	0.12	0.2616	0.0047	3.645	0.078	0.1011	0.0010	1644	18	91
34.1	345	63	0.43	0.1790	0.0033	2.503	0.064	0.1014	0.0016	1650	29	64
46.1	415	41	0.93	0.2257	0.0042	3.157	0.098	0.1014	0.0023	1651	42	80
53.1	479	219	0.42	0.2047	0.0036	2.872	0.083	0.1018	0.0021	1656	40	73
40.1	259	50	1.42	0.2845	0.0051	4.028	0.136	0.1027	0.0027	1673	50	97
36.1	224	97	0.24	0.2630	0.0057	3.740	0.104	0.1031	0.0016	1681	28	90
45.1	395	141	1.92	0.1900	0.0032	2.703	0.100	0.1032	0.0032	1682	58	67
14.1	1032	323	0.02	0.2249	0.0027	3.206	0.043	0.1034	0.0006	1686	10	78
58.1	820	185	0.14	0.2011	0.0026	2.891	0.051	0.1043	0.0011	1701	20	69
43.1	420	86	0.50	0.2233	0.0033	3.211	0.065	0.1043	0.0013	1702	23	76
9.1	620	206	0.14	0.2047	0.0027	2.948	0.047	0.1044	0.0008	1705	15	70
8.1	439	63	0.04	0.2902	0.0040	4.190	0.068	0.1047	0.0007	1709	13	96
3.1	60	34	0.29	0.3011	0.0086	4.370	0.168	0.1053	0.0024	1719	42	99
22.1	198	49	0.16	0.2946	0.0065	4.284	0.113	0.1055	0.0013	1722	23	97
24.1	171	81	0.13	0.3284	0.0058	4.844	0.106	0.1070	0.0012	1749	20	105
30.1	400	96	0.58	0.2501	0.0061	3.695	0.105	0.1072	0.0013	1752	22	82
56.1	29	13	1.17	0.3317	0.0222	4.903	0.493	0.1072	0.0072	1752	128	105
20.1	213	127	0.91	0.2886	0.0055	4.266	0.135	0.1072	0.0025	1753	42	93
60.1	180	143	0.46	0.3131	0.0061	4.630	0.165	0.1072	0.0030	1753	51	100
4.1	226	73	0.10	0.3256	0.0051	4.817	0.109	0.1073	0.0016	1754	27	104
25.1	229	85	0.44	0.2610	0.0045	3.862	0.086	0.1073	0.0013	1754	23	85
50.1	327	85	0.22	0.2772	0.0052	4.104	0.096	0.1074	0.0013	1756	22	90
61.1	321	83	0.92	0.2892	0.0049	4.283	0.103	0.1074	0.0016	1756	28	93
33.1	439	125	0.10	0.2784	0.0046	4.124	0.078	0.1074	0.0008	1757	14	90
6.1	176	41	0.11	0.2866	0.0052	4.249	0.113	0.1075	0.0019	1758	32	92
7.1	269	108	0.06	0.2909	0.0047	4.314	0.094	0.1076	0.0014	1759	23	94
21.1	348	102	0.23	0.2368	0.0034	3.512	0.063	0.1076	0.0010	1759	17	78
38.1	130	31	3.04	0.2158	0.0064	3.215	0.269	0.1081	0.0081	1767	143	71
15.1	99	38	0.10	0.3379	0.0065	5.038	0.119	0.1082	0.0012	1769	21	106
32.1	115	45	0.40	0.3246	0.0106	4.845	0.196	0.1083	0.0022	1771	37	102
52.1	537	146	<0.01	0.2467	0.0035	3.687	0.083	0.1084	0.0018	1773	30	80
44.1	206	82	0.39	0.2967	0.0049	4.444	0.106	0.1086	0.0017	1777	28	94
29.1	269	169	0.40	0.3179	0.0066	4.766	0.114	0.1087	0.0011	1778	18	100
2.1	224	45	0.11	0.3323	0.0069	4.985	0.118	0.1088	0.0010	1779	18	104
26.1	182	102	0.28	0.3140	0.0066	4.714	0.135	0.1089	0.0019	1781	31	99
35.1	237	113	0.36	0.3280	0.0061	4.930	0.113	0.1090	0.0012	1783	20	103
51.1	124	43	0.59	0.2830	0.0060	4.258	0.186	0.1091	0.0039	1785	66	90
19.1	350	84	0.08	0.3108	0.0046	4.680	0.079	0.1092	0.0007	1787	12	98
23.1	197	167	0.13	0.3132	0.0066	4.717	0.120	0.1092	0.0013	1787	22	98
54.1	261	91	1.07	0.2862	0.0050	4.311	0.116	0.1092	0.0020	1787	34	91
57.1	266	119	0.10	0.2791	0.0046	4.212	0.086	0.1095	0.0011	1790	18	89
18.1	162	42	0.07	0.3022	0.0050	4.566	0.090	0.1096	0.0010	1793	17	95
1.1	221	82	<0.01	0.3031	0.0051	4.592	0.092	0.1099	0.0010	1798	17	95
11.1	251	180	0.04	0.3187	0.0069	4.841	0.115	0.1102	0.0008	1802	13	99
42.1	196	72	0.27	0.3012	0.0064	4.574	0.123	0.1102	0.0015	1802	25	94
13.1	279	83	0.19	0.2936	0.0043	4.459	0.079	0.1102	0.0009	1802	16	92
16.1	201	176	0.02	0.3089	0.0064	4.697	0.111	0.1103	0.0010	1804	17	96
39.1	207	56	0.45	0.2994	0.0060	4.557	0.114	0.1104	0.0014	1806	24	94
12.1	138	27	<0.01	0.3121	0.0064	4.763	0.120	0.1107	0.0014	1811	22	97
47.1	177	69	0.55	0.3138	0.0076	4.794	0.155	0.1108	0.0021	1812	34	97
28.1	272	79	1.16	0.1914	0.0141	3.393	0.450	0.1285	0.0131	2078	191	54

Table 4 Summary of SHRIMP U–Pb zircon results for 6607/5-2, 4172.0 m. f^{206}% denotes the percentage of ^{206}Pb that is common Pb. For zircons >800 Ma, correction for common Pb made using the measured $^{204}Pb/^{206}Pb$ ratio. For zircons <800 Ma, correction for common Pb made using the measured Th/U ratio and calculated $^{208}Pb^{*}$ content. 100% denotes a concordant analysis.

Grain.spot	U (ppm)	Th (ppm)	f^{206}%	$^{206}Pb/^{238}U$	$\pm 1\sigma$	$^{207}Pb/^{235}U$	$\pm 1\sigma$	$^{207}Pb/^{206}Pb$	$\pm 1\sigma$	Preferred age	±	Conc.
74.1	183	101	5.99	0.0139	0.0006	0.11	0.03	0.0580	0.0147	89	4	–
69.1	527	342	3.13	0.0148	0.0006	0.11	0.01	0.0524	0.0046	95	4	–
125.1	75	45	9.12	0.0151	0.0005	0.09	0.03	0.0447	0.0140	97	3	–
17.1	6018	1376	0.36	0.0158	0.0006	0.10	0.00	0.0474	0.0007	101	4	–
104.1	53	33	7.16	0.0156	0.0005	0.14	0.04	0.0652	0.0166	103	3	–
116.1	2894	1274	19.42	0.0189	0.0004	0.10	0.03	0.0392	0.0100	121	3	–
7.1	1365	452	0.51	0.0281	0.0011	0.19	0.01	0.0480	0.0013	179	7	–
48.1	1280	1581	−0.01	0.0352	0.0014	0.28	0.01	0.0571	0.0013	223	9	–
58.1	600	391	0.91	0.0365	0.0015	0.27	0.01	0.0541	0.0014	231	9	–
66.1	77	57	8.62	0.0368	0.0017	0.21	0.09	0.0421	0.0173	233	10	–
76.1	949	586	0.25	0.0371	0.0015	0.27	0.01	0.0537	0.0012	235	9	–
67.1	84	159	3.89	0.0410	0.0017	0.33	0.05	0.0575	0.0080	259	11	–
33.1	68	44	7.27	0.0425	0.0017	0.24	0.03	0.0412	0.0052	268	11	–
38.1	722	577	0.37	0.0427	0.0017	0.32	0.02	0.0539	0.0013	270	11	–
2.1	253	216	0.13	0.0464	0.0019	0.36	0.02	0.0559	0.0020	292	12	–
19.1	102	59	2.82	0.0489	0.0020	0.37	0.03	0.0552	0.0030	308	12	–
83.1	217	219	1.26	0.0501	0.0020	0.36	0.03	0.0521	0.0027	315	13	–
22.1	327	163	1.3	0.0506	0.0021	0.38	0.02	0.0549	0.0017	318	13	–
9.1	205	307	1.22	0.0562	0.0023	0.42	0.03	0.0544	0.0033	353	14	–
108.1	96	77	1.71	0.0569	0.0012	0.41	0.04	0.0526	0.0054	357	7	–
139.1	547	281	0.68	0.0579	0.0012	0.43	0.02	0.0545	0.0016	363	7	–
105.1	36	3	5.07	0.0617	0.0013	0.43	0.05	0.0502	0.0062	386	8	–
151.1	235	111	0.93	0.0624	0.0013	0.47	0.02	0.0549	0.0022	390	8	–
138.1	429	240	16.94	0.0627	0.0015	0.46	0.10	0.0536	0.0109	392	9	–
141.1	1382	546	−0.04	0.0639	0.0013	0.49	0.01	0.0561	0.0008	399	8	–
155.1	306	232	−0.1	0.0679	0.0014	0.54	0.02	0.0581	0.0022	423	8	–
97.1	119	48	0.71	0.0688	0.0014	0.52	0.03	0.0547	0.0029	429	9	–
18.1	153	86	3.49	0.0688	0.0028	0.51	0.03	0.0542	0.0022	429	17	–
47.1	170	59	1.44	0.0707	0.0029	0.55	0.03	0.0568	0.0015	440	17	–
161.1	147	85	1.97	0.0737	0.0015	0.55	0.03	0.0545	0.0031	458	9	–
80.1	243	179	1.68	0.0737	0.0030	0.55	0.03	0.0543	0.0017	458	18	–
149.1	305	133	1.86	0.0738	0.0015	0.60	0.02	0.0590	0.0020	459	9	–
90.1	600	532	2.08	0.0746	0.0016	0.50	0.05	0.0487	0.0047	464	9	–
34.1	220	121	2.05	0.0756	0.0031	0.49	0.04	0.0473	0.0032	470	18	–
65.1	101	50	2.07	0.0807	0.0033	0.65	0.07	0.0581	0.0056	500	20	–
52.1	585	149	1.51	0.0563	0.0023	0.49	0.02	0.0626	0.0017	693	57	51
70.1	482	114	0.96	0.1109	0.0045	1.00	0.05	0.0657	0.0013	798	41	85
131.1	1546	1055	4.01	0.0408	0.0008	0.37	0.02	0.0665	0.0030	822	97	31
4.1	967	1564	5.71	0.0361	0.0015	0.33	0.02	0.0671	0.0035	840	113	27
124.1	428	176	0.43	0.1360	0.0028	1.29	0.04	0.0686	0.0016	885	49	93
96.1	231	520	−0.02	0.1251	0.0026	1.19	0.03	0.0688	0.0011	893	33	85
31.1	356	178	0.53	0.1597	0.0065	1.54	0.07	0.0700	0.0010	927	30	103
35.1	200	280	0.62	0.1263	0.0051	1.24	0.06	0.0710	0.0018	959	54	80
157.1	699	71	0.15	0.1675	0.0034	1.72	0.04	0.0746	0.0007	1056	19	95
152.1	266	82	1.07	0.1869	0.0039	1.93	0.08	0.0748	0.0026	1064	73	104
75.1	81	36	1.69	0.1861	0.0076	1.94	0.13	0.0756	0.0035	1085	97	101
143.1	338	62	0.05	0.1765	0.0036	1.85	0.05	0.0760	0.0010	1094	27	96
113.1	73	81	0.05	0.1615	0.0034	1.69	0.06	0.0761	0.0020	1098	53	88
77.1	1717	994	1.25	0.0386	0.0016	0.41	0.02	0.0762	0.0019	1101	51	22
24.1	89	64	1.17	0.1750	0.0071	1.86	0.11	0.0772	0.0027	1127	70	92
91.1	426	242	0.2	0.1844	0.0038	1.97	0.05	0.0774	0.0011	1132	28	96
159.1	420	228	1.29	0.2107	0.0043	2.30	0.07	0.0793	0.0018	1179	44	105
148.1	413	239	0.85	0.1393	0.0029	1.55	0.05	0.0806	0.0020	1211	51	69
130.1	381	208	0.47	0.2133	0.0044	2.37	0.07	0.0806	0.0016	1211	40	103
56.1	181	53	0.4	0.1860	0.0076	2.07	0.10	0.0807	0.0017	1215	42	90
26.1	228	97	0.45	0.1868	0.0076	2.16	0.10	0.0837	0.0013	1286	30	86
103.1	300	143	0.14	0.2230	0.0046	2.58	0.08	0.0840	0.0017	1291	39	101
147.1	27	22	0.49	0.1978	0.0045	2.29	0.15	0.0840	0.0049	1292	118	90
132.1	117	54	1.12	0.2781	0.0059	3.29	0.16	0.0859	0.0035	1336	80	118
41.1	506	180	0.97	0.2015	0.0082	2.40	0.10	0.0864	0.0010	1347	21	88
126.1	148	128	0.21	0.2006	0.0042	2.41	0.07	0.0873	0.0018	1366	40	86

(Continued)

Table 4 Continued.

Grain.spot	U (ppm)	Th (ppm)	f^{206}%	$^{206}Pb/^{238}U$	±1σ	$^{207}Pb/^{235}U$	±1σ	$^{207}Pb/^{206}Pb$	±1σ	Preferred age	±	Conc.
25.1	329	165	0.79	0.2355	0.0096	2.85	0.12	0.0877	0.0011	1376	23	99
150.1	55	35	1.17	0.2414	0.0053	2.96	0.19	0.0889	0.0052	1401	116	99
45.1	174	101	0.55	0.2573	0.0105	3.30	0.15	0.0931	0.0012	1490	25	99
99.1	338	158	0.05	0.2557	0.0053	3.30	0.08	0.0937	0.0008	1503	16	98
14.1	79	29	0.65	0.2523	0.0103	3.31	0.16	0.0952	0.0021	1533	43	95
153.1	680	288	0.06	0.2239	0.0046	2.94	0.07	0.0954	0.0006	1535	12	85
13.1	661	85	0.09	0.2611	0.0106	3.43	0.14	0.0954	0.0005	1536	10	97
154.1	99	73	0.31	0.2875	0.0061	3.88	0.13	0.0979	0.0025	1585	48	103
123.1	295	172	0.12	0.2797	0.0058	3.83	0.10	0.0993	0.0012	1612	23	99
1.1	106	89	0.71	0.2773	0.0113	3.81	0.20	0.0996	0.0027	1617	52	98
62.1	273	84	0.3	0.2634	0.0107	3.62	0.15	0.0997	0.0008	1618	15	93
115.1	205	107	0.5	0.3122	0.0065	4.32	0.13	0.1004	0.0018	1632	33	107
72.1	877	525	0.16	0.1565	0.0064	2.17	0.09	0.1007	0.0005	1638	9	57
137.1	81	48	0.54	0.2642	0.0056	3.69	0.15	0.1014	0.0031	1650	58	92
71.1	93	67	0.79	0.3094	0.0127	4.37	0.21	0.1024	0.0019	1669	34	104
50.1	623	483	6.97	0.1308	0.0053	1.85	0.10	0.1028	0.0028	1676	51	47
88.1	100	58	0.53	0.2970	0.0064	4.25	0.17	0.1037	0.0032	1691	57	99
162.1	196	78	0.98	0.2808	0.0058	4.02	0.12	0.1038	0.0021	1693	37	94
133.1	877	689	1.52	0.0861	0.0018	1.24	0.04	0.1045	0.0021	1706	37	31
46.1	319	205	0.26	0.3025	0.0123	4.36	0.18	0.1046	0.0007	1707	11	100
127.1	967	920	1.24	0.1460	0.0030	2.11	0.05	0.1047	0.0014	1709	24	51
119.1	236	117	5.74	0.2387	0.0051	3.45	0.22	0.1048	0.0061	1711	111	81
134.1	146	59	0.47	0.2921	0.0061	4.23	0.13	0.1050	0.0021	1715	38	96
64.1	196	110	0.39	0.2881	0.0117	4.20	0.18	0.1058	0.0010	1728	18	94
110.1	487	125	0.65	0.1682	0.0034	2.46	0.07	0.0106	0.0016	1732	27	58
20.1	387	108	0.19	0.2910	0.0118	4.28	0.18	0.1066	0.0006	1742	10	95
163.1	243	72	0.13	0.2687	0.0055	4.02	0.10	0.1086	0.0013	1775	23	86
107.1	177	167	0.36	0.3276	0.0068	4.91	0.14	0.1088	0.0020	1779	33	103
114.1	95	35	0.18	0.3209	0.0068	4.82	0.15	0.1089	0.0022	1782	37	101
30.1	905	711	0.41	0.1874	0.0076	2.83	0.12	0.1094	0.0006	1789	9	62
158.1	1207	193	0.84	0.2503	0.0051	3.78	0.09	0.1097	0.0008	1794	14	80
16.1	776	275	0.32	0.2258	0.0092	3.43	0.14	0.1100	0.0006	1800	9	73
140.1	432	138	0.12	0.3222	0.0066	4.90	0.11	0.1102	0.0007	1803	12	100
86.1	140	44	0.04	0.2761	0.0112	4.20	0.18	0.1103	0.0009	1804	14	87
142.1	266	38	0.21	0.3233	0.0067	4.93	0.12	0.1107	0.0012	1810	20	100
144.1	206	39	0.09	0.3036	0.0064	4.68	0.12	0.1119	0.0013	1830	22	93
168.1	112	51	0.54	0.3240	0.0069	5.01	0.15	0.1121	0.0021	1834	35	99
15.1	247	142	1.58	0.2305	0.0094	3.56	0.16	0.1121	0.0017	1834	28	73
29.1	253	86	0.31	0.3192	0.0130	4.94	0.21	0.1122	0.0011	1836	17	97
89.1	249	113	0.53	0.2584	0.0053	4.01	0.11	0.1125	0.0016	1840	26	81
89.1	249	113	0.53	0.2584	0.0053	4.01	0.11	0.1125	0.0016	1840	26	81
10.1	177	103	0.22	0.3331	0.0136	5.17	0.22	0.1126	0.0009	1841	14	101
40.1	198	168	0.31	0.3279	0.0134	5.11	0.22	0.1130	0.0008	1848	14	99
146.1	589	226	1.41	0.2586	0.0053	4.05	0.10	0.1136	0.0015	1857	23	80
120.1	275	149	0.12	0.2428	0.0050	3.81	0.10	0.1139	0.0014	1862	23	75
78.1	226	85	0.38	0.3380	0.0138	5.31	0.23	0.1139	0.0011	1863	17	101
68.1	262	135	0.15	0.2824	0.0115	4.45	0.19	0.1143	0.0006	1869	10	86
8.1	252	121	2.07	0.3003	0.0122	4.74	0.21	0.1144	0.0016	1871	26	90
3.1	239	227	0.18	0.3314	0.0135	5.25	0.22	0.1148	0.0008	1877	13	98
84.1	343	234	0.28	0.3349	0.0136	5.32	0.22	0.1151	0.0008	1882	12	99
32.1	452	339	0.49	0.2366	0.0096	3.76	0.16	0.1152	0.0008	1883	13	73
164.1	332	165	0.02	0.3269	0.0067	5.20	0.12	0.1153	0.0008	1885	13	97
98.1	304	151	0.21	0.3178	0.0066	5.05	0.12	0.1153	0.0011	1885	17	94
106.1	1655	3351	1.45	0.0610	0.0012	0.97	0.03	0.1154	0.0018	1885	28	20
54.1	418	232	1.76	0.1569	0.0064	2.50	0.11	0.1155	0.0017	1887	27	50
102.1	89	47	0.23	0.3423	0.0073	5.45	0.17	0.1155	0.0022	1888	35	101
56.1	188	156	0.45	0.3134	0.0128	5.00	0.21	0.1156	0.0011	1889	18	93
93.1	1550	101	0.06	0.3132	0.0064	5.00	0.11	0.1158	0.0004	1893	6	93
95.1	458	155	0.3	0.2217	0.0045	3.55	0.09	0.1162	0.0014	1899	21	68
165.1	477	437	0.68	0.1252	0.0026	2.01	0.05	0.1162	0.0018	1899	28	40
136.1	427	225	1.19	0.2524	0.0052	4.09	0.11	0.1174	0.0018	1917	27	76
122.1	268	119	0.27	0.3171	0.0066	5.16	0.12	0.1181	0.0012	1927	18	92
51.1	282	153	0.31	0.2878	0.0117	4.69	0.20	0.1182	0.0008	1929	13	85

(Continued)

Table 4 Continued.

Grain.spot	U (ppm)	Th (ppm)	f^{206}%	$^{206}Pb/^{238}U$	±1σ	$^{207}Pb/^{235}U$	±1σ	$^{207}Pb/^{206}Pb$	±1σ	Preferred age	±	Conc.
36.1	314	127	0.15	0.3170	0.0129	5.25	0.22	0.1200	0.0009	1956	13	91
42.1	527	200	0.91	0.3360	0.0137	5.56	0.23	0.1201	0.0008	1957	12	95
61.1	598	497	0.19	0.3281	0.0133	5.44	0.22	0.1203	0.0004	1960	6	93
11.1	255	160	0.31	0.3581	0.0146	5.97	0.25	0.1210	0.0008	1971	12	100
6.1	255	103	0.1	0.3329	0.0135	5.57	0.23	0.1214	0.0007	1977	10	94
94.1	141	49	0.19	0.3419	0.0072	5.75	0.15	0.1221	0.0016	1986	24	95
100.1	245	198	0.24	0.3487	0.0073	5.89	0.14	0.1225	0.0012	1992	18	97
53.1	167	70	0.51	0.3319	0.0135	5.66	0.24	0.1236	0.0012	2009	17	92
23.1	141	66	0.52	0.3199	0.0130	5.46	0.24	0.1238	0.0016	2011	23	89
85.1	194	118	0.05	0.3500	0.0143	6.01	0.25	0.1246	0.0006	2023	9	96
81.1	66	87	0.57	0.3531	0.0145	6.08	0.31	0.1248	0.0031	2026	45	96
59.1	227	153	0.02	0.3618	0.0147	6.28	0.26	0.1259	0.0006	2041	8	98
28.1	211	110	0.44	0.2472	0.0101	4.29	0.18	0.1259	0.0013	2041	18	70
101.1	313	145	0.06	0.3690	0.0076	6.53	0.15	0.1283	0.0009	2075	13	98
145.1	3743	8450	8	0.0275	0.0006	0.49	0.02	0.1304	0.0042	2104	58	8
166.1	391	244	2.06	0.1949	0.0040	3.61	0.11	0.1345	0.0025	2158	33	53
92.1	179	109	0.08	0.3997	0.0084	7.48	0.19	0.1357	0.0016	2173	21	100
118.1	119	170	0.46	0.2882	0.0061	5.61	0.17	0.1411	0.0026	2241	32	73
128.1	336	328	0.59	0.2124	0.0044	4.21	0.10	0.1437	0.0017	2272	20	55
57.1	355	248	2.05	0.3257	0.0133	6.56	0.29	0.1462	0.0017	2302	20	79
73.1	322	67	4.36	0.2288	0.0093	4.72	0.22	0.1497	0.0025	2343	29	57
63.1	228	91	0.29	0.4187	0.0171	9.05	0.38	0.1568	0.0010	2422	11	93
37.1	59	31	0.53	0.4279	0.0176	9.40	0.42	0.1594	0.0019	2449	21	94
87.1	279	62	0.29	0.4235	0.0172	9.61	0.40	0.1646	0.0008	2503	8	91
21.1	473	360	0.65	0.4416	0.0180	10.15	0.42	0.1667	0.0007	2524	7	93
117.1	595	523	1.52	0.1255	0.0026	2.92	0.08	0.1687	0.0024	2544	24	30
82.1	487	350	0.32	0.2758	0.0112	6.54	0.27	0.1719	0.0008	2577	7	61
79.1	1163	109	2.05	0.4017	0.0163	9.59	0.40	0.1732	0.0007	2589	7	84
43.1	401	317	0.4	0.3824	0.0156	9.41	0.39	0.1784	0.0007	2638	6	79
49.1	368	563	0.16	0.5010	0.0204	12.35	0.51	0.1788	0.0007	2641	6	99
111.1	122	94	0.26	0.4867	0.0104	12.24	0.31	0.1824	0.0020	2675	19	96
167.1	176	50	0.2	0.5051	0.0106	12.86	0.30	0.1846	0.0014	2695	12	98
129.1	163	229	0.12	0.5123	0.0107	13.06	0.30	0.1849	0.0013	2697	12	99
112.1	598	105	0.02	0.5025	0.0103	13.04	0.28	0.1882	0.0006	2726	5	96
5.1	240	108	0.2	0.4754	0.0194	12.38	0.51	0.1889	0.0008	2733	7	92
12.1	70	40	0.32	0.5309	0.0217	14.14	0.60	0.1931	0.0014	2769	12	99
44.1	287	69	0.14	0.4739	0.0193	13.36	0.55	0.2045	0.0007	2863	6	87
60.1	206	135	2.82	0.5626	0.0230	16.29	0.70	0.2100	0.0019	2905	15	99
27.1	163	115	0.56	0.5941	0.0242	21.85	0.91	0.2667	0.0011	3287	7	91

Table 5 Summary of SHRIMP U–Pb zircon results for 6610/3-1, 2300.0 m. f^{206}% denotes the percentage of ^{206}Pb that is common Pb. For zircons > 800 Ma, correction for common Pb made using the measured $^{204}Pb/^{206}Pb$ ratio. For zircons < 800 Ma, correction for common Pb made using the measured Th/U ratio and calculated $^{208}Pb^*$ content. 100% denotes a concordant analysis.

Grain.spot	U (ppm)	Th (ppm)	f^{206}%	$^{206}Pb/^{238}U$	±1σ	$^{207}Pb/^{235}U$	±1σ	$^{207}Pb/^{206}Pb$	±1σ	Preferred age	±	Conc.
56.1	2207	806	9.73	0.0299	0.0009	0.21	0.02	0.0499	0.0044	190	6	–
2.1	711	291	0.95	0.0319	0.0010	0.27	0.01	0.0606	0.0022	202	6	–
85.1	1785	283	1.11	0.0385	0.0012	0.31	0.01	0.0576	0.0016	243	7	–
52.1	1024	479	0.9	0.0430	0.0013	0.33	0.02	0.0563	0.0017	272	8	–
73.1	1318	615	1.14	0.0500	0.0015	0.39	0.02	0.0562	0.0013	315	9	–
1.1	1177	718	10.38	0.0507	0.0015	0.36	0.03	0.0511	0.0039	319	9	–
51.1	1292	558	1.53	0.0528	0.0016	0.41	0.02	0.0564	0.0014	332	10	–
59.1	1293	681	0.62	0.0539	0.0016	0.44	0.01	0.0591	0.0008	338	10	–
38.1	1013	420	1.32	0.0546	0.0016	0.43	0.02	0.0566	0.0016	342	10	–
27.1	1678	749	0.36	0.0576	0.0017	0.45	0.02	0.0567	0.0007	361	11	–
45.1	1135	643	0.52	0.0583	0.0018	0.49	0.02	0.0616	0.0008	365	11	–
54.1	113	2	4.88	0.0605	0.0019	0.32	0.08	0.0379	0.0098	378	12	–
44.1	1024	333	2.85	0.0647	0.0020	0.60	0.03	0.0677	0.0019	404	12	–
3.1	164	63	1.7	0.0651	0.0020	0.46	0.05	0.0513	0.0049	406	12	–
70.1	373	248	0.44	0.0681	0.0021	0.52	0.02	0.0554	0.0018	425	12	–
11.1	440	268	−0.04	0.0684	0.0021	0.55	0.02	0.0589	0.0009	426	12	–
14.1	416	701	0.69	0.0686	0.0021	0.51	0.02	0.0540	0.0015	428	12	–

(*Continued*)

Table 5 Continued.

Grain.spot	U (ppm)	Th (ppm)	f^{206}%	$^{206}Pb/^{238}U$	$\pm 1\sigma$	$^{207}Pb/^{235}U$	$\pm 1\sigma$	$^{207}Pb/^{206}Pb$	$\pm 1\sigma$	Preferred age	±	Conc.
25.1	451	275	0.31	0.0708	0.0021	0.55	0.03	0.0568	0.0025	441	13	–
22.1	281	88	0.94	0.1400	0.0042	1.31	0.05	0.0678	0.0017	863	52	98
39.1	178	243	1.31	0.1402	0.0042	1.38	0.08	0.0712	0.0031	962	91	88
26.1	255	101	1.35	0.1623	0.0049	1.62	0.07	0.0722	0.0020	991	56	98
13.1	291	101	0.32	0.1924	0.0058	2.03	0.07	0.0766	0.0012	1110	31	102
76.1	577	286	0.2	0.1907	0.0057	2.05	0.07	0.0781	0.0006	1149	16	98
17.1	1093	69	0.36	0.1022	0.0031	1.14	0.04	0.0812	0.0007	1227	17	51
75.1	372	142	0.35	0.1865	0.0056	2.12	0.07	0.0824	0.0011	1254	26	88
69.1	78	51	1.23	0.2525	0.0077	3.02	0.16	0.0866	0.0034	1352	77	107
21.1	349	158	0.28	0.1876	0.0056	2.27	0.07	0.0878	0.0009	1378	19	80
66.1	1176	299	1.7	0.0795	0.0024	0.97	0.04	0.0888	0.0016	1400	36	35
83.1	102	76	0.36	0.2319	0.0070	2.86	0.11	0.0895	0.0018	1415	38	95
40.1	299	154	0.35	0.2344	0.0071	2.89	0.10	0.0895	0.0013	1416	28	96
29.1	66	25	0.68	0.2479	0.0076	3.10	0.15	0.0908	0.0031	1443	66	99
30.1	207	176	0.39	0.2549	0.0077	3.22	0.11	0.0916	0.0010	1459	20	100
20.1	695	223	0.28	0.1541	0.0046	1.99	0.06	0.0937	0.0007	1501	14	62
5.1	68	23	1.18	0.1416	0.0044	1.83	0.13	0.0939	0.0055	1507	115	57
49.1	449	222	0.27	0.2314	0.0070	3.00	0.10	0.0940	0.0008	1509	15	89
84.1	470	156	22.74	0.2659	0.0081	3.45	0.22	0.0941	0.0049	1510	101	101
16.1	1133	167	0.21	0.1517	0.0046	1.97	0.06	0.0942	0.0005	1511	10	60
34.1	227	67	0.64	0.1367	0.0041	1.81	0.07	0.0960	0.0021	1548	41	53
43.1	255	210	0.29	0.2688	0.0081	3.58	0.12	0.0965	0.0009	1557	17	99
72.1	481	159	0.21	0.1860	0.0056	2.48	0.08	0.0968	0.0021	1563	15	70
4.1	522	117	0.63	0.1826	0.0055	2.45	0.08	0.0972	0.0016	1570	18	69
55.1	93	60	1.19	0.2702	0.0082	3.64	0.14	0.0978	0.0010	1582	42	97
47.1	123	105	0.49	0.2745	0.0083	3.71	0.13	0.0980	0.0010	1586	31	99
6.1	218	178	0.24	0.2704	0.0082	3.66	0.12	0.0982	0.0007	1591	19	97
80.1	335	160	0.64	0.2658	0.0080	3.67	0.12	0.1000	0.0012	1625	19	94
65.1	265	135	0.2	0.2802	0.0084	3.89	0.12	0.1006	0.0007	1635	13	97
74.1	202	178	0.45	0.2545	0.0077	3.53	0.12	0.1006	0.0012	1635	22	89
77.1	701	721	0.28	0.2140	0.0064	2.98	0.09	0.1011	0.0006	1644	11	76
18.1	1057	1666	0.17	0.2042	0.0061	2.85	0.09	0.1011	0.0004	1645	8	73
67.1	644	129	0.08	0.2802	0.0084	3.91	0.12	0.1013	0.0004	1648	7	97
63.1	672	301	0.23	0.1978	0.0059	2.79	0.09	0.1024	0.0006	1667	10	70
19.1	565	141	0.19	0.1793	0.0054	2.54	0.08	0.1028	0.0007	1676	12	63
53.1	139	41	0.65	0.2397	0.0073	3.41	0.13	0.1031	0.0020	1681	36	82
64.1	460	191	0.26	0.2719	0.0082	3.87	0.12	0.1031	0.0007	1681	12	92
24.1	758	243	0.5	0.1416	0.0043	2.02	0.06	0.1036	0.0008	1689	15	51
33.1	3096	1305	0.17	0.0960	0.0029	1.39	0.04	0.1049	0.0004	1713	7	34
57.1	239	198	0.37	0.2681	0.0081	3.89	0.13	0.1054	0.0010	1721	17	89
60.1	502	130	0.15	0.2985	0.0090	4.35	0.14	0.1056	0.0005	1725	9	98
10.1	674	282	0.27	0.1934	0.0058	2.85	0.09	0.1069	0.0007	1747	13	65
7.1	232	93	0.98	0.2780	0.0084	4.10	0.14	0.1071	0.0015	1751	26	90
31.1	372	130	0.5	0.2667	0.0080	3.94	0.13	0.1072	0.0009	1753	16	87
48.1	91	25	1.11	0.2645	0.0080	3.92	0.16	0.1074	0.0024	1755	41	86
61.1	292	79	0.86	0.2404	0.0072	3.57	0.12	0.1076	0.0014	1759	23	79
81.1	32	8	1.31	0.3053	0.0095	4.53	0.25	0.1077	0.0045	1761	78	98
46.1	225	78	1.32	0.2820	0.0085	4.21	0.15	0.1082	0.0016	1768	27	91
62.1	209	62	0.36	0.2783	0.0084	4.15	0.14	0.1081	0.0011	1768	18	90
68.1	226	85	0.31	0.2943	0.0089	4.40	0.14	0.1084	0.0009	1772	15	94
82.1	373	168	0.27	0.2897	0.0087	4.34	0.14	0.1086	0.0007	1776	11	92
36.1	238	106	0.15	0.3160	0.0095	4.73	0.16	0.1087	0.0011	1777	19	100
8.1	326	79	0.24	0.2987	0.0090	4.48	0.14	0.1087	0.0008	1778	13	95
28.1	165	39	0.36	0.3165	0.0096	4.76	0.16	0.1090	0.0012	1783	20	99
9.1	307	84	0.19	0.3038	0.0091	4.58	0.14	0.1092	0.0007	1787	12	96
71.1	240	115	0.25	0.2927	0.0088	4.42	0.14	0.1096	0.0010	1792	17	92
79.1	335	151	0.08	0.3176	0.0096	4.80	0.15	0.1096	0.0007	1792	11	99
35.1	514	317	0.19	0.2862	0.0086	4.33	0.14	0.1098	0.0006	1796	10	90
42.1	660	284	0.05	0.2991	0.0090	4.53	0.14	0.1098	0.0004	1797	7	94
37.1	684	354	0.14	0.2990	0.0090	4.57	0.14	0.1108	0.0004	1812	7	93
41.1	443	276	0.01	0.3215	0.0097	4.91	0.15	0.1108	0.0004	1813	7	99
32.1	102	48	0.3	0.3148	0.0095	4.82	0.17	0.1110	0.0015	1815	24	97
58.1	118	54	0.41	0.3012	0.0091	4.61	0.16	0.1109	0.0016	1815	26	94
23.1	302	109	0.1	0.3311	0.0100	5.09	0.16	0.1116	0.0006	1826	10	101
78.1	304	138	−0.04	0.2922	0.0088	4.52	0.14	0.1123	0.0006	1837	10	90
15.1	36	27	0.71	0.3291	0.0102	5.24	0.22	0.1155	0.0029	1887	46	97
50.1	306	115	0.3	0.3351	0.0101	8.42	0.26	0.1822	0.0010	2673	9	70

Table 6 Summary of SHRIMP U–Pb zircon results for 6707/10-1, 3002.8 m. f^{206}% denotes the percentage of ^{206}Pb that is common Pb. For zircons >800 Ma, correction for common Pb made using the measured $^{204}Pb/^{206}Pb$ ratio. For zircons <800 Ma, correction for common Pb made using the measured Th/U ratio and calculated $^{208}Pb^*$ content. 100% denotes a concordant analysis.

Grain.spot	U (ppm)	Th (ppm)	f^{206}%	$^{206}Pb/^{238}U$	±1σ	$^{207}Pb/^{235}U$	±1σ	$^{207}Pb/^{206}Pb$	±1σ	Preferred age	±	Conc.
16.1	372	195	0.16	0.0142	0.0003	0.101	0.008	0.0516	0.0040	91	2	–
13.1	147	87	<0.01	0.0151	0.0004	0.155	0.016	0.0744	0.0074	97	2	–
10.1	98	67	1.47	0.0155	0.0004	0.117	0.020	0.0547	0.0091	99	3	–
1.1	156	131	<0.01	0.0441	0.0008	0.355	0.032	0.0585	0.0050	278	5	–
26.1	200	71	<0.01	0.0463	0.0008	0.366	0.014	0.0574	0.0018	292	5	–
25.1	1116	1221	<0.01	0.0554	0.0007	0.442	0.007	0.0579	0.0005	347	4	–
54.1	528	365	<0.01	0.0703	0.0010	0.542	0.032	0.0560	0.0031	438	6	–
43.1	383	99	0.17	0.0707	0.0010	0.568	0.016	0.0583	0.0013	440	6	–
35.1	71	55	0.53	0.0727	0.0021	0.507	0.051	0.0505	0.0047	453	13	–
5.1	666	101	0.16	0.0775	0.0010	0.817	0.018	0.0765	0.0013	481	6	–
27.1	327	190	0.01	0.0802	0.0013	0.618	0.018	0.0559	0.0013	497	7	–
11.1	186	179	0.40	0.0911	0.0014	0.707	0.037	0.0563	0.0027	562	8	121
42.1	607	225	0.12	0.0962	0.0015	0.799	0.019	0.0603	0.0010	592	9	97
7.1	227	30	1.16	0.1038	0.0021	0.972	0.033	0.0680	0.0017	636	12	73
29.1	131	75	<0.01	0.1404	0.0028	1.334	0.041	0.0689	0.0015	897	44	95
2.1	197	75	<0.01	0.1597	0.0023	1.531	0.039	0.0695	0.0013	915	40	104
56.1	351	22	0.18	0.1382	0.0018	1.334	0.029	0.0700	0.0012	928	34	90
48.1	279	160	0.02	0.1551	0.0026	1.520	0.036	0.0711	0.0011	959	30	97
37.1	274	122	0.02	0.1841	0.0026	1.905	0.033	0.0750	0.0006	1069	16	102
24.1	341	99	0.05	0.2007	0.0026	2.150	0.036	0.0777	0.0007	1139	19	104
15.1	347	23	0.32	0.2085	0.0032	2.305	0.056	0.0802	0.0013	1202	33	102
47.1	295	182	0.01	0.2265	0.0042	2.547	0.052	0.0816	0.0006	1235	14	107
40.1	301	137	0.03	0.2512	0.0033	3.038	0.054	0.0877	0.0009	1376	20	105
38.1	196	43	0.02	0.2368	0.0036	2.896	0.055	0.0887	0.0009	1398	18	98
36.1	203	99	0.11	0.2198	0.0042	2.720	0.079	0.0898	0.0018	1421	38	90
14.1	259	154	0.02	0.2485	0.0038	3.096	0.069	0.0904	0.0013	1433	28	100
41.1	480	345	0.10	0.2469	0.0034	3.139	0.057	0.0922	0.0010	1472	20	97
52.1	269	119	0.04	0.2576	0.0040	3.289	0.059	0.0926	0.0007	1480	15	100
19.1	993	331	0.03	0.2309	0.0030	2.967	0.042	0.0932	0.0004	1491	8	90
23.1	704	246	0.04	0.2581	0.0033	3.343	0.047	0.0939	0.0004	1507	9	98
34.1	22	14	<0.01	0.2026	0.0233	2.626	0.401	0.0940	0.0082	1508	175	79
33.1	196	92	0.05	0.2569	0.0059	3.376	0.089	0.0953	0.0010	1534	20	96
17.1	690	184	<0.01	0.2576	0.0034	3.446	0.049	0.0970	0.0004	1568	7	94
45.1	122	64	0.27	0.2970	0.0061	3.993	0.102	0.0975	0.0013	1577	25	106
44.1	2829	173	2.37	0.2916	0.0035	4.010	0.057	0.0998	0.0006	1620	12	102
6.1	213	137	<0.01	0.2867	0.0046	3.969	0.116	0.1004	0.0023	1632	42	100
18.1	268	181	0.01	0.2981	0.0062	4.279	0.095	0.1041	0.0006	1699	10	99
39.1	340	240	<0.01	0.3113	0.0041	4.549	0.069	0.1060	0.0006	1731	11	101
12.1	985	67	0.05	0.2603	0.0030	3.856	0.050	0.1075	0.0005	1757	9	85
20.1	1088	210	0.02	0.3204	0.0040	4.779	0.066	0.1082	0.0005	1769	9	101
55.1	484	118	0.04	0.3111	0.0047	4.709	0.078	0.1098	0.0006	1796	10	97
4.1	494	307	0.09	0.2713	0.0038	4.137	0.070	0.1106	0.0008	1809	14	86
31.1	809	88	0.02	0.2930	0.0038	4.539	0.097	0.1124	0.0017	1838	28	90
22.1	268	58	<0.01	0.2947	0.0038	4.583	0.067	0.1128	0.0006	1845	10	90
53.1	52	54	0.16	0.3490	0.0091	5.439	0.169	0.1130	0.0016	1849	26	104
60.1	161	111	0.02	0.3292	0.0061	5.154	0.114	0.1135	0.0012	1857	19	99
57.1	314	108	0.03	0.3064	0.0061	4.841	0.105	0.1146	0.0007	1874	11	92
46.1	214	172	0.07	0.3281	0.0054	5.219	0.098	0.1154	0.0009	1886	14	97
51.1	277	86	0.20	0.3110	0.0053	4.965	0.104	0.1158	0.0012	1892	19	92
49.1	242	147	0.07	0.3396	0.0046	5.429	0.084	0.1159	0.0007	1894	11	100
32.1	190	170	<0.01	0.3544	0.0050	5.676	0.109	0.1162	0.0013	1898	21	103
58.1	132	44	0.26	0.3340	0.0049	5.367	0.102	0.1166	0.0012	1904	19	98
3.1	96	23	0.01	0.3638	0.0069	5.914	0.146	0.1179	0.0016	1924	25	104
8.1	218	53	0.08	0.3407	0.0062	5.546	0.115	0.1181	0.0009	1927	14	98
21.1	132	78	0.09	0.3513	0.0059	5.871	0.129	0.1212	0.0015	1974	22	98
59.1	126	25	0.03	0.3831	0.0100	6.470	0.185	0.1225	0.0011	1993	16	105
50.1	392	69	0.14	0.3890	0.0051	6.972	0.099	0.1300	0.0005	2098	7	101
9.1	296	241	0.01	0.5019	0.0100	11.806	0.244	0.1706	0.0006	2564	5	102
28.1	166	49	0.06	0.4826	0.0112	11.826	0.295	0.1777	0.0012	2632	11	97
30.1	370	294	0.11	0.3874	0.0078	9.520	0.230	0.1782	0.0020	2637	19	80

Table 7 Summary of SHRIMP U–Pb zircon results for sample W4567, Barremian, Store Koldewey. f^{206}% denotes the percentage of ^{206}Pb that is common Pb. For zircons >800 Ma, correction for common Pb made using the measured $^{204}Pb/^{206}Pb$ ratio. For zircons <800 Ma, correction for common Pb made using the measured Th/U ratio and calculated $^{208}Pb^*$ content. 100% denotes a concordant analysis.

Grain.spot	U (ppm)	Th (ppm)	f^{206}%	$^{206}Pb/^{238}U$	$\pm 1\sigma$	$^{207}Pb/^{235}U$	$\pm 1\sigma$	$^{207}Pb/^{206}Pb$	$\pm 1\sigma$	Preferred age	±	Conc.
3.1	13	< 1	1.19	0.0623	0.0032	0.451	0.043	0.0525	0.0039	390	19	–
49.1	395	5	0.12	0.0647	0.0008	0.488	0.011	0.0547	0.0009	404	5	–
14.1	70	< 1	0.18	0.0660	0.0014	0.473	0.029	0.0520	0.0028	412	8	–
25.1	136	1	0.04	0.0666	0.0009	0.512	0.012	0.0557	0.0009	415	6	–
32.2	987	254	1.22	0.1265	0.0014	1.621	0.026	0.0929	0.0010	1487	20	52
32.1	148	26	0.07	0.1341	0.0043	1.801	0.069	0.0974	0.0017	1574	33	52
43.1	325	71	0.14	0.2390	0.0036	3.238	0.076	0.0983	0.0016	1591	31	87
9.1	129	23	0.08	0.2845	0.0054	3.894	0.089	0.0993	0.0011	1610	20	100
6.1	135	74	0.07	0.2500	0.0040	3.508	0.070	0.1018	0.0010	1657	19	87
26.1	245	69	0.02	0.1853	0.0026	2.606	0.045	0.1020	0.0008	1660	15	66
1.1	139	89	<0.01	0.2433	0.0035	3.438	0.064	0.1025	0.0011	1670	20	84
2.1	77	32	0.02	0.2530	0.0050	3.582	0.091	0.1027	0.0014	1673	26	87
19.1	190	48	0.03	0.2693	0.0034	3.838	0.060	0.1034	0.0008	1685	15	91
13.1	695	25	0.07	0.1740	0.0023	2.483	0.038	0.1035	0.0006	1688	11	61
46.1	68	40	<0.01	0.2740	0.0049	3.939	0.112	0.1043	0.0021	1701	37	92
15.1	12	1	0.02	0.1365	0.0057	1.966	0.200	0.1045	0.0092	1705	171	48
37.1	346	109	0.17	0.1757	0.0026	2.536	0.046	0.1047	0.0009	1709	16	61
42.1	82	43	<0.01	0.2996	0.0056	4.326	0.107	0.1047	0.0015	1709	26	99
50.1	103	54	0.12	0.2985	0.0048	4.314	0.104	0.1048	0.0017	1711	30	98
36.1	173	41	0.08	0.3010	0.0042	4.358	0.078	0.1050	0.0010	1715	18	99
23.1	53	40	<0.01	0.2932	0.0053	4.265	0.163	0.1055	0.0033	1723	59	96
35.1	145	71	<0.01	0.2358	0.0039	3.546	0.077	0.1091	0.0013	1784	23	77
34.1	1579	1074	3.46	0.0838	0.0010	1.273	0.025	0.1101	0.0016	1802	27	29
10.1	929	595	0.04	0.2188	0.0024	3.342	0.039	0.1108	0.0004	1812	6	70
44.1	638	48	0.18	0.2833	0.0044	4.371	0.096	0.1119	0.0015	1831	25	88
5.1	599	222	0.57	0.1619	0.0019	2.509	0.040	0.1124	0.0011	1838	17	53
38.1	1052	342	1.48	0.1924	0.0021	2.998	0.074	0.1130	0.0023	1848	38	61
8.1	420	127	0.01	0.2430	0.0028	3.801	0.052	0.1135	0.0007	1855	11	76
41.1	323	242	<0.01	0.3006	0.0035	4.744	0.070	0.1145	0.0009	1871	14	91
45.1	530	203	0.02	0.3037	0.0035	4.795	0.061	0.1145	0.0005	1873	8	91
31.1	821	248	0.13	0.2987	0.0036	4.741	0.062	0.1151	0.0005	1882	7	90
30.1	459	126	0.15	0.2637	0.0030	4.207	0.058	0.1157	0.0007	1891	12	80
29.1	382	112	0.07	0.3461	0.0040	5.576	0.079	0.1169	0.0008	1909	12	100
48.1	949	160	0.08	0.3354	0.0036	5.404	0.061	0.1169	0.0003	1909	5	98
7.1	189	33	<0.01	0.2307	0.0034	3.721	0.065	0.1170	0.0009	1911	14	70
18.1	387	113	<0.01	0.3003	0.0037	4.850	0.068	0.1171	0.0006	1913	9	89
33.1	454	213	<0.01	0.3045	0.0040	4.951	0.077	0.1179	0.0009	1925	13	89
12.1	289	67	0.07	0.2876	0.0042	4.687	0.082	0.1182	0.0010	1929	15	85
22.1	428	131	0.04	0.3015	0.0037	4.913	0.065	0.1182	0.0005	1929	7	88
17.1	220	62	<0.01	0.3182	0.0038	5.223	0.079	0.1191	0.0009	1942	14	92
20.1	314	150	0.02	0.3171	0.0046	5.214	0.090	0.1193	0.0009	1945	13	91
47.1	249	58	<0.01	0.3183	0.0046	5.252	0.084	0.1197	0.0006	1951	9	91
16.1	242	48	0.06	0.3222	0.0042	5.319	0.085	0.1197	0.0009	1952	14	92
4.1	472	22	0.03	0.3366	0.0042	5.569	0.078	0.1200	0.0006	1957	8	96
24.1	280	46	1.14	0.3311	0.0050	5.483	0.135	0.1201	0.0021	1958	32	94
39.1	439	262	0.03	0.3400	0.0040	5.637	0.072	0.1203	0.0005	1960	7	96
28.1	220	66	0.01	0.3366	0.0046	5.588	0.085	0.1204	0.0007	1962	10	95
27.1	270	115	<0.01	0.3130	0.0038	5.212	0.072	0.1208	0.0006	1968	9	89
40.1	334	129	0.01	0.3485	0.0042	5.832	0.078	0.1214	0.0005	1977	7	98
21.1	203	40	0.01	0.3369	0.0049	5.666	0.090	0.1220	0.0006	1985	9	94
11.1	108	73	0.36	0.5144	0.0076	14.176	0.239	0.1999	0.0013	2825	10	95

The two age spectra (Fig. 6) are closely comparable, both being dominated by Proterozoic zircons between 1600–1800 Ma, and in particular by a major cluster of zircons at c. 1780–1790 Ma. Both also contain a number of Caledonian-age zircons (400–450 Ma), although these are better represented in the sample from 6610/3–1. The zircon population in the sample from 6610/3–1 is slightly more diverse than in 6507/2–2, containing a cluster of grains around c. 1400–1500 Ma and a small number of grains between 900 and 1200 Ma.

The strong similarity between the two zircon populations confirms the mineralogical evidence for a common provenance. The zircon evidence

Table 8 Summary of SHRIMP U–Pb zircon results for sample W4337, Cenomanian, Hold with Hope. f^{206}% denotes the percentage of ^{206}Pb that is common Pb. For zircons >800 Ma, correction for common Pb made using the measured $^{204}Pb/^{206}Pb$ ratio. For zircons <800 Ma, correction for common Pb made using the measured Th/U ratio and calculated $^{208}Pb^*$ content. 100% denotes a concordant analysis.

Grain.spot	U (ppm)	Th (ppm)	f^{206}%	$^{206}Pb/^{238}U$	±	$^{207}Pb/^{235}U$	±	$^{207}Pb/^{206}Pb$	±	Preferred age	±	Conc.
15.1	128	0	0.50	0.0660	0.0024	0.482	0.037	0.0530	0.0034	412	14	–
55.1	349	316	2.66	0.0693	0.0013	0.592	0.042	0.0620	0.0041	432	8	–
22.1	112	110	0.69	0.0798	0.0031	0.608	0.077	0.0553	0.0064	495	19	–
45.1	244	125	0.26	0.1544	0.0028	1.488	0.044	0.0699	0.0015	924	45	100
11.1	346	85	0.03	0.1744	0.0027	1.762	0.033	0.0733	0.0007	1022	19	101
59.1	721	156	0.04	0.1740	0.0024	1.786	0.043	0.0744	0.0014	1053	38	98
8.1	197	219	0.21	0.1910	0.0034	1.980	0.052	0.0752	0.0013	1074	35	105
54.1	604	262	<0.01	0.1917	0.0034	2.026	0.041	0.0766	0.0006	1112	16	102
52.1	352	82	0.10	0.1895	0.0041	2.016	0.058	0.0772	0.0013	1125	34	99
16.1	151	92	0.30	0.1592	0.0030	1.709	0.067	0.0779	0.0025	1143	65	83
42.1	164	60	0.05	0.2055	0.0049	2.228	0.086	0.0786	0.0022	1162	55	104
23.1	191	64	<0.01	0.2220	0.0039	2.545	0.067	0.0831	0.0014	1272	34	102
43.1	525	127	0.16	0.1712	0.0043	1.991	0.065	0.0844	0.0015	1301	35	78
5.1	471	165	0.12	0.2329	0.0048	2.808	0.103	0.0874	0.0024	1370	54	99
2.1	206	122	0.01	0.2323	0.0045	2.816	0.066	0.0879	0.0010	1380	21	98
29.1	37	14	1.23	0.2936	0.0144	3.824	0.336	0.0945	0.0063	1517	132	109
19.1	557	294	0.04	0.2795	0.0052	3.666	0.076	0.0951	0.0007	1531	13	104
48.1	125	109	0.22	0.3001	0.0062	4.098	0.125	0.0990	0.0020	1606	38	105
20.1	321	128	<0.01	0.3032	0.0060	4.144	0.093	0.0991	0.0009	1607	17	106
30.1	134	71	0.02	0.3032	0.0073	4.165	0.150	0.0996	0.0024	1617	46	106
41.1	123	82	0.16	0.3021	0.0089	4.172	0.154	0.1002	0.0019	1627	35	105
25.1	282	73	0.09	0.2950	0.0066	4.086	0.120	0.1005	0.0017	1633	31	102
21.1	141	69	<0.01	0.2965	0.0083	4.139	0.137	0.1012	0.0015	1647	27	102
31.1	353	181	0.05	0.2859	0.0044	3.991	0.072	0.1013	0.0008	1647	14	98
51.1	119	83	0.10	0.3017	0.0061	4.217	0.138	0.1014	0.0023	1650	43	103
38.1	108	61	0.28	0.2484	0.0066	3.499	0.160	0.1022	0.0035	1664	64	86
35.1	318	58	0.34	0.2764	0.0061	3.984	0.147	0.1046	0.0028	1707	51	92
9.1	355	100	0.10	0.3025	0.0064	4.366	0.110	0.1047	0.0012	1709	21	100
3.1	45	21	0.76	0.2571	0.0106	3.720	0.244	0.1050	0.0049	1714	88	86
27.1	119	56	0.14	0.2841	0.0065	4.133	0.156	0.1055	0.0029	1724	51	94
32.1	388	231	0.07	0.3105	0.0053	4.559	0.098	0.1065	0.0012	1740	20	100
50.1	339	192	0.09	0.3159	0.0048	4.695	0.092	0.1078	0.0012	1762	20	100
46.1	622	156	<0.01	0.2660	0.0038	4.017	0.071	0.1095	0.0010	1792	17	85
44.1	614	64	0.01	0.2799	0.0042	4.242	0.075	0.1099	0.0008	1798	14	89
12.1	367	113	0.26	0.2743	0.0041	4.177	0.090	0.1105	0.0016	1807	26	87
26.1	290	72	0.32	0.2132	0.0050	3.286	0.109	0.1118	0.0023	1829	38	68
4.1	208	101	0.04	0.3114	0.0059	4.801	0.113	0.1118	0.0013	1829	21	96
13.1	231	388	0.14	0.3386	0.0060	5.276	0.120	0.1130	0.0014	1848	22	102
7.1	249	68	0.19	0.3232	0.0062	5.076	0.123	0.1139	0.0014	1863	22	97
34.1	835	670	0.04	0.2435	0.0032	3.853	0.066	0.1148	0.0011	1877	17	75
37.1	222	168	0.03	0.3368	0.0057	5.353	0.118	0.1153	0.0014	1884	22	99
36.1	179	62	0.08	0.3418	0.0067	5.454	0.138	0.1157	0.0016	1891	25	100
24.1	159	35	0.16	0.3397	0.0081	5.439	0.173	0.1161	0.0021	1898	33	99
56.1	248	99	0.21	0.3552	0.0057	5.691	0.118	0.1162	0.0013	1898	20	103
57.1	442	80	0.49	0.2646	0.0039	4.262	0.082	0.1168	0.0013	1908	20	79
18.1	454	325	0.03	0.3307	0.0050	5.378	0.097	0.1179	0.0010	1925	15	96
17.1	205	68	0.07	0.2873	0.0055	4.693	0.119	0.1185	0.0017	1933	26	84
60.1	383	162	0.13	0.3445	0.0064	5.628	0.136	0.1185	0.0016	1933	24	99
39.1	273	77	0.02	0.3272	0.0053	5.347	0.121	0.1185	0.0016	1934	25	94
47.1	416	227	0.03	0.3420	0.0058	5.598	0.118	0.1187	0.0012	1937	19	98
10.1	433	72	0.02	0.3370	0.0053	5.519	0.099	0.1188	0.0008	1938	12	97
28.1	261	54	0.08	0.3600	0.0063	5.903	0.121	0.1189	0.0010	1940	15	102
6.1	152	146	0.01	0.3537	0.0093	5.885	0.194	0.1207	0.0020	1966	30	99
33.1	182	103	0.21	0.2574	0.0068	4.310	0.160	0.1214	0.0028	1978	42	75
49.1	262	122	0.10	0.3657	0.0066	6.441	0.149	0.1277	0.0016	2067	22	97
14.1	94	38	<0.01	0.4114	0.0109	7.608	0.275	0.1341	0.0029	2153	38	103
53.1	481	72	<0.01	0.3556	0.0099	6.800	0.246	0.1387	0.0027	2211	35	89
40.1	228	76	0.15	0.3088	0.0054	6.278	0.150	0.1475	0.0021	2317	25	75
58.1	64	53	0.01	0.6892	0.0187	20.363	2.384	0.2143	0.0237	2938	191	115
1.1	190	32	<0.01	0.5613	0.0141	16.897	0.629	0.2184	0.0053	2969	40	97

Table 9 Summary of SHRIMP U–Pb zircon results for sample W4346, Santonian, Hold with Hope. f^{206}% denotes the percentage of ^{206}Pb that is common Pb. For zircons >800 Ma, correction for common Pb made using the measured $^{204}Pb/^{206}Pb$ ratio. For zircons <800 Ma, correction for common Pb made using the measured Th/U ratio and calculated $^{208}Pb^*$ content. 100% denotes a concordant analysis.

Grain.spot	U (ppm)	Th (ppm)	f^{206}%	$^{206}Pb/^{238}U$	±1σ	$^{207}Pb/^{235}U$	±1σ	$^{207}Pb/^{206}Pb$	±1σ	Preferred age	±	Conc.
50.1	325	4	<0.01	0.0650	0.0009	0.496	0.013	0.0553	0.0011	406	6	–
13.1	36	1	0.30	0.0654	0.0026	0.498	0.040	0.0553	0.0035	408	16	–
22.1	128	58	0.27	0.0693	0.0011	0.537	0.039	0.0563	0.0039	432	7	–
39.1	412	331	<0.01	0.0700	0.0009	0.545	0.020	0.0565	0.0018	436	6	–
45.1	397	100	3.92	0.1058	0.0013	1.019	0.019	0.0698	0.0009	648	8	–
20.1	344	25	0.10	0.1560	0.0021	1.472	0.029	0.0684	0.0009	881	27	106
17.1	292	84	0.08	0.1506	0.0025	1.437	0.029	0.0692	0.0007	905	20	100
46.1	822	321	<0.01	0.1511	0.0016	2.031	0.026	0.0975	0.0005	907	9	58
3.1	657	8	0.02	0.1616	0.0018	1.565	0.020	0.0703	0.0004	936	10	103
40.1	320	66	0.16	0.1861	0.0024	1.885	0.033	0.0735	0.0008	1027	21	107
19.1	47	28	<0.01	0.1582	0.0043	1.625	0.079	0.0745	0.0028	1055	77	90
41.1	16	10	0.86	0.18	0.0052	1.778	0.209	0.071	0.0078	1076	28	113
18.1	395	111	1.33	0.1534	0.0022	1.605	0.062	0.0759	0.0026	1092	70	84
14.1	476	69	0.02	0.1831	0.0025	1.916	0.031	0.0759	0.0005	1092	14	99
26.1	37	27	0.05	0.1905	0.0042	2.156	0.084	0.0821	0.0024	1247	59	90
37.1	49	27	0.33	0.2681	0.0064	3.271	0.108	0.0885	0.0018	1393	39	110
34.1	936	93	<0.01	0.2389	0.0026	2.931	0.038	0.0890	0.0005	1404	12	98
21.1	199	59	0.42	0.2138	0.0029	2.695	0.052	0.0914	0.0011	1456	23	86
27.1	379	135	0.23	0.2406	0.0030	3.040	0.046	0.0916	0.0007	1460	14	95
1.1	260	129	<0.01	0.2536	0.0040	3.281	0.060	0.0938	0.0007	1504	14	97
47.1	83	152	<0.01	0.2654	0.0042	3.480	0.101	0.0951	0.0021	1530	43	99
38.1	552	182	0.06	0.2055	0.0027	2.799	0.046	0.0988	0.0009	1601	17	75
35.1	226	92	0.12	0.2794	0.0040	3.829	0.069	0.0994	0.0009	1613	18	99
29.1	365	44	<0.01	0.2896	0.0038	4.003	0.066	0.1003	0.0009	1629	16	101
7.1	142	80	<0.01	0.2659	0.0037	3.684	0.061	0.1005	0.0008	1633	14	93
16.1	345	242	0.09	0.2762	0.0033	3.827	0.056	0.1005	0.0007	1634	13	96
44.1	129	128	0.13	0.2815	0.0047	3.908	0.078	0.1007	0.0009	1636	17	98
42.1	367	272	1.36	0.2805	0.0032	3.896	0.137	0.1007	0.0032	1637	60	97
25.1	99	43	0.02	0.3070	0.0053	4.353	0.086	0.1028	0.0008	1676	14	103
23.1	39	41	<0.01	0.3009	0.0080	4.291	0.168	0.1034	0.0027	1686	48	101
30.1	352	109	0.04	0.2922	0.0035	4.221	0.057	0.1048	0.0005	1710	9	97
4.1	331	138	<0.01	0.3164	0.0040	4.687	0.069	0.1074	0.0007	1757	11	101
49.1	426	137	0.02	0.2152	0.0025	3.253	0.047	0.1096	0.0008	1793	14	70
6.1	150	40	0.09	0.3253	0.0053	4.981	0.092	0.1110	0.0008	1817	13	100
2.1	604	182	0.00	0.2531	0.0030	3.914	0.050	0.1122	0.0004	1835	7	79
12.1	482	95	0.05	0.2781	0.0034	4.323	0.058	0.1128	0.0005	1844	8	86
43.1	102	34	0.15	0.3090	0.0046	4.860	0.097	0.1141	0.0013	1865	21	93
32.1	244	70	<0.01	0.2365	0.0033	3.752	0.060	0.1151	0.0008	1881	12	73
33.1	277	188	0.02	0.3244	0.0043	5.210	0.079	0.1165	0.0007	1903	11	95
8.1	978	92	0.01	0.3348	0.0036	5.398	0.062	0.1169	0.0003	1910	5	98
48.1	301	113	<0.01	0.3066	0.0053	5.053	0.113	0.1195	0.0015	1949	22	88
5.1	338	52	0.01	0.3373	0.0042	5.570	0.078	0.1198	0.0006	1953	9	96
11.1	303	45	0.06	0.3575	0.0052	5.950	0.096	0.1207	0.0007	1967	10	100
31.1	160	31	0.02	0.3590	0.0049	5.986	0.098	0.1210	0.0009	1970	13	100
9.1	528	115	<0.01	0.3425	0.0041	5.735	0.077	0.1215	0.0006	1978	9	96
10.1	261	118	0.06	0.3443	0.0043	5.778	0.082	0.1217	0.0007	1982	10	96
28.1	299	72	0.02	0.3542	0.0049	5.964	0.094	0.1221	0.0007	1988	10	98
15.1	41	19	0.20	0.4061	0.0108	7.699	0.248	0.1375	0.0021	2196	27	100
36.1	470	181	0.06	0.2614	0.0037	5.515	0.086	0.1530	0.0008	2380	8	63
24.1	47	28	<0.01	0.5294	0.0107	13.983	0.346	0.1916	0.0023	2756	20	99

strongly supports derivation of K1 detritus from the adjacent part of the Scandinavian landmass, a terrain that comprises late Precambrian metasediments and components of the trans-Scandinavian igneous belt, intruded by granites during the Caledonian Orogeny. The predominant group of c. 1780–1790 Ma detrital zircons is consistent with a source within the trans-Scandinavian igneous belt, which forms a significant part of the outcrop on the adjacent Norwegian landmass (Fig. 1). The wider scatter associated with the younger Proterozoic components (900–1780 Ma) could represent direct erosion of a diverse Proterozoic hinterland, but is probably better interpreted in terms of second generation zircon detritus, eroded from pre-existing sediments which themselves were the repository of

Table 10 Summary of SHRIMP U–Pb zircon results for sample W4470, Cenomanian, Geographical Society Ø. $f^{206}\%$ denotes the percentage of ^{206}Pb that is common Pb. For zircons > 800 Ma, correction for common Pb made using the measured $^{204}Pb/^{206}Pb$ ratio. For zircons < 800 Ma, correction for common Pb made using the measured Th/U ratio and calculated $^{208}Pb^*$ content. 100% denotes a concordant analysis.

Grain.spot	U (ppm)	Th (ppm)	$f^{206}\%$	$^{206}Pb/^{238}U$	$\pm 1\sigma$	$^{207}Pb/^{235}U$	$\pm 1\sigma$	$^{207}Pb/^{206}Pb$	$\pm 1\sigma$	Preferred age	±	Conc.
37.1	126	214	1.77	0.0255	0.0006	0.143	0.033	0.0407	0.0092	162	4	–
48.1	475	147	0.25	0.0338	0.0004	0.239	0.009	0.0513	0.0018	214	3	–
3.1	308	157	4.68	0.0371	0.0008	0.259	0.082	0.0506	0.0158	235	5	–
46.1	108	100	1.92	0.0391	0.0008	0.248	0.032	0.0460	0.0057	247	5	–
58.1	394	541	0.97	0.0395	0.0006	0.000	0.000	0.0000	0.0000	250	4	–
18.1	286	291	0.83	0.0396	0.0005	0.278	0.021	0.0509	0.0037	250	3	–
42.1	399	151	1.29	0.0401	0.0006	0.295	0.017	0.0533	0.0028	254	4	–
50.1	77	67	2.23	0.0417	0.0011	0.275	0.037	0.0479	0.0061	263	7	–
35.1	289	398	3.82	0.0439	0.0007	0.258	0.039	0.0427	0.0064	277	4	–
7.1	79	91	7.30	0.0447	0.0012	0.229	0.050	0.0372	0.0079	282	7	–
30.1	131	198	16.11	0.0450	0.0013	0.266	0.072	0.0429	0.0114	284	8	–
47.1	543	804	0.34	0.0452	0.0006	0.327	0.017	0.0524	0.0026	285	3	–
32.1	396	401	< 0.01	0.0454	0.0007	0.361	0.017	0.0576	0.0025	286	4	–
60.1	211	199	2.45	0.0475	0.0009	0.000	0.000	0.0000	0.0000	299	5	–
59.1	599	188	0.41	0.0479	0.0006	0.000	0.000	0.0000	0.0000	302	4	–
41.1	1245	510	0.18	0.0486	0.0006	0.343	0.008	0.0513	0.0010	306	3	–
6.1	388	187	4.07	0.0514	0.0007	0.378	0.014	0.0534	0.0017	323	4	–
63.1	637	135	3.08	0.0547	0.0007	0.000	0.000	0.0000	0.0000	343	4	–
22.1	1770	39	1.41	0.0632	0.0015	0.492	0.020	0.0565	0.0017	395	9	–
21.1	82	0	1.29	0.0633	0.0015	0.502	0.057	0.0575	0.0063	396	9	–
49.1	248	175	0.42	0.0648	0.0009	0.469	0.019	0.0525	0.0019	405	5	–
33.1	211	1	0.26	0.0662	0.0011	0.509	0.016	0.0557	0.0014	413	7	–
28.1	158	68	0.59	0.0690	0.0010	0.515	0.028	0.0541	0.0028	430	6	–
36.1	236	82	0.85	0.0734	0.0010	0.596	0.020	0.0589	0.0017	456	6	–
10.1	267	100	0.90	0.0745	0.0013	0.557	0.025	0.0542	0.0021	463	8	–
40.1	126	96	0.88	0.0888	0.0015	0.659	0.037	0.0538	0.0028	549	9	–
31.1	26	14	9.93	0.1189	0.0032	0.919	0.146	0.0561	0.0086	724	19	–
12.1	754	648	11.11	0.1203	0.0023	2.076	0.212	0.1252	0.0122	732	13	–
17.1	257	133	0.13	0.1837	0.0029	1.886	0.036	0.0745	0.0007	1054	18	103
19.1	96	64	0.69	0.2527	0.0043	3.451	0.113	0.0991	0.0026	1607	49	90
54.1	263	109	0.34	0.2772	0.0044	3.820	0.109	0.1000	0.0022	1623	41	97
56.1	395	209	0.08	0.2861	0.0039	3.988	0.071	0.1011	0.0010	1644	18	99
55.1	969	168	0.25	0.1871	0.0023	2.610	0.042	0.1012	0.0010	1646	18	67
4.1	119	56	0.18	0.2924	0.0040	4.103	0.080	0.1018	0.0012	1657	23	100
13.1	681	161	0.02	0.2957	0.0038	4.151	0.056	0.1018	0.0003	1657	6	101
26.1	40	24	0.55	0.3175	0.0082	4.482	0.179	0.1024	0.0028	1668	52	107
2.1	224	61	0.23	0.2919	0.0036	4.183	0.088	0.1039	0.0016	1695	29	97
52.1	91	66	1.17	0.3258	0.0074	4.774	0.187	0.1063	0.0031	1736	54	105
61.1	177	69	0.82	0.3009	0.0055	4.438	0.142	0.1070	0.0026	1748	45	97
14.1	66	14	0.41	0.3163	0.0060	4.670	0.133	0.1071	0.0020	1750	35	101
8.1	521	42	0.78	0.2412	0.0031	3.651	0.063	0.1098	0.0011	1796	18	78
57.1	46	45	1.06	0.3288	0.0101	5.020	0.234	0.1108	0.0035	1812	58	101
44.1	48	24	0.47	0.3558	0.0108	5.586	0.214	0.1139	0.0023	1862	36	105
9.1	590	1119	0.44	0.3185	0.0038	5.006	0.075	0.1140	0.0009	1864	14	96
53.1	86	41	0.67	0.3356	0.0110	5.278	0.230	0.1141	0.0028	1865	45	100
45.1	138	65	0.09	0.3457	0.0049	5.503	0.089	0.1154	0.0008	1887	12	102
20.1	224	79	0.14	0.3411	0.0045	5.437	0.080	0.1156	0.0006	1889	10	100
27.1	51	56	0.15	0.3425	0.0101	5.529	0.188	0.1171	0.0016	1912	25	99
51.1	348	214	0.03	0.3386	0.0040	5.524	0.074	0.1183	0.0006	1931	9	97
34.1	203	90	0.55	0.3089	0.0046	5.043	0.095	0.1184	0.0012	1932	18	90
15.1	50	31	0.02	0.3425	0.0088	5.609	0.159	0.1188	0.0011	1938	17	98
39.1	96	90	0.02	0.3397	0.0054	5.562	0.110	0.1188	0.0012	1938	18	97
16.1	312	273	0.69	0.2685	0.0031	4.467	0.064	0.1207	0.0009	1966	13	78
1.1	431	216	0.11	0.3589	0.0047	5.975	0.097	0.1207	0.0010	1967	14	101
5.1	112	71	1.25	0.3747	0.0078	6.324	0.169	0.1224	0.0018	1992	26	103
43.1	96	77	0.65	0.3842	0.0074	6.710	0.166	0.1267	0.0017	2052	24	102
23.1	52	7	0.50	0.3532	0.0062	5.608	0.189	0.1151	0.0031	2226	36	86
62.1	431	160	1.05	0.3535	0.0072	7.640	0.260	0.1568	0.0039	2421	43	81
11.1	60	39	0.44	0.5153	0.0099	12.837	0.319	0.1807	0.0024	2659	23	101
29.1	104	64	0.20	0.5079	0.0110	13.005	0.324	0.1857	0.0019	2704	17	98
38.1	88	173	0.06	0.5249	0.0077	13.561	0.225	0.1874	0.0012	2719	10	100
25.1	132	62	0.14	0.5395	0.0079	16.175	0.270	0.2174	0.0014	2962	11	94

Table 11 Summary of SHRIMP U–Pb zircon results for sample S3920, Turonian, Traill Ø. $f^{206}\%$ denotes the percentage of ^{206}Pb that is common Pb. For zircons > 800 Ma, correction for common Pb made using the measured $^{204}Pb/^{206}Pb$ ratio. For zircons < 800 Ma, correction for common Pb made using the measured Th/U ratio and calculated $^{208}Pb^*$ content. 100% denotes a concordant analysis.

Grain.spot	U (ppm)	Th (ppm)	$f^{206}\%$	$^{206}Pb/^{238}U$	$\pm 1\sigma$	$^{207}Pb/^{235}U$	$\pm 1\sigma$	$^{207}Pb/^{206}Pb$	$\pm 1\sigma$	Preferred age	±	Conc.
48.1	368	274	0.51	0.0471	0.0008	0.325	0.018	0.0501	0.0025	297	5	–
44.1	608	372	< 0.01	0.0690	0.0011	0.535	0.022	0.0563	0.0020	430	7	–
39.1	499	570	0.33	0.0693	0.0015	0.508	0.023	0.0532	0.0019	432	9	–
36.1	358	432	0.10	0.0736	0.0014	0.555	0.048	0.0547	0.0045	458	8	–
47.1	501	84	< 0.01	0.0989	0.0019	0.880	0.025	0.0645	0.0012	608	11	80
23.1	76	52	0.24	0.1053	0.0046	0.917	0.099	0.0632	0.0059	645	27	91
19.1	405	34	0.02	0.1437	0.0022	1.411	0.028	0.0712	0.0008	963	23	90
14.1	462	152	< 0.01	0.1837	0.0030	1.862	0.050	0.0735	0.0014	1027	40	106
53.1	440	213	0.14	0.1721	0.0027	1.753	0.045	0.0739	0.0014	1037	38	99
43.1	54	44	1.40	0.1748	0.0077	1.523	0.204	0.0632	0.0076	1039	42	145
55.1	58	85	0.11	0.1697	0.0058	1.754	0.090	0.0750	0.0026	1067	71	95
21.1	81	27	0.35	0.1854	0.0061	1.937	0.105	0.0758	0.0030	1089	80	101
41.1	196	63	0.15	0.2031	0.0054	2.132	0.088	0.0761	0.0022	1099	58	109
37.1	643	308	0.04	0.1724	0.0028	1.813	0.035	0.0763	0.0007	1102	18	93
2.1	241	74	0.19	0.2096	0.0048	2.218	0.067	0.0768	0.0013	1115	34	110
10.1	431	310	< 0.01	0.1782	0.0031	1.898	0.040	0.0772	0.0008	1127	20	94
60.1	481	230	0.05	0.1960	0.0030	2.122	0.041	0.0785	0.0008	1160	20	99
1.1	153	61	0.37	0.1979	0.0106	2.318	0.214	0.0849	0.0058	1164	57	89
16.1	176	71	0.09	0.2066	0.0039	2.252	0.057	0.0791	0.0012	1174	29	103
54.1	425	177	0.04	0.1881	0.0034	2.056	0.045	0.0793	0.0008	1180	21	94
11.1	17	9	< 0.01	0.2010	0.0106	2.492	0.201	0.0899	0.0049	1181	57	83
31.1	302	251	0.04	0.2166	0.0031	2.404	0.052	0.0805	0.0012	1209	28	105
58.1	213	109	0.11	0.1816	0.0054	2.017	0.076	0.0805	0.0016	1210	39	89
25.1	149	42	0.35	0.2116	0.0049	2.242	0.124	0.0768	0.0036	1238	26	111
26.1	133	46	0.14	0.2100	0.0053	2.386	0.083	0.0824	0.0017	1255	41	98
34.1	63	42	< 0.01	0.1828	0.0059	2.083	0.108	0.0826	0.0031	1260	74	86
17.1	81	29	0.45	0.2188	0.0050	2.466	0.126	0.0818	0.0035	1276	26	103
7.1	178	179	0.04	0.2077	0.0037	2.438	0.066	0.0851	0.0015	1319	35	92
24.1	647	360	0.03	0.2319	0.0031	2.793	0.051	0.0873	0.0010	1368	21	98
38.1	184	321	0.19	0.2337	0.0044	2.981	0.082	0.0925	0.0017	1478	34	92
40.1	330	154	0.09	0.2759	0.0043	3.597	0.077	0.0946	0.0012	1519	25	103
18.1	719	225	0.02	0.2496	0.0040	3.284	0.058	0.0954	0.0006	1536	11	94
13.1	682	47	0.01	0.2776	0.0037	3.676	0.055	0.0960	0.0005	1548	10	102
22.1	379	34	0.03	0.2839	0.0052	3.892	0.089	0.0994	0.0012	1614	22	100
20.1	177	151	0.18	0.2889	0.0055	3.961	0.105	0.0995	0.0016	1614	30	101
32.1	122	144	< 0.01	0.2948	0.0084	4.064	0.169	0.1000	0.0027	1624	51	103
27.1	335	167	0.02	0.2165	0.0035	3.037	0.069	0.1017	0.0014	1656	26	76
8.1	589	292	0.28	0.1764	0.0029	2.480	0.053	0.1019	0.0012	1660	23	63
52.1	32	23	0.97	0.2939	0.0135	3.868	0.374	0.0955	0.0076	1661	67	108
28.1	89	113	< 0.01	0.3061	0.0095	4.329	0.156	0.1026	0.0015	1672	28	103
29.1	383	101	0.05	0.2571	0.0046	3.659	0.077	0.1032	0.0010	1683	17	88
5.1	289	248	0.02	0.3083	0.0056	4.443	0.110	0.1045	0.0016	1706	28	102
33.1	17	29	0.99	0.2732	0.0185	3.984	0.361	0.1058	0.0055	1727	99	90
9.1	151	96	0.39	0.3356	0.0080	4.970	0.161	0.1074	0.0021	1756	35	106
30.1	296	70	0.00	0.2829	0.0046	4.229	0.081	0.1084	0.0009	1773	15	91
59.1	192	98	< 0.01	0.3153	0.0110	4.757	0.255	0.1094	0.0040	1790	68	99
56.1	537	225	0.01	0.3238	0.0045	4.907	0.085	0.1099	0.0010	1798	16	101
51.1	436	258	0.02	0.2843	0.0056	4.448	0.109	0.1135	0.0014	1856	23	87
46.1	1142	813	< 0.01	0.2956	0.0049	4.688	0.083	0.1150	0.0005	1881	8	89
4.1	413	90	0.11	0.3257	0.0048	5.204	0.099	0.1159	0.0012	1894	19	96
15.1	425	107	0.10	0.2787	0.0039	4.512	0.074	0.1175	0.0008	1918	13	83
50.1	221	178	0.07	0.3592	0.0084	5.910	0.169	0.1193	0.0017	1946	25	102
35.1	1266	326	< 0.01	0.2732	0.0032	4.498	0.065	0.1194	0.0009	1948	13	80
3.1	161	50	< 0.01	0.3680	0.0071	6.133	0.155	0.1209	0.0017	1969	25	103
45.1	158	72	0.06	0.3631	0.0088	6.189	0.212	0.1236	0.0026	2009	38	99
49.1	120	63	0.57	0.1519	0.0059	2.815	0.175	0.1344	0.0060	2156	79	42
12.1	63	43	0.13	0.4681	0.0181	10.003	0.474	0.1550	0.0036	2402	40	103
42.1	272	65	0.07	0.4832	0.0085	10.878	0.213	0.1633	0.0011	2490	11	102
57.1	607	233	0.02	0.4460	0.0060	10.547	0.188	0.1715	0.0017	2573	17	92
6.1	188	32	< 0.01	0.4892	0.0111	11.864	0.296	0.1759	0.0014	2615	14	98

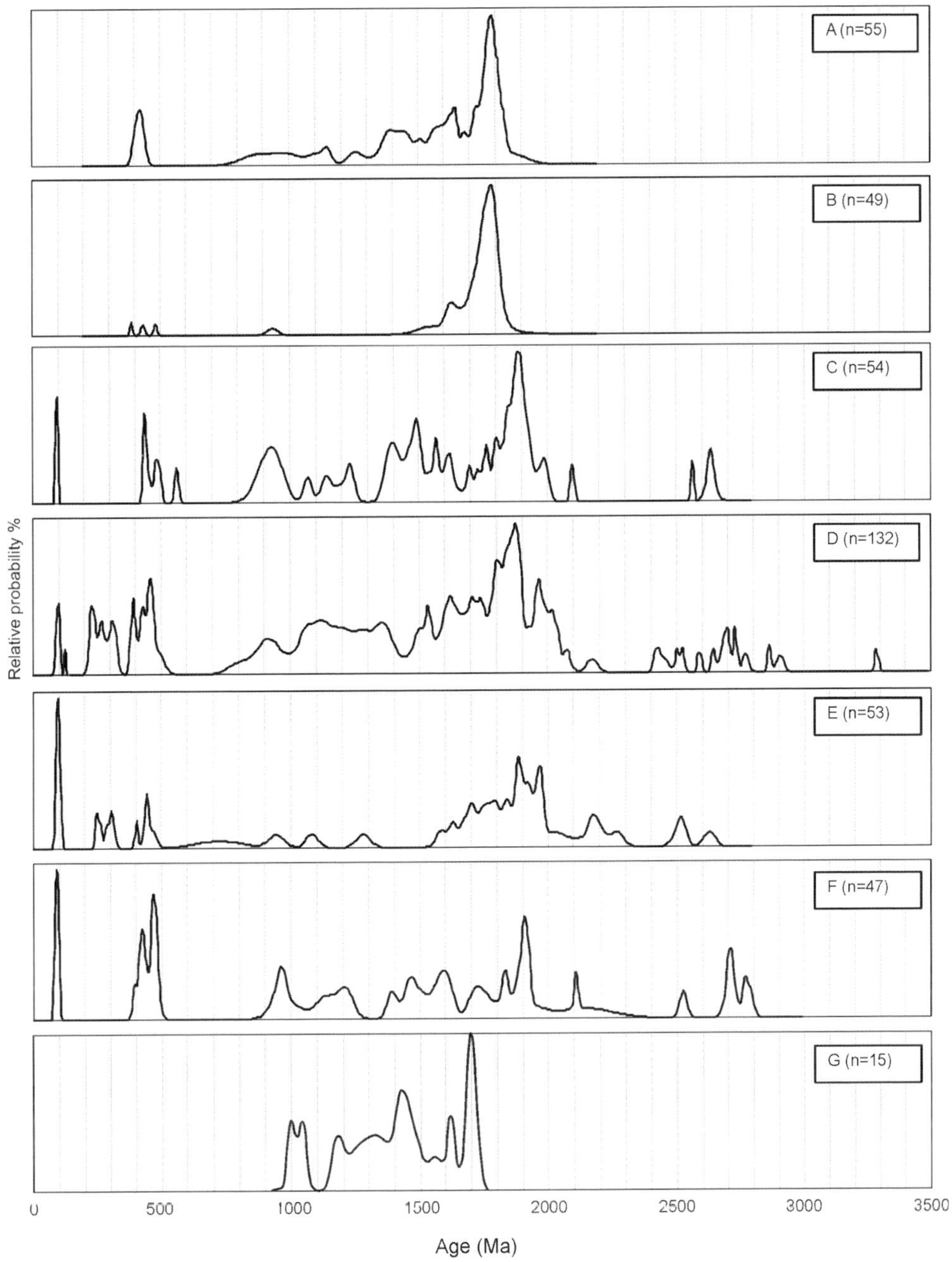

Fig. 6 Relative probability diagrams showing distribution of detrital zircon ages in K1 and K2 sandstones. n = number of zircons >80% concordant. A = 6610/3-1, 2300.0 m (K1). B = 6507/2-2, 3281.0 m (K1). C = 6707/10-1, 3002.8 m (K2). D = 6607/5-2, 4172.0 m (K2). E = 6507/2-2, 2830.1 m (K2). F = 6505/10-1, 3711.6 m (K2). G = Metasediments of the Seve Nappes (Williams and Claesson, 1987), adapted from Watt and Thrane (2001).

zircons eroded from the primary Proterozoic source regions. The metasediments of the adjacent Caledonian fold belt are late Precambrian and are inferred to contain Proterozoic detrital zircons, by analogy with the findings of a study of related metasediments in the Seve Nappes (part of the Caledonian fold belt north of 64°N in Sweden), which are dominated by Proterozoic-derived detritus (Williams and Claesson, 1987), as shown in Fig. 6. The Palaeozoic zircons are interpreted as representing erosional products of Caledonian granites, although one zircon of a possible Caledonian origin (6610/3-1, grain 54.1) has a very low Th/U ratio suggesting derivation from high-grade metamorphic rocks.

K2 sandstones

Four sandstones with K2 mineralogy have been analysed using SHRIMP: well 6707/10-1, 3002.8 m (Maastrichtian), well 6607/5-2, 4172.0 m (Early Campanian), well 6507/2-2, 2830.1 m (Turonian)

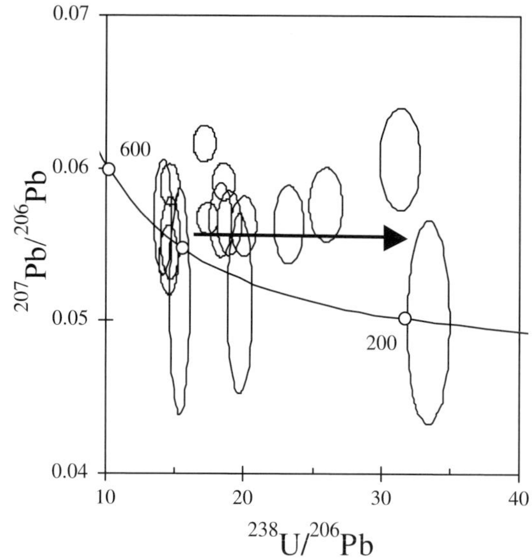

Fig. 7 Expansion of the Phanerozoic part of the concordia diagram for the K1 sample from 6610/3-1 (2300.0 m), showing that many of the zircons with apparent ages younger than the Caledonian (400–450 Ma) are likely to have suffered Pb-loss. The ages of these apparently young zircons are therefore probably unreliable. Arrow shows predicted Pb-loss trajectory for zircons with crystallisation ages of 450 Ma.

and well 6505/10-1, 3711.6 m (Coniacian). Data from the 6607/5-2 sample have been previously described in brief by Morton and Grant (1998). The zircon age populations in sandstone type K2 are diverse and complex (Fig. 6). Because of this complexity, a large zircon age data set was collected for the sample from well 6607/5-2, to ensure that relatively low-abundance components of the source of this sample were identified. With 168 detrital zircons dated, of which 132 have concordant or near-concordant measured ages, we can be 95% confident that the largest provenance component that has missed detection has a frequency of just 2.7% in the sample (Dodson et al., 1988). The age spectrum obtained is therefore a realistic representation of K2 provenance, including terrains that contributed small amounts of detritus. Although there are some differences in detail, the four zircon age spectra have the same set of elements, confirming that they share a common provenance. The most important components of the age structure are:

Archaean: all samples contain a small but significant Archaean component, best represented in the sample from well 6607/5-2, probably because of the larger number of zircons analysed from this sample. Most of the Archaean grains in the sample from well 6607/5-2 range from 2400 Ma to 3000 Ma, with a single zircon dated as 3287 Ma. Those from the other samples have a more limited age range, between 2500 Ma and 2700 Ma.

Early Proterozoic: all the four samples contain a major group of zircons between 1800–2100 Ma, peaking at c. 1900 Ma.

Early-Mid Proterozoic: zircons between 1500–1800 Ma are found in all the four samples, without showing any clear structure. The youngest sample (Maastrichtian, well 6707/10-1) also contains a group of zircons dated as between 1400–1500 Ma. This component has not been identified in the other samples.

Mid-Late Proterozoic: zircons dated between 900–1200 Ma (corresponding to the Grenvillian/Sveconorwegian orogenic event) occur in all the four samples. They appear to be more significant in the younger sandstones, being scarce in the Turonian sample (well 6507/2-2), and relatively common in the Maastrichtian (well 6707/10-1).

Palaeozoic: zircons dated between 390 Ma and 460 Ma (corresponding to the Caledonian orogeny) form a small but significant component of all the four samples.

Permo-Triassic: all the K2 samples contain zircons with Permo–Triassic apparent ages (c. 230–300 Ma). However, these are scarce in 6505/10-1 and 6707/10-1, and could be interpreted as older zircons that have suffered Pb loss, as described for the K1 sample from Well 6610/3-1 (Fig. 7). In view of this uncertainty, they are not shown on the relative probability plots (Fig. 6). By contrast, wells 6507/2-2 and 6607/5-2 contain distinct and relatively large clusters of zircons with Permo–Triassic apparent ages, many of which fall on the concordia curve. In these wells, therefore, we believe there is a genuine contribution from Permo-Triassic rocks (Fig. 6). The Permo–Triassic zircons are invariably intricately zoned and have high Th/U ratios, and are therefore interpreted as being of igneous, rather than high-grade metamorphic origin.

Mid-Cretaceous: all the four samples contain a significant number of zircons dated at c. 100 Ma (late Albian). There is some doubt as to whether these represent coeval air-fall volcanism, since although they apparently pre-date the analysed sediments (the oldest being c. 90 Myr old), the ages are associated with relatively large error bars. Although some of the zircons are near-euhedral, many display evidence for abrasion and rounding, and have evidently undergone sedimentary transport.

Detrital zircons in sandstone type K2 have a wider-ranging and more complex age structure than found in type K1. They are similar only in the predominance of Proterozoic detritus and the evidence for Caledonian-age components in the provenance. The most important differences are the presence of Mid-Cretaceous, Permo-Triassic, Early Proterozoic (c. 1900 Ma) and Archaean detrital zircons. These features confirm the mineralogical evidence for a major difference in source between the K1 and K2 sandstones. Furthermore, the similarity in the zircon age spectra from the four K2 samples provides support for a common provenance, as inferred from the heavy mineral data.

The combination of the Archaean and the Early Proterozoic zircons in the K2 age spectrum rules out the immediately adjacent part of Scandinavia as a feasible source area. There are exposures of reworked Archaean rocks on the Lofoten Islands (Fig. 2), interpreted by Jacobsen and Wasserburg, 1978 to have a 2600 Ma age. However, these are not spatially associated with Svecofennian basement that could have provided the pre-1800 Ma zircons. In any case, the oldest dates associated with the Svecofennian are c. 1900 Ma (Gaal and Gorbatschev, 1987), yet a significant number of Early Proterozoic zircons in the K2 samples predate 1900 Ma.

By contrast, the detrital zircon ages in sandstone type K2 offer compelling evidence of a match to the Greenland area. The Greenland basement, 400 km north of the Cretaceous outcrops on Traill Ø and Geographical Ø has recently been shown to represent an Early Proterozoic crust-forming event at about 1900–2000 Ma (Kalsbeek et al., 1993; Thrane, 2002). This terrain therefore provides a match for the distinctive c. 1900 Ma zircon population in the K2 sediment. It is possible that this terrain also provided the Archaean zircons, from Archaean rocks reworked in the Early Proterozoic, but this is not proven. A single occurrence of reworked Archaean rocks in NE Greenland has been reported just to the north of Store Koldewey by Nutman and Kalsbeek (1994), who present evidence for a major igneous and metamorphic event at about 1965 Ma. An alternative candidate source for the Archaean zircons lies immediately to the SW of Traill Ø (Thrane, 2002), where there are both Archaean basement rocks and Archaean protoliths reworked in the Proterozoic (Fig. 1). Zircons in metasediments of the Caledonian fold belt in East Greenland have been measured in SHRIMP studies by Strachan et al. (1995) and Watt et al. (2000). Although they differ in detail, detrital zircon ages in the Smallefjord supracrustal sequence, migmatites from the Krummedal supracrustal sequence and the Stauning Alper migmatite zone and the Nathorst Land Group (Eleanore Bay Supergroup) cover a wide range from c. 1900 Ma to c. 1000 Ma, together with younger Neoproterozoic zircons believed to have formed during the Grenvillian. East Greenland was therefore capable of providing detrital zircons directly from Archaean and Early Proterozoic (1800–2100 Ma) basement, together with a wide range of later Proterozoic zircons from metasediments in the Caledonian fold belt, and Palaeozoic zircons formed during the Caledonian Orogeny.

East Greenland

In order to test the hypothesis that the K2 sandstones were derived from the west, a comprehensive heavy mineral study on Cretaceous sandstones along the East Greenland margin has been initiated. This study has identified major differences in mineralogy, both stratigraphically and regionally, indicating the interplay of several different sediment sources. Further information on the source areas for the various mineralogical types has been acquired using SHRIMP. To date, five samples from four different areas have been analysed, ranging from Store Koldewey in the north, to Hold with Hope, Geographical Society Ø and Traill Ø in the south. The stratigraphic positions of the five samples analysed by SHRIMP are shown in Fig. 8, with locations shown in Fig. 2.

Store Koldewey

Detrital zircons have been analysed from a Barremian sample from Store Koldewey (W4567). In mineralogical terms, this sample is characterised by very high garnet : zircon and apatite : tourmaline (both GZi and ATi > 90), together with low rutile : zircon ($RuZi < 10$), low monazite : zircon ($MZi = 2.1$) and very low chrome spinel : zircon ($CZi = 0.7$). It has a distinctive garnet population (Fig. 9) dominated by high-Ca, high-Mg types, with subordinate low-Ca, high-Mg types. Garnets of this type indicate supply from a hinterland dominated by high-grade basic gneisses (including eclogites). Brueckner et al. (1998) have identified eclogites within the gneiss terrain to the north of Store Koldewey, between Dove Bugt and Holm Land, and this region is therefore a possible candidate source for the Barremian sandstones in Store Koldewey.

The age structure shown by sample W4567 is comparatively simple, with two main components and two subsidiary peaks (Fig. 10). Most of the

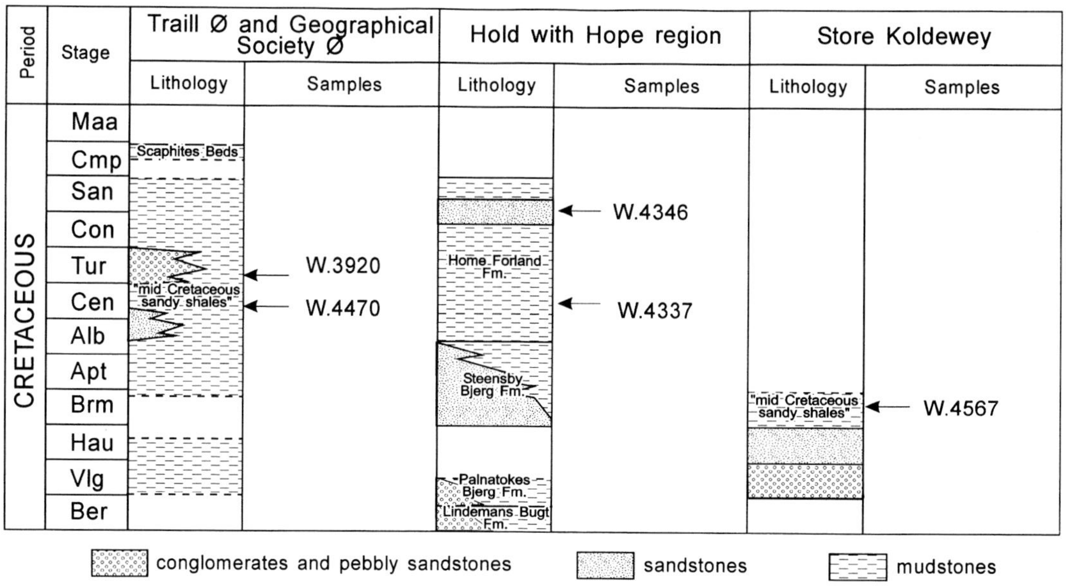

Fig. 8 Stratigraphy of the Cretaceous in East Greenland, showing the distribution of sandstones and the stratigraphic position of samples analysed in this study. Adapted from Donovan (1953, 1955), Kelly et al. (1998), Nøhr-Hansen (1993) and Surlyk (1978).

grains are >80% concordant, and those that are >20% discordant can be readily tracked back to the two main age groups. The two main components consist of a dominant group peaking at c. 1960 Ma and a second major group peaking at 1680 Ma. Apart from a single Archaean grain (2825 Ma) and four Caledonian zircons (390–415 Ma), the entire population falls between 1400 Ma and 2000 Ma.

Hold with Hope

Two samples from Hold with Hope have been analysed by SHRIMP, one of Cenomanian age (W4337) and one of Santonian age (W4346). The Cenomanian sample has low garnet (GZi < 5), moderate apatite : tourmaline and rutile : zircon (ATi = 42, RuZi = 5), low monazite : zircon (MZi = 0), and relatively high chrome spinel : zircon (CZi = 9). Its most distinctive feature is the presence of abundant clinopyroxene of augite-ferroaugite and pigeonite compositions, indicating supply from basaltic-andesitic igneous lithologies. The Santonian sample, by contrast, lacks clinopyroxene, and is characterised by an assemblage rich in garnet, low in apatite, monazite and chrome spinel, with moderate rutile (GZi = 84, RuZi = 43, ATi < 1, MZi = 2, CZi = 0). The garnets in the Santonian sample are almost exclusively composed of low Mg varieties with a range of Ca contents (Fig. 9), typical of an amphibolite-facies metasedimentary source.

The zircon age spectra from the two samples (Fig. 10) are similar, both being dominated by grains between c. 1600 Ma and 2000 Ma, with distinct peaks at c. 1630–1640 Ma and 1930–1980 Ma. They also contain a distinct group of zircons between c. 900–1200 Ma, corresponding to the Grenvillian/Sveconorwegian orogenic event, together with small numbers of Archaean and Palaeozoic (Caledonian) grains and a scattering of Proterozoic zircons between 1200–1600 Ma. Since the two spectra are similar, the data do not provide any indications as to the age of the igneous (basaltic-andesitic) source that supplied the clinopyroxene in the Cenomanian sample.

Geographical Society Ø

The Geographical Society Ø sample discussed in this chapter (W4470) is from the Cenomanian (Fig. 8). It has high garnet (GZi c. 90), with moderate apatite and rutile (ATi = 62, RuZi = 53), relatively low monazite (MZi c. 5) and relatively high chrome spinel (CZi c. 11). The garnet population (Fig. 9) is dominated by low Ca, high Mg types, typical of a high-grade, granulite-facies, metasedimentary or charnockitic source, although there are some lower Mg types suggesting incorporation of amphibolite-facies metasediments.

The age spectrum from W4470 (Fig. 10) is remarkable in that there are a large number of Phanerozoic grains. This includes a small but distinct group of Caledonian zircons, five of which fall between 395–413 Ma and four between 430–477 Ma. More significant, however, is the presence of a large number of post-Caledonian zircons, 24 of the 81 analysed grains falling into the 153–343 Ma

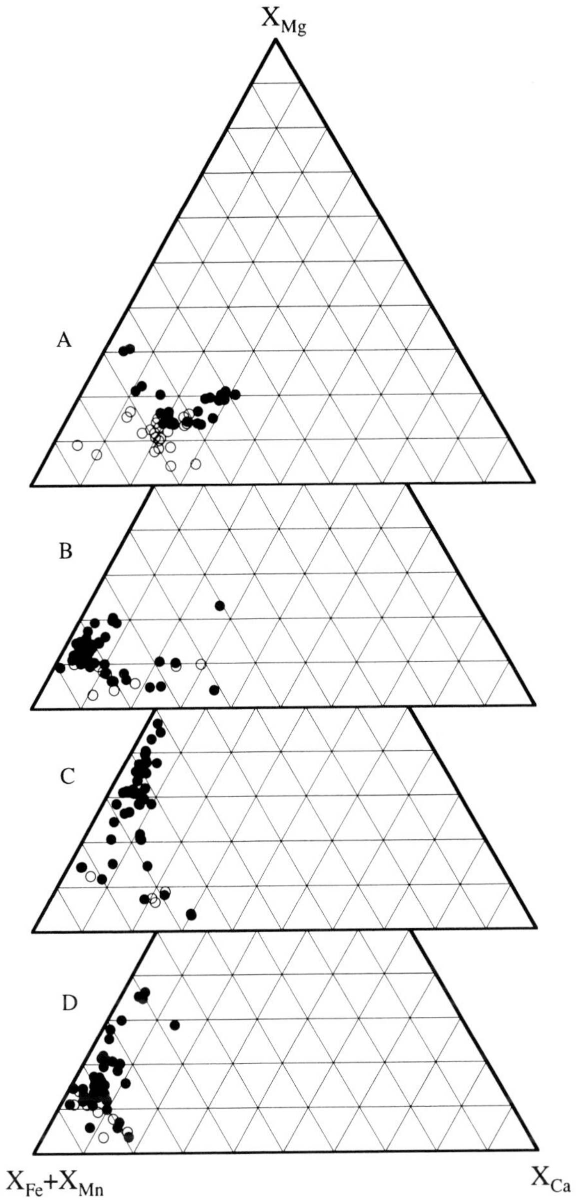

Fig. 9 Garnet compositions from East Greenland Cretaceous sandstones with SHRIMP zircon age data. Sample W4337 (Cenomanian, Hold with Hope) is garnet-poor and no geochemical data are available. A = W4567 (Barremian, Store Koldewey) B = W4346 (Santonian, Hold with Hope) C = W4470 (Cenomanian, Geographical Society Ø) D = W3920 (Turonian, Traill Ø) X_{Fe}, X_{Mg}, X_{Ca}, X_{Mn} = molecular values of Fe, Mg, Ca and Mn respectively, calculated on the basis of 24 oxygens, and normalised to total Fe + Mg + Ca + Mn, as recommended by Droop and Harte (1995). All Fe calculated as Fe^{2+}. ● – garnets with $X_{Mn} < 5\%$, ○ – garnets with $X_{Mn} > 5\%$.

range. As discussed in connection with the K1 age spectra, the reliability of the ages of some of the Phanerozoic zircons is questionable because some may have suffered Pb-loss. However, in view of the large number of zircons with post-Caledonian apparent ages and the fact that most of these fall precisely on the concordia curve, the majority of the apparent ages are considered reliable. The spectrum shows particular peaks at c. 250 Ma (early Triassic), c. 280 Ma (mid-Permian) and c. 300 Ma (latest Carboniferous). The significance of the scarce younger zircons (153 Ma, 162 Ma) is uncertain, and consequently, these have not been included on the age spectrum. Since all of these young zircons show zoning, many are euhedral-subhedral, and all have high Th/U ratios, the data are interpreted as providing evidence for post-Caledonian felsic igneous activity.

The Cenomanian sample also includes Precambrian zircons, most of which lie in the 1500–2050 Ma range (Fig. 10). This large group has a similar bimodal distribution to that seen in the Store Koldewey and Hold with Hope samples. The older peak, at c. 1950 Ma, is virtually identical to that seen in the other two samples, and the other peak, at c. 1650 Ma, is very similar to that found in the Hold with Hope sample and c. 30 Ma younger than that found in Store Koldewey. In addition, there is an Archaean-earliest Proterozoic population (2421–2962 Ma) and a small number of zircons between c. 900–1200 Ma (Grenvillian/Sveconorwegian).

Traill Ø

The Turonian sample (W3920) from Hold with Hope has relatively high garnet : zircon (GZi c. 75), moderate rutile : zircon (RuZi c. 57), low apatite tourmaline (ATi = 9), and very low monazite: zircon and chrome spinel : zircon (MZi and CZi both 0). The garnet assemblage is rich in the low Mg, variable Ca component suggesting derivation from amphibolite-facies metasediments, with a subordinate high Mg, low Ca group derived from granulite-facies metasediments or charnockites (Fig. 9).

The zircon age spectrum in W3920 (Fig. 10) is distinctive in containing a dominant group between c. 950–1300 Ma (broadly equating to the Grenvillian/Sveconorwegian). This contrasts with the other samples from East Greenland, which have comparatively few Grenvillian-age zircons. The other components of the spectrum are comparable to those present in the other East Greenland samples. There is a major group between 1500 Ma and 2000 Ma, although in this case the main peak in this range is at c. 1680 Ma, with zircons between 1900–2000 Ma being comparatively scarce. The sample also contains a small number of Archaean-earliest Proterozoic (2402–2619 Ma) and Caledonian-age grains. There is also a single, apparently concordant, grain dated as 297 Ma.

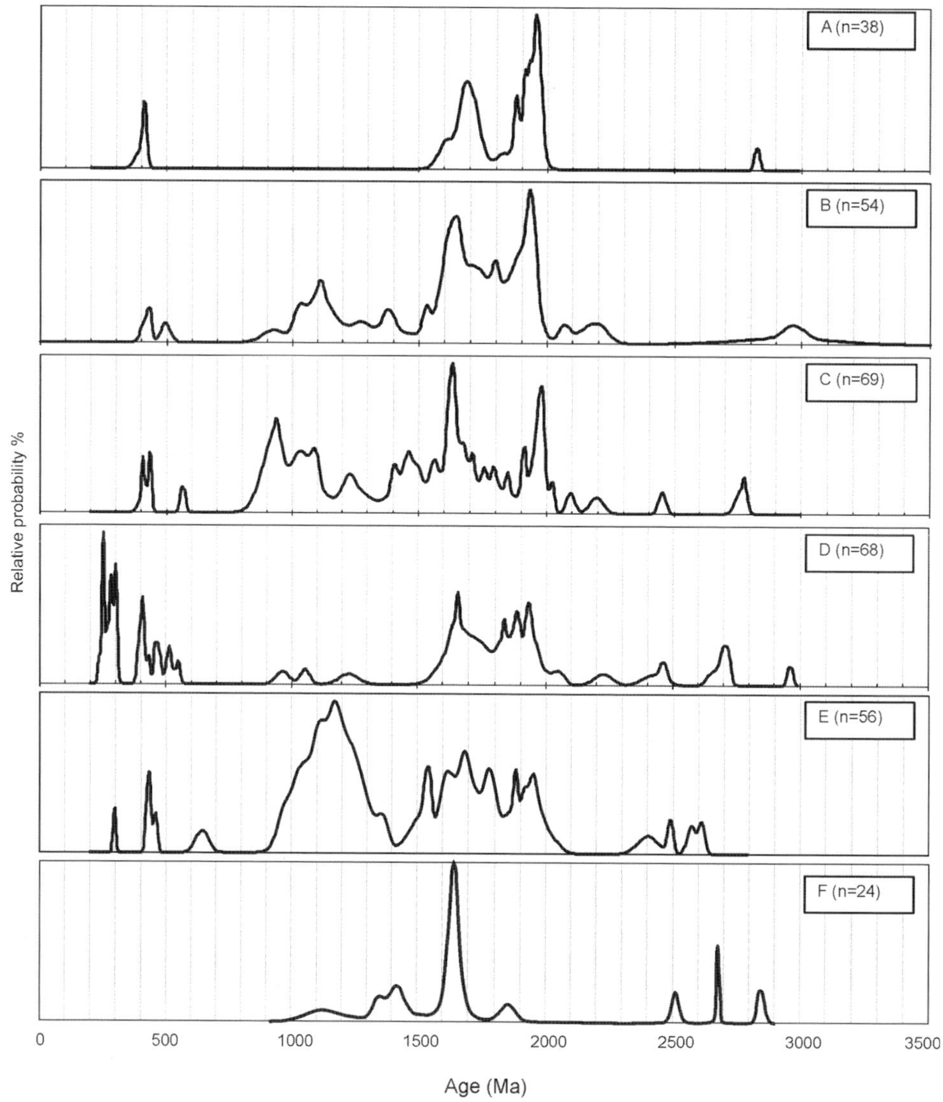

Fig. 10 Relative probability diagrams showing distribution of detrital zircon ages from East Greenland Cretaceous sandstones. A = W4567 (Barremian, Store Koldewey). B = W4337 (Cenomanian, Hold with Hope). C = W4346 (Santonian age, Hold with Hope). D = W4470 (Cenomanian, Geographical Society Ø). E = W3920 (Turonian, Traill Ø). F = Nathorst Land Group (Eleanore Bay Supergroup) sample 445335 (data from Watt et al., 2000).

Discussion

Zircon age data from the Cretaceous sandstones of the East Greenland margin has identified the involvement of six distinct crust-forming events, one in the Archaean, two in the Early Proterozoic, one in the late Mid-Proterozoic, one in the Palaeozoic and one in the Permo-Triassic.

The Archaean is represented in all the samples, although it is least abundant in the most northern sample, from Store Koldewey. This in accordance with the distribution of Archaean basement in the region (Escher and Pulvertaft, 1995), which dies out towards the north (Fig. 1). However, even in the north of the study region where Archaean basement is absent, some Palaeoproterozoic gneisses have high Tdm model ages (> 3 Ga, Kalsbeek et al., 1993), indicating the local incorporation of precursor Archaean material. Therefore, the single Archaean grain in the Store Koldewey sample could either have been derived from Archaean crust or from inherited Archaean material within the Palaeoproterozoic terrain.

Kalsbeek et al. (1993) and Nutman and Kalsbeek (1994) have identified an Early Proterozoic (> 1600 Ma) terrain in NE Greenland in the vicinity of Store Koldewey on the basis of Rb–Sr and Sm–Nd dating of orthogneisses. These orthogneisses contain zircons with SHRIMP U–Pb ages of 1974 ± 17 Ma, 1764 ± 20 Ma and 1739 ± 11 Ma, believed to represent two phases of crustal melting and granite generation. The detrital zircon age data from the Barremian sample of Store Koldewey indicate that this terrain was the main source of sediment, and show that no other major geological events affected this area. Thus, the outline geological history proposed by Kalsbeek et al. (1993) is confirmed by the detrital zircon geochronology. The four grains between 390–415 Ma indicate a minor degree of Caledonian reworking, which was also found by Kalsbeek et al. (1993).

The bimodal Early Proterozoic pattern seen both in the basement of NE Greenland and in the Barremian sample of Store Koldewey can also be detected further south, in Hold with Hope, Geographical Society Ø and Traill Ø, with all the samples showing distinct age clusters between 1900–2000 Ma and between 1630–1680 Ma, and derivation from Early Proterozoic basement can be inferred. However, zircons corresponding to the younger of these two groups could alternatively have been derived from the Precambrian metasediments. The majority of zircons in the Krummedal and Stauning Alper migmatites and in Nathorst Land Group metasediments lie in the 1500–1700 Ma range (Watt et al., 2000), with a strong peak at c. 1650 Ma that corresponds closely to the peak seen in samples from Hold with Hope and the Geographical Society Ø.

The influence of the Grenvillian orogenic event (c. 900–1200 Ma) can be recognised in the samples from Geographical Society Ø, Traill Ø and Hold with Hope, and is especially strong in the Turonian sample from Traill Ø. This is in accord with the view of Escher and Pulvertaft (1995), who indicate that the basement in this area was affected by Grenvillian events. There is no evidence of Grenvillian activity to the north, in the Store Koldewey area. This accords to the view of Kalsbeek et al. (1993), who consider that Grenvillian activity was essentially confined to the area south of 74°N.

As with the Grenvillian, the effects of the Caledonian Orogeny appear to have been minor, but widely distributed. Their age distribution appears to be bimodal, with all four from Store Koldewey falling in the 390–415 Ma bracket, as do two from Hold with Hope and five from Geographical Society Ø. In addition, two zircons from Hold with Hope, three from Traill Ø and four from the Geographical Society Ø, fall in the 430–477 Ma range. This supports an earlier evidence for polyphase Caledonian events in East Greenland. Strachan et al. (1995) identified zircon growth at 445 ± 10 Ma in the Smallefjord sequence, whereas Kalsbeek et al. (1993) recognised a later event at 404 ± 6 Ma in the basement gneisses of Dove Bugt.

Possibly the most surprising feature of the zircon age study was the identification of a major group of post-Caledonian zircons in the Cenomanian sample from Geographical Society Ø. The spectrum shows particular peaks at c. 250 Ma (early Triassic), c. 280 Ma (mid-Permian) and c. 300 Ma (latest Carboniferous). The main peaks in this age distribution correspond to major rift events in East Greenland. Late Carboniferous-Early Permian rifting resulted in pronounced fault-block-rotations and was followed by regional uplift and peneplanation (Surlyk et al., 1986). Early Triassic rifting is indicated by growth faults on Wegner Halvø (Seidler, 2000) and a sharp increase in water-loaded basement subsidence rate (Price and Whitham, 1997). However, at present, there is little evidence to suggest that these rift events were associated with magmatism, although Permian lamprophyre dykes have been identified in southern Scoresby Land by Stemmerik and Sørensen (1980). The detrital zircon age data suggest that Permo-Triassic rift-related magmatism may have been more significant than is presently recognised.

Provenance links between East Greenland and K2 sandstones

The above discussion shows that there are similarities in mineralogy and zircon age data between the K2 sandstones in the Vøring Basin and parts of the East Greenland Cretaceous succession.

One of the characteristic features of the K2 sandstones is the prevalence of low Mg, variable Ca garnet, in association with minor amounts of high Mg, low Ca garnet (Fig. 5). Sandstones characterised by low Mg, variable Ca garnet assemblages are also found in the East Greenland Cretaceous, such as the Santonian of Hold with Hope and the Turonian of Traill Ø (Fig. 9). Furthermore, the East Greenland Cretaceous also contains sandstones with abundant high Mg, low Ca garnets, similar to those forming the subsidiary group in the K2 assemblages. Such assemblages are seen, for example, in the Cenomanian of Geographical Society Ø (Fig. 9). By contrast, garnet geochemistry rules out a link between the source of the Barremian sandstones of Store Koldewey and that of K2, with the Barremian sandstones being dominated by high Mg, high Ca garnets derived from high-grade orthogneisses, which are distinctly different to those in K2.

The zircon age data also suggest links between the two areas. The critical factor distinguishing K1 and K2 sandstones is the co-occurrence of Archaean and early Proterozoic (1800–2100 Ma) zircons in K2. Both of these groups occur in all the East Greenland samples analysed to date. Most of the other components of the K2 age spectra also occur in the East Greenland samples, including the Early Proterozoic (1500–1800 Ma), which is found in all the samples; the Mid-Proterozoic (Grenvillian, 900–1200 Ma), which is particularly well-developed in the Turonian of Traill Ø, the Caledonian (390–460 Ma), which occurs in all samples, and the Permo-Triassic (c. 230–300 Ma), which is present in the Cenomanian of Geographical Society Ø.

Using the mineralogical and the zircon age data in combination, constraints can be placed on the likely potential entry points for K2 sandstones into the Vøring Basin. The limited diversity of the zircon age spectrum shown by the Barremian sample from Store Koldewey rules out a link with K2, supporting the evidence from the garnet data outlined above. Furthermore, the dominance of the Grenvillian-age group (950–1300 Ma) in Turonian sample W3920 argues against derivation of K2 from the Traill Ø region. The predominance of high Mg, low Ca garnet in the Cenomanian of Geographical Society Ø contrasts with the garnets found in K2, although these garnets form a subsidiary component of the K2 garnet assemblage. The closest match to K2 in terms of both mineralogy and zircon age data therefore appears to be Hold with Hope, since the Santonian sample from this region has a similar garnet assemblage and a zircon population with similar age groupings. A minor influence from the source that supplied the Cenomanian sample from Geographical Society Ø also appears to be required, given the presence of high Mg, low Ca garnet and Permo-Triassic zircons in the K2 sandstones.

Despite the overall similarities, K2 sandstones have some features that cannot be matched with the East Greenland Cretaceous samples described in this chapter. The most notable mineralogical difference is the higher rutile:zircon in the Hold with Hope, Geographical Society Ø and Traill Ø samples compared with K2. In terms of the zircon age spectra, none of the analysed East Greenland samples contain Mid-Cretaceous zircons, and the cluster of zircon ages at c. 1600–1700 Ma seen in all East Greenland samples is very subdued in the K2 spectra.

Consideration of the Cretaceous evolution of East Greenland provides additional evidence for the region being the source of the K2 sandstones and a possible explanation as to the lack of a direct match between sandstone types in the two areas. The rifted topography that dominated the basin architecture during the Early Cretaceous would have acted as a trap for sediment derived from Greenland and prevented significant eastward sediment transport (Whitham et al., 1999). With the end of rifting in the Late Cretaceous, sediment would have covered the rifted seafloor topography, created a shelf break margin in East Greenland and smoothed out the basin floor. This would have favoured the transport of sediment into the centre of the Vøring Basin. The temporal distribution of K2 sandstones broadly supports the inferences provided by the Greenland observations since they do not occur in the pre-Turonian succession (Morton and Grant, 1998). After rifting in the Mid-Albian, thermal subsidence of the rift flanks would have occurred, causing the burial of many of areas of East Greenland basement that are currently exposed, some of which would have been the source of Pre-Cenomanian sediment. Rift-flank subsidence may also have captured depositional systems sourced in the interior of Greenland and previously excluded from the basin during rifting. This may explain the subdued nature of the 1600–1700 Ma peak in the K2 sandstones, since this is the main peak found in the Proterozoic supracrustal successions that separate the crystalline basement of interior Greenland from the Phanerozoic sedimentary basins, along the East Greenland margin (Watt et al., 2000). Reduced input from Proterozoic metasediments such as the Eleanore Bay Supergroup would account for the absence of a specific cluster of zircons dated between 1600–1700 Ma in the K2 sandstones. A change in the drainage pattern may also account for the absence of the c. 100 Ma peak in the East Greenland sandstones, although another explanation might be input from Cretaceous igneous centres located within the Vøring Basin. Another point that should be borne in mind is that the K2 sandstones in the Vøring Basin represent an average of a large number of individual point sources in East Greenland.

This chapter, therefore, provides firm isotopic evidence for deriving K2 sandstones from East Greenland. This conclusion is shared by Fonneland et al. (2004), and was also reached independently through seismic mapping work (Vergara et al., 2001). However, there are other potential sources also. One possibility is that material with K2 characteristics derived from East Greenland found its way to Mid Norway, prior to the Cretaceous, perhaps during the Jurassic or Triassic. Reworking of such material from fault-block crests and from the basin flanks would generate sandstones with K2 mineralogy and zircon age spectra. Another potential source that might have supplied Early Proterozoic and Archaean zircons to the Vøring Basin region is the area east of the Caledonian front (Fig. 1), which consists of Svecofennian basement. Although today there are no rivers in the Mid Norway region that drain from this region into the Atlantic, it is possible that during periods of post-rift thermal subsidence, rivers to the east of the Caledonian front were captured and flowed westward, bringing material of Archaean and Early Proterozoic age into the Vøring Basin.

Concluding remarks

Earlier work on the heavy mineral and mineral chemical characteristics of Late Cretaceous sandstones from the Trøndelag Platform, Halten Terrace and Vøring Basin identified two distinct sandstone types, K1 and K2. These two sandstone types indicate the involvement of two different source regions (Morton and Grant, 1998). Additional SHRIMP U–Pb dating of zircons from sandstone types K1 and K2 has validated this mineralogical differentiation, and has confirmed that the sediment was supplied from two different source regions with different geological histories. A similar conclusion was reached by Fonneland et al. (2004), on the basis of a combined U–Pb and Pb–Pb dating study in the same area.

The K1 sandstones, which occur in wells along the eastern margin of the basin, have zircon age spectra that can be tied back to the adjacent Scandinavian landmass. Most of the zircons fall into a relatively narrow age group peaking at c. 1780–1790 Ma, implying derivation from the Trans-Scandinavian Igneous Belt. Other components of the zircon spectra, in conjunction with the heavy mineral evidence, indicate contributions from the Precambrian metasediments forming the Caledonian nappes, and granites.

The K2 sandstones have much more complex zircon age spectra, with components that cannot be readily tied back to the adjacent parts of Scandinavia, most notably the co-occurrence of Archaean and Early Proterozoic groups. The K2 sandstones share many zircon age characteristics with the Cretaceous sandstones exposed in East Greenland, most notably those from the Hold with Hope, Geographical Society Ø and Traill Ø areas, and clear links between their provenances can be established. Derivation of K2 sandstones from East Greenland is therefore considered most probable. A link in provenance between Greenland and the Vøring Basin is of great importance for petroleum exploration because it predicts thickening sandstone packets and increasing net : gross ratios away from the Mid-Norwegian coastline. This is contrary to initial impressions, when looking at a map of the present day Norwegian Sea.

Certain questions remain to be answered, however. It remains possible that some of the K2 sandstones were not sourced directly from East Greenland, but through recycling of Jurassic sediment of ultimate Greenland origin. Another possibility is that some sediment was shed from east of the Caledonian front, from the Svecofennian domain. The zircon age data from both K2 sandstones and from the Geographical Society Ø area of East Greenland has identified a magmatic episode in the Permo-Trias, the geological evidence for which is scanty, but may be related to contemporaneous rifting events. The data also provide evidence for a Mid-Cretaceous magmatic event, but as with the Permo-Trias, the locations of the igneous centres responsible remain unknown.

Acknowledgements

The authors are grateful for the financial support of BP, ConocoPhillips, ChevronTexaco, ExxonMobil, Shell and Statoil. Clive Johnson, Simon Kelly, Kenn Nielsen, Caroline Pickles, Simon Price and Dominic Strogen are thanked for their assistance and valuable company in the field. The work also has benefited from the reviews of Hege Fønneland, Hans Amundsen and Rónadh Cox.

References

Brekke, H., 2000. The tectonic evolution of the Norwegian Sea Continental Margin with emphasis on the Vøring and Møre Basins. In: Nøttvedt, A. et al. (eds.), Dynamics of the Norwegian Margin. Geological Society, London, Special Publication, 167: 327–378.

Brekke, H., Dahlgren, S., Nyland, B. and Magnus, C., 1999. The prospectivity of the Vøring and Møre basins on the Norwegian Sea continental margin. In: Fleet, A.J. and Boldy, S.A.R. (eds), Petroleum Geology of Northwest Europe: Proceedings of the 5th Conference. Geological Society of London, 261–274.

Brueckner, H.K., Gilotti, J.A. and Nutman, A.P., 1998. Caledonian eclogite-facies metamorphism of Early Proterozoic protoliths from the North-East Greenland Eclogite Province. Contributions to Mineralogy and Petrology, 130: 103–120.

Claoué-Long, J.C., Compston, W., Roberts, J. and Fanning, C.M., 1995. Two Carboniferous ages: a comparison of SHRIMP zircon dating with conventional zircon ages and 40Ar/39Ar analysis. In: Berggren, W.A., Kent, D.V. and Hardenbol, J. (eds). Geochronology, time scales and global stratigraphic correlation. SEPM Special Publication, 54: 3–21.

Dodson, M.H., Compston, W., Williams, I.S. and Wilson, J.F., 1988. A search for ancient detrital zircons in Zimbabwean sediments. Journal of the Geological Society, London, 145: 977–983.

Donovan, D.T., 1953. The Jurassic and Cretaceous stratigraphy and palaeontology of Traill Ø, East Greenland. Meddelelser om Grønland, 111(4): 1–150.

Donovan, D.T., 1955. The stratigraphy of the Jurassic and Cretaceous rocks of Geographical Society Ø, East Greenland. Meddelelser om Grønland, 103(9): 1–60.

Droop, G.T.R. and Harte, B., 1995. The effect of Mn on the phase relations of medium grade pelites: constraints from natural assemblages on petrogenetic grid topology. Journal of Petrology, 36: 1549–1578.

Escher, J.C. and Pulvertaft, T.C.R., 1995. Geological map of Greenland, 1:2,500,000. Geological Survey of Greenland, Copenhagen.

Fonneland, H.C., Lien, T., Martinsen, O.J., Pedersen, R.B. and Kosler, J., 2004. Onshore and offshore provenance studies: a key to understanding the deposition of deepmarine sandstones in the Norwegian Sea. Sedimentary Geology, 164: 147–156.

Gaal, G. and Gorbatschev, R., 1987. An outline of the Precambrian evolution of the Baltic Shield. Precambrian Research, 35: 15–52.

Henry, D.J. and Guidotti, C.V., 1985. Tourmaline as a petrogenetic indicator mineral: an example from the staurolite-grade metapelites of NW Maine. American Mineralogist, 70: 1–15.

Jacobsen, S.B. and Wasserburg, G.J., 1978, Interpretation of Nd, Sr and Pb isotope data from Archaean migmatites in Lofoten-Vesteralen, Norway. Earth and Planetary Science Letters, 41: 245–253.

Kalsbeek, F., Nutman, A.P. and Taylor, P.N., 1993. Palaeoproterozoic basement province in the Caledonian fold belt of North-East Greenland. Journal of Precambrian Research, 63: 163–178.

Kittilsen, J.E., Olsen, R.R., Marten, R.F., Hansen, E.K. and Hollingsworth, R.R., 1999. The first deepwater well in Norway and its implications for the Upper Cretaceous play, Vøring Basin. In: Fleet, A.J. and Boldy, S.A.R. (eds) Petroleum Geology of Northwest Europe: Proceedings of the 5th Conference. Geological Society, London, 275–280.

Ludwig, K.R., 1999. Isoplot, a geochronological toolkit for Microsoft Excel. Berkeley Geochronology Center, Special Publication 1a.

Maync, W., 1949. The Cretaceous between Kuhn Island and Cape Franklin (Gauss Peninsula), northern East Greenland. Meddelelser om Grønland, 133(2): 1–291.

Morton, A.C. and Grant, S., 1998. Cretaceous depositional systems in the Norwegian Sea: heavy mineral constraints. Bulletin of the American Association of Petroleum Geologists, 82: 274–290.

Morton, A.C. and Hallsworth, C.R., 1994. Identifying provenance-specific features of detrital heavy mineral assemblages in sandstones. Sedimentary Geology, 90: 241–256.

Morton, A.C. and Hallsworth, C.R., 1999. Processes controlling the composition of heavy mineral assemblages in sandstones. Sedimentary Geology, 124: 3–29.

Nøhr-Hansen, H., 1993. Dinoflagellate cyst stratigraphy of the Barremian to Albian, Lower Cretaceous, North-East Greenland. Grønlands Geologiske Undersøgelse Bulletin, 166: 171 pp.

Nutman, A.P. and Kalsbeek, F., 1994, Search for Archaean basement in the Caledonian fold belt of North-East Greenland. Grønlands Geologiske Undersøgelse Rapport, 162: 129–134.

Price, S.P. and Whitham, A.G., 1997. Exhumed hydrocarbon traps in East Greenland: analogs for the Lower-Middle Jurassic play of NW Europe. Bulletin of the American Association of Petroleum Geologists, 81: 196–221.

Scott, R.A., 2000. Mesozoic-Cenozoic evolution of East Greenland: implications of a reinterpreted continent-ocean boundary location. Polarforschung, 68: 83–91.

Seidler, L., 2000. Incised submarine canyons governing new evidence of Early Triassic rifting in East Greenland. Palaeogeography, Palaeoclimatology and Palaeoecology, 161: 267–293.

Shanmugam, G., Lehtonen, L.R., Straume, T., Syvertsen, S.E., Hodgkinson, R.J. and Skibeli, M., 1994. Slump and debris-flow dominated upper slope facies in the Cretaceous of the Norwegian and northern North Sea (61–67°N): implications for sand distribution: Bulletin of the American Association of Petroleum Geologists, 78: 910–937.

Spencer, A., Birkelund, Ø., Knag, G. and Fredsted, R., 1999. Petroleum systems of the Atlantic margin of northwest Europe. In: Fleet, A.J. and Boldy, S.A.R. (eds.), Petroleum Geology of Northwest Europe: Proceedings of the 5th Conference. Geological Society, London, 231–246.

Strachan, R.A., Nutman, A.P. and Friderichsen, J.D., 1995. SHRIMP U-Pb geochronology and metamorphic history of the Smallefjord sequence, NE Greenland Caledonides. Journal of the Geological Society, London, 152: 779–784.

Stemmerik, L. and Sørensen, M., 1980. Upper Permian dykes in southern Scoresby Land. East Greenland. Grønlands Geologiske Undersøgelse Rapport, 100, 108.

Surlyk, F., 1978. Submarine fan sedimentation along fault scarps on tilted fault blocks (Jurassic-Cretaceous boundary, East Greenland). Grønlands Geologiske Undersøgelse, Bulletin, 128: 1–108.

Surlyk, F., Hurst, J.M., Piasecki, S., Rolle, F., Scholle, P.A., Stemmerik, L. and Thomsen, E., 1986. The Permian of the western margin of the Greenland Sea — a future exploration target. In: Halbouty M.T. (Ed.), Future petroleum provinces of the World. American Association of Petroleum Geologists, Memoir, 40: 629–659.

Thrane, K., 2002. Relationships between Archaean and Palaeoproterozoic crystalline basement complexes in the southern part of the East Greenland Caledonides: an ion microprobe study. Precambrian Research, 113: 19–42.

Vergara, L., Wreglesworth, I., Trayfoot, M. and Richardson, G., 2001. The distribution of Cretaceous and Paleocene deep-water reservoirs in the Norwegian Sea basins. Petroleum Geoscience, 7: 395–408.

Vischer, A., 1943. Die postdevonische tektonik von ostgrönland zwischen 74° und 75°N. Meddelelser om Grønland, 133(1): 1–194.

Watt, G.R., Kinny, P.D. and Frederichsen, J.D., 2000. U-Pb geochronology of Neoproterozoic and Caledonian tectonothermal events in the East Greenland Caledonides. Journal of the Geological Society, London, 157: 1031–1048.

Watt, G.R. and Thrane, K., 2001. Early Neoproterozoic events in East Greenland. Precambrian Research, 110: 165–184.

Whitham, A.G., Price, S.P., Koraini, A.M. and Kelly, S.R.A., 1999. Cretaceous (post-Valanginian) sedimentation and rift events in the NE Greenland (71–77°N). In: Fleet, A.J. and Boldy, S.A.R. (eds.), Petroleum Geology of Northwest Europe: Proceedings of the 5th Conference. Geological Society, London, 325–336.

Williams, I.S. and Claesson, S., 1987, Isotopic evidence for the Precambrian provenance and Caledonian metamorphism of high grade paragneisses from the Seve Nappes, Scandinavian Caledonides. II. Ion microprobe U-Th-Pb. Contributions to Mineralogy and Petrology, 97: 205–217.

The Norwegian Sea during the Cenozoic

Sverre Henriksen, Christine Fichler, Arne Grønlie, Tormod Henningsen, Inger Laursen, Helge Løseth, Dag Ottesen and Ian Prince

Based on 2D seismic surveys covering the entire Norwegian Sea (250 000 km^2), selected 3D surveys and an extensive well database, the Cenozoic depositional history for the area has been reconstructed. Interpretation of this large database has made possible a regional overview of, and a new insight into the Cenozoic depositional systems. Significant amounts of sediments were fed to the Norwegian Sea during the Cenozoic, while, apart from a thin Quaternary cover, no Cenozoic sediments are preserved onshore. This is interpreted to be the result of several phases of uplift and erosion of the mainland during this period. The sedimentary filling of the basins is interpreted in a sequence stratigraphic context, aiming towards a dynamic understanding of the depositional history. In the Palaeocene, extensional tectonics prevailed and the Norwegian Sea received sediments from uplifted land areas, both to the east and the west. The input of sediments to the deeper parts of the basin were to some degree determined by the intersection of NW–SE trending lineaments intersecting with older structural features on the shelf. With the onset of sea floor spreading in the Eocene, the tectonic regime changed from extensional to compressional. Extrusion of basaltic lavas dominated the western land areas, while a major transgressive event resulted in the deposition of shaly sediments on the eastern continental shelf. Large parts of Scandinavia were probably flooded during this time period. A deltaic system constituting the 'Molo Formation' was deposited all along the eastern Norwegian Sea margin, as a response to regional uplift of the Norwegian mainland. Difficulties in seismic ties and the sparse well control have made the actual age of the Molo Formation a subject for discussion. Both Oligocene and Early Pliocene ages have been suggested. New seismic correlations presented in this chapter suggest that the Molo Formation is Early Pliocene in age. Erosional channels with possible fluvial drainage patterns suggest subaerial exposure over large parts of the continental shelf during the Miocene. Prograding shelf geometries within Middle to Late Miocene sediments support this theory. An unconformity in the Miocene is associated with a strong compressional event leading to flexural doming and inversion of older depocentres on the shelf. Basin scale tectonic movements are the possible causes for both, the unconformity and the compressive movements. An Early Pliocene flooding event shifted the locus of sedimentation in an eastward direction, and the Molo Formation was the first sedimentary unit deposited onto this surface. A marked shift in the prograding style occurred in mid Pliocene, and Late Pliocene/Pleistocene glacial sediments prograded westward as continental ice sheets expanded onto the shelf. Once glacial conditions were established on the shelf, the glacial drainage pattern followed bedrock boundaries and older structural features in the subsurface.

Introduction

The early stage of exploration in the Norwegian Sea was restricted mainly to Jurassic targets in the Halten/ Dønna Terrace and Nordland Ridge areas (Fig. 1). Here, reservoir sandstones of Cenozoic age are rare and this play was consequently not considered very prolific. However, the hydrocarbon potential of the early Cenozoic is well-documented in the North Sea, west of Shetland and also in later years by the Ormen Lange gas find in the Norwegian Sea. Sandy deposits of early Cenozoic age are also found in the northern North Sea, at the Selje High, Gossen High, Frøya High, in the Helgeland Basin and the Vestfjorden Basin.

Following the opening of the Møre and Vøring Basins for exploration in 1995 (Norwegian 15th round), new exploration opportunities have been identified within the Cenozoic play in the Norwegian Sea between 63°N and 68°N (Fig. 1). The Ormen Lange Dome well (6305/5-1) was the first well in this area to have the Lower Cenozoic (Palaeocene) as its prime target. Apart from questions related directly to prospectivity, the changing Cenozoic depositional systems in relation to tectonics and basin physiography have exerted major control on burial history and thereby on the timing of formation, migration and trapping of hydrocarbons in the area.

One of the main objectives of this chapter is to establish a sequence stratigraphic framework for

Onshore–Offshore Relationships on the North Atlantic Margin
edited by B. Wandas et al. NPF Special Publication 12, pp. 111–133,

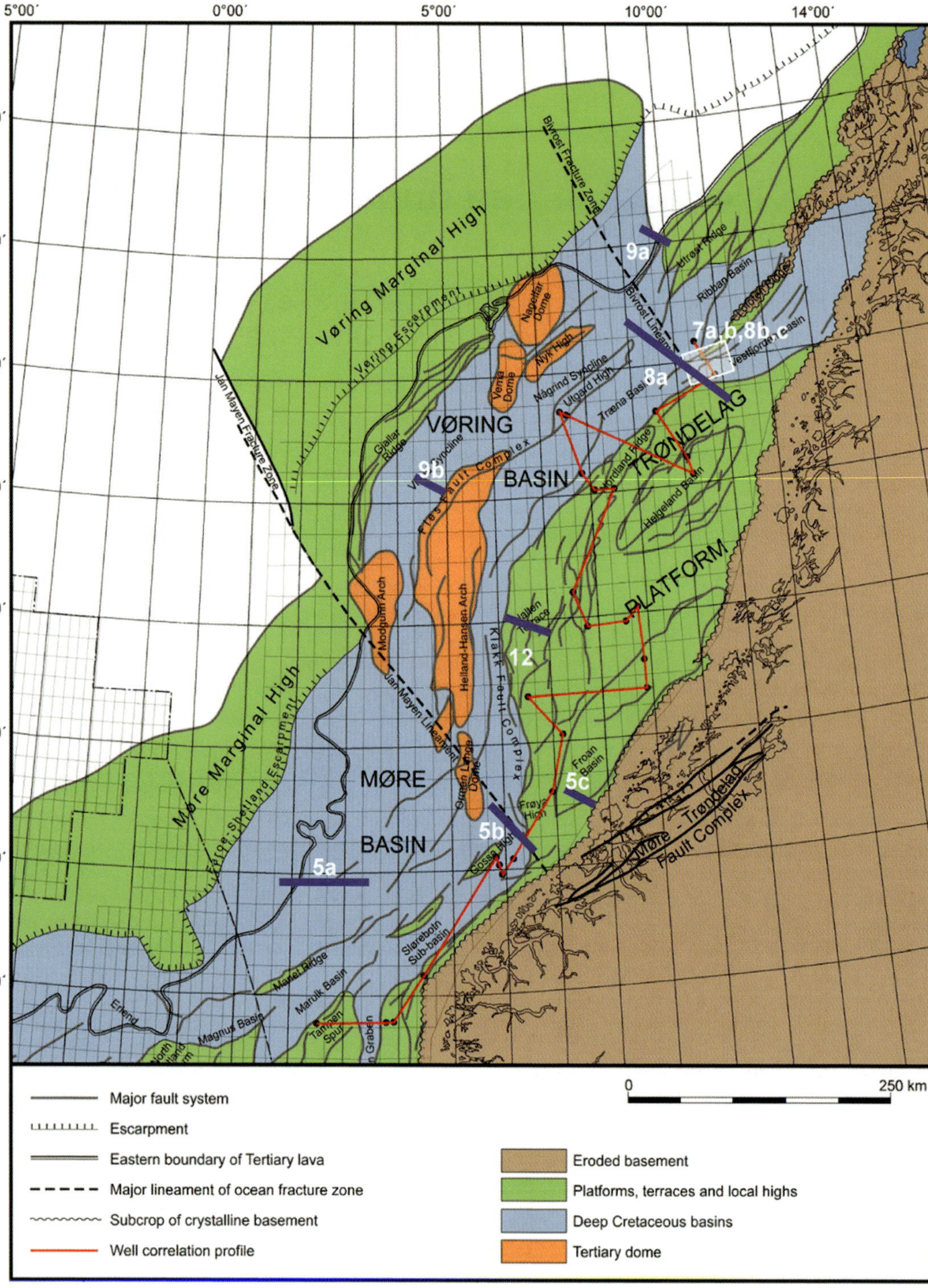

Fig. 1 Main structural elements of the Norwegian Sea (simplified from Blystad et al., 1995), and location of wells used in the correlation panel (Fig. 2), and location of figures marked by corresponding numbers. 3D area (Figs. 7a, b and 8b, c) in northern part of study area marked by white square.

the Norwegian Sea. By viewing the sequence development along depositional dip and strike in a basinwide perspective we also aim to create a dynamic approach to understanding the basin infill and stratigraphy in the region.

In this work, an extensive 2D seismic database covering most of the platform and basin areas in the Norwegian Sea, between 63° and 68° N, has been used in the interpretation of strata relationships and in construction of regional maps. A large number of surveys of different vintage have been used and altogether, several thousand kilometres of seismic data have been interpreted. The 3D seismic cubes from the Ormen Lange Dome, the

Helland-Hansen Arch, Nyk High/Vema Dome and the Nordland area, respectively, have been interpreted.

Palynological and micropalaeontological data from selected wells were used in a chronostratigraphic review and correlation of the Cenozoic succession, some of which has been grouped together in a correlation panel (Fig. 1). These wells have also been subject to palaeo-environmental interpretations.

Tectonics and geological framework

The main outline of the geological history of this area has been extensively covered by several authors, more recently; (Brekke and Riis, 1987; Blystad et al., 1995, Dore and Lundin, 1997; Lundin and Dore, 1997; Dore and Lundin, 1996; Brekke et al., 1999; Brekke, 2000), and is also considered to be common knowledge and will not be repeated in detail here, except when considered to have direct relevance to the Cenozoic.

Tectonism and magmatism, along the Møre and Vøring Margins, during NE Atlantic continental break-up lasted for a period of 15–20 My (Million years), from the onset of faulting in the Campanian–Maastrichtian to final continental separation at the Palaeocene–Eocene transition (Skogseid et al., 1992). The final stages of continental separation were accompanied by deposition of large volumes of lavas, mostly emplaced subaerially, as well as abundant sill intrusion in the basins adjoining the marginal highs (Fig. 1).

In the Norwegian Sea, extensional faulting took place along the Møre Marginal High, Fles Fault Complex, Gjallar Ridge, Nyk High, Utgard High and Utrøst Ridge (Fig. 1). In the west, igneous uplift affected the Møre and Vøring Marginal Highs and in the east the Norwegian landmass was uplifted, as indicated by increased clastic input.

On a larger scale, three major fault trends; NE–SW, N–S and NW–SE, define the overall basement grain and structural geometry of the area (Aanstad et al., 1981, Bucovics, 1984). This inherited Caledonian basement grain has to a large extent determined the later development of Mesozoic and Cenozoic basins and highs in the Norwegian Sea area (Aanstad et al., 1981; Blystad et al., 1995; Doré et al., 1997; Olesen et al., 1997). Together, the NW–SE trending lineaments and the main Caledonian NE–SW structural grain determine the width of the continental margin. In fact, Mosar et al. (2002) defines a large part of the onshore mountain belt as a part of the continental margin. In addition, movement along the lineaments, such as the Bivrost and the Jan Mayen lineaments and corresponding fracture zones have probably had major influence on the sedimentation throughout the Cenozoic.

With the onset of sea floor spreading in the North Atlantic, the tectonic regime changed from being mainly extensional to becoming mainly compressional. Several phases of compressional movements during the Eocene to Miocene period led to inversion and formation of structural domes in the Vøring Basin (e.g. Ormen Lange Dome, Helland-Hansen Arch and the Vema Dome).

The compressive movements in the Møre Basin seem to have been concentrated along the Jan Mayen Lineament, where the Ormen Lange Dome and a number of other similar domes are situated en èchelon along the lineament. Also in the Slørbotn Sub-basin and along the western rim of the Møre Basin, there are signs of inverted dome structures. However, we agree with Brekke et al. (1999) and Brekke (2000) in that there are few signs of compression in the central parts of the basin, and thus, large parts of the Møre Basin have subsided throughout the Cenozoic (Brekke, 2000).

Cenozoic stratigraphy

Two major surfaces are important in establishing a stratigraphic framework in a basin: (1) the erosional unconformity and (2) the downlap surface. Both surfaces appear as disconformities on the seismic section, but the processes involved in creation of the surfaces are significantly different. The erosional unconformity represents a true time stratigraphic break and represents the time extent of eroded sediments during a relative sea-level fall. It thus represents a sequence boundary and transport of sediments in a basinward direction (e.g. Posamentier and Allen, 1999).

The downlap surface represents a starvation surface produced during the time of transgression that subsequently forms a surface on which prograding clinoforms downlap. The corollary third surface, the transgressive surface, exists immediately beneath the downlap surface. It is formed during a transgression as the high-energy (wave-dominated) nearshore facies transgresses the shelf after a lowstand situation. This causes minor erosion and sediment starvation basinward of the transgressive beach. This also implies a landward shift in the locus of sedimentation and formation of a marine flooding surface (MFS) (Loutit et al., 1988). In practice, the transgressive surface becomes difficult to identify on seismic data and is generally regarded as constituting the basal part of the downlap surface. Sediment starvation of the outer shelf and deep basin will prevail during transgression, maximum highstand and even after turnaround and onset of highstand progradation. The entire time interval may be present, but due to the low sedimentation rates it is highly condensed and only detectable by high-resolution data. The inferred hiatus associated with the condensed section may thus be only apparent and not a true-time stratigraphic break.

The two types of surfaces have significant bearing on the stratigraphic architecture and the geological implications associated with each of them may be utilised in reconstruction of palaeogeography and prediction of lithology (c.f. Wilgus et al., 1988; Weimer and Posamentier, 1993).

The stratigraphic breaks and marine flooding surfaces observed from wells have been integrated with the horizon interpretation from the seismic database and a general stratigraphy for the Norwegian Sea is proposed (Figs. 2 and 3). On the Norwegian Sea continental shelf, the Cenozoic sediments comprise seven main seismic units (Fig. 3). Above the base Tertiary unconformity both Lower and Upper Palaeocene show a marked landward thickness increase. Over the shelf variable thickness of Palaeocene sediments are deposited. In the Vøring and Møre Basins, there are local depocentres containing sandy sediments. Later compressive movements inverted some of these depocentres. The Ormen Lange Dome is a good example of such a depocentre (Fig. 4). The Eocene also shows a marked landward thickness increase (Fig. 3). The internal reflection pattern of the Eocene is distorted by numerous small faults, confined to this package. However, an overall westward prograding reflection pattern can also be indicated for this unit. The distribution of Oligocene sediments on the inner shelf is uncertain. A high-angle sigmoid prograding unit with a deltaic appearance (Fig. 3), was given an Oligocene age by Eidvin et al., (1998). This is a matter of discussion and Henriksen and Weimer (1996) have suggested that the deltaic unit equally well may be Early Pliocene in age. The Oligocene becomes extremely thin over the shelf, but the deeper basins also seem to have been a locus for deep-marine sedimentation in the Oligocene. The Miocene is found as a thin wedge of low-angle westward prograding clinoforms on the middle and outer continental shelf. Also for the Miocene, a certain expansion in a basinward direction is indicated. Finally, a thick wedge of Late Pliocene and Pleistocene, mainly glacial, sediments is deposited on the shelf. This unit is recognised by a series of low-angle lateral persistent clinoforms (Fig. 3). On the inner shelf, the clinoforms merge and become truncated at the top by an upper regional unconformity (URU) defining the base of the Quaternary in the Barents Sea (Vorren et al., 1992). On the outer shelf, this relation is less obvious and Base Quaternary could by any of the westward dipping clinoforms (Fig. 3).

Palaeocene depositional system

Thick accumulations of Palaeocene sediments are found in both the Vøring and Møre Basins (Fig. 4). Sedimentary thickness reaches more than 750 m TWT in the Naglfar Dome and in the Vigrid and Någrind synclines of the northern Vøring Basin. Substantial elongate depocentres follow the trend of the Vestfjord Basin and its intersection with the Bivrost Lineament (Fig. 4). Additionally, an increase in the thickness of Palaeocene strata is observed in the Northern Helland Hansen Arch/Rås Basin. Evidently, the Naglfar Dome and the Helland Hansen Arch both represent inverted Palaeocene depocentres. In the Møre Basin, increased thickness of Palaeocene strata coincides with the intersection of two major long-lived structural elements: The Jan-Mayen Lineament (JML) and the Møre-Trøndelag Fault Zone (Fig. 4). The Ormen Lange Dome represents an inversion of this depocentre (Doré and Lundin, 1997).

Thick sediment accumulations in the western Norwegian Sea, e.g. in the central part of the Møre Basin, locally show downlap in an eastward direction, indicating a western source area for Palaeocene sediments (Figs. 4, 5a). The Palaeocene strata again show a marked thickness increase along the Møre Margin towards the Norwegian mainland (Fig. 5b). Thick accumulations of both the Lower and Upper Palaeocene strata indicate that the mainland acted as a source area for sedimentation throughout the entire Palaeocene (Fig. 5c). The input of Palaeocene sediments to the Norwegian Sea basins thus occurred from both eastern and western source areas and the main entry points for sediments were determined by the interaction of several structural elements.

The large-scale basin physiography has also had a major influence on the locus of Palaeocene

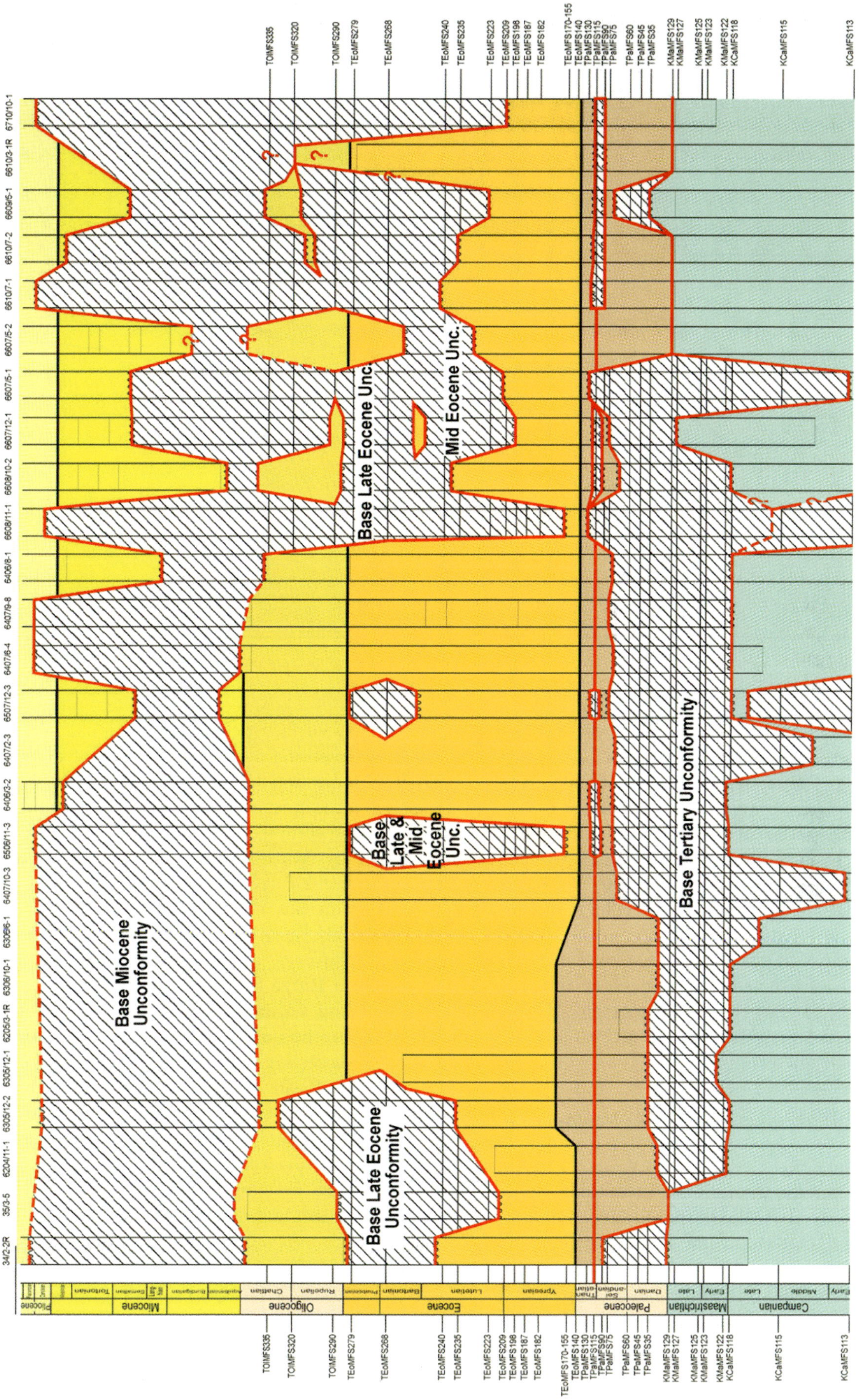

Fig. 2 Norwegian Sea regional chronostratigraphy for the Cenozoic. See Fig. 1 for location of wells.

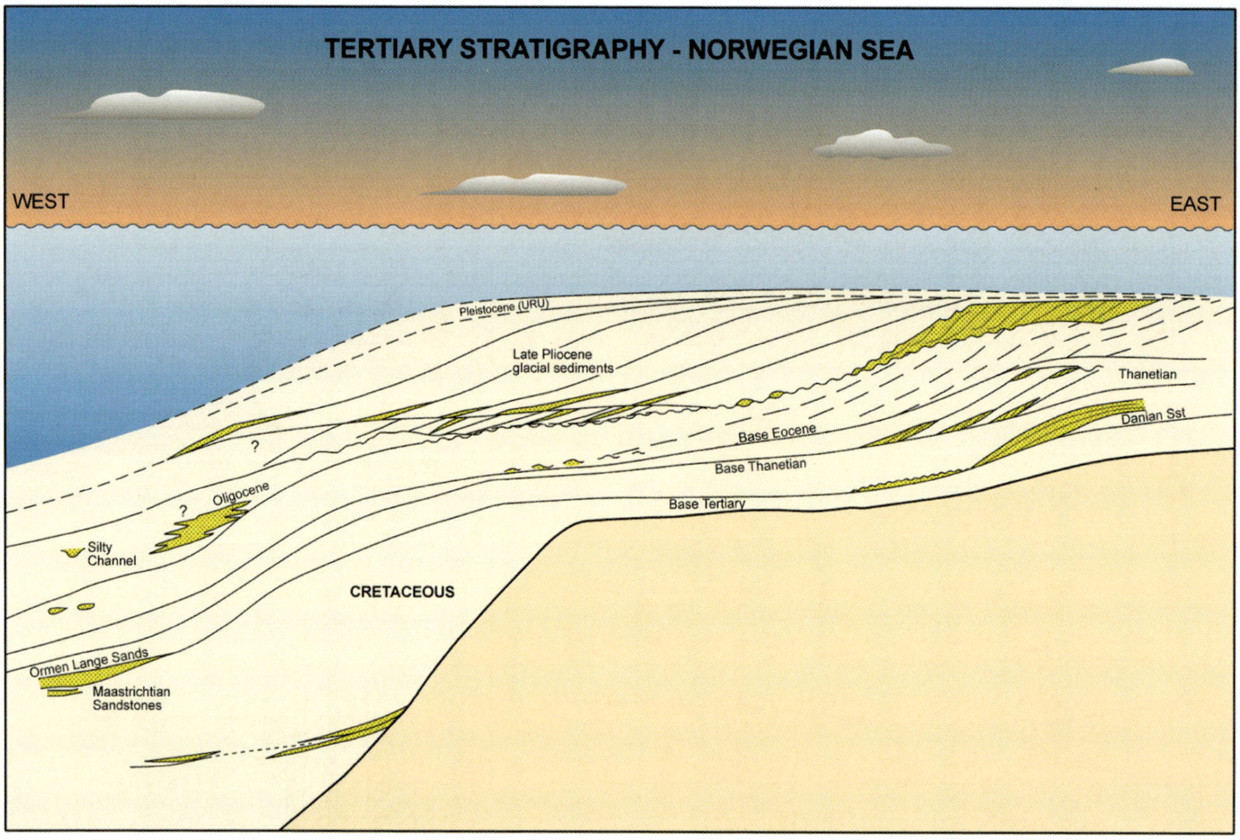

Fig. 3 Schematic and generalised cross-section over the Norwegian Sea shelf. Important sedimentary units are projected onto the section for illustration purposes.

depocentres and distribution of lithologies in the region. Like today, the Norwegian Sea was recognised by a narrow shelf and relatively steep slopes at the Lofoten and Møre Margins (Fig. 1). These areas have little capacity to store sediments in a landward position and the steep slopes provide effective bypass routes for sediments to the deep marine basins. The intervening Trøndelag Platform has been a wide and gently dipping shelf area throughout most of the Cenozoic. Consequently, large amounts of sediments may have been trapped in prograding sequences on the shelf. These sediments have largely been removed by the Late Cenozoic uplift and erosion.

Seismic stratigraphy of the Lower Palaeocene

Exploration wells along the Norwegian margin clearly show that the Danian is the most sand-prone Palaeocene interval. At the Møre Margin, southeast of the Ormen Lange Dome, the seismic signature of the Danian sandstones appear as high amplitude, semi-parallel reflections with good continuity (Fig. 5b). The internal reflections of the Danian downlap the base Tertiary surface along the inner margin (Fig. 5b), and this surface, therefore, defines a marine flooding surface. At the same time, the base Tertiary surface defines an erosional unconformity representing a variable amount of missing stratigraphic section (Fig. 2). Evidently, the base Tertiary is a composite surface, recording a number of events. The composite nature of this surface was also noted by Martinsen et al., 1999. The base Palaeocene break is recorded in most of the wells in the Norwegian Sea. The extent of the break is, however, highly variable. The largest break is found over the platform areas to the east and on the Nordland Ridge (structural highs). Because of the prograding nature of the Palaeocene on the Møre Margin, the base Palaeocene break is downlapped by successively younger sediments. In the western Møre Basin, there is also a clear downlap relation to the east, suggesting that the Palaeocene sediments above the break also become progressively younger towards the west. Eventually, the two progradings systems flatten out and meet near the middle of the Møre Basin.

Seismic stratigraphy of the Upper Palaeocene

Upper Palaeocene strata are generally thin over the platform areas, but a trend of landward thickening is evident over most of the study area. This is best expressed in the area between 63°N and

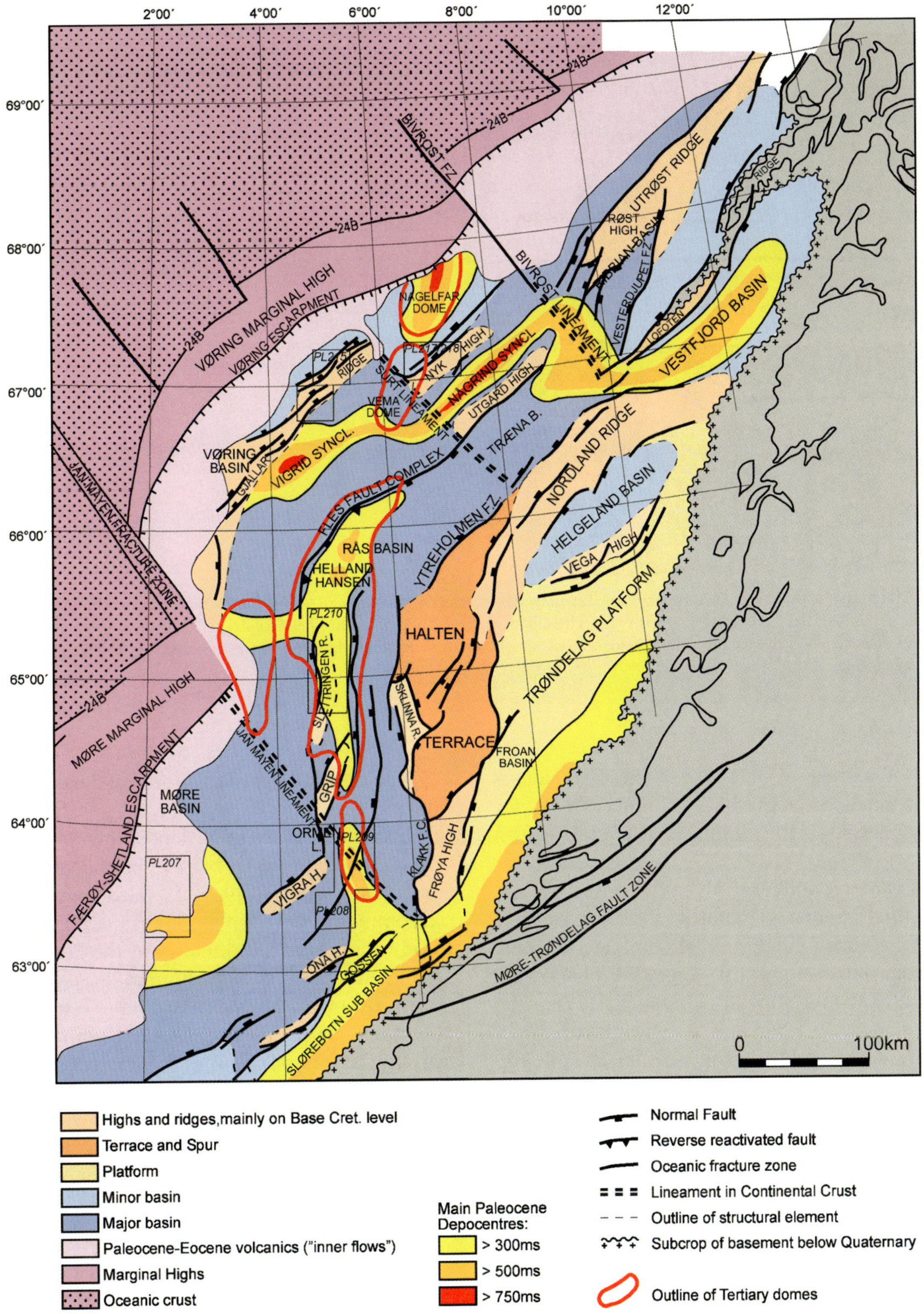

Fig. 4 Main Palaeocene depocentres of the Norwegian Sea. (Basemap modified from Blystad et al., 1995).

64°N, where the maximum thickness of the Upper Palaeocene strata reaches 500 m TWT (Fig. 5c). In this area, the Upper Palaeocene strata constitute a complex depositional system characterised by a series of prograding clinoforms (Fig. 5c). Some of these clinoforms are marked unconformities defined by onlap and erosional truncation of underlying strata. The stratal stacking pattern between these unconformities reveals all components of a depositional sequence as defined by Vail et al. (1977)

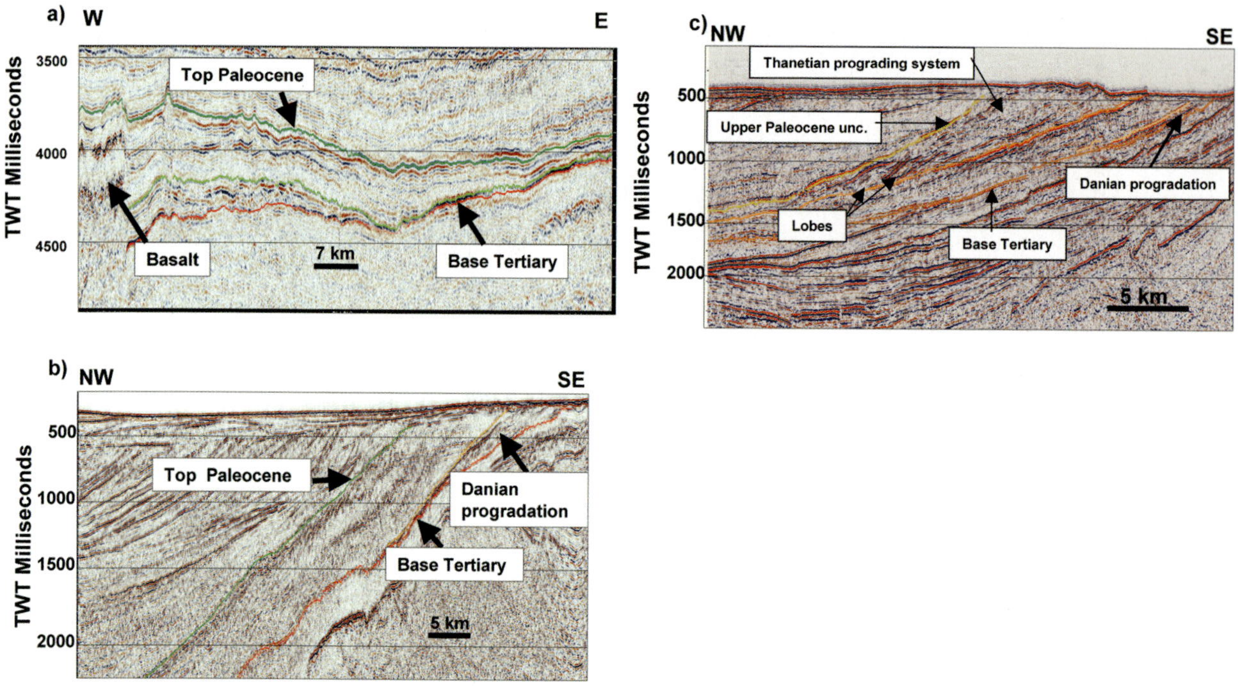

Fig. 5 (a) Internal Palaeocene reflectors showing downlap in an easterly direction in the western Møre Basin. (b) Danian prograding depositional system at the Trøndelag Platform. (c) Thanetian prograding pattern at the Trøndelag Platform. Note the well-developed unconformities with associated lobes at clinoform toes. Note also the small-scale prograding geometries near the top of the sequences. See Fig. 1 for location of figure.

and Van Wagoner et al. (1988). These sequences are interpreted to represent a higher frequency sequence development within an overall lower frequency system.

The downlap configuration at the lower boundary of the Thanetian defines a major marine flooding surface. Slightly below this flooding surface, the biostratigraphy suggests a shallowing in depositional environment. This suggests that the lower Palaeocene (Danian) is terminated by a sequence boundary. Seismic evidence of top Danian erosion is not clear but this event correlates with renewed input of siliciclastic sediments after deposition of chalk in Late Cretaceous and Early Tertiary in the North Sea (Berggren et al., 1995; Hardenbol et al., 1998). The erosional vacuity (if at all present) associated with the formation of this sequence boundary must have been situated eastward of the Tertiary subcrop line.

The break was most probably of short duration and a flooding surface (Tpal MFS90; Fig. 2) is situated directly above the inferred break. This flooding is of major significance and is recorded in all wells in the Norwegian Sea. Progressively, younger sediments downlap the flooding surface in a basinward direction and the condensed section thus records an increasing amount of time towards the west.

The overall gradual loss of accommodation during the Late Palaeocene (i.e. relative sea-level fall) suggests a regional depositional system developing in response to broad epeirogenetic uplift of the Norwegian mainland. This overall loss of accommodation eventually resulted in the formation of a sequence boundary in the late Palaeocene (Thanetian). This depositional break is evident, both from wells and seismic data (Figs. 2 and 5c). It is noteworthy that the Thanetian shelf edge prograded several tens of kilometres farther westwards (Fig. 5c). This may explain why this break, although shorter in duration, appears to be present even in the deeper part of the basins, where there is a continuous record of deposition over the K/T boundary (Fig. 2).

No significant basin floor fan or other lowstand deposits are recorded above this Late Palaeocene depositional break. Instead, a thin wedge of onlapping strata occur above the unconformity.

The seismic observations pertaining to wells with reviewed biostratigraphy, has enabled a palaeogeographic reconstruction of the Palaeocene in the Norwegian Sea (Fig. 6). Notable is the input of sediments from both the eastern (Norwegian mainland) and the western (Greenland/marginal highs) source areas, during this time period (Fig. 6).

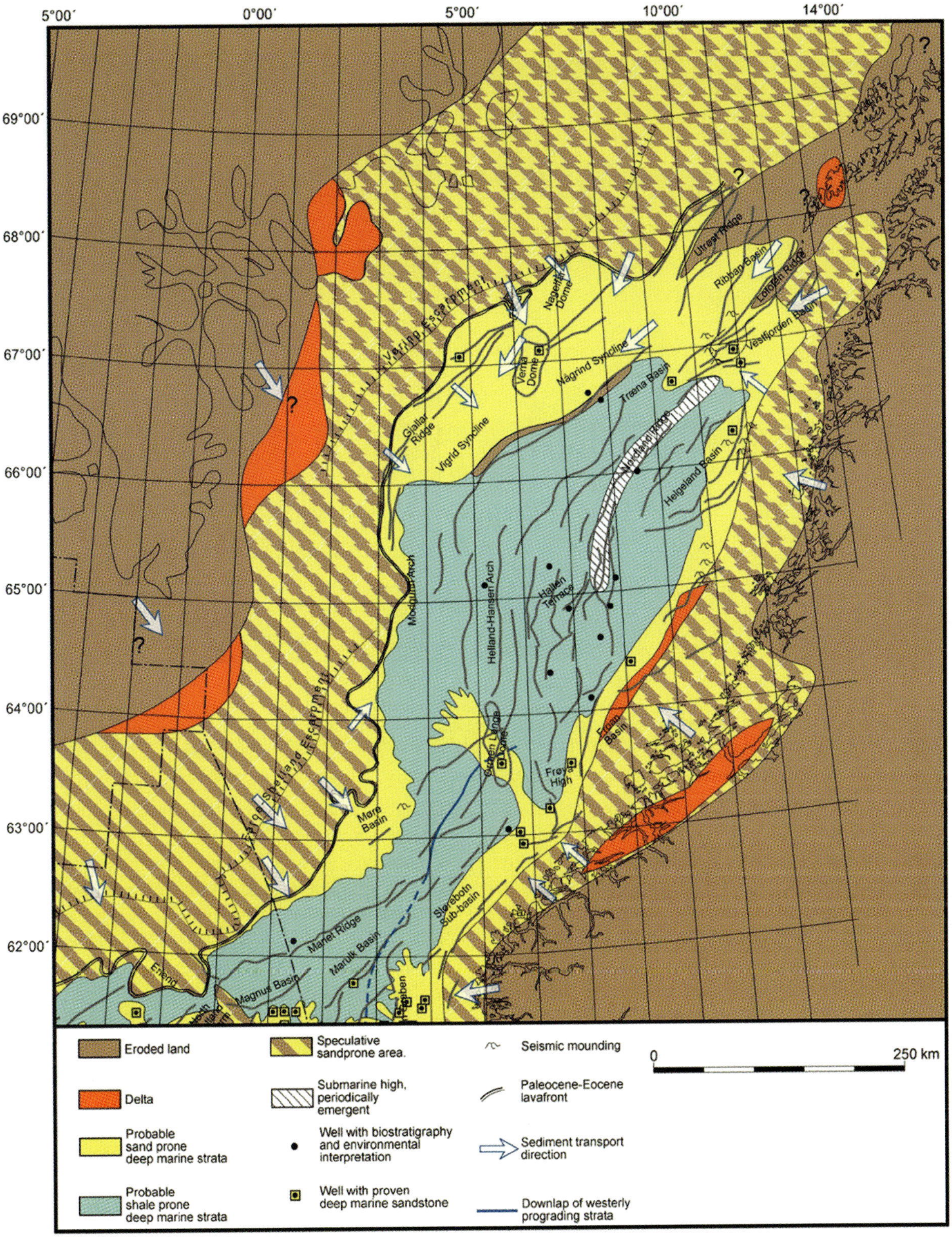

Fig. 6 Palaeocene palaeogeography and depositional environment in the Norwegian Sea.

Eocene depositional systems

The Møre and Vøring Marginal Highs were totally dominated by extrusion of basaltic lavas during the Eocene (c.f. Skogseid et al., 1992). The lavas are believed to have been subaerially emplaced (c.f. Mutter, 1984; Eldholm et al., 1984; Skogseid et al., 2000). Landwards the top Palaeocene/base Eocene surface also defines a major downlap surface (Figs. 3, 7a). This downlap surface is recognised

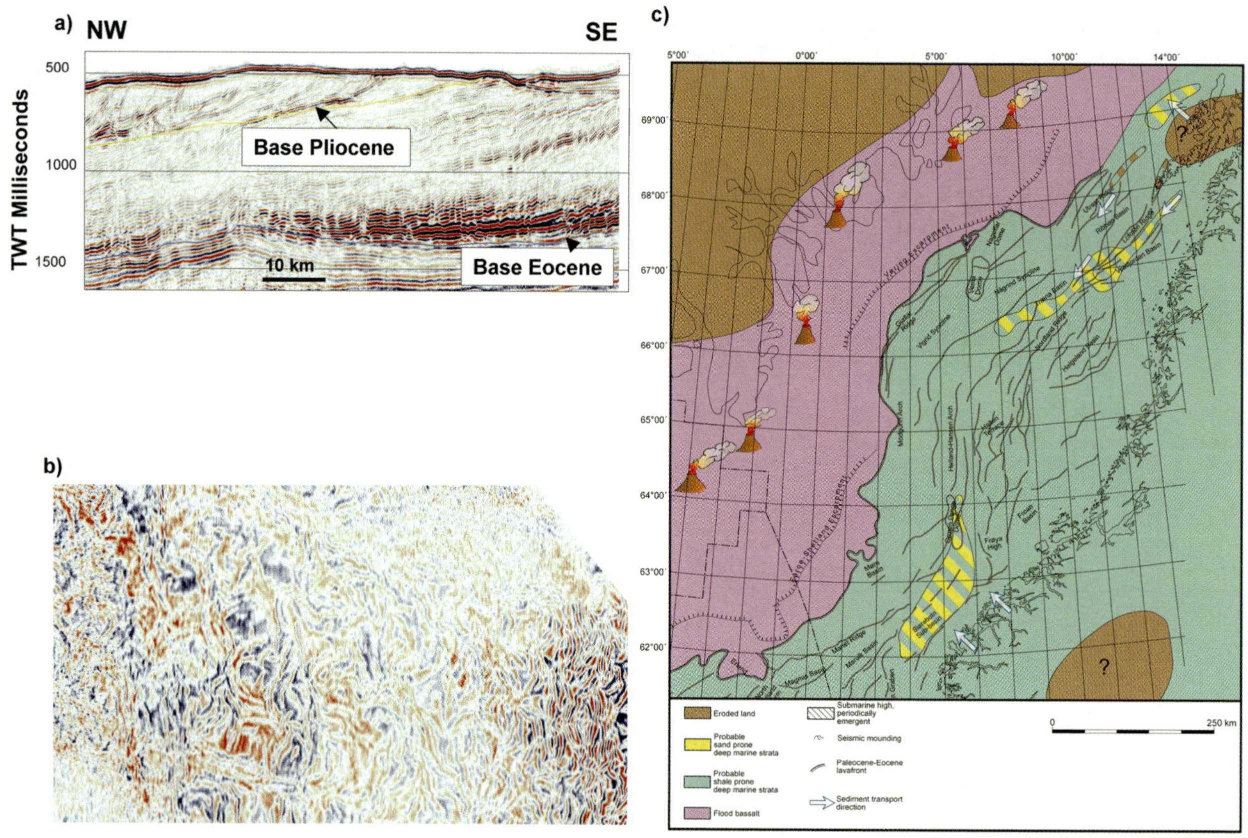

Fig. 7 (a) Typical seismic appearance of the Eocene succession in the Norwegian Sea. Note the numerous faults cutting trough the stratigraphy. Note also that the units below and above the Eocene is practically undisturbed by this deformation. (b) 3D time-slice of the Eocene succession southwest of the Lofoten Islands. Note the chaotic to polygonal pattern arising from the numerous faults in the formation. (c) Eocene palaeogeography and depositional environment in the Norwegian Sea. See Fig. 1 for location of figure.

over the entire Norwegian Sea area, as well as in the North Sea (Jordt et al., 1995; Brekke, 2000), and represents a major marine flooding surface. From the biostratigraphic analysis in the wells, it is evident that the base Eocene flooding surface (TEoMFS170–155) most likely represents one of the most condensed sections within the Cenozoic era (Fig. 2). The regional extent of this surface indicates that this flooding probably was associated with major basin subsidence, most likely controlled by tectonic movement related to continental break-up (Brekke et al., 1999; Brekke, 2000). The early sea-floor spreading in the North Atlantic also induced a compressional regime in the Norwegian Sea, which initiated the inversion of the main Palaeocene depocentres (Doré and Lundin, 1996).

The prograding nature of the Eocene succession suggests an overall loss of accommodation throughout the epoch. Two major depositional breaks in the form of erosional vacuities at mid Eocene level also suggest formation of sequence boundaries and possibly subaerial conditions in a landward direction. In some areas, the mid Eocene break and the base Late Eocene break merge and become one surface (Fig. 2). In these cases, multiple events of relative rise and fall of the sea level are recorded over one horizon.

However, the Eocene succession generally consists of fine-grained marine sediments (shale) throughout the entire Norwegian Sea area (Dalland et al., 1988; Brekke et al., 1999, 2001) (Fig. 2). Only locally, the sands are of rather poor reservoir quality encountered. This suggests that potential subaerial conditions occurred way landward of the site of deposition. Sediments from Late Palaeocene/Early Eocene are preserved onshore in Denmark, in central and northern Sweden and in Finland (Heilmann-Clausen et al., 1985; Hirvas and Tynni, 1976). Both the lithology and fossil content of these sediments point to a deep marine depositional environment. The transgression associated with the base Eocene flooding surface was thus probably widespread, and large parts of Fennoscandia were flooded during this event. The actual extent of this flooding is difficult to assess, but we suggest that only the most high-standing areas of Scandinavia were above sea level (Fig. 7c). As for the Palaeocene, the sandiest portions of the Eocene are found where the palaeo-shelf is inferred to have been at its narrowest (Fig. 7c).

Seismic stratigraphy of the Eocene

The seismic stratigraphic resolution of the Eocene is generally poor. This is mainly due to the internal deformation of the sedimentary package. Numerous small faults with different directions and throws seem to cut the internal stratigraphy (Figs. 7a, b). The deformation is concentrated to the Eocene unit, leaving both the underlying and the overlying sedimentary packages relatively undisturbed (Fig. 7a, b). Similar fault patterns have been interpreted to result from early diagenesis and water escape from high porous shales (Cartwright, 1996; Dewhurst et al., 1999).

Oligocene depositional systems

Oligocene sediments are generally thin over the platform areas to the east. In places, it is below seismic resolution, or is totally missing. New biostratigraphic analyses from selected wells suggest that there are no significant sedimentary breaks within the Oligocene, and that the succession, in places, is more or less complete (Fig. 2). The stratigraphic thinning is therefore, in part, related to low sedimentation rates and condensation. Additionally, the unit is bounded at base, and top, by the base late Eocene and base Miocene unconformities respectively. At some locations the erosion associated with the base Miocene unconformity cuts into, and removes the entire Oligocene succession. Occasionally, it also removes most of the Eocene, leaving only small remnants of Eocene sediments (Fig. 2). In these cases, the Lower Eocene, the base Upper Eocene and the Miocene unconformity merge and define one surface. Intervening flooding surfaces also coalesce in this erosional vacuity, resulting in a true multi-story composite surface.

The unconformities separate basinward and the Oligocene succession is an overall fine-grained unit recognised by semi-transparent and parallel seismic facies patterns. Both the wells and the seismic facies suggest a deep marine depositional environment dominated by hemipelagic sedimentation.

Miocene depositional systems

Miocene sediments in the Norwegian Sea are found to be a relatively thin wedge (~400 m TWT) on the middle and the outer continental shelf (Fig. 3). The succession is fine-grained, with several sandy intervals. In these areas, the Miocene deposition is interpreted to have occurred in a marine shelf setting. Inferred shallowing of the basin margins is supported by identification of an early to middle Miocene hiatus over the entire Norwegian Sea continental shelf (Eidvin and Riis, 1989, 1991, 1992; Eidvin et al., 1993; Gradstein and Backstrøm, 1996). An Early to Middle Miocene event of regional uplift of the Norwegian mainland and inner shelf areas was also suggested by Jordt et al. (1995), Brekke (2000) and Løseth and Henriksen (in press).

The compressive movements in the mid Miocene led to renewed vertical movements of the large arches and domes in the Norwegian Sea. These movements enhanced the basin relief induced by compressive movements in the Eocene/Oligocene and resulted in the formation of a Mid Miocene unconformity over these structures (Brekke et al., 2000). The unconformity appears to be erosional, but its formation is not quite clear (Brekke et al., 2000). Miocene sediments above and below the unconformity are interpreted as deep marine, and it is thus likely that the unconformity formed by submarine erosion during a stage of shallowing of the basin. Bruns et al. (1998) and Laberg et al. (2002) have described the Mid- and Late Miocene depositional systems to result from contour currents. The ocean currents and palaeobathymetry may in part have controlled the lack of strata or thin succession over structural highs compared with the thicker accumulations in the basins.

Seismic stratigraphy of the Miocene

The Miocene on the continental shelf is recognised by a series of low angle westward prograding clinoforms on the middle and outer continental shelves (Figs. 3, 8a). These clinoforms downlap the base Miocene surface regionally and thus, most likely define a major marine flooding surface (Fig. 8a). In some locations, the internal Miocene reflections onlap the basal surface in a landward direction (Fig. 8a). In these locations, the base Miocene also defines an unconformity. This unconformity can be mapped landwards and merges with the base Pliocene surface (Fig. 8a). In an area of 3D seismic coverage SW of the Lofoten area, there are identified marked channel geometries on this surface (Fig. 8b). Mapping of these features reveals a network of channels with a possible dendritic drainage pattern (Figs. 8c, d). These stratigraphic relations point to the possibility of subaerial exposure and fluvial drainage over the inner shelf in the Miocene. The onlap towards the

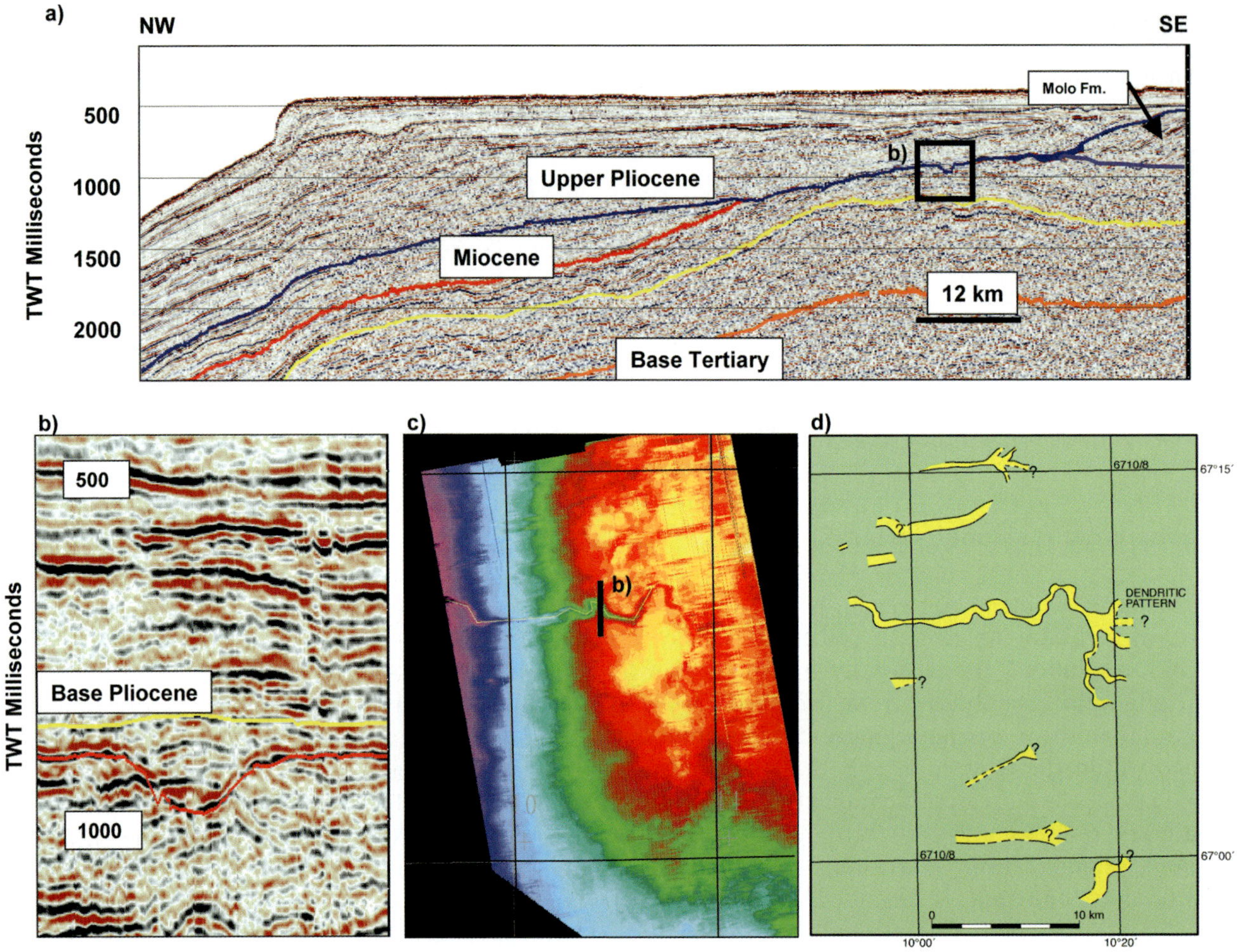

Fig. 8 (a) Seismic transect across the continental shelf SW of Lofoten illustrating the stratal relationships of the Miocene succession in the area. Note the low-angle westward prograding style, the onlap relation in a landward direction and the erosional relief east of the Miocene subcrop-line (encircled area) (See Fig. 1 for location of figure). (b) Seismic section through channel system at base Miocene/Pliocene level. (c) Time depth map of channel system. Location of (b) marked. (d) Interpretation of channel system. Note the dendritic pattern in the eastern reaches of the channel system.

base Miocene surface could thus be interpreted as a coastal onlap. The subcrop-line of the Miocene sediments then closely defines a palaeo-coastline. Indications of shallow marine, or marginal marine, conditions are also found at the western flank of the Utrøst Ridge. Here, a distinct, prograding geometry interpreted as a delta is identified (Fig. 9a). The inferred fluvial drainage pattern, together with the westerly position of the palaeo-coastline, suggests that large parts of the continental shelf was subaerially exposed during the Miocene. It may be stated that the highest degree of subaerial exposure of the Norwegian shelf probably occurred during the early-mid Miocene. This statement is also in accordance with previous reconstructions from the Northern North Sea (Martinsen et al., 1999). From the observations and arguments above, it is clear that the base Miocene surface most likely represents both a sequence boundary and a marine flooding surface and is a compound surface.

In the deeper basins, the Miocene succession is recognised by a semi-transparent, parallel and continuous reflection pattern. At several locations along the flanks of intra-basin highs, the reflection pattern shows 'clinoform' geometries, apparently climbing in a landward direction (Fig. 9b). These geometries are in favour of the interpretation of these deposits as contourite drifts. Laberg et al. (2002), suggested that these sediments were brought into the basin by northward flowing ocean currents. We will however, not rule out the possibility of a more local provenance by winnowing of the shelf and structural highs.

A wide range of depositional environments is inferred for the Miocene of the Norwegian Sea. These observations are summarised in a palaeogeographic map for the study area (Fig. 9c).

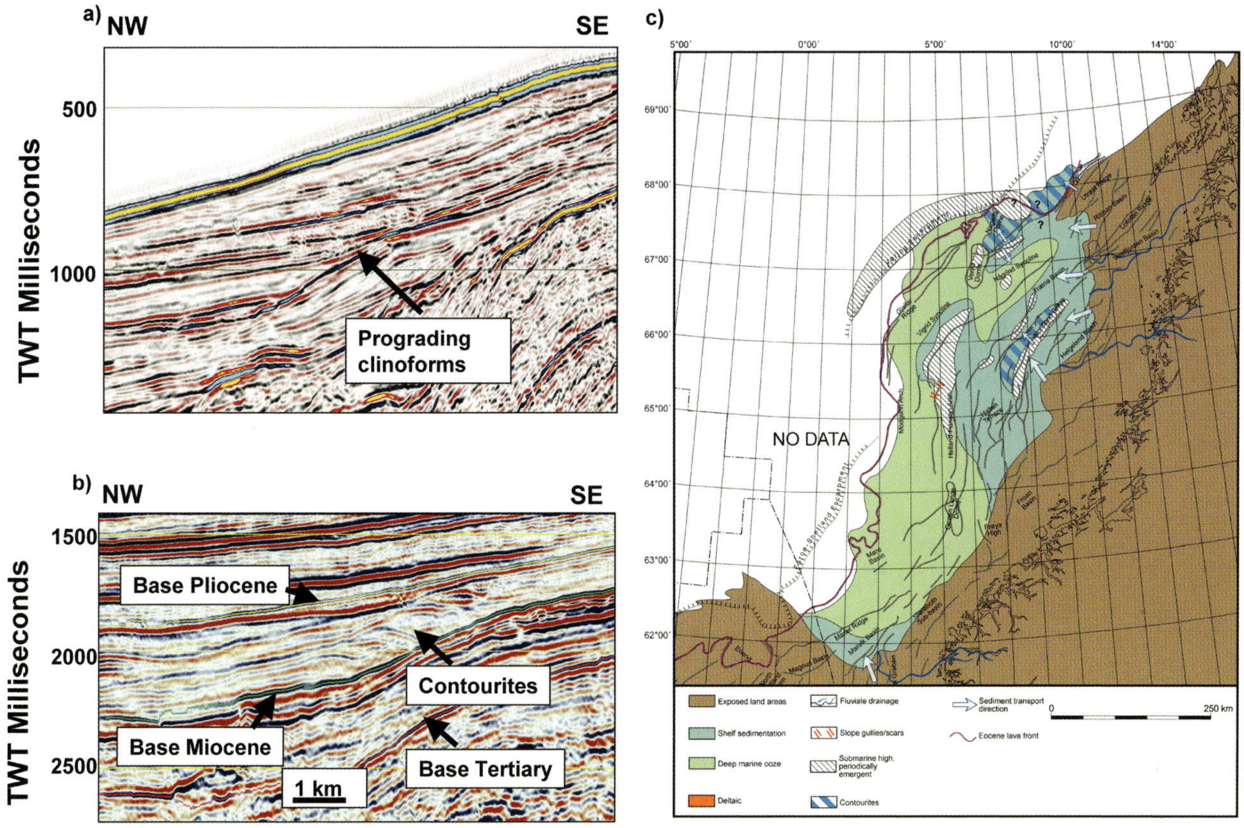

Fig. 9 (a) Well-developed prograding facies interpreted as a deltaic system on the western flank of the Utrøst Ridge. (b) Seismic section from the western flank of the Helland Hansen Arch illustrating 'clinoforms' climbing in an updip direction (c) Miocene palaeogeography and depositional pattern in the Norwegian Sea. See text for discussion. (See Fig 1. for location of figure).

The Molo Formation

A depositional system with a pronounced deltaic appearance called the Molo Formation is deposited all along the eastern margin of the Norwegian Sea (Figs. 3, 10a). A marked downlap surface defines the base of the formation. Westward, this downlap surface appears to merge with the base of the late Plio-Pleistocene prograding wedges. Eastwards, the deltaic system becomes truncated by the base Quaternary unconformity (Figs. 3, 10b, c). The Quaternary succession is very thin, and in places, the base Quaternary merges with the sea floor. The Molo Formation is terminated westward by a marked unconformity (Fig. 10b). This unconformity merges with the Base Pliocene downlap surface, and in places with the Miocene unconformity and the top Eocene unconformities (Fig. 8a). Because of this apparent merging of a number of surfaces and lack of a good well control, the actual age of the Molo Formation has been a subject of discussion.

Micropalaeontological studies from sidewall cores in well 6607/10-3 within the deltaic succession attribute an Oligocene age (Eidvin et al., 1998). Based on stratigraphic position and regional considerations, Henriksen and Weimer (1996) proposed that the deltaic system could equally well be Early Pliocene in age. New seismic ties further south along the margin strongly suggest that the eastern wedge of the Miocene succession underlies the distal toe of the deltaic complex (Figs. 10b, c). New biostratigaphic analysis from nearby wells document finding of early Pliocene type fossils (Eidvin et al., personal communication) This interpretation is much in favour of an Early Pliocene age, as suggested by Henriksen and Vorren (1996) and Henriksen and Weimer (1996) and Løseth and Henriksen (in press).

Regardless of age, the regional distribution of the deltaic system points to deposition related to a regional geological event. Our primary interpretation is that the deltaic systems were deposited in response to broad epeirogenic uplift of the Norwegian mainland. Prior, or concomitant, to uplift, there was a major flooding event over the shelf areas to the east and the base Pliocene thus defines a major marine flooding surface. Compared with the major regression during the Miocene, a significant shift in the locus of sedimentation is associated with the base Pliocene flooding. After this major transgression, a relative drop in sea level

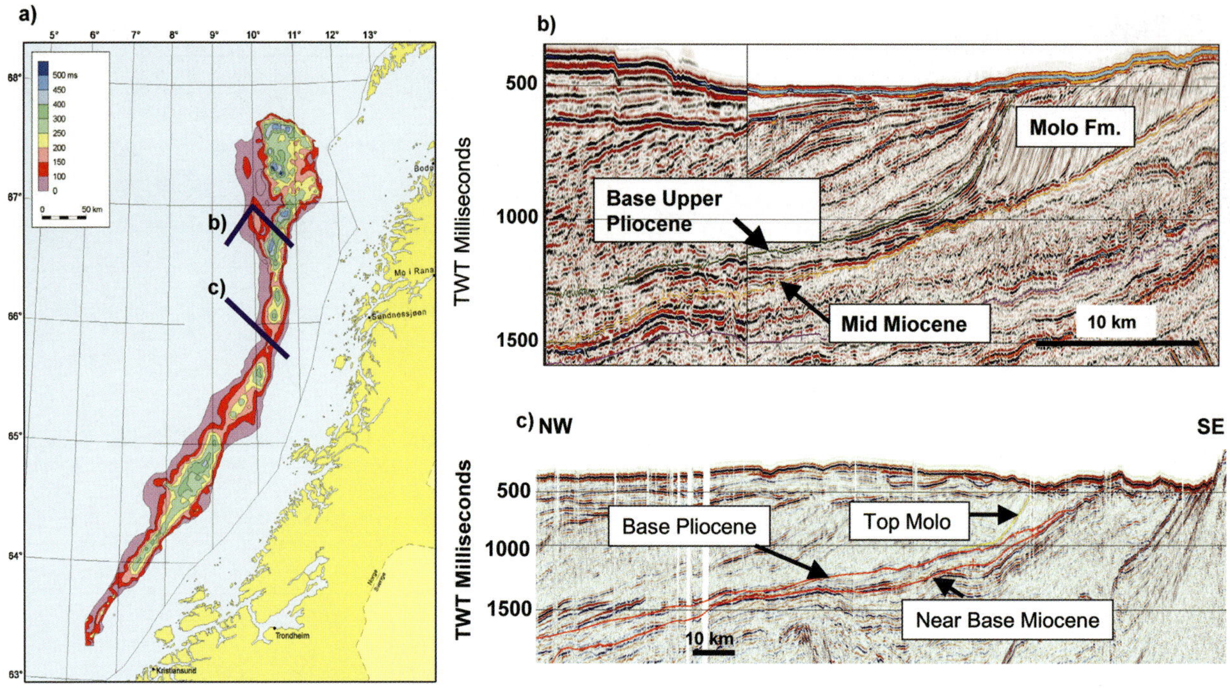

Fig. 10 (a) Time isopach map the Molo Formation in the Norwegian Sea. Contour interval 50 ms TWT. Note the regional distribution and coast-parallel depositional pattern of the system. (Location of (b) and (c) marked). (b) and (c) Seismic composite lines indicating that the Miocene succession lies stratigraphically below the deltaic 'Molo' Formation.

occurred and the first sedimentation onto this surface is represented by the clinoforms of the Molo Formation.

Due to the shallow burial depth, the seismic resolution is excellent, and a number of internal unconformities and depositional packages can be resolved (Fig. 10b). The sequence stratigraphy and the sequence stacking patterns of the Molo Formation in the northern part of the study area are described in detail by Henriksen and Weimer (1996).

Late Plio-/Pleistocene depositional systems

A thick wedge of glacial/glaciomarine sediments is deposited over most of the study area during the late Pliocene and Pleistocene (Figs. 3 and 8a). In the largest depocentre, SW of Lofoten, about 1.8 s TWT (~2000 m) of sediments are deposited (Henriksen and Vorren, 1996). The whole wedge is believed to have been deposited during the past 2.4 My (Eidvin et al., 1998), and considerable sedimentation rates are implied (Riis and Fjeldskaar, 1992; Henriksen and Vorren, 1996).

The deposition of the wedge has been related to regional uplift of the Scandinavian mainland, climatic deterioration and glacial expansion over the continental shelf. The resulting sediments could not, by far, be accommodated on the shelf. This resulted in an overall prograding depositional style throughout the Pliocene (Figs. 3 and 8). During deposition of the unit, the shelf-edge migrated more than 100 km westward (Henriksen and Vorren, 1996).

Once glacial conditions were established on the shelf, the ice moved in separate ice flows (Ottesen et al., 2002). The fast-flowing ice followed, and excavated, the large glacial troughs transversing the shelf (e.g. Trænadjupet and Sklinnadjupet troughs) (Fig. 11). Stagnant ice covered the shallower banks as Haltenbanken, Trænabanken and the Lofoten area (Fig. 11).

Plio-/Pleistocene seismic stratigraphy

The internal reflection pattern of the late Pliocene and Pleistocene wedge is recognised by low-angle (1–2°) laterally persistent clinoforms (Figs. 3 and 8). These clinoforms downlap the base Pliocene surface regionally and may be viewed as a continuation of the high-angle clinoforms of the Molo Formation. The marked shift in prograding style from early to late Pliocene may signify the turnover from peri-glacial to glacial conditions on the Norwegian Sea continental shelf (Henriksen and Vorren, 1996).

Each of the prograding clinoforms defines an internal unconformity surface with a certain facies

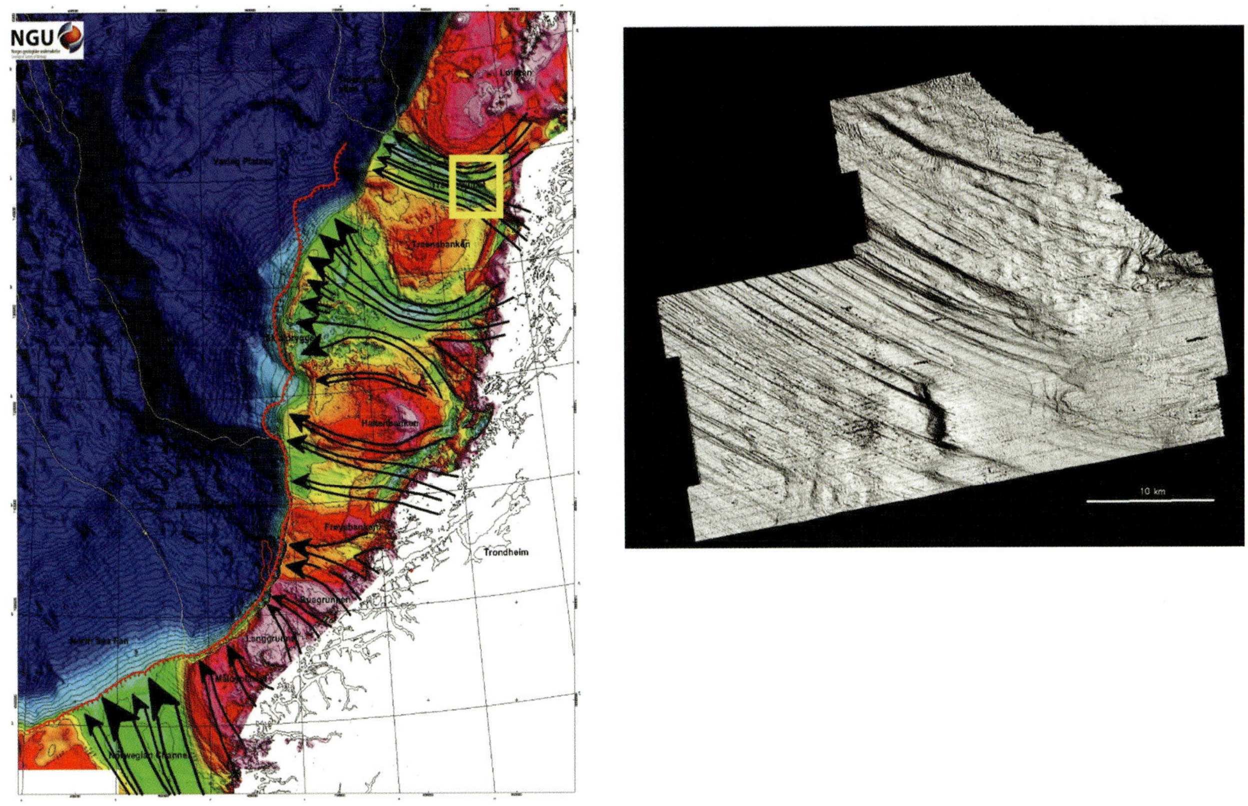

Fig. 11 Bathymetry of the present day sea floor in the Norwegian Sea with reconstructed ice flow pattern (Ottesen et al., 2002). Enlarged image from the mouth of Vestfjorden.

association attached to it (Fig. 12a). Although the observed reflection pattern is quite similar to what is observed in sequences controlled by relative sea-level changes (i.e. eustasy + tectonics) (e.g. Van Wagoner et al., 1988), we favour an interpretation where this facies association is attributed to the glacial depositional regime on the shelf (Fig. 12b). The internal clionforms top-lap, merges, and in places, become truncated by one, or several unconformities up-dip (Figs. 3, 8 and 10). These unconformities were probably eroded during subsequent glacial expansions over the shelf. Several events of glacial advances and retreats occurred over the inner shelf and the unconformities merges and become one upper regional unconformity (URU) Vorren et al. (1992). In these areas, the upper regional unconformity defines the base of the Quaternary succession. Basinward, however, it becomes difficult to decide which of the upper unconformities that actually defines the base of the Quaternary.

Also, the sea floor has a distinct reflectivity signature. In the bathyal troughs, once occupied by the fast-flowing ice streams, there are a number of striations, or flutes (Fig. 11). These flutes result from the movement of grounded ice on the shelf. The direction of ice flow can easily be deduced from the orientation of the flutes (Fig. 11). Apparently, the glacial drainage followed bedrock boundaries and older structural features in the subsurface.

Sequence stratigraphy

The Base Palaeocene, the Late Palaeocene, the Late Eocene and base Miocene sedimentary breaks are all interpreted to represent major sequence boundaries (Fig. 13a). The URU, although formed by glacial processes, may also be defined as a sequence boundary.

These surfaces represent basin wide unconformities and significant basinward shifts in the locus of sedimentation is associated with these surfaces. A number of unconformities are identified within the individual units. These surfaces of erosion represent higher-order sequence boundaries within a lower-order frequency system. Subaerial exposure is indicated for the Late Palaeocene, the Miocene and the Top Molo sequence boundaries. Formation of the other sequence boundaries is probably also associated with subaerial exposure but proximal parts of these depositional systems, and thus evidence for continental conditions have been removed by the Late Cenozoic glacial erosion.

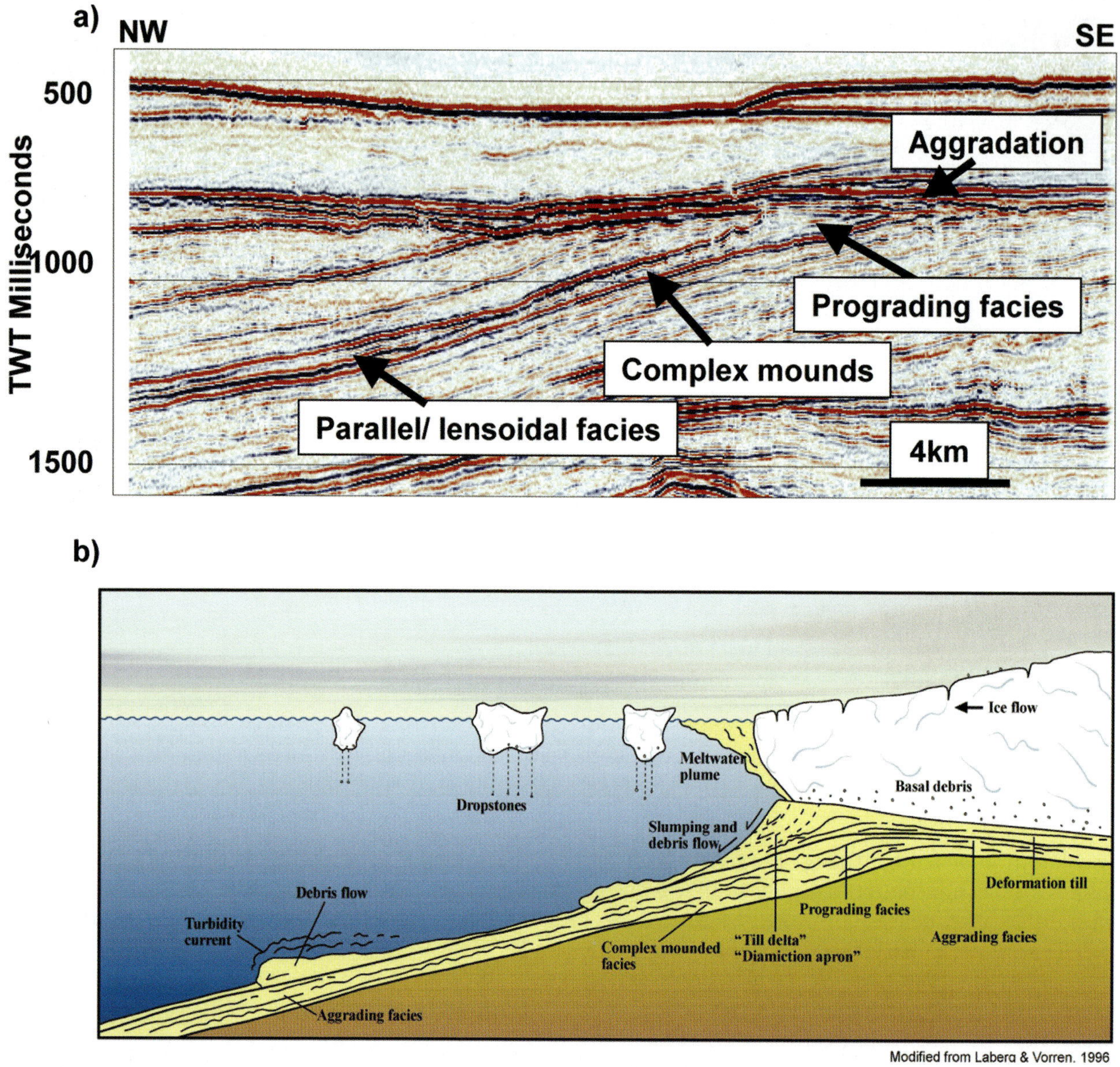

Fig. 12 (a) Stratal relationships and seismic facies associated with the internal sequences of the Late Pliocene. (See Fig. 1 for location of figure). (b) Ice sheet depositional model for the sequences within the late Pliocene on the Norwegian Sea shelf. Sediments are eroded and carried from the continent and inner shelf areas, principally as basal debris in grounded ice sheets and are deposited on the continental shelf as deformation tills and on the continental slope as marine diamictons. From Henriksen and Vorren (1996).

The base Tertiary, base Late Palaeocene, base Early Eocene and base late Pliocene are all interpreted as major marine flooding surfaces (Fig. 13a). These surfaces probably extended far eastward, and large parts of the Norwegian mainland, possibly also large parts of Scandinavia were flooded during these events.

Systems tracts

Based on the overall stratigraphic framework, the Cenozoic succession may be subdivided into mega-scale systems tracts, which are contemporaneous depositional systems deposited during a given part of one complete cycle of fall and rise of eustatic sea level (Brown and Fischer, 1977; Mitchum and Van Wagoner, 1991).

The definition of each systems tract in this study is based on its specific position within the succession, its internal reflection pattern and its relation to bounding surfaces (Fig. 13b). As documented in the description of each depositional system, it is evident that the mega-scale systems tracts may be subdivided into a number of higher frequency sequences. This has been described in detail for the Molo Formation by Henriksen and Weimer (1996) and for the Late Pliocene succession by Henriksen and Vorren (1996). These high frequency

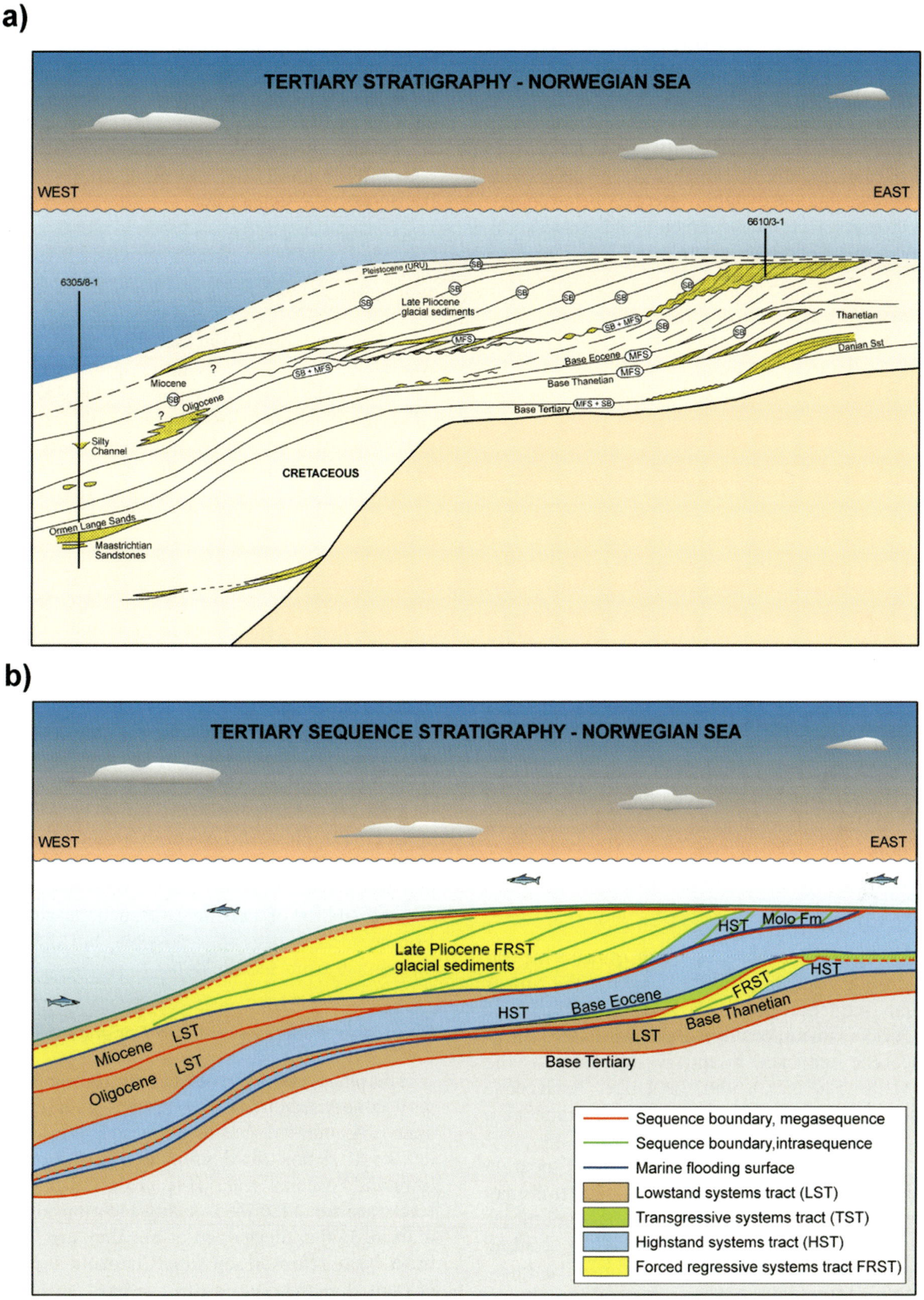

Fig. 13 General Cenozoic stratigraphy in the Norwegian Sea interpreted in a sequence stratigraphic context. Sequence boundaries (SB) and marine flooding surfaces (MFS) labelled.

sequences constitute the internal stacking pattern of the mega-scale systems tracts (Figs. 13a, b). It is not within the scope of this chapter to do a detailed analysis of these high frequency events. However, recognition and integration of these high frequency sequences and their variability in a mega-scale hierarchy along the margin will be an important step in revealing the dynamic interplay between sea-level changes, tectonic movements, sediment supply and variable basin physiography.

Palaeocene

Lower Palaeocene

The Lower Palaeocene (Danian) is interpreted as a lowstand systems tract (Fig. 13). Sandy turbidite fans in the Møre Basin (Ormen Lange sands) constitute the basin floor fans and the rock record in these areas roughly equals the erosional vacuity represented by the base Tertiary unconformity. The sandy prograding facies on the Møre Margin is slightly younger than the basin floor fans and is interpreted to represent the lowstand prograding complex. A semi-transparent overall shaly package in late- early Palaeocene is interpreted to represent the transgressive and highstand systems tract. Based on the observed shallowing and renewed input of siliciclastic material, an upper Danian sequence boundary is inferred. The Lower Palaeocene, thus, constitutes a complete depositional sequence of approximately 3 million years in duration.

Upper Palaeocene

Also, the Upper Palaeocene constitutes a depositional sequence, bounded below by the upper Danian sequence boundary and the intra-upper Palaeocene unconformity. Basinward the intra-Palaeocene sequence boundary merges with the TPaMFS115 (Fig. 2), and the Upper Palaeocene sequence is thus also deposited during a time span of approximately 3 million years.

The two Palaeocene sequences define separate third-order sequences. These cycles of events may be grouped together and viewed as one mega-sequence. The localised Thanetian prograding wedge on the Trøndelag Platform may be interpreted as a late highstand/forced regressive systems tract (Fig. 13). It represents renewed input of sediment to the basin margin and a major basinward shift in the locus of sedimentation. Within this mega-sequence scale the internal sequence boundaries of the Thanetian represent regressive surfaces of marine erosion (Figs. 3, 5c). A lowstand situation is expected above the Upper Palaeocene sequence boundary, but as stated, no obvious lowstand deposits are identified above this unconformity. This may be explained by the short duration of the depositional break. The thin onlapping wedge may constitute a minor lowstand wedge and a possible transgressive systems tract.

Sequence 2: Upper Palaeocene–late Eocene

Following Upper Palaeocene lowstand, a major condensation and a marine flooding surface are defined at the near base Eocene level (TEoMFS170 – 155), and a major transgression is associated with this flooding, The most dominant part of the Eocene mega-sequence in the Norwegian Sea is thus inferred to be a highstand systems tract (Fig. 13). A late Eocene erosive event interpreted to represent a mega-scale sequence boundary with an increasing amount of section missing in an eastward position defines the top of the sequence. An overall loss of accommodation and erosion of early highstand deposits is associated with the Eocene progradation. A forced regressive systems tract is thus likely to have developed towards the end of highstand. Poor seismic resolution and deep erosion makes it difficult to single out this systems tract. This interpretation is in line with the interpretation from the Møre Basin by Martinsen et al. (1999). However, north of about 65°N in the Norwegian Sea, an intra-Eocene sequence boundary becomes increasingly more important (Fig. 2). The possibility of two mega-sequences within the Eocene succession is thus possible, but difficult to distinguish from seismic data.

Sequence 3: Oligocene (Late Eocene–Miocene)

The Oligocene sequence is very thin and only a very condensed interval is present over the shelf areas. A marked basinward thickness increase, defines a depositional complex, constituting the lowstand systems tract (Fig. 13). Other systems tracts are not identified at any location along the margin. If they developed at all, they are likely to have been removed by late Cenozoic uplift and erosion.

Sequence 4: Miocene to Quaternary

The base Miocene unconformity is interpreted as a mega-scale sequence boundary. The Miocene clinoforms probably represent a lowstand basin margin progradation, where successively larger shelf areas became subaerially exposed (Fig. 13). The interpreted deltaic features on the Nordland Ridge and on the Utrøst High, probably represent the final stage of this progradation, before turn-around and transgression of the shelf. The flooding associated with this transgression extends far landward and constitutes the top Miocene/base Pliocene flooding surface. The first sediments deposited onto this surface are the clinoforms of the Molo Formation. If one assumes that the topset segment of the Miocene deltas at the Utrøst Ridge and the clinoforms of the Molo Formation, respectively were at, or close to, sea level during formation, then the sea level rise associated with this flooding is about 400 m. This is way out of range of all eustatic sea level curves, and the formation of the flooding surface must be tectonically enhanced.

The strongly progradational nature of the Molo Formation as well as the Late Pliocene result from an overall loss of accommodation, and relative sea level fall during deposition. Consequently, both these prograding wedges may be interpreted as forced regressive systems tract. The easternmost parts of the Molo Formation (largely eroded by base Quaternary) may constitute the highstand systems tract of this mega-sequence (Fig. 13). We thus disagree with Martinsen et al., 2000 who included the Late Pliocene in the highstand systems tract, but we are open to define the Molo Formation as one individual sequence within a lower-order mega-sequence.

The Late Pliocene prograding system is inferred to have occurred in association with glacial expansion onto the continental shelf (Henriksen and Vorren, 1996), and is thus different from prograding systems driven by relative sea level changes alone. All the same, the glacial expansion onto the shelf must have been associated with a relative sea level fall. Similarly, the base Quaternary defines a major sequence boundary, even though it is formed by glacial processes.

Sequence 5: Quaternary–Recent

The sedimentary cover above the upper regional unconformity represents the final stage of glacial expansion over the shelf. This represents the lowermost sea level during the glacial cycle, and sediments deposited during this stage may be viewed as a lowstand systems tract (LST) (Fig. 13). Also within this succession there is a high frequency signal and several glacial cycles are recorded above the URU (Olsen, 1997 and Sejrup et al., 2000). After the last glacial maximum around 20 K year ago, the glaciers melted away relatively rapidly (Vorren et al., 1983). Sediments released during de-glaciation and the concomitant sea level rise may be viewed as a transgressive systems tract (TST). The continental shelf is currently in a highstand situation and receives very little sediments from the continent (Fig. 13). Sediments eroded from the mainland by fluvial run-off are largely trapped in the fjord basins inland.

Basin physiography and control on sequence development

The variable basin physiography with narrow shelf and steep slopes in the north and south and the intervening wide and gently dipping Trøndelag Platform have exerted major control on the sequence development throughout the Cenozoic (Fig. 1). Presuming that the sedimentation rates were fairly constant, this is largely a question about accommodation. On the low accommodation shelf and steep slopes, the bypass of sediments will result in vertical stacking of turbidite fans at toe of slope and basin floor (Fig. 14). Because of the high relief, these turbidite fans may be totally detached from the shelf systems and form true up-dip pinch out. Additionally, the run-out distance for these turbidite fans is expected to be large. In the Norwegian Sea the sandiest portion of both the Palaeocene and Eocene deep marine systems are found where the palaeo-shelf is inferred to have been narrowest and steepest.

On the broad high-accommodation shelf, stacking of continental and marginal-marine sediments is expected. If the sedimentation rates exceed the accommodation on the shelf, prograding sequences with clinoform relief roughly at scale with the water depth on the shelf will result (Fig. 14). Basin floor fans may be associated with each of the internal sequence boundaries within the prograding systems. These fans are however to be expected of a single cycle with a relatively short run-out distance and with a high risk for up-dip connection to incised valleys and shallow marine deposits.

For the Palaeocene and Eocene, such deposits are likely to have been deposited on the landward part of the Trøndelag Platform. With an exception of the Thanetian, these prograding shelf systems, and associated basin floor fans, are removed by the late Cenozoic uplift and erosion.

By the Early Pliocene most of the basin relief on the inner shelf was filled in by early Tertiary sediments. Consequently, the entire Molo Formation prograded over a broad, low-relief platform,

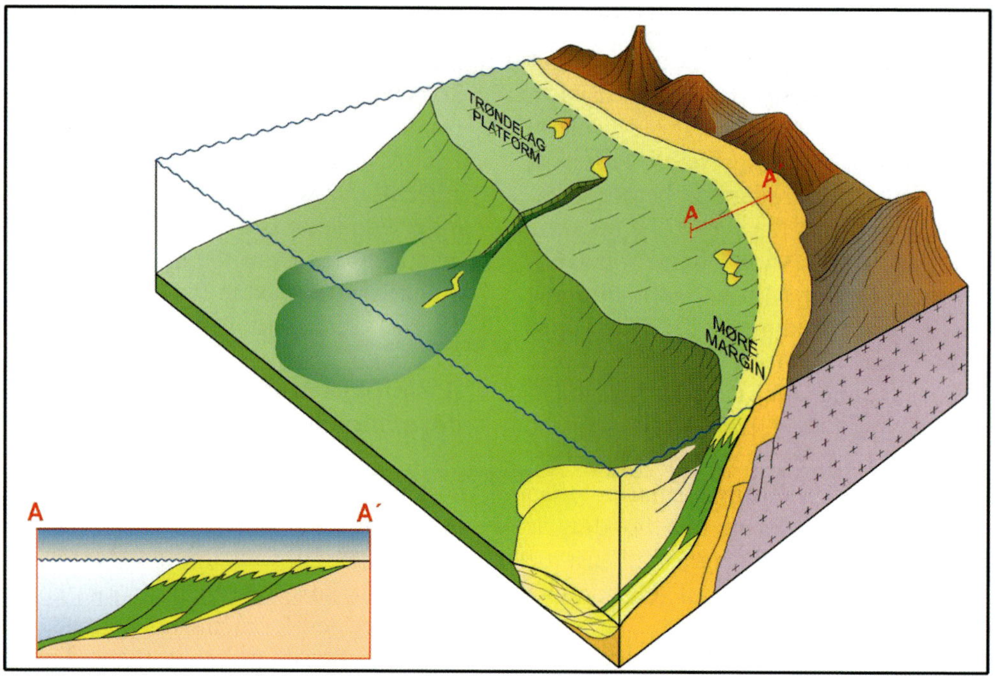

Fig. 14 Conceptual model for along strike variablility in mega-sequence development and lithology-distribution in a basin.

and basin floor fans associated with these clinoforms are all small and single cycle. The same largely applies for the Late Pliocene and Pleistocene, but here, increased basin subsidence result in larger clinoforms and basin floor fans. The fans are, however, still single cycle.

Conclusions

The stratigraphic development of the Cenozoic can be illustrated by a schematic back-stripping across the Møre Basin (Fig. 15). Both Lower and Upper Palaeocene progradational shelf systems were deposited along the eastern and the western margin of the Norwegian Sea. These sediments were deposited in response to regional epeirogenic uplift of the mainland as well as major rifting and uplift of the Møre and Vøring Marginal Highs. The main entry points for Palaeocene sediments seem to be determined by the intersection of major NW–SE trending lineaments associated with long-lived Caledonian basement structures, e.g. erosional products from the Norwegian mainland were fed into the basin along the Vestfjord–Lofoten trend and brought further out and into the Vøring Basin, via the intersection with the Bivrost Lineament. Similarly, the Jan Mayen Lineament trapped and brought sediments derived from the shelf areas close to the Møre-Trøndelag Fault Zone into the Møre Basin.

Mainland-derived Upper Palaeocene depositional systems show a westward emplacement relative to the Lower Palaeocene systems. Several erosional unconformities, interpreted to be subaerial exposure surfaces were identified within the Upper Palaeocene section.

During the Eocene, the western Norwegian Sea was totally dominated by the extrusion of basaltic lavas. Major basin subsidence to the east resulted in flooding and deposition of predominantly fine-grained sediments over large areas during this time period. It may be speculated that large parts of Fennoscandia was flooded during the Eocene. The western land areas gradually cooled, underwent thermal subsidence and became uneffective as sediment source areas by the Oligocene.

Due to condensation, the Oligocene is extremely thin over the shelf. In the deeper basins, relatively thick successions of hemi-pelagic ooze was deposited. The succession is interpreted to constitute a lowstand systems tract of a mega scale depositional sequence.

A phase of regional uplift and subaerial exposure with fluvial incision of the shelf probably occurred in the early Miocene. Possible coastal onlap relations and inferred shelf and deltaic progradations suggest shallow-marine conditions over the shelf in Miocene times. In the Vøring Basin, deep marine conditions prevailed. A mid Miocene phase of compressive movements gave renewed growth of intra-basin highs. The increased relief, possibly in conjunction with major oceanographic changes,

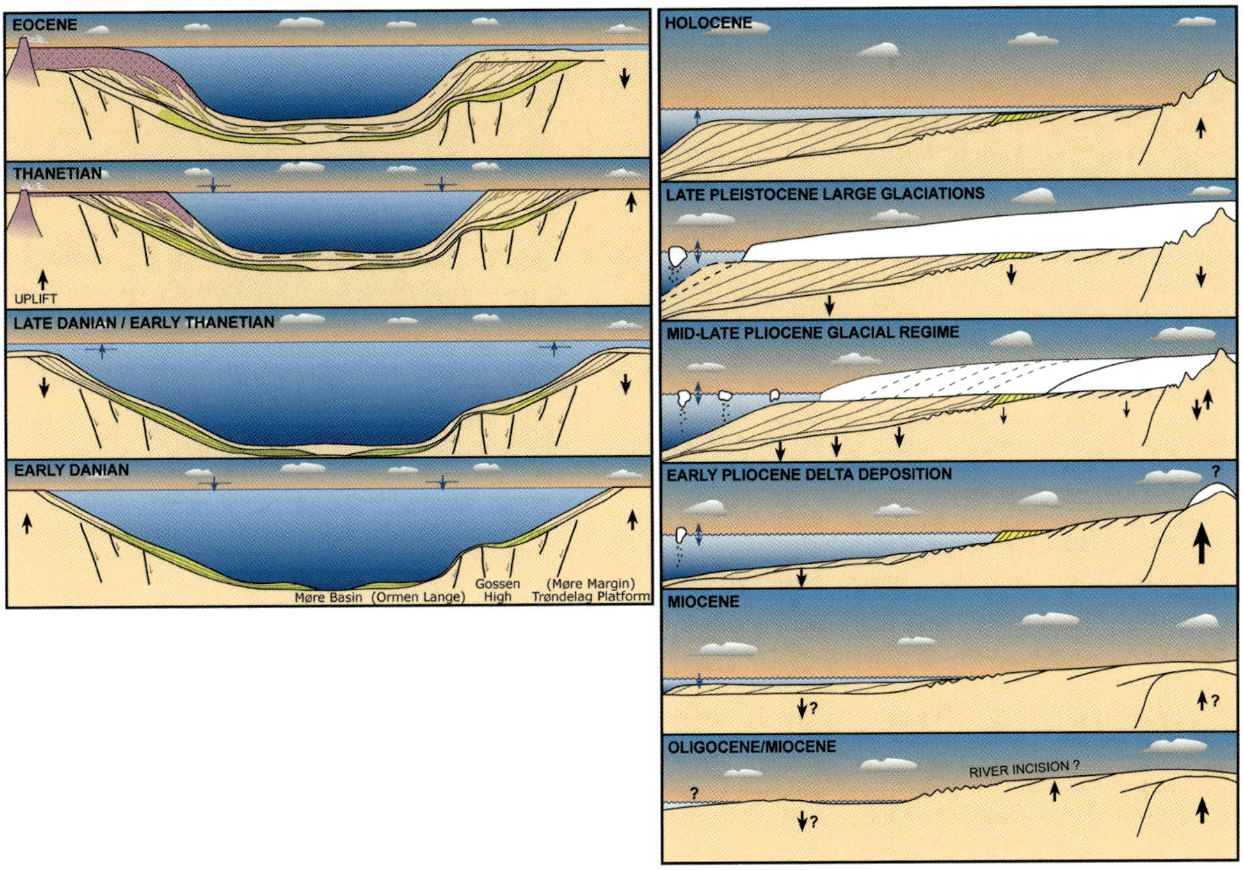

Fig. 15 Backstripped geoseismic profile across the Norwegian Sea, illustrating the sequence development and inferred vertical movements of the basin and basin flanks through the Cenozoic. See text for discussion.

gave favourable conditions for reworking of sediments by contourite currents.

After this lowstand situation, a major flooding of the shelf occurred. This base Pliocene flooding surface was probably tectonically enhanced and associated with uplift of the basin margin and increased subsidence of the basin. After flooding, and as a response to uplift, the early Pliocene Molo Formation was deposited onto this surface.

The marked shift in style of progradation from the early to late Pliocene prograding wedges signify a fundamental change in the sedimentary environment in the area and are interpreted to represent the turnover from a peri-glacial to a glacial regime on the shelf. The deposition of the late Plio-Pleistocene sequences probably reflects a gradual climatic deterioration and regional advances of major ice sheets across the continental shelf. The movement of grounded ice on the shelf left behind a pattern of glacial striations, or flutes. The direction of ice movement can easily be deduced from the flute pattern, and it appears that the main ice streams, at least partly, followed bedrock boundaries and structural features in the subsurface.

Acknowledgements

The authors wish to thank Statoil ASA for the permission to publish this work. We offer our sincere thanks to Norsk Hydro AS, Eni Norge AS (3D survey, DTV 2000 in the PL259 licence), TGS Nopec AS, Fugro Geoteam AS, Amerada Hess Norge AS for allowing us to use and publish the seismic data owned by these companies. Lars Reistad at Statoil has made a tremendous effort in drafting the figures used for illustrations. We are also grateful to Mai-Britt Mørk, Gavin Lewis, Elisabeth Eide and Tom Bugge for their insightful comments on an earlier version of the manuscript.

References

Aanstad, K.M., Gabrielsen, R.H., Hagevang, T., Ramberg, I.B. and Torvanger, O., 1981. Correlation of offshore and onshore structural features between 62°N and 68°N. Proceedings, Norwegian Symposium on Exploration, Bergen 1981, NSE/11. Norwegian Petroleum Society (NPF), pp. 1–25.

Berggren, W.A, Kent, D.V., Swisher III, C.C. and Aubry, M.-P., 1995. A Revised Cenozoic Geochronology and chronostratigraphy. In: W.A. Bergren, D.V. Kent, M.-P. Aubry and J. Hardenbol (Editors), Geochronology, time scales and global stratigraphic correlation: Tulsa, SEPM Spec. Publ., 54: 129–212.

Blystad, P., Brekke, H., Færseth, R.B., Larsen, B.T., Skogseid, J. and Tørudbakken, B., 1995. Structural elements of the Norwegian continental shelf. Part II The Norwegian Sea Region. NPD Bull., 8: 45 pp.

Brekke, H. and Riis, F., 1987. Tectonics and basin evolution of the Norwegian shelf between 62°N and 72°N. Nor. J. Geol. (NGT), 67: 295–322.

Brekke, H., Dahlgren, S., Nyland, B. and Magnus, C., 1999. The prospectivity of the Vøring and Møre basins on the Norwegian Sea margin. In: A.J. Fleet and S.A.R. Boldy (Editors), Petroleum Geology of Northwest Europe: Proceedings from the 5th Conference. Geological Society, London, pp. 261–274.

Brekke, H. 2000. The tectonic evolution of the Norwegian Continental Margin with emphasis on the Vøring and Møre Basins. In: A. Nøttvedt et al. (Editors), Dynamics of the Norwegian Margin. Geol. Soc., London, Spec. Publ., 167: 327–378.

Brekke, H., Sjulstad, H.I., Magnus, C. and Williams, R.W., 2001. Sedimentary environments offshore Norway—an overview. In: O.J. Martinsen and T. Dreyer (Editors), Sedimentary environments Offshore Norway—Palaeozoic to Recent. Norwegian Petroleum Society (NPF) Special Publication 10, Elsevier, Amsterdam, pp. 7–37.

Brown, L.F. and Fisher, W.L. (1977). Seismic-Strategraphic Interpretation of Depositional Systems: Examples from Brazilian Rift and Pull-Apart Basins. In: C.E. Payton (ed.), Seismic Stratigraphy—Applications to Hydrocarbon Exploration. AAPG Memoir, 26: 213–248.

Bruns, P., Dullo, W.-C., Hay, W.W., Frank, M. and Kubik, P., 1998. Hiatuses on the Vøring Plateau: sedimentary gaps or preservation artefacts? Mar. Geol. 145: 61–84.

Bukovics, C. and Ziegler, P.A., 1985. Tectonic development of the Mid-Norway continental margin. Mar. Petrol. Geol., 2: 2–22.

Cartwright, J.A., 1996. Polygonal fault systems: a new type of fault structure revealed by 3D seismic data from the North Sea basin. Am. Assoc. Petrol. Geol., Studies in Geology, 42: 225–230.

Cooper, A.K., Barett, P.J., Hinz, K., Traube, V., Leitchenkov, G. and Stagg, H.M.J., 1991. Cenozoic prograding sequences of the Antarctic continental margin: a record of glacio-eustatic and tectonic events. Mar. Petrol. Geol., 102: 175–213.

Dalland, A., Worsley, D. and Ofstad, K,. 1988. A lithostratigraphic scheme for the Mesozoic and Cenozoic succession offshore mid- and northern Norway. The Norwegian Petroleum Directorate, Stavanger, NPD Bull., 4: 65 pp.

Dewhurst, D.N., Cartwright, J.A. and Lonergan, L., 1999. The development of polygonal fault systems by syneresis of colloidal sediments. Mar. Petrol. Geol.. 16: 793–810.

Doré, A.G. and Lundin, E.R., 1996. Cenozoic compressional structures on the NE Atlantic margin: nature, origin and potential significance for hydrocarbon exploration. Petrol. Geosci. 2: 299–311.

Doré, A.G., Lundin, E.R., Fichler, C. and Olesen, O., 1997. Patterns of basement structure and reactivation along the NE Atlantic margin. J. Geol. Soc., London, 154: 85–92.

Doré, A.G., Lundin, E.R., Birkeland, Ø., Eliassen, P.E. and Jensen, L.N., 1997. The NE Atlantic Margin: implications of late Mesozoic and Cenozoic events for Hydrocarbon prospectivity. Petrol. Geosci., 3: 117–131.

Eidvin, T. and Riis, F., 1989. Nye dateringer av de tre vestligste borehullene i Barentshavet. Resulateter og konsekvenser for den tertiære hevning. NPD Contrib., 27: 44 pp.

Eidvin, T. and Riis, F., 1991. En biostratigrafisk analyse av tertiære sedimenter på kontinentalmarginen utenfor Midt-Norge, med hovedvekt på øvre pliocene vifteavsetninger. NPD Contrib., 29, 44 pp.

Eidvin, T. and Riis, F., 1992. En biostratigrafisk og seismostratigrafisk analyse av tertiære sedimenter i nordlige deler av Norskerenna, med hovedvekt på øvre pliocene vifteavsetninger. NPD Contrib., 32, 40 pp.

Eidvin, T., Riis, F., Beyer, C. and Rundberg, Y., 1993. Et multidisiplinært, stratigrafisk studie av neogene sedimenter i sentrale deler av Nordsjøen (Ekofiskfeltet). NPD Contrib., 34, 73 pp.

Eidvin, T., Brekke, H., Riis, F. and Renshaw, D.K., 1998. Cenozoic stratigraphy of the Norwegian continental shelf, 64°N–68°N. Norsk Geol. Tidsskr., 78 (2): 125–151.

Eidvin, T, Smelror, M. and Bugge, T., In press. New age of the Molo Formation (new) on the mid Norwegian Continental shelf. Nor. J. Geol. (NGT).

Eldholm, O., Sundvor, E, Myhre, A, Faleide, J.I., 1984. Cenozoic evolution of the continental margin off Norway and western Svalbard. Petroleum Geology of the North European Margin. Norwegian Petroleum Society (NPF), Graham and Trotman, London, pp. 3–8.

Gradstein, F.M. and Bäckstrøm, S.A., 1996. Cainozoic biostratigraphy and paleobathymetry , northern North Sea and Haltenbanken. Nor. J. Geol. (NGT), 76: 2–32.

Hardenbol, J., Thierry, J., Farley, M.B., Jacquin, T., De Graciansky, P.-C. and Vail, P.R., 1998. Mesozoic and Cenozoic Sequence Chronostratigraphic framework of European Basins. Tulsa. SEPM Spec. Publ., 60: 3–13.

Heilmann-Clausen, Nielsen, O.B. and Gersner, F., 1985. Lithostratigraphy and depositional environment in the Upper Palaeocene and Eocene of Denmark. Bull, Geol. Soc. Den., 33: 287–323.

Henriksen, S. and Vorren, T.O., 1996. Late Cenozoic sedimentation and uplift history on the mid-Norwegian continental shelf. Global Planet. Change, 12: 171–199.

Henriksen, S. and Weimer, P., 1996. High Frequency depositional sequences and Stratal Stacking Patterns in Tertiary coastal Deltas, Mid Norwegian continental shelf. AAPG Bull, 80 (12): 1867–1895.

Hirvas, H. and Tynni, R., 1976. Tertiary clay deposits at Savukoski, Finnish Lapland, and observations of Tertiary microfossils, preliminary report. Geol. Soc., Fin., 28: 22–40.

Jordt, H., Faleide, J.I., Bjørlykke, K. and Maged, T.I., 1995. Cenozoic sequence stratigraphy of central and northern North Sea Basin: tectonic development, sediment distribution and provenance areas. Mar. Petrol. Geol., 12 (8): 845–879.

Laberg, J.S., Vorren, T.O., Mienert, J., Kvalstad, T.J., Haflidason, H., Bryn, P. and Lien, R., 2002. The Trænadjupet Slide offshore Norway: Physical and Geotechnical properties of the failed sediments—implication for the slope stability. In: A. Hurst (Editor), Onshore-Offshore Relationships on the Nordic Atlantic Margin. NGF Abstracts and Proceedings 2, 2002 of the Norwegian Petroleum Society (NPF) and Norwegian Geological Society (NGF) Conference, October 7–9 Trondheim, pp. 114–115.

Loutit, T.S., Hardenbol, J., Vail, P.R. and Baum, G.R., 1988. Condensed sections: The key to age determination and correlation of continental margin sequences. In: C.K. Wilgus et al. (Editors), Sea Level Changes-An Integrated Approach. SEPM Spec. Publ., 42:183–213.

Lundin, E.R. and Doré, A.G., 1997. A tectonic model for the Norwegian passive margin with implications for the NE Atlantic: Early Cretaceous to break-up. J. Geol. Soc., London, 154: 545–550.

Løseth, H. and Henriksen, S., 2005. A mid to Late Miocene compression phase along the Norwegian passive margin. In: A.G. Doré and B. Vining (Editors), Petroleum Geology: North-West

Europe and Global Perspectives—Proceedings of the 6th Conference, Petroleum Geology Conferences Ltd. Geological Society, London.

Martinsen, O.J., Bøen, F., Charnock, M.A. and Mangerud, G., 1999. Cenozoic development of the Norwegian Margin 60–64°N: sequences and sedimentary response to variable basin physiography and tectonic setting. In: A.J. Fleet and S.A.R. Boldy (Editors), Petroleum Geology of Northwest Europe: Proceedings from the 5th Conference. Geological Society, London, pp. 293–304.

Mitchum, R.M. and Van Wagoner, J.C., 1991. High-frequency sequences and their stacking patterns. Sequence stratigraphic evidence of high frequency eustatic cycles: Sedimentary Geology, 70: 131–160.

Mosar, J., Eide, E.A., Osmundsen, P.T., Sommaruga, A. and Torsvik, T.H., 2002. Greenland–Norway separation: A geodynamic model for the North Atlantic. Norwegian Journal of Geology, 82: 281–298.

Mutter, J.C., Talwani, M. and Stoffa, P., 1984. Evidence of Thick Oceanic Crust Adjacent to the Norwegian Margin. J. Geophys. Res., 89 (B1): 483–502.

Olesen, O., Torsvik, T.H., Tveten, E., Zwaan, K.B., Løseth, H. and Henningsen, T., 1997. Basement structure of the continental margin in the Lofoten-Lopphavet area, northern Norway: Constraints from potential field data, on-land mapping and palaeomagnetic data. Norsk Geol. Tidsskr., 77: 15–30.

Olsen, L., 1997. Rapid shifts in glacial extension characterise a new conceptual model for glacial variations during the mid and late Weichselian in Norway. Nor. Geol. Surv., Bull, 433: 54–55.

Ottesen, D., Dowdeswell, J.A., Rise, L., Rokoengen, K. and Henriksen, S., 2002. Large-scale morphologic evidence for past ice-stream flow on the mid-Norwegian continental margin. In: Dowdeswell, J.A. and Cofaigh, C. (eds.), Glacier-Influenced Sedimentation on High-Latitude Continental Margins. Geological Society, London, Special Publications, 203, 245–258.

Posamentier, W. and Allen, G., 1999. Siliciclastic Sequence Stratigraphy: Concepts and Applications, SEPM Concepts in Sedimentology and Paleontology 7: 204 pp.

Posamentier, W. and Vail, P.R., 1988. Eustatic controls on clastic deposition II—sequences and systems tract models. In: C.K. Wilgus et al. (Editors), Sea Level Changes-An Integrated Approach. SEPM Spec. Publ., 42: 55–95.

Riis F. and Fjeldskaar, W., 1992. On the magnitude of the late Tertiary and Quaternary erosion and its significance for the uplift of Scandinavia and the Barents Sea. In: R.M. Larsen et al. (Editors), Structural and Tectonic Modeling and its application to Petroleum Geology. Norwegian Petroleum Society (NPF), Special Publication 1, Elsevier, Amsterdam, pp. 163–185.

Sejrup, H.P., Larsen, E., Landvik, J., King, E.L., Haflidason, H. and Nesje, A., 2000. Quaternary glaciations in southern Fennsoscandia: evidence from southwestern Norway and northern North Sea region. Quaternary Sci. Rev., 19: 667–685.

Skogseid, J., Pedersen, T., Eldholm, O. and Larsen, B.T., 1992. Tectonism and magmatism during NE Atlantic continental break-up: the Vøring Margin. In: B.C. Storey, T. Alabaster and R.J. Pankhurst (Editors), Magmatism and the Causes of Continental Break-up, Geol. Soc. London, Spec. Publ., 68: 305–320.

Skogseid J., Planke, S., Faleide, J.I., Pedersen, T., Eldholm, O. Neverdal, F., 2000. NE Atlantic continental rifting and volcanic margin formation. In: A. Nøttvedt et al. (Editors), Dynamics of the Norwegian Margin. Geol. Soc., London, Spec. Publ., 167: 295–326.

Vail. P.T., Mitchum, R.M., Todd, R.G., Widmier, J.M., Thomson, S., Sangree, J.B., Bubb, J.N. and Hatlelid, W.G., 1977. Seismic stratigraphy and global changes in sea level, parts I-II: overview. In: C.E. Payton (Editor), Seismic stratigraphy—application to hydrocarbon exploration: AAPG Mem., 26: 51–212.

Van Wagoner, J.C., Mitchum, R.M. Jr., Posamentier, H.W., Vail, P.R., Sarg, J.F., Loutit, T.S. and Hardenbol, J., 1988. An overview of the fundamentals of the sequence stratigraphy and key definitions. In: C.K. Wilgus et al. (Editors), Sea Level Changes-An Integrated Approach. SEPM Spec. Publ., 42: 39–45.

Vorren, T.O., Edvardsen, M., Hald, M. and Thomsen, E., 1983. Deglaciation of the Continental Shelf off Southern Troms, North Norway. Norg. Geol. Unders. Bull., 380, 173–187.

Vorren T.O., Rokoengen,K., Bugge, T. and Larsen, O.A., 1992. Kontinentalsokkelen, Tykkelsen på kvartære sediementer, 1:3 mill. Nasjonalatlas for Norge, kartblad 2.3.9. Statens Kartverk.

Vergara; L., Wreglesworth, I., Trayfoot, M. and Richardsen, G., 2001. The distribution of Cretaceous and Palaeocene deepwater reservoirs in the Norwegian Sea basins. Petrol. Geosci., 7 (4): 395–408.

Weimer, P. and Posamentier, H.W. (eds.), 1993. Siliciclastic Sequence Stratigraphy. Recent developments and applications. AAPG Memoir, 58, 492 pp.

Wilgus, C.K., Hastings, B.S., Posamentier, H., Van Wagner, J., Ross, C.A. and Kendall, C.G. St C. (eds.), 1988. Sea level changes—an integrated approach. SEPM Special Publication, 42, 407 pp.

Upper Cretaceous basin-floor fans in the Vøring Basin, Mid Norway shelf

Erik Fjellanger, Finn Surlyk, Lee C. Wamsteeker and Torill Midtun

Recent deep-water wells drilled in the Vøring Basin, situated on the outermost Mid Norway shelf, have revealed new insights into the presence of extensive Upper Cretaceous basin-floor fan deposits, including intervals of stacked, laterally continuous, sheet-like sandstones, up to 1 km thick.

The first exploration well aiming for this play was drilled by BP on the Nyk High. BP made a gas discovery in excellent quality reservoir sandstones, thought to represent the middle to proximal part of a Campanian basin-floor system. Correlation of cores, logs and reflection seismic profiles between the wells drilled on the Nyk High and the Vema Dome, 25 km apart, shows a remarkable continuity of intervals of stacked turbidite sandstones separated by mudstones.

A Maastrichtian deep-water reservoir section was also drilled by the Nyk well, and was cored by a well on the Gjallar Ridge. Palaeogeographic reconstructions and provenance studies show that the sediments for both the fans were most probably derived from the Greenland craton and its palaeo-shelf, prior to opening of the Norwegian–Greenland Sea.

During the Late Cretaceous, the outer Vøring Basin was an active rift basin. Bathymetric features controlled by the main bounding faults and transfer zones of the rift basin are interpreted to have had a strong affect on the sediment distribution. The Norwegian–Greenland Sea provides an outstanding example of the progressive evolution of deep-water clastic systems reflecting changes in tectonic style and bathymetry. The Albian–Cenomanian was characterised by erosion of the inherited Jurassic topography and the formation of small immature clastic systems in the pre-existing basins. During Turonian–Coniacian time, a smoother sea-floor topography developed with more regional subsidence and erosion of the large Nordland Ridge during tectonic episodes. The deep-water facies were still heterogeneous but the depositional systems were larger systems and possessed a higher degree of connectivity and reservoir quality in the single-cycle reservoirs. During Campanian–Maastrichtian times, large hinterland areas of the East Greenland mainland became subject to erosion and sourced extensive sheet-like basin-floor fan systems, up to 1 km thick, forming multi-cycle reservoirs.

Introduction

The aim of this study is to present a synthesis for the structural and sedimentological evolution of the Vøring Basin on the Mid Norway shelf in order to improve the understanding of the Campanian–Maastrichtian basin-floor plays of the basin. The temporal evolution from small and immature Lower Cretaceous deep-water clastic systems over more extensive mid-Cretaceous single cycle fans to Campanian–Maastrichtian multi-cycle fan systems is highlighted.

Palaeogeographic reconstructions of the margins of the Norwegian–Greenland Sea emphasize the Late Cretaceous central position of the Vøring Basin, relative to the Norwegian and Greenland mainlands. Interpretation of the structural evolution is based on regional seismic interpretation and mapping of the Upper Cretaceous and Cenozoic intervals. Core interpretation shows that the dominant process active during transport and deposition of the main reservoir facies was that of concentrated density flows (in the sense of Mulder and Alexander, 2001). Depositional environments and palaeogeography of the deep-water plays are inferred from the integration of the sedimentological results and linking the wells by log correlation and biostratigraphic data, supplemented by seismic interpretations and the structural setting.

Exploration History

The Campanian–Maastrichtian deep-water plays of the Vøring Basin were first drilled on the Utgard High by wells 6607/5-1 and -2 operated by Esso in 1987 and 1992, respectively (Fig. 1). The targets

Onshore–Offshore Relationships on the North Atlantic Margin
edited by B. Wandas et al. NPF Special Publication 12, pp. 135–164,
Published by Elsevier B.V., Amsterdam © Norwegian Petroleum Society (NPF), 2005

were at the time interpreted as being of Jurassic age, but were later shown to represent distal parts of Lower–Middle Campanian (Nise Formation) deep-water deposits. In 1993, IKU cored three shallow wells in the Vøring Basin, two on the Nyk High and one on the Naglfar Dome. The cores proved the presence of Campanian–Maastrichtian deep-water sandstones in this part of the Vøring Basin. In 1997, BP drilled well 6707/10-1 on the Nyk High and discovered gas in central parts of a Lower–Middle Campanian basin-floor fan system (Fig. 1; Kittelsen et al., 1999). The sandstones in the Nyk well are of very good reservoir quality. The well also penetrated deep-water sandstones of

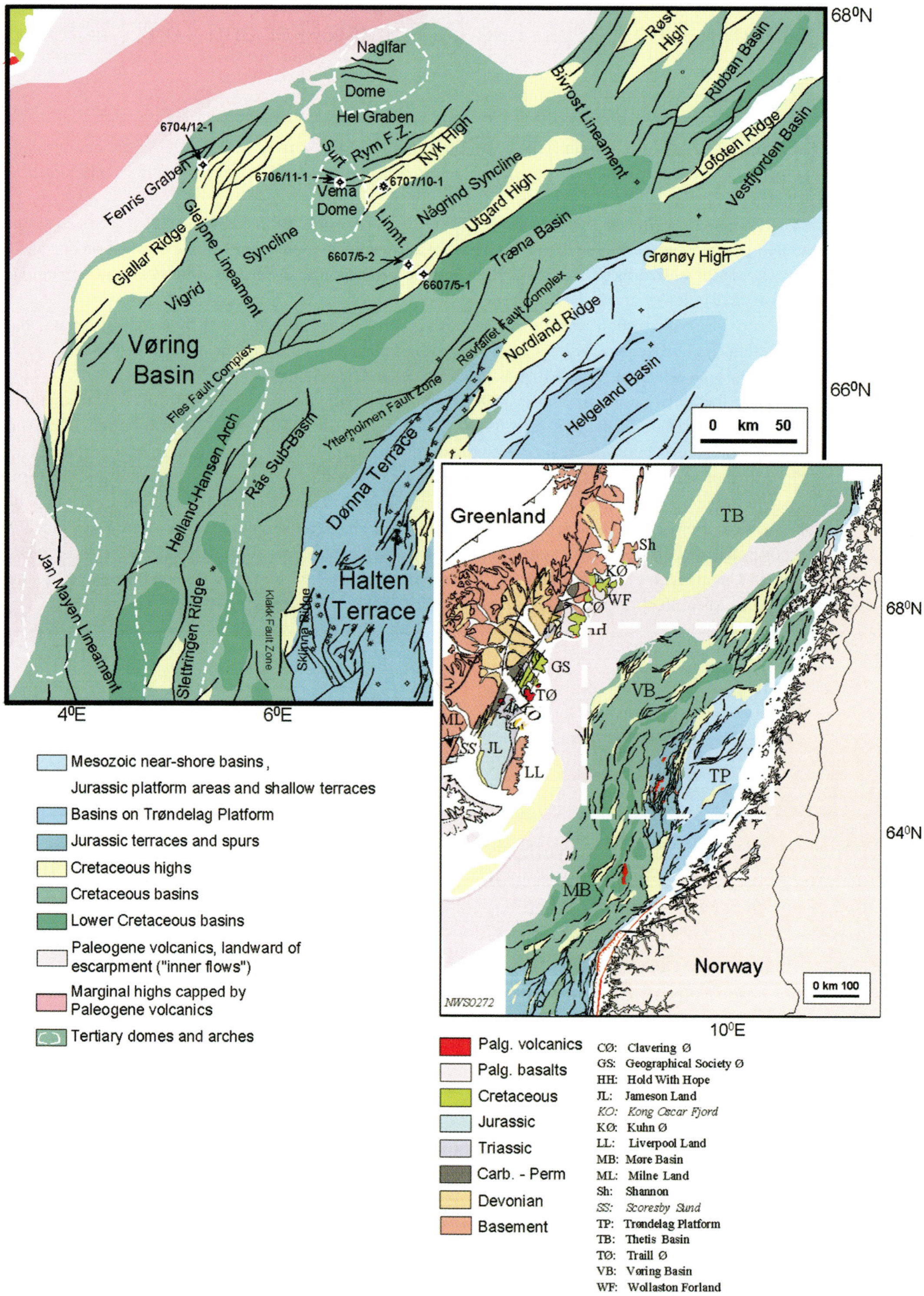

Fig. 1 Tectonic elements map of the Vøring Basin, Norwegian Sea. Modified after Blystad et al. (1995). Palg. = Palaeogene.

Maastrichtian age above the Campanian target, but they were not successfully cored. In 1997–98, Statoil drilled well 6706/11-1 on the Vema Dome. The same good quality Lower–Middle Campanian reservoir sandstones were penetrated, but hydrocarbons were not encountered. During the summer of 1999, Saga Petroleum completed the drilling of well 6704/12-1 on the Gjallar Ridge. The well showed the presence of a sand-rich Upper Cretaceous interval, 750 m of which were of Late Campanian–Early Maastrichtian age (Springar Formation) and 250 m of Early–Middle Campanian age (Nise Formation). The well was dry.

Structural setting and basin evolution

The main structural elements of the Vøring Basin are the results of progressive overprinting of several tectonic extensional and subsequent compressional events (Fig. 1; e.g. Gabrielsen et al., 1984; Bukovics and Ziegler, 1985; Brekke and Riis, 1987; Surlyk, 1990; Ziegler, 1990; Blystad et al., 1995; Doré et al., 1999; Brekke et al., 2001). A generalised lithostratigraphic scheme based on Dalland et al. (1988) and Gradstein et al. (1994) is modified and applied to the Vøring Basin to give an overview of the major depositional units on the Mid Norway shelf (Fig. 2).

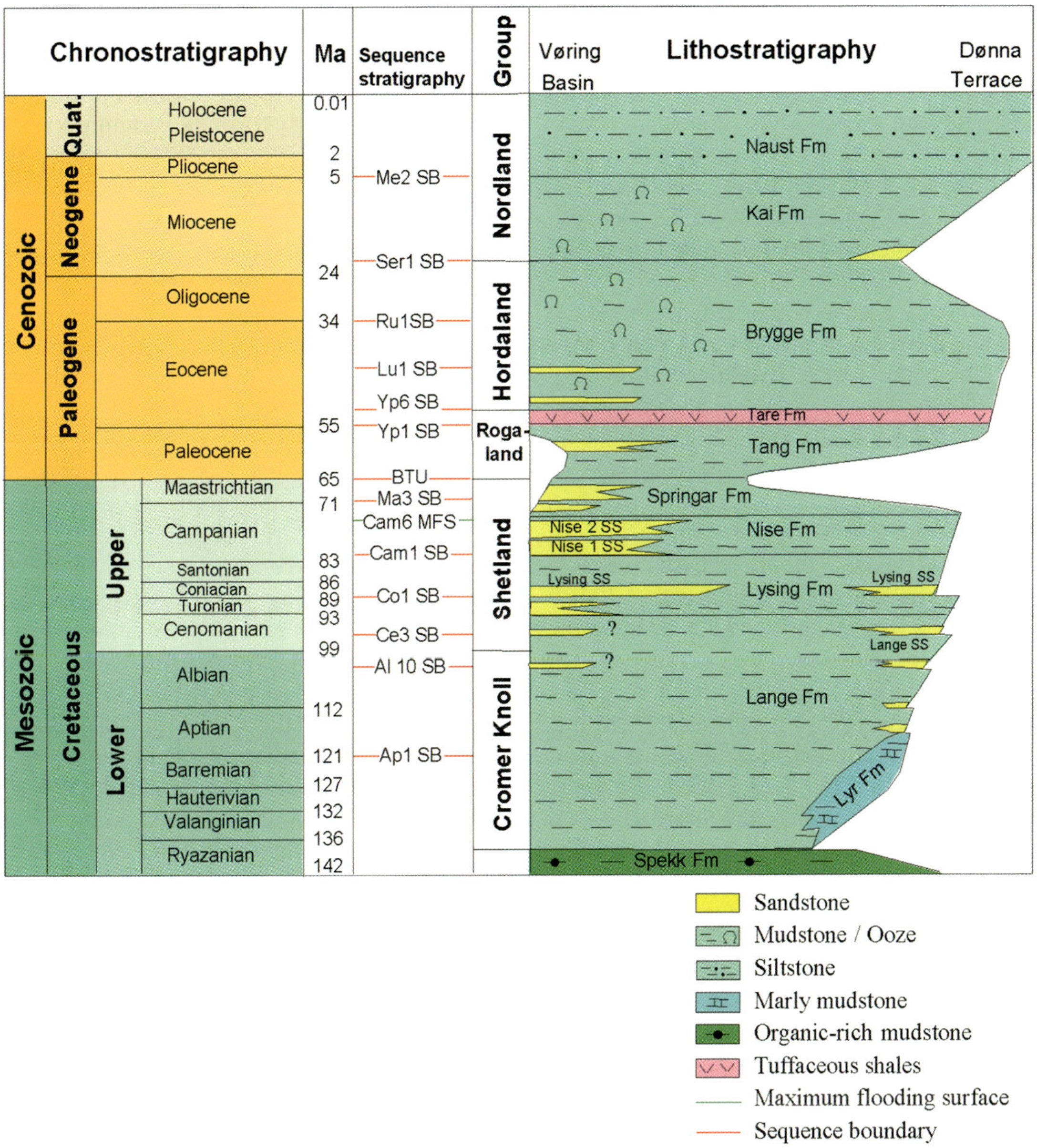

Fig. 2 Stratigraphy of the Cretaceous–Cenozoic succession of the Vøring Basin. Modified from Dalland et al. (1988) and Gradstein et al. (1994).

The overall NE–SW structural grain of the basin is interpreted as inherited from the original Caledonian suture system (Bukovics and Ziegler, 1985; Doré et al., 1997). This structural grain largely controlled the orientation of the axes of successive Cretaceous–Paleogene rift episodes, during which the axes of rifting became progressively focused between Greenland and Norway, so that the axes on the Norwegian side stepped towards the W–NW (Bukovics and Ziegler, 1985). In the northern part of the Vøring Basin, the axes of the west-stepping rifting follow a NE–SW orientation (Fig. 1). These trends change in the southern part of the basin, with a transition to a more N–S orientation that extends into the northern Møre Basin. In the southern Møre basin, the structural grain shifts back to the NE–SW direction. The primary NE–SW trend of the basin is intersected by a series of NW–SE oriented fault lineaments (Blystad et al., 1995). These are interpreted to be discontinuities in the main rift faults and trend obliquely to the main fault direction. The lineaments had a significant effect on Cretaceous–Paleogene sedimentary patterns and structural geometry (Doré et al., 1999; this chapter).

The dominant NE–SW oriented basement grain was reactivated during the Early Cretaceous rift phase along the North Atlantic margin (e.g. Lundin and Doré, 1997), which opened the Møre and Træna Basins and the Rås Sub-basin. The Klakk Fault Zone bounds the east side of an Early Cretaceous rift in the N–S trending Rås Sub-basin and the Slettringen Ridge bounds the west side. Towards the northeast, the Rås Sub-basin narrows and merges with the NE–SW oriented Træna Basin, bounded to the east by the Ytterholmen Fault Zone and to the west by the Utgard High and Fles Fault Complex (Fig. 1). The Early Cretaceous rift is interpreted to continue north of the Træna Basin into the Ribban and Vestfjorden Basins (Blystad et al., 1995).

In Early Cretaceous times, the Nordland and Sklinna Ridges became significant topographic highs, situated east of the Træna Basin and the Rås Sub-basin. The Halten and Dønna Terraces, flanking the highs, saw very low sedimentation rates and condensation throughout the Early Cretaceous (Figs. 3–5; Bukovics and Ziegler, 1985; Skogseid et al., 1992a). Structural highs were eroded and immature sediments were deposited in the restricted basins, between the highs along the Møre–Rås–Træna rift axis (Fig. 3A). The structural highs provided a limited provenance area for sands deposited in the restricted basins (Fig. 3B). The sandstones show poorly developed reservoir characteristics and connectivity, as demonstrated by wells drilled in the Agat area (Fig. 3A; Shanmugam et al., 1994). Similar characteristics are seen for age equivalent sub-marine, conglomeratic, avalanche breccias of the Rold Bjerge Formation in North-East Greenland (Surlyk and Noe-Nygaard, 2001).

During Albian time, major basin-bounding faults were reactivated, and the Halten and Dønna Terraces developed into truly open seaways (Brekke and Riis, 1997; Brekke et al., 1999, 2001). Subsequent tectonic events continued through the Cenomanian and Turonian, each contributing new pulses of erosional products into the basin (Fig. 5). Coarse clastic sediments from the rejuvenated Nordland Ridge footwall were shed into the basin and form the Lange Formation sandstones. A tectonic event occurred in the Late Turonian–Coniacian and resulted in the formation of an unconformity forming an important sequence boundary (Co1 SB, Figs. 2, 4 and 5). By this time, the pre-existing intra-basin relief was effectively infilled, and most of the tilted fault blocks and structural highs became draped by sediments (Figs. 4 and 5). Hence, the Lysing Formation sandstones overlying and onlapping the Co1 sequence boundary have a less restricted distribution than the previously deposited sandstones. The Nordland Ridge acted as a provenance area of sufficient size for the Lysing Formation to develop as a 50–100 m thick deep-water fan succession, covering the Dønna Terrace (Fig. 4). The deep-water sandstones are recognised on seismic sections as a single-cycle reservoir resting on the Co1 SB (Fig. 5). Recycling of the Palaeozoic–Mesozoic sand-rich rocks generally provides good to moderate reservoir properties to the Lysing Formation sandstones. The deep-water deposits comprise a mixture of turbidites, debrites and heterolithic sandstones and mudstones (Fig. 6), which suggest a short travel distance for the gravity flow deposits. However, the Lysing deposits are substantially more mature and connected than the Lower Cretaceous sediments. In North-East Greenland, a similar improvement in original reservoir properties is seen passing from the Lower Cretaceous Rold Bjerge Formation breccias over Upper Turonian–Lower Coniacian conglomerates of the Månedal Formation to the turbidite-dominated Vega Sund Formation sandstones of the assumed Coniacian age (Surlyk and Noe-Nygaard, 2001).

In Albian–Cenomanian time, the rift axis on the Norwegian side shifted to a more westward position (Lundin and Doré, 1997) and thick sedimentary successions were deposited in the widened rift basin

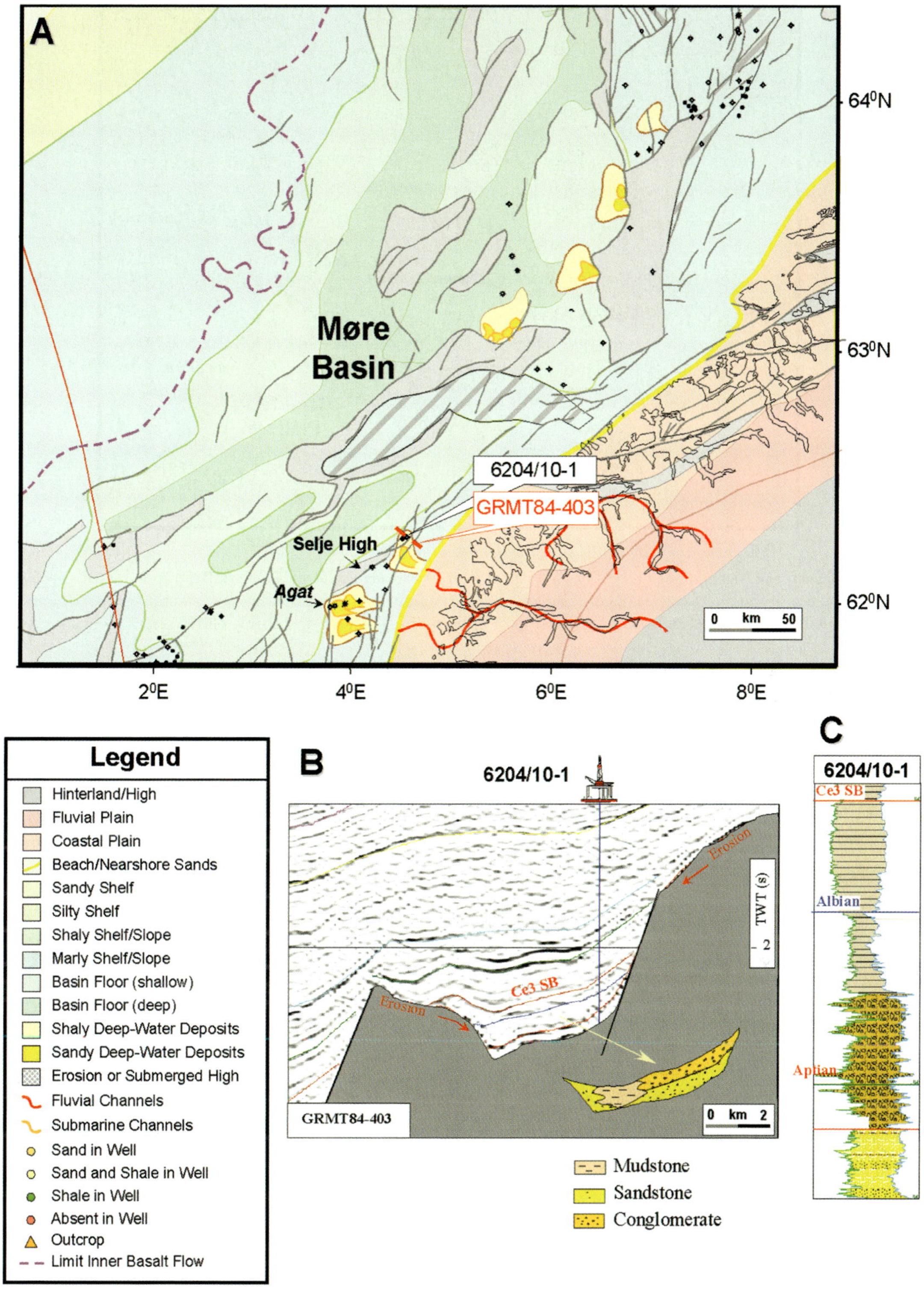

Fig. 3 (A) Aptian–Albian palaeogeography of the Møre Basin. Structural highs were exposed for erosion. (B) During the Early Cretaceous, clastic sediments were shed into restricted sedimentary basins. Fan lobes of limited extent and poor connectivity within the reservoir were deposited along the flanks of the highs. Shales and marls were deposited along and away from the highs. The deposits are thin and condensed due to the low sedimentation rate during the Early Cretaceous. (C) Well 6204/10-1 drilled an Aptian–Albian section consisting of poorly sorted sandstones, conglomerates and mudstones.

in the Någrind and Vigrid Synclines. The western rift shoulder was controlled by the Rym Fault Zone and the Gjallar Ridge, and is expressed by onlap and thinning onto the interpreted mid-Cenomanian unconformity on the Gjallar Ridge (Ce3 SB, on Fig. 2; Lundin and Doré, 1997). The Gjallar Ridge rift faults were periodically active throughout the Late Cretaceous, and the Fenris Graben was

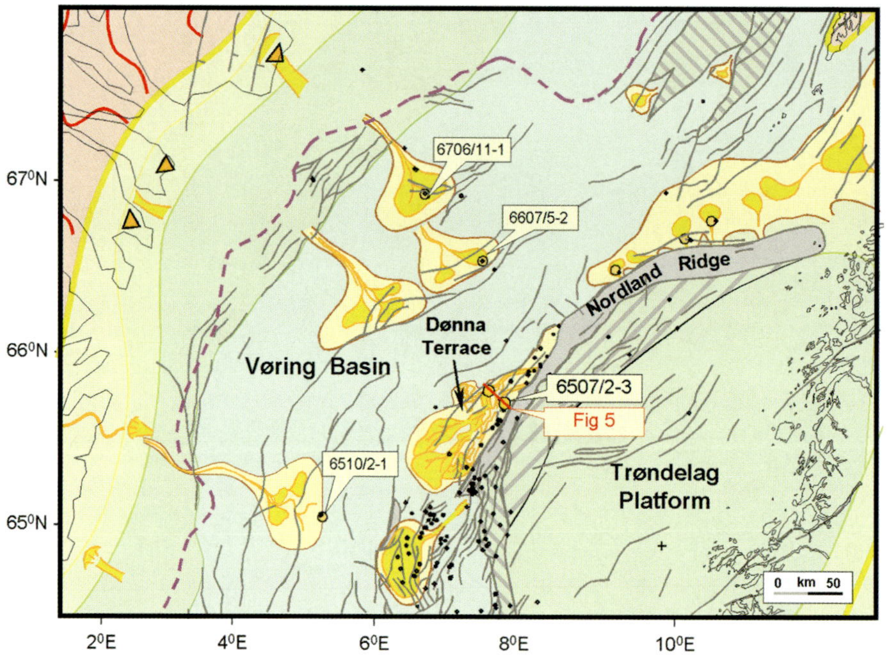

Fig. 4 Turonian–Coniacian palaeogeography of the Vøring Basin and the Trøndelag Platform. The Nordland Ridge acted as provenance for a deep-water fan system deposited on the Dønna Terrace. This fan system is well-documented by many wells, and shows a good connectivity. Less well-documented distal fan systems are penetrated by wells in the Vøring Basin. These are tentatively derived from a Greenland provenance. For legend, see Fig. 3.

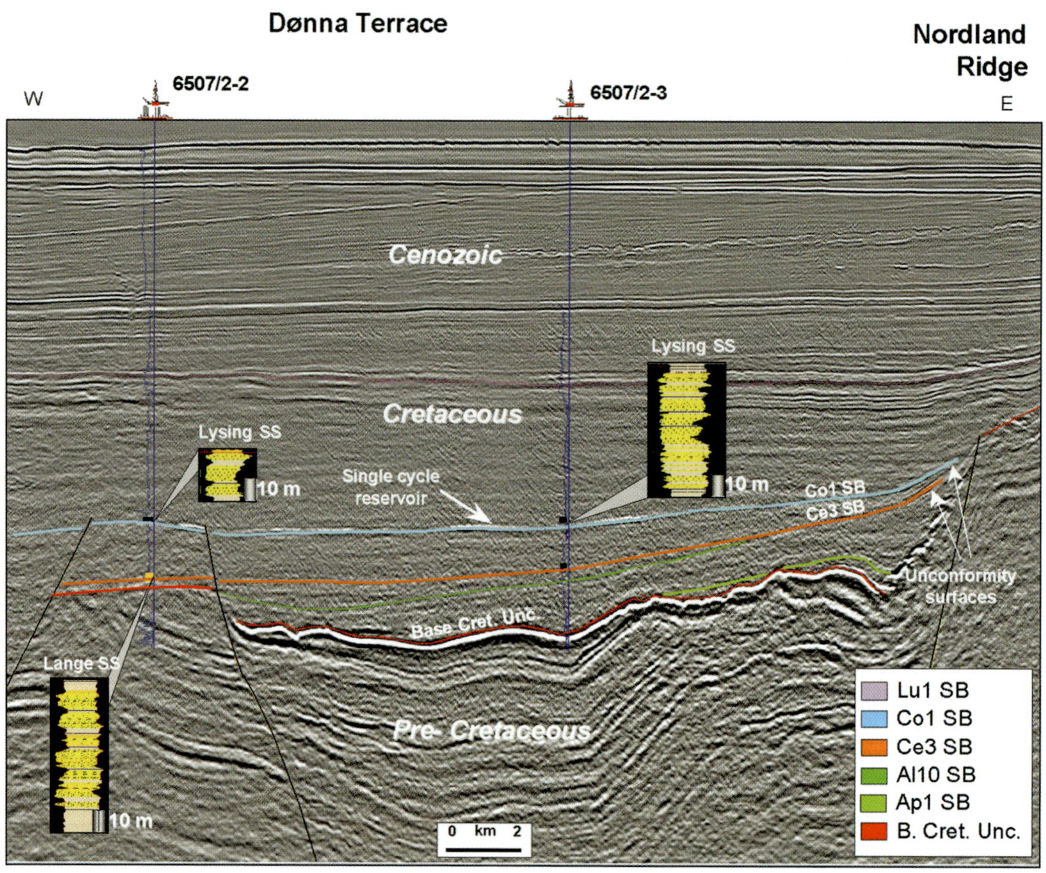

Fig. 5 Seismic section across the Dønna Terrace showing the Cretaceous reservoir intervals. The Lysing Formation sands resting on the Co1 SB show a single-cycle, patchy appearance due to variations in thickness, reservoir porosity and hydrocarbon content. The Lange Formation sands resting on the Ce3 SB are invisible on the seismic line at this scale, and are in general difficult to identify and map. The sequence boundaries are observed as unconformities close to the Nordland Ridge. This suggests a tectonic impact on the sandstone deposits onlapping and resting on these surfaces. The low sedimentation rates in the Early Cretaceous are indicated by the presence of a condensed Lower Cretaceous succession between the Ap1 SB and the Base Cretaceous Unconformity.

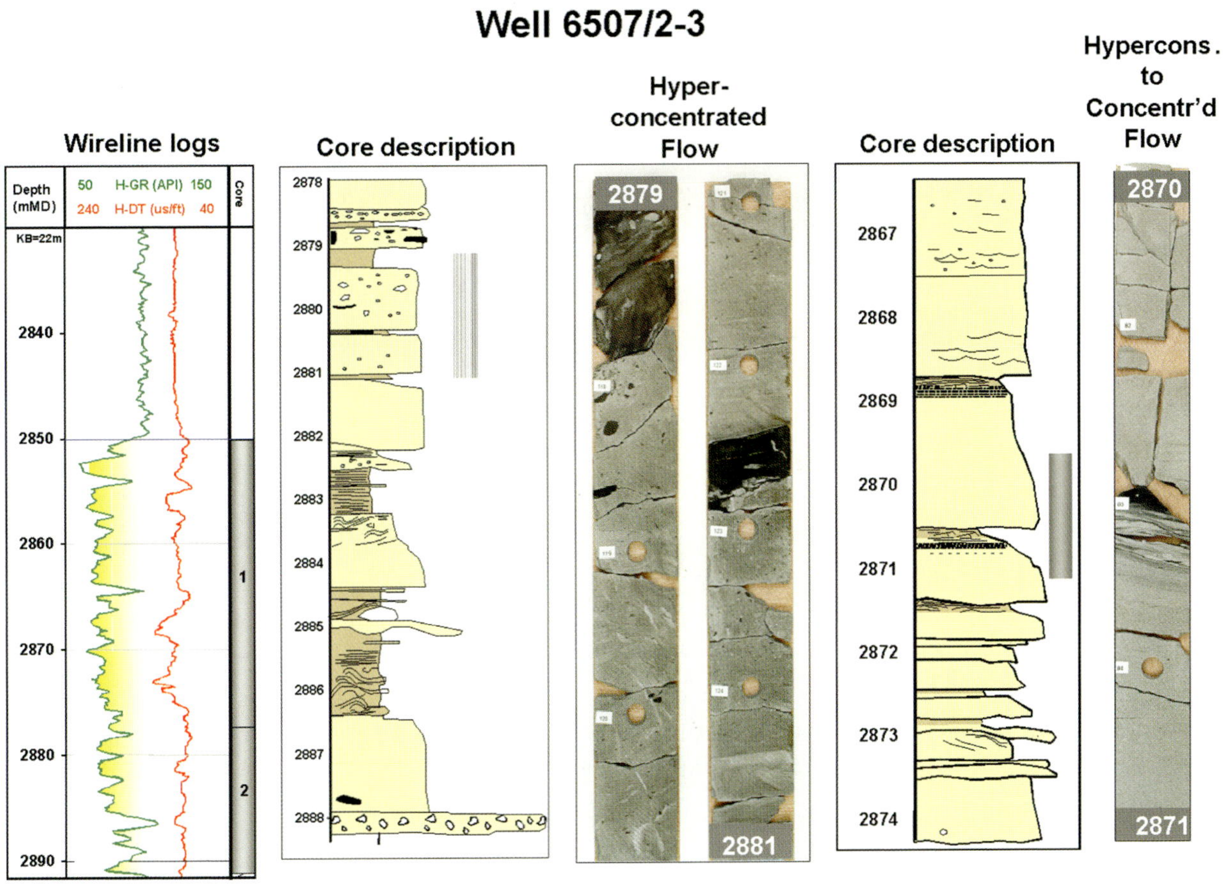

Fig. 6 Heterolithic composition of the Lysing Formation reservoir in Well 6507/2-3 on the Dønna Terrace. The immature, deep-water sediments are derived by different hyperconcentrated and concentrated flows from the nearby Nordland Ridge. The sand-rich deposits have good reservoir properties.

formed west of the Gjallar Ridge (Blystad et al., 1995). Expansion of the stratigraphic section in the Hel Graben, Någrind Syncline and Vigrid Syncline indicates continued Late Cretaceous extension and subsidence in those areas (Bukovics and Ziegler, 1985). The effect of the transfer zones on sedimentation was pronounced during the Late Cretaceous, with major changes in sand distribution across the Surt and Bivrost Lineaments. The lineaments are believed to have controlled the position of the main sediment entry points for the thick sheet-like basin-floor turbidite systems of the Nise and Springar Formations, deposited in the Vøring Basin during the Campanian–Maastrichtian (Fig. 7). North-East Greenland and its shelf are interpreted as the provenance area for these basin-floor fans (Morton and Grant, 1998; Surlyk and Noe-Nygaard, 2001). The large size and uniform nature of the Campanian–Maastrichtian basin-floor systems and their provenance stands in marked contrast to the less well-developed deep-water systems deposited during Early to mid-Cretaceous times in the Norwegian Sea area (Figs. 3 and 4).

Parts of the Vøring Basin were uplifted, faulted and eroded during the Maastrichtian, continuing into the Paleocene (Bukovics and Ziegler, 1985). Reactivation of Cretaceous normal faults occurred along the Utgard High and the Fles Fault Zone, with evidence of active erosion on the footwalls (Skogseid et al., 1992a). Significant uplift and extensional faulting occurred along the Nyk High, resulting in the formation a 100 km long NE–SW trend of rotated, east-dipping fault blocks. This faulting is interpreted to have taken place primarily during latest Cretaceous–Paleocene times, mainly involving an earlier non-faulted Cretaceous succession (Kittelsen et al., 1999). Uplift, erosion and reactivation of earlier faults occurred along the western margin of the Gjallar Ridge and in the Fenris Graben (Skogseid et al., 1992a). The NW–SE trending Bivrost, Surt and Gleipne Lineaments played a major role in this rift pulse, controlling significant changes in structural style across the lineaments (Fig. 1).

Continental separation was initiated in the Early Eocene, associated with intense volcanism, with formation of both extrusive and intrusive magmatic rocks, along the western margin of the basin (Eldholm et al., 1989; Skogseid et al., 1992b). The onset of continental separation in the Early

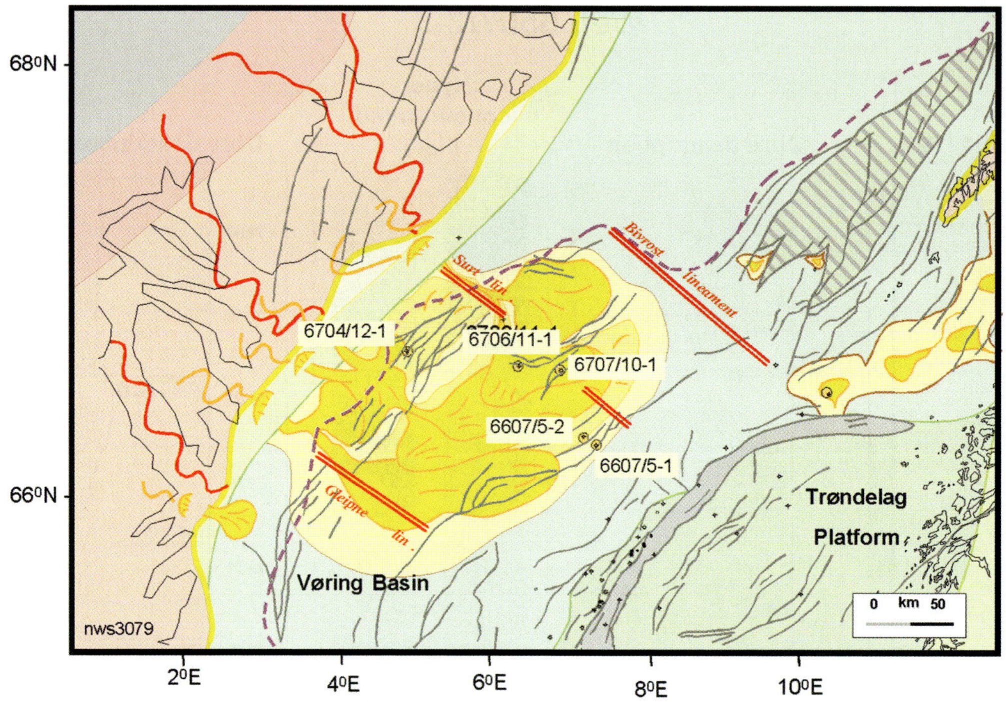

Fig. 7 Campanian palaeogeography of the Vøring Basin. A large basin-floor fan system is well-documented by five exploration wells and seismic mapping and as described in subsequent figures. A Greenland sediment provenance is suggested. For structural names and legend, see Figs. 1 and 3, respectively.

Eocene resulted in a change of the horizontal stress patterns in the Vøring Basin. The compressive stress is attributed to ridge-push forces from the opening Norwegian–Greenland Sea (Doré et al., 1999). A NW–SE extensional regime changed to one of primarily NW–SE compression and formation of Eocene–Miocene inversion structures in many areas of the Vøring Basin and inversion of Cretaceous–Paleogene structural lows (e.g. Doré and Lundin, 1996; Lundin and Doré, 2002). The hanging walls of the Fles Fault Zone and the Slettringen Ridge were thus inverted, resulting in the formation of the 200 km long Helland Hansen Arch (Sanchez-Ferrer et al., 1999). The hanging wall of the Rym Fault Zone (southern part of Hel Graben) underwent uplift and inversion, resulting in formation of the Vema Dome, where the northern Hel Graben was inverted to form the Naglfar Dome (Fig. 1) (Blystad et al., 1995).

Sedimentary facies and depositional processes

Sedimentological core description was performed for all the exploration wells drilled in the Vøring Basin, leading to a facies classification scheme that formed the basis for interpretation of sedimentary processes and depositional environments. Six main sedimentary facies groups were identified from the Vøring Basin cores (Fig. 8). A few beds belong to other facies but have not been included in the scheme. They are mentioned in the descriptive text wherever relevant. The generalised facies scheme covers all the studied cores but the wells are described independently as there are subtle differences in characteristics of similar facies between wells, mainly in lithology and degree of bioturbation.

The Campanian–Maastrichtian plays in the Vøring Basin were studied by facies analysis of cores from wells 6707/10-1, 6706/11-1, 6607/5-2 and 6704/12-1. The Lower–Middle Campanian depositional system is described on the basis of the first three wells, whereas the Maastrichtian system is based on the last well. Below follows a description and interpretation of the sedimentary facies and depositional environment recognised in the cores.

Well 6707/10-1 (Nyk High)

The Nyk cores cover 205 m of the Lower–Middle Campanian succession (3145–2967 m and 4145–4118 m) belonging to the Nise 1 sandstone and Nise 2 sandstone of the Nise Formation, respectively (Fig. 9). Six main facies groups are identified in the logged sections of the Nyk well. They are in turn, described from the coarsest to the finest grained.

Facies	Observations	Process	Description
C1	Massive sandstone or faint inverse graded sandstone	Hyper-concentrated density flow	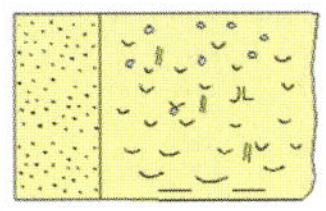
C2	Massive or top-only graded sandstone	Hyper-concentrated to concentrated density flow	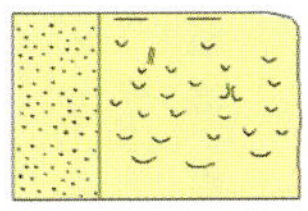
T1	Graded sandstone with Tabcde Bouma divisions	Concentrated density flow and turbidity flow	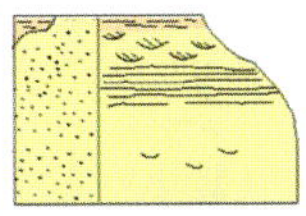
T2	Stratified sandstone	Concentrated density flow and turbidity flow modified by currents or waves	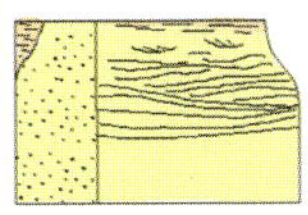
H1	Heterolithic sandstone and mudstone	Low density turbidity flow / bottom current	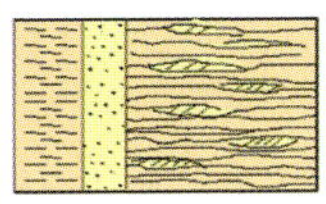
S1	Overturned / brecciated deposits	Slumping	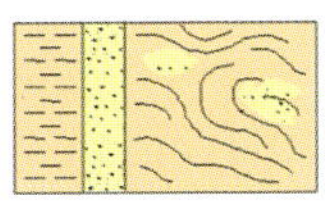

Core photo

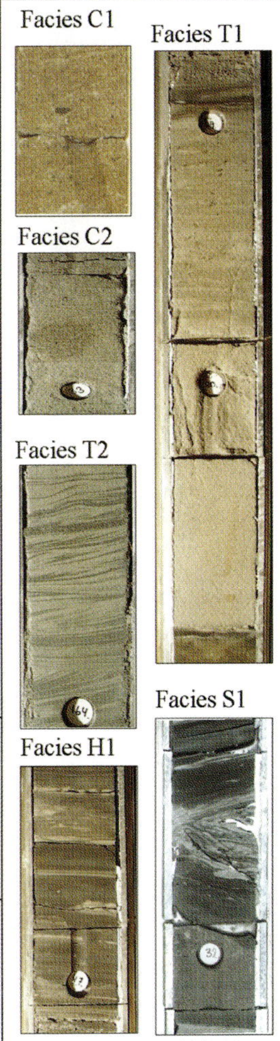

Fig. 8 Main sedimentary facies recognised in the Campanian–Maastrichtian basin-floor fan deposits in the Vøring Basin. Locations of core photographs are shown on Figs. 9–12.

Facies C1 and C2 — Massive sandstone

The sandstone is fine-grained and non-graded, except for the top 5–10 cm, which in most cases show rapid normal grading into a strongly bioturbated, heterolithic sandstone–mudstone (Figs. 8 and 9). Cross-lamination is in some cases preserved in the heterolithic part. The base of the beds is sharp and horizontal. The lower part of some beds shows a weak inverse grading. The sandstone beds are 15–280 cm thick, averaging 100 cm. Amalgamation of beds is common and is revealed by internal erosion surfaces, bioturbated clay-rich horizons, and change in size and style of water escape structures across a diffuse bedding plane. The sandstones show no primary sedimentary structures, but contain abundant water escape structures. They comprise gently curved, concave-up laminae in the lower part of the beds giving way upward to more strongly curved, classical dish structures and sub-vertical fluidisation pipes. Flow folds outlined by inclined and overturned water escape laminae are common. The upward succession from gently to more strongly curved dish structures with shorter wavelength and higher amplitude is commonly repeated in the thicker beds, suggesting the presence of cryptic amalgamation surfaces. Small clay chips are common in some beds, especially, in the upper 20 cm of the sandstone portion.

Beds showing a top-normal grading are transitional to Facies T1. Facies C2 dominates the interval from 3148–3125 m. It constitutes about a third of the thicker beds in the interval from 3083–3021 m, and most of the 2985–2975 m interval.

The processes active during sediment transport are difficult to interpret due to the lack of grading

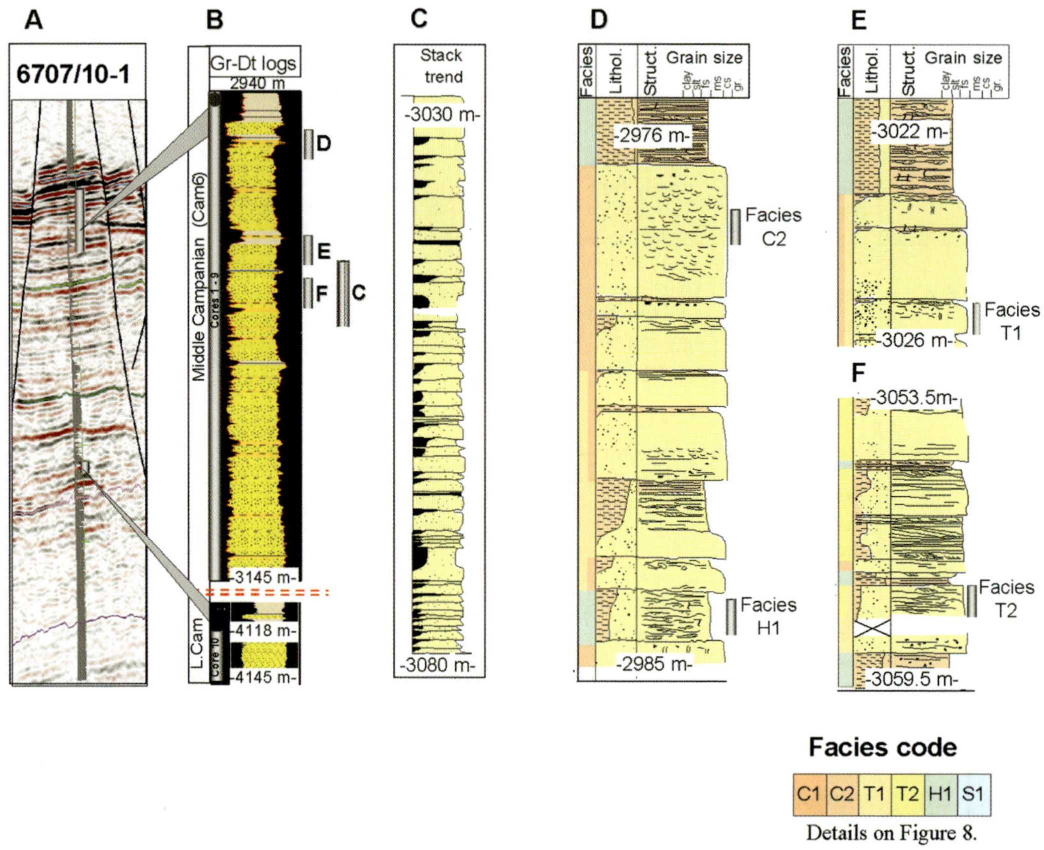

Fig. 9 Key sedimentary facies described from the cores in Well 6707/10-1 drilled on the Nyk High. (A) Seismic section at the well location of the Lower–Middle Campanian succession. (B) Wireline logs of the cored interval. (C) Simplified core description showing the lack of trends in the stacking of the sandy density-flow deposits. (D–F) Core descriptions showing the dominance of hyper-concentrated and concentrated density-flow deposits and examples of key facies (core photos on Fig. 8).

and primary sedimentary structures in the sandy part of the beds. The sharp base, massive texture and top normal grading indicate deposition from sediment gravity flows. The non-graded part was deposited by frictional freezing of hyper-concentrated density flows (in the sense of Mulder and Alexander, 2001) and the ubiquitous occurrence of water escape structures suggests that deposition was rapid. Grain-to-grain interaction generating dispersive pressure was an important particle support mechanism at least during the late stages of transport. The normally graded top interval indicates that dilution allowed the upper part of the flow to become fully turbulent and deposition was from a waning turbidity flow. The graded heterolithic top was in most cases colonised and bioturbated by a benthic fauna, indicating that the sea floor was oxygenated. The uniform nature of the beds and their vertical stacking in thick successions indicate that hyper-concentrated transitional to concentrated density flows (Mulder and Alexander, 2001) with a thin, upper, fully turbulent flow were the dominant transport mechanisms, and were active in the depositional system over long periods of time.

Erratic events or chaotic processes are thus unlikely as causative mechanisms, and triggering of the sediment gravity flows was a regularly recurring event. The stacking of numerous essentially identical beds and the gradual transition from the top-graded beds of Facies C2 to the normally graded Facies T1 sandstones suggest that transport was mainly by highly concentrated density flows. Flow distance was possibly too short to allow the development of a pervasive grading, or deposition from suspension may have continued for a protracted period of time beneath a steady flow. Kneller (1995) suggested that thick massive sands or thick sequences of climbing ripples could be deposited from depletive steady flows, which were sustained at relatively constant discharge for long periods. The thickness of the beds may thus not bear any relation to the thickness of the flows.

Facies T1 – Graded sandstone

This facies consists of fine-grained sandstone, which is normally graded throughout or in the top part only (Figs. 8 and 9). The beds vary in thickness between 25–190 cm, averaging 80 cm. They show an upward sequence of a massive division, followed by a planar-laminated, a cross-laminated and a very fine-grained planar-laminated division corresponding to the Tabcd turbidite divisions of Bouma (1962). The Ta division is non-graded in most cases, or the top few centimetres are graded, but some Ta divisions show strong grading throughout. The Tb division shows well-developed planar lamination and the Tc division is a cross-laminated heterolith, in a few cases with climbing ripples. All Bouma divisions are only rarely encountered in a single bed, and the most common sequence is Tac. The Ta division contains abundant water escape structures. These are mainly consolidation laminae, which may be over-folded, but dish-and-pillar structures are also common. Clay chips and organic debris occur in the middle to upper part of the Ta division in a number of beds. The upper part of the Tc division and the Td division are moderately bioturbated.

The grading and the presence of Bouma divisions show that transport and deposition of the lower Ta and Tb divisions were from concentrated density flows, whereas the overlying Tbcd divisions were deposited by suspension fall-out from the higher, more dilute and turbulent part of the flow forming a true, waning turbidity current. This interpretation is confirmed by the stacking of numerous similar beds, which excludes an erratic or chaotic mode of transport. The Ta divisions were deposited rapidly, as indicated by the ubiquitous water escape structures.

Facies T2 — Graded, cross-stratified sandstone

This facies is restricted to the 3058–3050 m interval (Figs. 8 and 9). It consists of fine-grained sandstone beds, 75–135 cm thick, averaging 112 cm. The sandstones are very rich in glauconite, and coaly detritus is also common. The beds have a sharp base, and the lower part is massive, similar to the sandstones of Facies C2 and the Ta divisions of Facies T1. They are weakly normally graded in the lower part and more strongly graded in the upper part. Irregular, widely spaced laminae occur in the lower part of some beds. The upper part is heterolithic and shows divergent lamination or low-angle, inclined stratification without a preferred dip direction. The lowest bed of Facies T2 shows a unidirectional lamination transitional to non-directional low-angle cross-stratification (Figs. 8 and 9). Laminae are gently curved concave-up or convex-up. Higher in the section both uni- and non-directional, low-angle cross-stratification occur. There are no clear signs of bioturbation, although a few beds have a mottled texture in the middle part.

The sandstones are interpreted as deposited from concentrated density flows. The mainly non-directional, low-angle nature of the cross-stratification is similar to hummocky cross-stratification normally interpreted as formed by deep-storm waves (Dott and Bourgeois 1982). The occurrence of this structure in deep-water density-flow sandstones is interpreted as caused by a wave field generated during flow by interference between the flow and the water layer above. The wave field will affect the flows and cause the formation of combined current and wave structures, which are commonly generated during spillover of subtle channels. The spillover flow widens out and becomes diluted as it is cut off from the main channel fairway, where it dies out rapidly. The non- and uni-directional laminae are generated during this final phase of the density flow (David Mohrig, pers. comm. 2001). It should be noted, however, that the lithology of the beds showing low-angle cross-stratification deviates from the other facies in their high content of glauconite, which may have been exported to the basinal areas from the outer shelf during major storms. The possibility of reworking of density-flow sands by deep-storm waves thus cannot be completely ruled out. Flood-generated delta-front sandstone lobes deposited from concentrated hyper-pycnal flows have commonly been mistaken for storm-dominated nearshore and shelf deposits because of the common occurrence of hummocky cross-stratification (Mutti et al., 2000). Combined-flow conditions generating this type of stratification are, however, claimed to be inherent to the dynamics of hyper-pycnal flood-generated flows (Mutti et al., 2000).

Facies H1 — Heterolithic sandstone–mudstone

The heterolithic facies comprises alternating layers of mudstone and current ripple cross-laminated fine-grained sandstone (Figs. 8 and 9). The beds are highly bioturbated, and well-defined burrows can commonly be distinguished. The heteroliths vary in thickness between 5–155 cm, averaging 27 cm, but most units are only 5–10 cm thick. Thicker units occur above 3023 m in the middle core, and above 2976 m in the upper core. The latter unit forms a drape of the sandstone-dominated succession.

The heteroliths reflect deposition of mud out of suspension under quiet conditions, alternating with sand deposition by bottom currents or possibly by dilute turbidity flows. The heterolithic units are generally thin and were probably deposited during relatively short time intervals between the sandy gravity flow events. The thicker heteroliths probably represent temporal abandonment of the coarse-grained gravity flow systems.

Facies S1 — Slumped and brecciated heterolithic sandstone–mudstone

This facies is represented by two bedsets. The first set consists of two beds, 20 and 30 cm thick, with clasts of heteroliths or mudstone 10–20 cm long, set in a heterolithic matrix. The clasts are siderite-impregnated in one of the beds. The beds represent heteroliths of Facies H1, which were mobilised by slumping and downslope redeposition. The second set consists of two beds, 65 and 70 cm thick, that show asymmetrical slump folds and scattered mudstone chips. They represent heteroliths that have undergone some down-slope slumping with associated folding of the layers, but without complete disruption as in the former two beds.

Well 6706/11-1 (Vema Dome)

The Vema cores cover 65 m (2304–2295 m, 2334–2307 m and 3137–3108 m) of the Lower–Middle Campanian interval, which belongs to the Nise Formation (Fig. 10). The first two cores are cut in the Nise 1 sandstones and the third core spans the Nise 2 sandstones. Six facies are identified in the measured core sections and are described, here from the coarsest to the finest grained.

Facies C1 and C2 — Massive sandstone

This facies consists of thick, massive, medium-grained sandstone beds. The base of the beds is sharp and the top is sharp or graded in the uppermost 5–10 cm, which pass into a 10 cm thick heterolith. In the latter case, the top-graded portion commonly shows cross-lamination (Fig. 10). The beds (or amalgamated beds) are 45–345 cm thick, averaging 160 cm. Organic detritus, mudstone clasts and chips are common and may be horizontally aligned. Dewatering structures are ubiquitous and include dish structures and occasionally

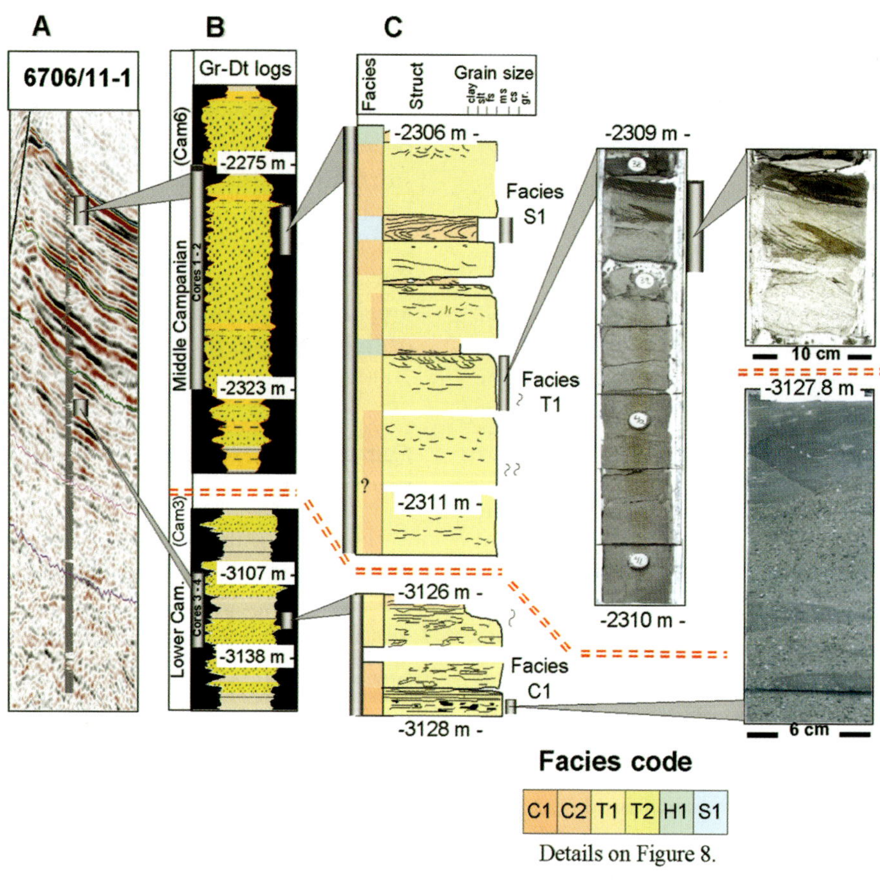

Fig. 10 Key sedimentary facies described from the cores in well 6706/11-1 drilled on the Vema Dome. (A) Seismic section at the well location of the Lower–Middle Campanian succession. (B) Wireline logs of the cored interval. (C) Core description showing that hyper-concentrated and concentrated density-flow deposits dominate the succession. Slumping is recorded in a fine-grained interval (core photograph on Fig. 8). Muddy debris-flow deposits are present in the lowermost, cored section.

pillars, wavy laminae, flow-folds, oblique pillars and injection structures. A 345 cm thick bed (or amalgamated beds) is low-angle cross-stratified in the top 25 cm and is transitional to Facies T2. Bioturbation is sparse to absent, and is mainly restricted to the graded heterolithic top of some beds. The top-graded type is transitional to the graded turbidite sandstones of Facies T1, but the latter in this well are strongly bioturbated throughout.

The non-graded beds with sharp boundaries are interpreted as representing transport and deposition from hyper-concentrated density flows. The massive, top-graded sandstones were deposited from hyper-concentrated transitional to concentrated density flows, probably by progressive aggradation under relatively sustained and steady flows rather than by instantaneous collapse of a particularly thick flow. This is supported by the presence of floating mudstone clasts at certain levels, which may represent an instantaneous position of the upward-moving bed surface. It is likely that a continuum existed during deposition hyper-concentrated to concentrated density flow and sustained steady-state turbidity current mechanisms.

Facies T1 — Graded, bioturbated sandstone

This facies accounts for about two thirds of the 3137–3108 m interval, whereas it is not represented in the 2323–2297 m interval. The beds are 5–90 cm thick, averaging 35 cm. They are graded throughout, and there is a rapid transition from the lower sandstone-dominated to the upper mudstone-dominated part, which forms about half the thickness of most beds. This is in marked contrast to the top-graded sub-facies of Facies C2, which have much thinner mudstone/heterolithic intervals. The beds are pervasively bioturbated and the density and diversity of trace fossils do not appear to change vertically through the beds. The burrows are mainly sub-horizontal. Bed boundaries are strongly modified by burrowing, but appear to have been sharp at the time of deposition. Organic detritus is common in most beds and glauconite also occurs. Slump folds can be discerned in some cases, in spite of the pervasive bioturbation.

Interpretation of the depositional process is far from straightforward. The grading suggests event deposition from turbidity flows. The pervasive, sub-horizontal bioturbation, on the other hand, would normally be taken to reflect slow rates of deposition with total, uniform mottling at all levels by infaunal, deposit feeding animals. Bioturbation is common in turbidites, but is normally most intense or even confined to the upper mud-rich portions or to amalgamation surfaces. The sea bottom was clearly well-oxygenated and hospitable to a diverse burrowing infauna. The stacking of many graded beds, each 20–30 cm thick, clearly suggests repetition of similar depositional events.

Facies T2 — Massive to cross-stratified sandstone

This facies is transitional to the top-graded sub-facies of Facies C2 and is restricted to a few thick beds. The beds are massive with dewatering structures and clay clasts. They are very similar to Facies C2 except for the top 20–30 cm, which show low-angle inclined stratification and unidirectional cross-lamination transitional to hummocky cross-stratification occurs in a few beds. The levels with inclined stratification contain abundant glauconite, organic detritus, mudstone clasts and chips, and rounded siderite clasts. The sandstone layers in the heterolithic tops of some of the beds show pinch-and-swell structures.

The massive nature of the bulk of the thick beds and the restriction of inclined stratification to the bed tops suggest transport and deposition from sediment gravity flows. Subsequent reworking by combined flow currents was associated with a wave field, which may have been generated at the transition zone between the gravity flow and the overlying water (see interpretation of Facies T2 for well 6707/10-1).

Facies H1 — Heterolithic sandstone–mudstone

Facies H1 consists of mudstone with laminae of siltstone or fine-grained sandstone. The facies is moderately bioturbated, and injection features, such as compaction-folded sandstone dykes occur. The facies forms packages, 10–85 cm thick, averaging 25 cm, which may show a fining-upward trend and a decrease in sandstone to mudstone ratio. Some of the thin sandstone beds show small-scale uni- or non-directional cross-stratification.

The facies represents deposition under fluctuating energy. The mudstone laminae were deposited from suspension under quiet conditions, whereas the sandstone laminae represent transport and deposition from bottom currents or relatively low-energy turbidity flows. The fining-upward packages are similar to the top part of Facies C2 and T1, and may have been deposited from dilute, fluctuating turbidity flows. Similarly, the cross-stratified beds are thin turbidites of Facies T2.

Facies S1 – Slump folded sandstone and mudstone

Some of the heterolithic sandstones and mudstones are commonly contorted and brecciated by post-depositional movements (Figs. 8 and 10).

Well 6607/5-2 (Utgard High)

The studied core includes a Lower Campanian (Nise Formation; Fig. 11) section, 28 m thick, from a depth of 4189–4161 m. Facies C1, C2, T1 and H1 were identified.

Facies C2 and C1 – Massive sandstone

The facies consists of massive, medium-grained sandstones, which are immature quartzose in composition, and contain abundant red feldspar, mica, glauconite, and mudstone clasts and chips. Green clasts up to 10 mm in size are highly conspicuous in many beds, and are tentatively referred to as glauconite. The beds (or amalgamated beds) vary in thickness from 5–700 cm, but most beds fall in the 5–40 cm range. The base and the top of the beds are sharp and commonly inclined, with dips up to 45°. The tops may, however, be bioturbated. Primary sedimentary structures are absent, whereas water escape structures are common. They include dish structures mainly in the lower part of the bed, slump or flow folds, and pillars, which may be densely spaced near the top of the bed. Most beds are non-graded but some show normal grading in their uppermost parts.

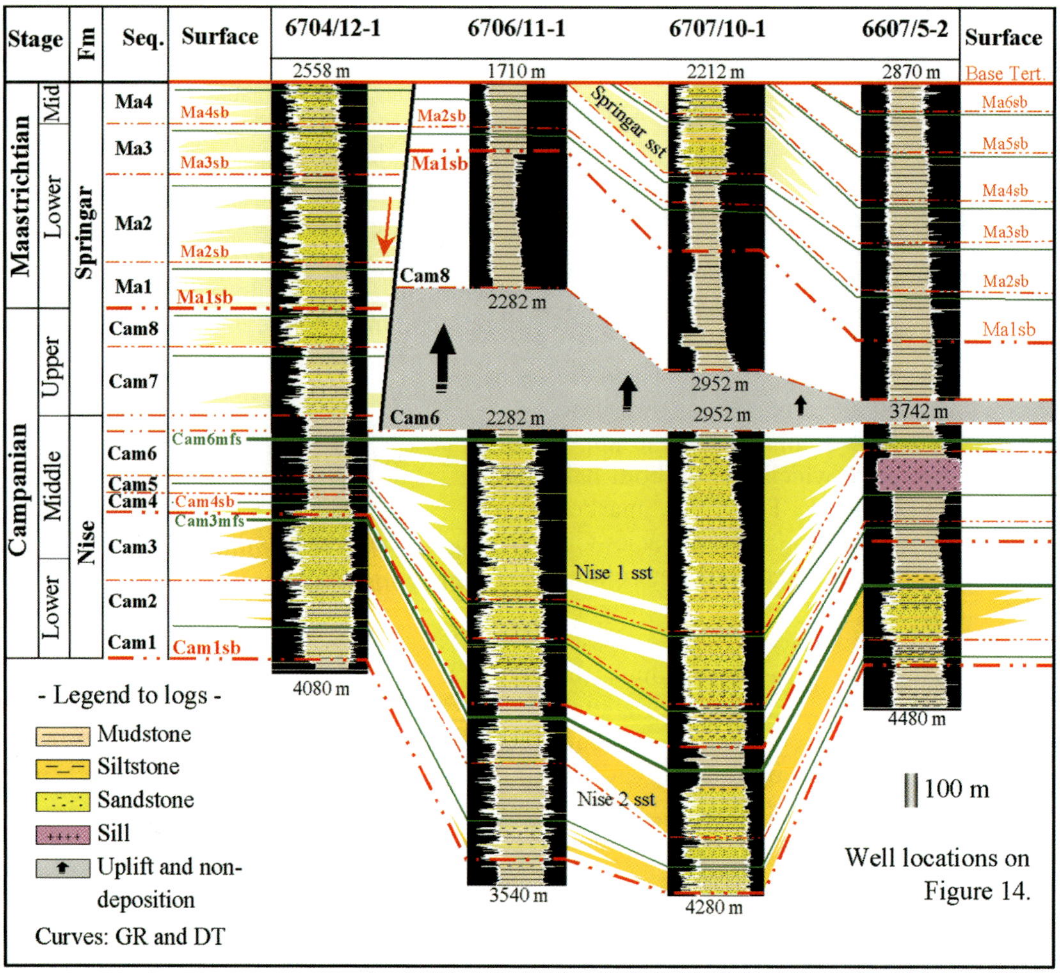

Fig. 11 Sequence stratigraphic log correlation of the Campanian–Maastrichtian sections from wells 6704/12-1, 6706/11-1, 6707/10-1 and 6607/5-2. Significant sandstone units (e.g. Cam3) are interpreted as lowstand deposits on top of their corresponding sequence boundaries, covered by shales deposited during transgression. However, a distinction from auto-cyclicity is not always evident. The sandstone packages, split into the Nise 1, Nise 2 and Springar sandstones, are interpreted as deposited during overall lowering of the sea level, and may be described as major lowstand systems.

The latter is transitional to the graded turbidite sandstones of Facies T1. Mudstone clasts and chips and quartz granules are common.

The sands were deposited by sediment gravity flows representing a spectrum from hyper-concentrated to concentrated density flows. The inclined bases may reflect deposition in scours or post-depositional loading. A 3 m-thick bed probably represents deposition by gradual aggradation of sand beneath a steady density current, and upward-migration of the depositional flow boundary that is dominated by grain hyper-concentration and hindered settling (cf. Kneller and Branney, 1995). Thus, the thickness need not bear any relation to the thickness of the parental current. The thin, graded tops of some of the thicker beds represent the eventual waning of sustained currents and associate development of full turbulence.

Facies T1 — Graded sandstone

The facies consists of fine- to medium-grained sandstone of similar lithology to Facies C2, grading upward into heterolithic sandstone–mudstone. The bed thickness varies from 10–140 cm, averaging 50 cm. The sandstone contains dewatering structures, including dish and pillar structures and slump or flow folds, some of which are isoclinal. The sandstones are weakly graded and occasionally even non-graded in the lower part. They may show planar lamination in the top part. The heterolithic part is strongly normal-graded, and shows convolute bedding or cross-lamination, commonly masked by bioturbation. The succession of structures corresponds well with the classical Ta, Tb and Tc turbidite divisions of Bouma (1962).

The graded bedding and the presence of Bouma divisions indicate that Facies T1 was deposited from concentrated density flow (Ta) and turbidity flows (Tb and Tc). The thick, massive Ta division with water-escape structures suggests rapid dumping from suspension or continued but still rapid deposition from sustained steady-state currents. The abundance of asymmetrical to isoclinal flow and slump folds indicates deposition on a slope with continued shearing and down-slope movement during and after deposition by the density flows.

Facies H1 — Heterolithic sandstone–mudstone

Facies H1 consists of heteroliths, mainly with equal proportions of sandstone and mudstone. The packages vary in thickness between 5–75 cm, averaging 67 cm. The mudstones and the sandstones form cm-thick beds, but structures are mottled by strong bioturbation. Well-preserved horizontal back-fill structures, 1 cm wide, occur in some beds. Most heterolithic packages fine upwards or show an upward decrease in sandstone to mudstone ratio. Slump folds are common and mm-wide sandstone injections have been observed.

The facies reflects alternating deposition from suspension under quiet conditions and pulses of traction current deposition. The strong bioturbation has destroyed the details of the cross-lamination. The graded or upward fining nature of most units suggests deposition from dilute turbidity flows, but bottom current transport and deposition is also possible. The asymmetrical slump or flow folds indicate deposition on a slope.

Well 6704/12-1 (Gjallar Ridge)

Two cores were cut in the Maastrichtian succession in well 6704/12-1, Core 1 from 2554–2571 m, and Core 2 from 2997–3005 m (Fig. 12). The topmost 3 m of Core 2 are of Paleocene age. Core 3 was cut at TD in Santonian mudstones and thin turbidites. Three main facies C2, T1 and H1 are identified in the Maastrichtian sections, in addition to an observation of Facies C1 (Figs. 8 and 12).

Facies C2 — Massive to top-graded sandstones

A few beds of this facies category were found in Core 1. They consist of medium- or fine-grained sandstone and are 0.7–1 m thick. The beds have sharp, slightly erosional bases and are non-graded, except for a grading of the topmost 0.1–0.3 m to fine or very fine-grained sandstone (Fig. 12D). Scattered fine pebbles occur in the lower part of one bed, and two beds in the upper part of the core are very rich in glauconite. The lower, non-graded part of the beds is massive without primary sedimentary structures. However, dish structures were observed in one bed, and mudstone clasts up to 3 cm occur (Figs. 8 and 12B). The topmost, normally graded part of one of the sandstones is plane-parallel laminated. Some beds show pervasive bioturbation and both sub-horizontal and sub-vertical burrows are present, but sub-horizontal traces predominate.

The massive, largely non-graded character of the lower part of the sandstone beds indicates deposition from hyper-concentrated flows, where the turbulence in the lower flow part was virtually suppressed by a high rate of sediment settling from the original turbulent suspension (cf. Lowe, 1982;

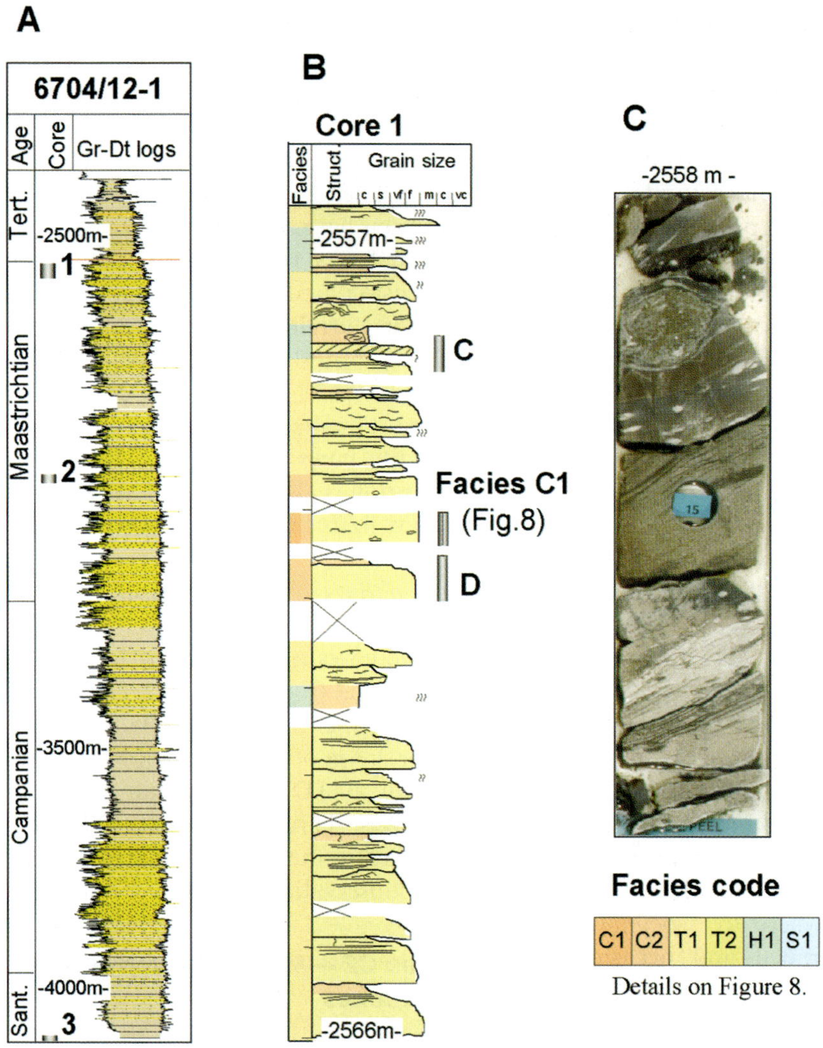

Fig. 12 Key sedimentary facies described from the cores in well 6704/12-1 drilled on the Gjallar Ridge. (A) Two Maastrichtian cores were cut. (B) The succession is dominated by density-flow deposits, where single beds are thinner and less sandy than for the Campanian core described on Figs. 9 and 10. Presence of pebbles suggests hyper-concentrated density-flow deposition (Facies C1) for some of the massive beds, see photo on Fig. 8. (C) Facies H1: heterolithic sandstone and mudstone including siderite nodule. (D) Facies C2: concentrated density-flow deposit with rapid grading in topmost part. Scale: Core width 10 cm.

Mulder and Alexander, 2001). The upper, normally graded and occasionally laminated part of the sandstones beds reflects development of full turbulence in the upper part of the flow.

Facies T1–Normally graded sandstones

Normally graded sandstones represent the most common facies in the cored intervals of well 6704/12-1 (Fig. 12B). The beds are 10–105 cm thick, averaging 50 cm. The sandstones are medium- or fine-grained in the lower part and grade gradually upward to very fine-grained sandstone or siltstone. Coarse and very coarse-grained sand grade grains occur scattered in the basal part of some beds. The lower part of the beds is commonly massive and some beds show inverse grading of the basal 2–3 cm. Rectangular or occasionally folded or contorted clasts of mudstone and less commonly very fine-grained sandstone may occur in the massive part. Vague dish structures are noted in a few beds.

The upper part of the beds may be parallel-laminated, locally defined by elongate horizontally oriented millimetre-sized organic particles, and occasionally succeeded by a ripple-cross lamination. The topmost, finest-grained portions of the beds may show parallel lamination. Some of the normally graded sandstones are totally bioturbated, mainly by sub-horizontal trace fossils, whereas other beds are devoid of burrows.

Most beds of facies T1 comprise the Tab Bouma sequences, but Tabc, Tc, Tac, Tacd and Td sequences are also present. The complete Bouma

Tabcde succession was observed only once. The normal grading and the presence of Bouma divisions suggest deposition from gradually decelerating concentrated density flows passing into turbidity flows. The massive Ta division reflects rapid deposition from suspension, whereas the parallel and ripple cross-laminated Tb, Tc and Td(e) divisions were formed by combined traction and suspension fall-out from the waning, more dilute part of the flow. Direct settling of sediment from suspension was occasionally associated with the formation of dish structures formed by upward migration and expulsion of pore water. The basal inverse grading developed when the rate of shear was at its maximum.

Strong burrowing characterises some turbidite sandstones in Core 2, whereas there is an almost complete lack of burrows in most of the overlying mudstones. This suggests that redeposition of sand by turbidity currents into deep water was associated with an intermittent increase in oxygen content at the sea floor. This was accompanied with colonisation of a short-lived infauna, which disappeared with the gradual return to poorly oxygenated conditions.

Facies H1 — Heterolithic sandstone–mudstone

This facies consists of mudstone or siltstone with beds and streaks of very fine-grained sandstones (Figs. 12B and C). The sandstones are up to 10 cm thick, but occur locally also as 15 mm-thick stringers and as isolated lenses of ripple-cross-laminated sandstone. The sandstones range from non-graded to the normally graded ones and show parallel lamination and/or ripple-cross-lamination, in some cases developed as climbing ripples. Fading ripples are also observed. Small load casts are present at the bases of some beds. The mudstones show a faint parallel lamination and siderite nodules are scattered through the units (Fig. 12C). Some heterolithic sandstones are totally bioturbated, and the density of burrowing apparently increases upward towards the top of the bed. Some thoroughly ripple-cross-laminated and parallel-laminated sandstones are virtually non-burrowed.

The mudstones of this facies were deposited from suspension under quiet conditions, whereas the sandstones probably represent the product of mainly dilute, low-energy turbidity flows. Periodically, the sediments seem to have been subjected to stronger bottom currents, resulting in thick current-winnowed units of ripple- or parallel-laminated, non-burrowed deposits. The lack of bioturbation probably reflects a relatively high sedimentation rate.

Depositional environments

The depositional environments of the Campanian–Maastrichtian deposits in the Vøring Basin are interpreted on the basis of the core analysis with support from seismic observations and wireline log interpretations (Fig. 13). The Lower–Middle Campanian (Nise Formation) deposits are described from cores cut in wells 6707/10-1 (Nyk High), 6706/11-1 (Vema Dome) and 6607/5-2 (Utgard High), while the Maastrichtian deposits (Springar Formation) are described from cores in well 6704/12-1 (Gjallar Ridge) (Fig. 11).

Lower–Middle Campanian system (Nise Formation sandstones)

The Lower–Middle Campanian succession cored on the Nyk High consists mainly of stacked, massive, normally graded sandstones deposited from hyper-concentrated and concentrated density flows passing into turbidity flows. The beds are highly uniform and of similar composition, thickness and grain size, typical of sheet-like fan systems. The great thickness of the succession suggests an aggradational sheet system. Generally random thickness trends are observed (Fig. 9c), indicating compensatory infill by lobe switching on the basin floor.

Most of the beds have non-erosional bases, typical of sheet deposition in a mid-fan position at the termination of the main fan channels. A higher degree of amalgamation suggests a position closer to wide, gentle channels. The middle part of the logged section is characterised by the presence of graded, uni- and non-directionally cross-stratified sandstones rich in glauconite interpreted as spillover deposits from subtle channel fairways (cf. Well 6707/10-1 Facies T2 description). Minor currents and waves at the flow–water interface probably generated the cross-stratification as the flows come to rest (David Mohrig, pers. comm. 2001). Intervals of m-thick, bioturbated mudstones (Facies H1) were deposited during periods of fan abandonment.

Stacked beds deposited from hyper-concentrated to concentrated density flows, similar to the Nyk High succession, also dominate the section encountered in the well on the Vema Dome. Slight differences in the sandstones include a higher

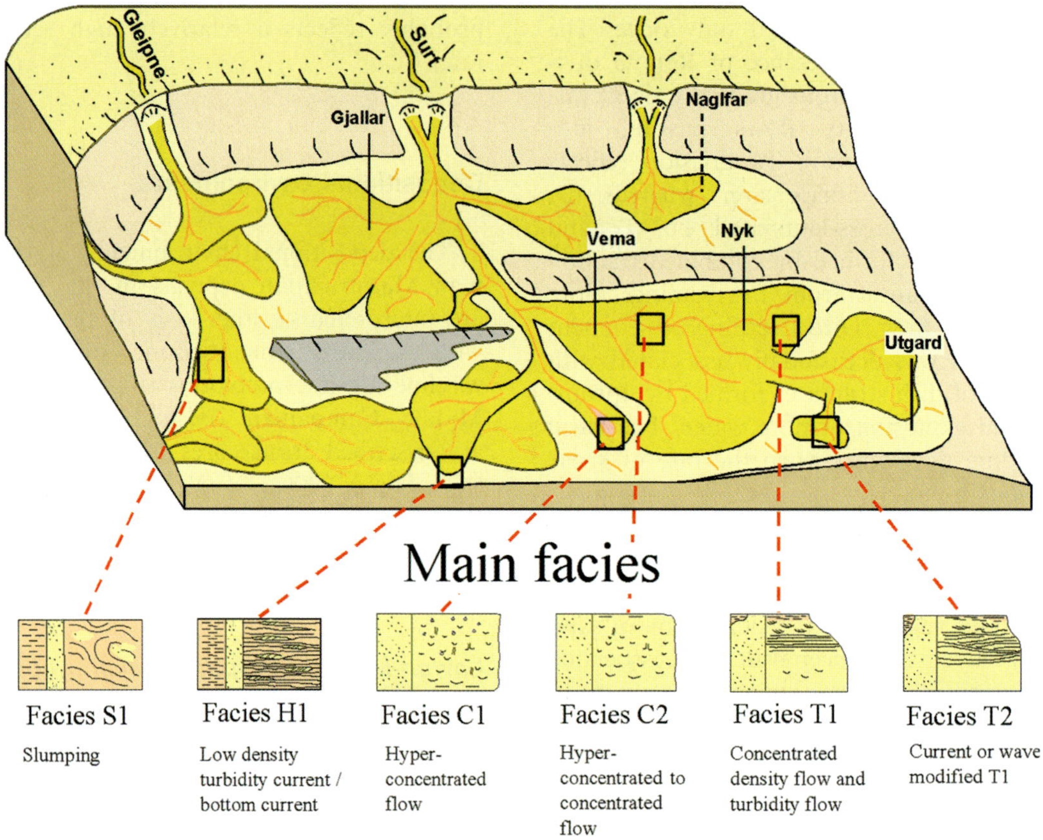

Fig. 13 Schematic illustration of the depositional environments of the Campanian–Maastrichtian fan systems in the Vøring Basin. Illustrations of typical locations for the six main facies are included. The main rift ridges are disrupted by fault displacements and fault polarity shifts along the main transfer zones. These lineaments acted as entry zones for sediments from the shelf, between and perpendicular to the fault ridges and into the basins. Note that Gjallar, Vema and Nyk were all in hanging-wall positions at the time of deposition, and received thick basin-floor deposits.

degree of amalgamation, and more abundant, occasionally contorted clay clasts, whereas grading of the sandstone beds is less well-developed in the Vema well. Bioturbated mudstones representing temporary fan abandonment also occur.

The Vema well (6706/11-1) is considered to be located in a basinal position similar to the Nyk well (6707/10-1), i.e. a mid-fan position of a basin-floor sheet sandstone succession, supported by the similar seismic character at the well locations (Figs. 14, 15 and 16).

The Lower Campanian succession cored on the Utgard High shows alternating graded and heterolithic sandstones with a thick unit of massive sandstones at the top. The thick units of hyper-concentrated to concentrated density-flow deposits seen in the Nyk–Vema area are not observed here (Figs. 11 and 17), and the Utgard High is therefore interpreted as representing a more distal position with respect to the provenance area than the Nyk and Vema Highs. However, major depositional fairways are still present, as suggested by the presence of amalgamated, massive sandstone beds at the top of the cored interval.

The lack of vertical trends and the very uniform nature of the sandstone beds in the Nyk–Vema area indicate that influx of sand was governed by one or a few relatively stable point-source feeder systems. Erratic processes, such as line-sourced debris flows and slides triggered from an over-steepened slope can be ruled out. The provenance for the Lower–Middle Campanian deposits cored in the Nyk and Vema wells was sandstones exposed in the source area or in a contemporaneous relatively shallow outer shelf environment. This is indicated by the generally well-sorted, fine-grained nature of most sandstones and the abundance of glauconite. A direct fluvio-deltaic source is less likely, but may have existed, if the erosion products from the hinterland had a relatively uniform grain composition. The depositional model involves a wide sandy shelf with formation of glauconite in the outer, deeper part. A rather steep slope, passing into a wide basin, flanked the shelf. It is suggested that channels or canyons dissected the shelf-slope break and were probably controlled by fault intersections or relay ramps. The channels tapped sand transported by shelf

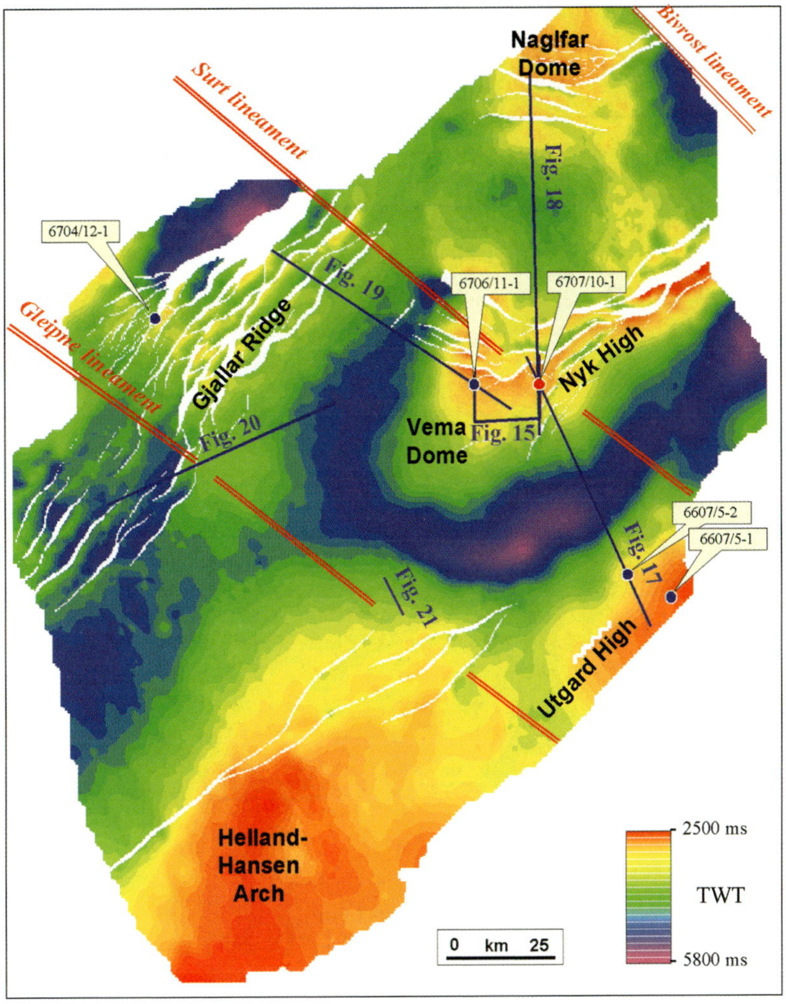

Fig. 14 Top Middle Campanian (Cam6 MFS) time structure map with locations of seismic sections in Figs. 15–21.

currents and the sand was transported down the channels by turbidity currents.

Upper Campanian — Lower Maastrichtian system (Springar Formation Sandstones)

The Maastrichtian succession was cut by two cores in well 6704/12-1 on the Gjallar Ridge; both cores are interpreted as being parts of the same syn-rift depositional system. The interval was also penetrated by well 6707/10-1, but no core was recovered. The section in Core 1 in well 6704/12-1 is sand-rich and dominated by beds of normally graded sandstones, interpreted as deposited from concentrated density flows transitional to turbidity flows (Facies T1). Sandstone beds commonly have few, if any, intercalations of fine-grained and muddy heteroliths (Facies H1). The section is rather monotonous, but intervals of slightly amalgamated sandstones occur. Signs of minor scouring and erosion are seen at the base of the sandstones. The same facies types dominate Core 2, and the beds form two well-defined thickening-and-thinning-up cycles. The sandstones are highly bioturbated with common mudstone clasts and slump-deformations.

The characteristics of Cores 1 and 2 suggest deposition in a basin-floor fan setting. A mid-to-outer fan position is suggested due to the uniform nature of the sandstones and rarity of clearly amalgamated sandstones and coarse-grained lags, typical for a more proximal, inner fan position. The succession is likely to have been affected by proximity to subtle channel fairways undergoing avulsion, as suggested by the varying thickness trend of the sandstones. Core 2 appears to represent a more axial position than Core 1, but the presence of silty intercalations suggests that the cores were not cut in an axial channel position.

The monotonous nature of the Upper Campanian–Lower Maastrichtian succession indicates influx of sediment from one or perhaps a few stable point-source feeder systems. However, the density-flow deposits at the Gjallar location

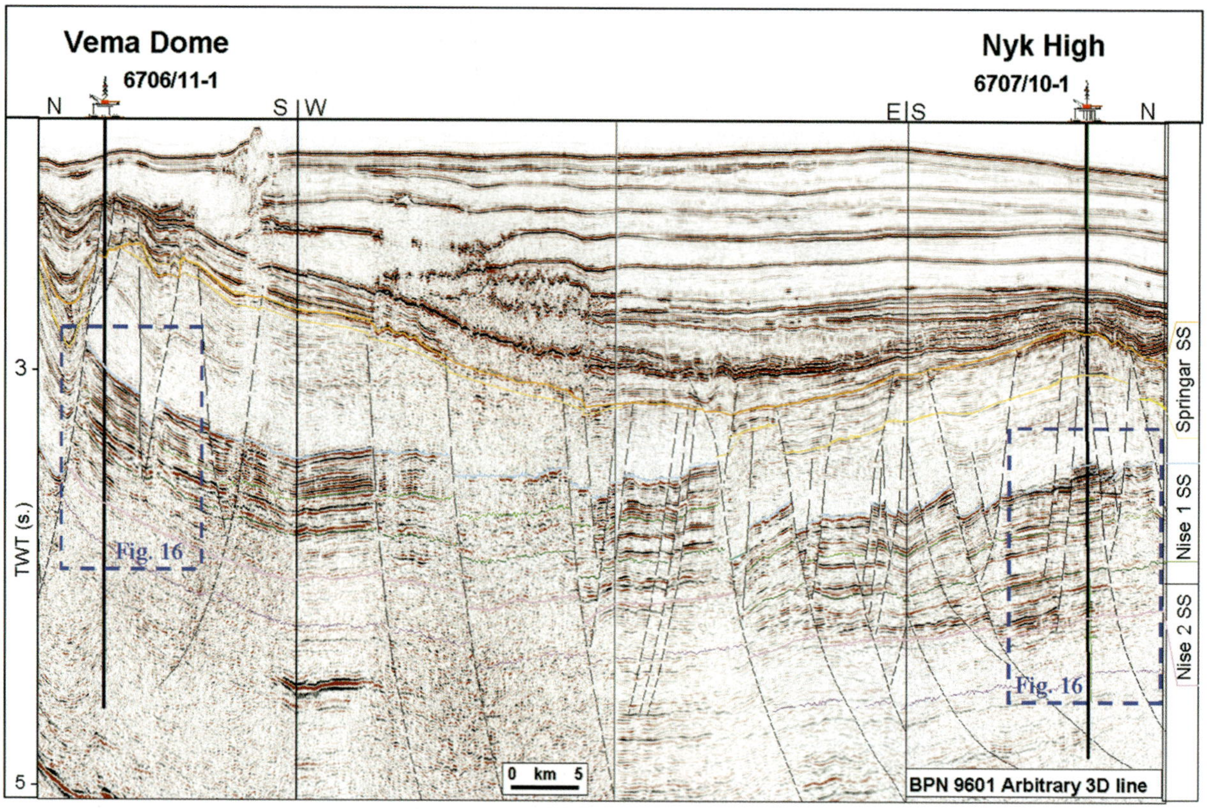

Me2 SB	Base Pliocene	Cam6 SB	Intra Nise 1 Fm
Ser1 SB	Top Oligocene	Cam4 SB	Base Nise 1 Fm
Ru1 SB	Top Eocene	Cam1 SB	Base Nise Fm
Yp1 SB	Top Tang	SaX	Intra Santonian
BTU	Base Tertiary Unc.	Co1 SB	Base Lysing Fm
Ma1 SB	Base Springer Fm	Ce3 SB	Intra Cenomanian
Cam6 MFS	Top Nise Fm	BCU	Base Cret. Unc.

Legend for Figures 15-21

Fig. 15 Seismic 3D (BPN9601) tie between wells 6706/11-1 and 6707/10-1. The Nise Formation deep-water sandstones show a remarkably strong seismic character, enclosed by transparent mudstones above and below. The multi-cycle, continuous character supports an interpretation of a large, stable basin-floor fan system. The Springar Formation sandstones are thinner and less visible on the seismic section. For location see Fig. 14.

are thinner and less sandy compared to the Lower Campanian system cored in the Nyk–Vema area.

Sequence stratigraphy

A sequence stratigraphic framework is established for the Campanian–Maastrichtian basin-floor fans of the Vøring Basin (Fig. 11). Maximum and minor flooding surfaces are identified on the basis of lithological changes, presence of condensed intervals and biostratigraphic data, whereas sequence boundaries are interpreted based on the presence of unconformities and abrupt downwards shifts in facies recognised on well logs and supported by observations from seismic sections. The sequences are named Cam 1, 2, 3 etc. after the age of the sequence boundary (Campanian) and their relative stratigraphic position (cf. Hardenbol et al., 1998).

Two Campanian (Nise 1 and 2 sandstones) and one, mainly Maastrichtian (Springar sandstones), lowstand systems are recognised (Fig. 11). They are characterised by downward shifts of facies, suggesting that proximal parts of the depositional system were exposed for erosion. Deep erosion of the East Greenland cratonic margin and

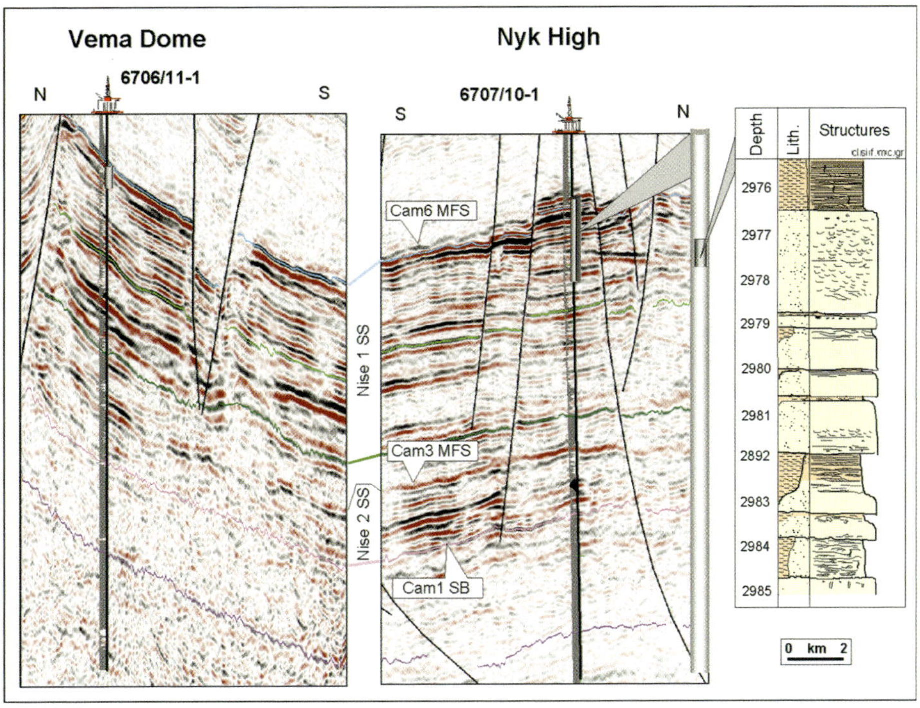

Fig. 16 Magnified seismic sections close to the 6706/11-1 and 6707/10-1 well bores. Strong seismic reflectivity is caused by impedance contrast between stacked sandstone packages and hemi-pelagic mudstone deposits. The seismic signature varies laterally, suggesting local variations in stacking pattern. Both amalgamated sandstone packages and mudstone intervals show transparent character. Some maximum flooding surfaces within mudstone intervals can be correlated between the wells (e.g. Cam3 MFS). Single sandstone beds covered by thin shales will not have the required resolution and impedance contrast to form a seismic reflection response. For location and legend see Figs. 14 and 15.

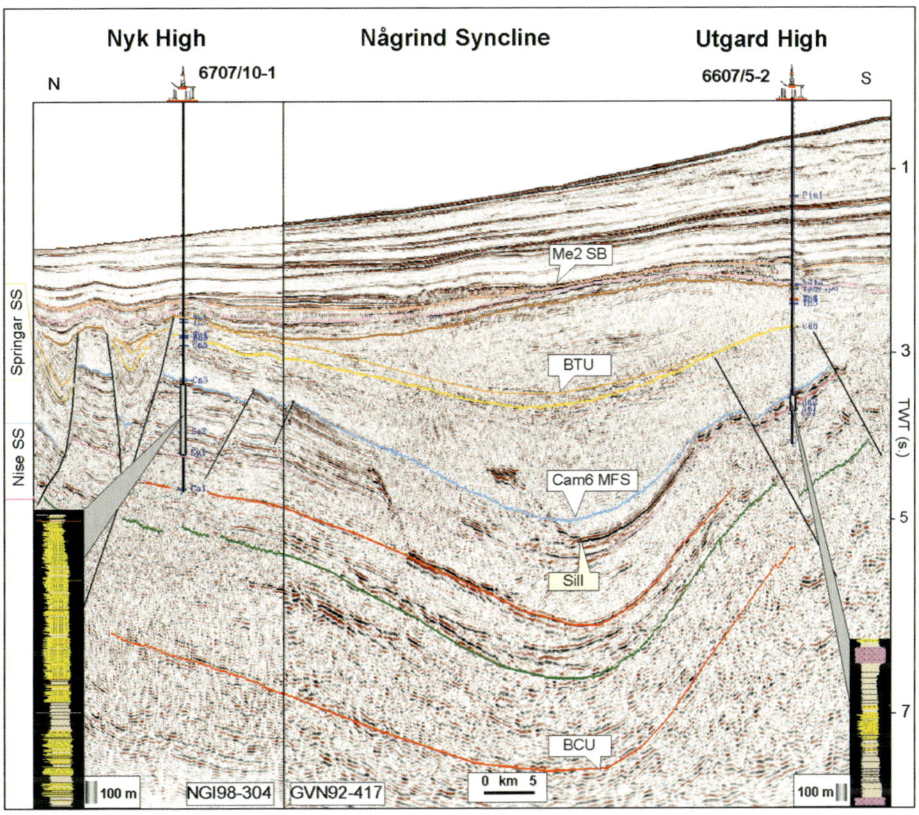

Fig. 17 Seismic section from the Nyk High to the Utgard High. The thick Nise Formation basin-floor fan system at the Nyk High thins into the Någrind Syncline and towards the Utgard High. The Springar Formation sandstones also thin southwards from the Nyk High and shale out before reaching the Utgard High. For location and legend, see Figs. 14 and 15, respectively.

possibly re-deposition from a sand-rich, exposed shelf is believed to have provided the main sediment source for the thick deep-water sheet-sand deposits in the Vøring Basin (Fig. 11).

The Lower–Middle Campanian system

The Campanian deep-water succession is divided into two major sandstone packages, Cam1–3 (Nise 2 sst) and Cam4–6 (Nise 1 sst) (Fig. 11). Cam1–3 is penetrated by wells on the Nyk, Vema, Gjallar and Utgard structures, and Cam3 comprises the most widespread of the Campanian sandstone units.

Cam1–3 is very sand-rich in the Nyk well, which is interpreted as being located in a central position of the basin-floor deposit. The succession is less sandy but fully developed in the neighbouring Vema well. Cam1–3 in well 6704/12-1 on the Gjallar Ridge shows a coarsening-up trend from a muddy basal part to a sand-rich Cam3 interval, interpreted as distally located with respect to the main sandstone fairway. On the Utgard High in well 6607/5-2, Cam1–2 is muddy, whereas Cam3 is very sand-rich, and is interpreted as onlapping the high.

The base of the lowermost Campanian basin-floor fan (Nise 2 SS) in wells 6707/10-1 and 6706/11-1 is interpreted as a sequence boundary, marked by a pronounced downward shift of facies. It is termed, Cam1 SB, and occurs at the Santonian–Campanian transition. The lowermost Campanian sandstones received a regional shale drape during Cam3 maximum flooding. Hence the Cam1–3 basin-floor fan is interpreted as a lowstand deposit, resting on the Cam3 SB, and is covered by the transgressive systems tract, defined between the Cam3 FS and MFS.

Cam4–6 constitutes the bulk of the sandy density flow-dominated deposits in the Vøring Basin and consists of about 900 m of stacked sandstone beds deposited from hyper-concentrated and concentrated flows in the Nyk–Vema area. The interval is reduced to 330 m in the Gjallar well, where it is sandstone-dominated in the lower part and muddy in the upper part. The interval is shale-dominated in well 6607/5-2 on the Utgard High, except for a thin turbidite unit in Cam6.

Cam4 SB is interpreted on the basis of a downward shift in facies from regional shale deposition of Cam3 MFS to basin-floor fan deposition of Cam4–6. No unconformity is detected at Cam4 SB, which may simply reflect the deep-water setting, and in general Cam4–6 follows the same depositional pattern as was established during Cam1–3 times. The massive sandstone package is interpreted as representing a second Campanian lowstand systems tract (Nise 1 sst), deposited during renewed exposure of the source areas without significant tectonic movements in the basin. Significant Campanian density-flow deposition in the Vøring Basin came to an end by Cam6 MFS time and mud deposition prevailed until the end of the Campanian in the Nyk–Vema area and on the Utgard High. In the Gjallar area, sporadic turbidity flows entered the basin during Cam6–7 times, until a new phase of major sand deposition took place during Late Campanian–Maastrichtian.

The Upper Campanian–Maastrichtian system

The Upper Campanian–Maastrichtian sandstone deposits were drilled by the Gjallar and Nyk wells with a base defined by Cam8 SB in well 6704/12-1 on the Gjallar Ridge (Fig. 11). This sequence boundary coincides with an unconformity seen on seismic lines on top of the Nise Formation across the Gjallar Ridge and on the Nyk High. However, the Cam8 SB and the Cam6 MFS, drowning the Nise Formation, cannot be seismically distinguished and are regarded as two merged surfaces on the seismic sections (Figs. 15–18). The Maastrichtian sandstone deposits are truncated both in the Gjallar and the Nyk well, and are faulted out in the Vema well. Hence the age of the termination of sand deposition is not known and Ma4 in the Gjallar well represents the youngest sandstones recorded in any of these wells.

The thickest Upper Campanian–Maastrichtian sandstone succession was drilled on the Gjallar Ridge, where it rests on Cam8 SB and comprises 680 m of density flow-dominated deposits of Cam8–Ma4 age (Springar Formation sandstones). At the Nyk well location muds were deposited from Cam8 to Ma3 and sands from Ma3 to Ma5. Thus sand deposition was not always contemporaneous on the Nyk High and the Gjallar Ridge, suggesting that multiple entry points for sand were present, each dominating at different times dependant on provenance, drainage and local tectonics. The predominantly Maastrichtian deep-water lowstand sandstones are interpreted as deposited during renewed exposure of the North-East Greenland margin combined with tectonic movements along the rift margins of the Vøring Basin.

Palaeogeography

The palaeogeographic setting of the Campanian–Maastrichtian deep-water sandstones in the Vøring Basin is interpreted by integrating the seismic character, mapping and the structural framework with the sedimentology and sequence stratigraphy described earlier.

Seismic Character

The Lower–Middle Campanian (Nise Formation) and Upper Campanian–Maastrichtian (Springar Formation) depositional systems identified on well logs are also recognised by seismic interpretation (Figs. 14–21). The lateral extent of the Campanian–Maastrichtian sandstones has been mapped in the Vøring Basin based on their high reflectivity appearance on seismic sections.

The Lower–Middle Campanian fan system (Nise Formation sandstones)

The high reflectivity of the Campanian Nise Formation sandstones on seismic sections is particularly well-documented in the Nyk High and Vema Dome areas. A seismic 3D multi-line between wells 6706/11-1 and 6707/10-1 displays the thick, Lower–Middle Campanian high reflectivity reservoir successions penetrated by the two wells, situated 25 km apart (Figs. 14 and 15). The distinct seismic reflectivity supports the interpretation of the depositional environment as a sheet-like, widespread basin-floor depositional system without significant channel incision developed at this location. Thick mudstones with a transparent seismic character enclose the sand-rich, density flow-dominated system (Fig. 15).

Magnified displays around the well bores document how the acoustic impedance contrast between sandstone and mudstone packages causes the strong seismic reflectivity (Fig. 16). In some cases the thickest, amalgamated sandstone packages, indicative of an axial channel position, show a relatively transparent seismic character that varies laterally. Occasionally, mudstone layers within the turbidite section are also seismically transparent and can be followed between the two wells. The thickest mudstones are interpreted to represent regional flooding events, supported by biostratigraphic data. Cam3 MFS represents one of these events. Some of the thinner mudstone layers may represent temporary abandonment of fan deposition during lobe switching and are expected to have a smaller lateral extent.

The Lower–Middle Campanian Nise Formation sandstones in the Nyk–Vema areas can be followed seismically into the Någrind and Vigrid Synclines, where the strong seismic reflectivity dims out to the southeast and south. The dim-out is interpreted as a shale-out of the fan succession and a stratigraphic thinning of the Lower–Middle Campanian interval, confirmed by a seismic tie to well 6607/5-2 on the Utgard High (Fig. 17).

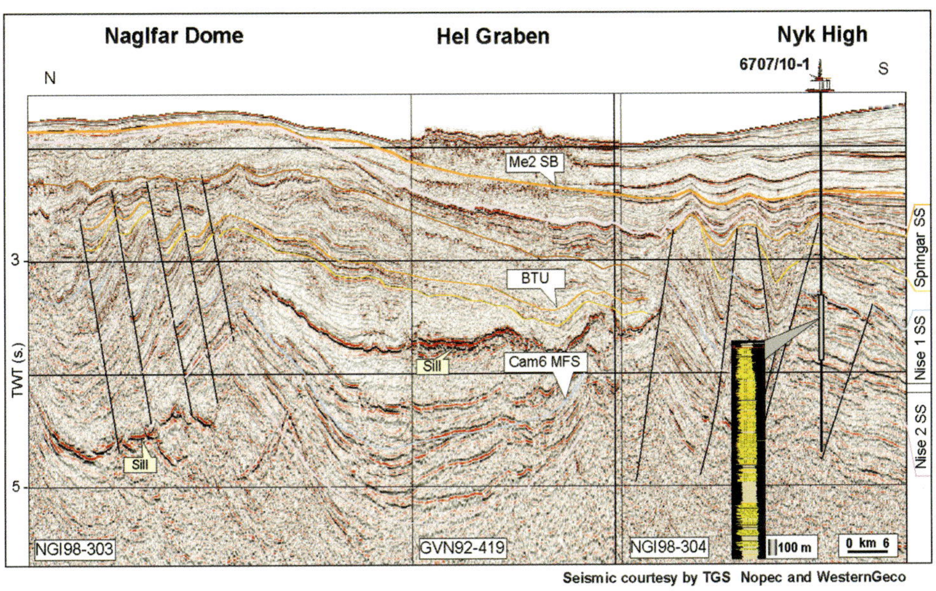

Fig. 18 Seismic section from the Naglfar Dome to the Nyk High. The Nise Formation sandstones preserve a strong seismic signature across the Naglfar Dome but the succession is thinner than on the Nyk High. The Springar Formation sandstones are probably present at several intervals in the Naglfar area similar to the Springar Formation on the Gjallar Ridge. Sills mark the seismic character in the Hel Graben. For location and legend, see Figs. 14 and 15, respectively.

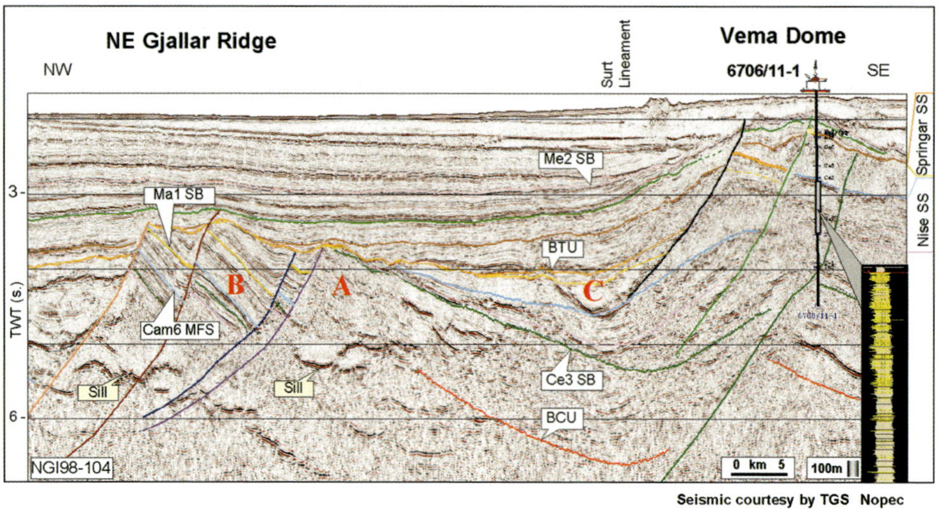

Fig. 19 Seismic section from the Gjallar Ridge to the Vema Dome. The Springar and Nise sandstones are seen onlapping the outermost fault block of the Gjallar Ridge (at A). Strong seismic response is seen within the western fault blocks on the Gjallar Ridge (at B), which can be tied to Campanian–Maastrichtian sandstones in the offset well 6704/12-1. The seismic signature is reduced across the Surt Lineament (at C). For location and legend, see Figs. 14 and 15, respectively.

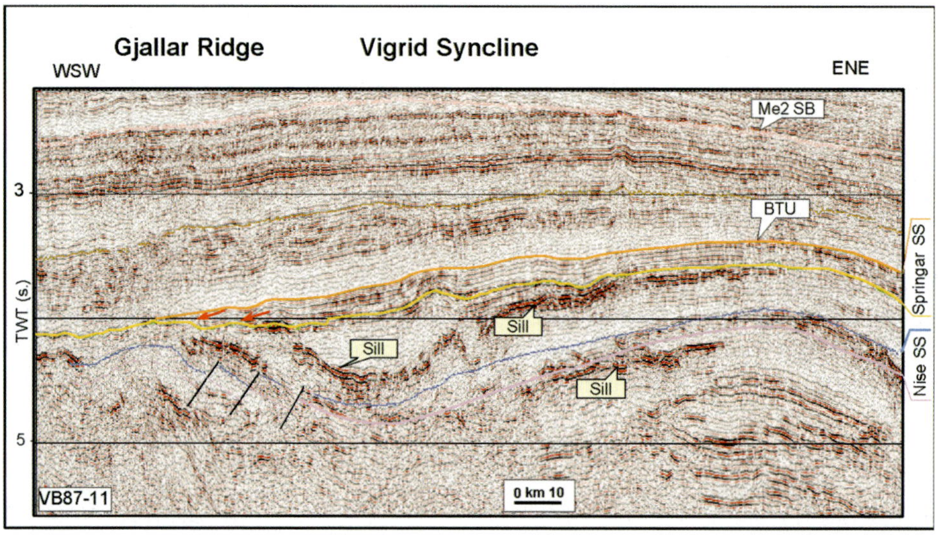

Fig. 20 Seismic section across the Vigrid Syncline. The Springar Formation sandstones show a distinct seismic response tied to sandstone intervals in well 6704/12-1. The Nise Formation sandstones have a less distinct seismic response, being disrupted by sill intrusions. For location and legend, see Figs. 14 and 15, respectively.

Lithology and biostratigraphic data from the Utgard well confirm that the 900 m thick Lower–Middle Campanian interval in Nyk is reduced to 100 m in this area, and it is mainly the Cam3 interval that extends over these distances. Volcanic sills intruded during the Paleogene onset of sea-floor spreading have modified the seismic character close to the Utgard well.

A high-reflectivity sequence is observed in the Hel Graben, and based on seismic character it is correlated with the Lower–Middle Campanian sandstones outside the Hel Graben. However, the seismic reflectivity indicates that the sandstone interval is thinner in the Hel Graben than farther south in the Nyk area. Continued depositional thinning is expected northwards to the Naglfar Dome (Fig. 18). The thickening of the turbidite succession from the Hel Graben to the Nyk–Vema area suggests that more space was available for accommodation of sediments in the Nyk–Vema area. At the time of deposition, the Nyk–Vema area is therefore interpreted to have been in a structurally deeper and more basinward position than the Naglfar–Hel area. The fault system separating the Nyk High and the Hel Graben probably acted as a rift-bounding fault zone, down-throwing to the south during the Campanian. However, this structural model is only vaguely supported by seismic interpretation (Osmundsen et al., 2002). In the Early Eocene, the fault system acted as a rift zone

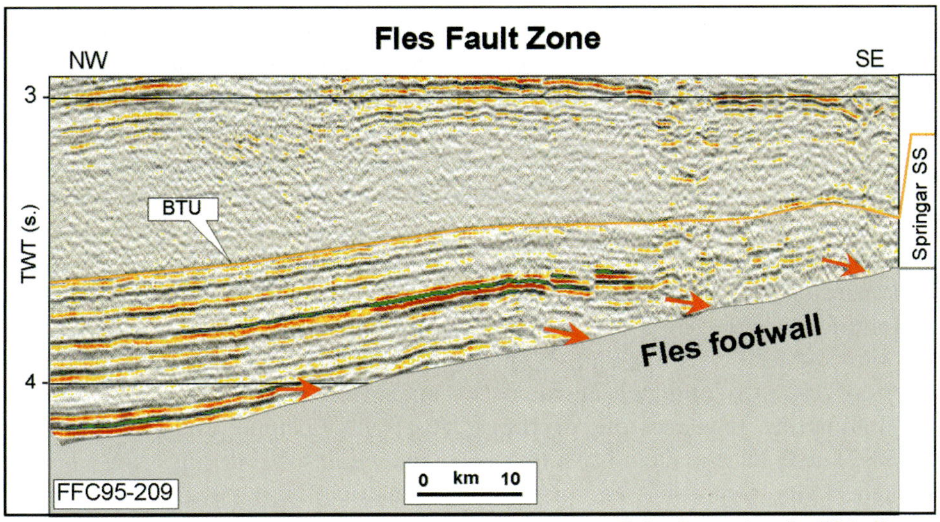

Fig. 21 Seismic section showing Springar Formation sandstones onlapping the Fles footwall. For location and legend, see Figs. 14 and 15, respectively.

down-throwing to the north, bringing the Nyk High–Vema Dome to its present-day, structurally shallower position relative to the Hel Graben (Fig. 18).

During Late Cretaceous times, the Gjallar Ridge was down-thrown to the northwest. This is documented by the onlap of Lower–Middle Campanian deposits onto the mid-Cenomanian unconformity from south to north onto the southernmost rift block of the Gjallar Ridge (Fig. 19 at A). The high seismic reflectivity representing the Campanian sandstones appears as sub-parallel reflectors to the northwest of this master fault (Fig. 19 at B); the units show faint syn-sedimentary thickening into the hanging-wall side of the faults that down-throw to the northwest. Well 6704/12-1 drilled on the Gjallar Ridge suggests that the high reflectivity deposits in Gjallar are coeval and younger than the Nyk–Vema fan system.

It is likely that the Surt and Gleipne Lineaments acted as fairways for deep-water density flows during Late Cretaceous times. The lineaments are associated with disruptions in the continuity of the Gjallar Ridge, and were conceivably suitable fairways for sediment transport and deposition perpendicular to the fault direction, and also formed sediment entry points to the deeper parts of the basin. The seismic character of the Upper Cretaceous succession along the Surt and Gleipne Lineaments is disturbed by the numerous faults and fractures (Fig. 19 at C). However, significant reflectivity in the Upper Cretaceous interval is recorded on both sides of the lineaments, i.e. on the northern Gjallar Ridge to the west and the Hel Graben–Vema Dome to the east. Seismic ties to wells indicate that the Surt Lineament was the most active sediment fairway during the Early–Middle Campanian, while the Gleipne Lineament was the most pronounced sediment transport route during latest Campanian–Maastrichtian time.

Strong seismic reflectivity can be followed along-strike in the tilted fault blocks of the Gjallar Ridge, which at the time of deposition were in a hanging-wall or graben position, relative to the southeastern main fault of the Gjallar Ridge (Fig. 19). The Gjallar well proved that the sandstones deposited on the northwestern side of the Gjallar Ridge were of predominantly Campanian–Maastrichtian age (Springar Formation sandstones), but the well also penetrated a 250 m thick Lower–Middle Campanian Nise Formation sandstone interval (Fig. 12). No cores were cut, but log responses indicate a muddier reservoir of poorer quality than for the time-equivalent interval in the Nyk and Vema wells.

The highly reflective seismic character associated with sandy density-flow deposits in the Gjallar well continues southwestwards, but dims out southwest of the Gleipne Lineament. Sills tend to affect the seismic character, but in general the seismic dim-out is evident on several sections. This indicates a reduction in thickness of the Lower–Middle Campanian Nise Formation reservoir towards the southwest.

The Upper Campanian–Maastrichtian fan system (Springar Formation sandstones)

Following a period of mud deposition during mid-Campanian time, a new basin-floor fan system developed in the Vøring Basin, starting with the Cam7 sequence (Fig. 11). The Upper Campanian–Maastrichtian fan system appears to be less sand-rich, but its extent is broadly similar to the Lower–Middle Campanian basin-floor deposits.

The Gjallar well penetrated 600 m of Upper Campanian–Maastrichtian Springar Formation sandstones. The uppermost, cored section is dominated by normally graded turbidites. These were less massive than the Lower–Middle Campanian (Nise Formation) hyper-concentrated and concentrated flow sandstones dominating the reservoir of the Nyk and Vema wells. The Gjallar well confirms that the seismic high-reflectivity response seen on the Gjallar Ridge is correlative to a sandy interval below the base-Palaeocene unconformity. However, the contrast between the sandstone interval and the under- and overlying mudstones is less distinct here than between the Nise Formation sandstones and the surrounding mudstones in the Vema Dome and Nyk High areas. High reflectivity is also recorded in mudstone-dominated intervals with occasional turbidites. Hence, based on seismic character, the reservoir extent is less well-defined for the Upper Campanian–Lower Maastrichtian interval than for the Lower–Middle Campanian interval.

The Springar Formation fan system drilled by the Gjallar well can be correlated across the Gleipne Lineament, and the interval thickens into the lineament (Fig. 22). This observation supports that the Gleipne Lineament was an active sediment fairway during the latest Campanian–Maastrichtian, equivalent to the importance of the Surt Lineament as a sediment fairway during the Early–Middle Campanian.

The Nyk well penetrated 150 m of Springar Formation sandstones, but they were eroded at the

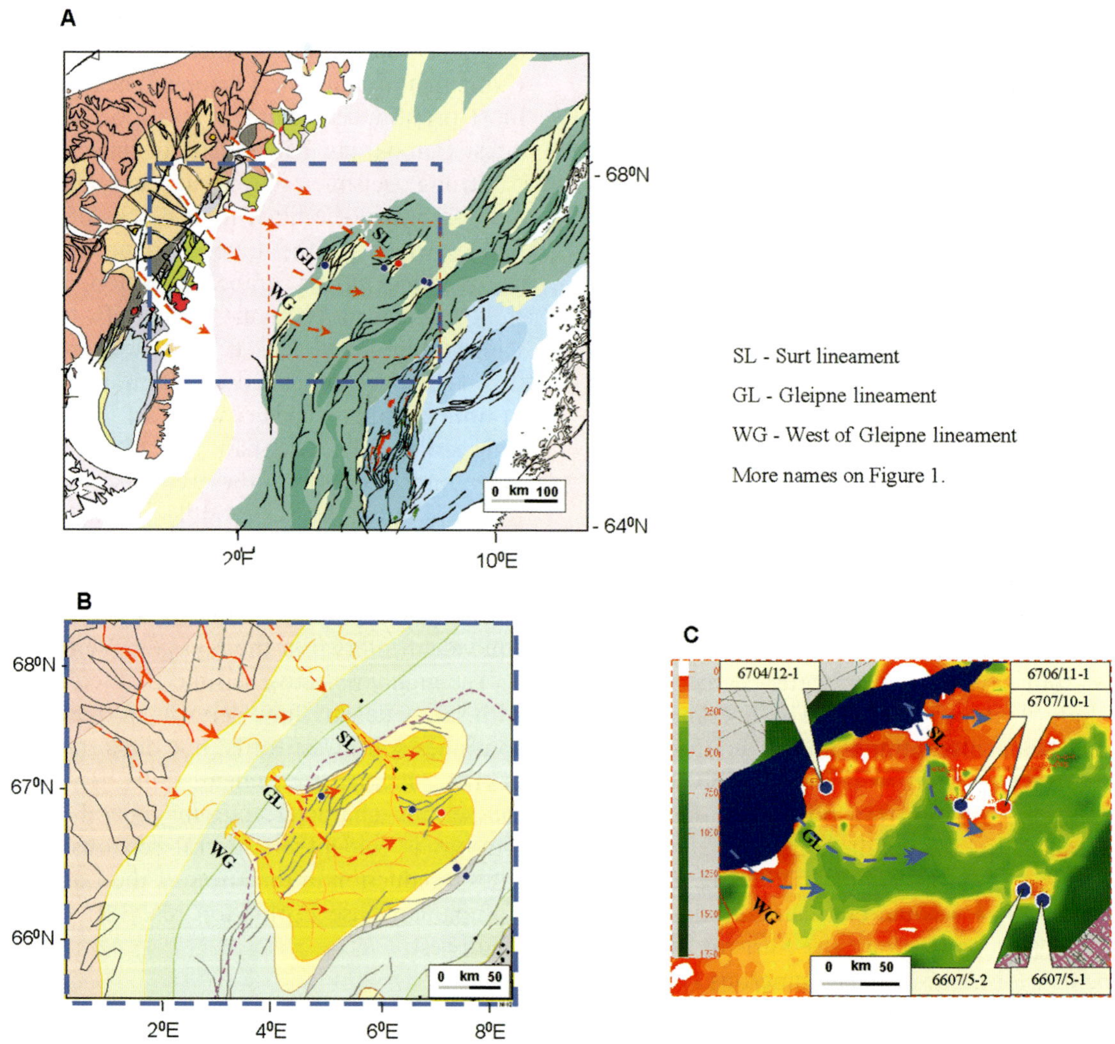

Fig. 22 (A) Structural element map of the Vøring Basin with Greenland restored back to its pre-drift position in Paleocene. Possible sediment fairways are indicated. For legend see Fig. 1. (B) Palaeogeography of the Springar Formation basin-floor fans in the Vøring Basin including likely sediment entry zones. For legend see Fig. 3. (C) Springar Formation isochron map of the Vøring Basin. Thickest sediments are mapped close to the fault lineaments. The thin interval in the Nyk–Vema area is due to later erosion.

Vema well location (Figs. 11 and 15). Also in the Nyk–Vema area, the Upper Campanian–Lower Maastrichtian succession shows less reflectivity than the Lower–Middle Campanian succession. The seismic reflectivity patterns can be correlated northwards into the Hel Graben and across the Naglfar Dome. The distinct seismic character becomes weaker and covers a thinner interval northwards, suggesting that the sandstones are less well-developed here than on the Nyk High.

The seismic reflectivity patterns can be followed into the Vigrid Syncline as a well-defined seismic package of sub-parallel reflectors (Fig. 20). This succession is seen onlapping both westward and eastward against the flanks of the Vigrid Syncline. These onlaps occur against the Gjallar Ridge, and the Fles Fault complex, respectively (Fig. 21). The thinning towards footwall blocks suggests that the Maastrichtian sandstones are syn-tectonic. On the Gjallar Ridge the Maastrichtian sandstones form syn-tectonic wedges within the tilted fault blocks, and several small unconformities are detected within the Campanian–Maastrichtian interval. These tectonic episodes are precursors to the rifting, tilting and erosion that took place in Late Maastrichtian — Paleocene time, prior to the onset of sea-floor spreading (Skogseid et al., 1992a).

Structural reconstructions

A structural elements map of the North Atlantic margin forms the base for development of geological models for deep-water depositional systems in the Vøring and Møre Basins (Fig. 22A). The position of Greenland has been restored relative to the Norwegian shelf close to its break up position in the Paleogene. Combined with the seismic signature of the deep-water fans (Figs. 15–21), the map suggests that East Greenland and its palaeo-shelf acted as a source area for the Campanian–Maastrichtian basin-floor fans deposited in the Vøring Basin (Fig. 22B). The restored map shows that the Vøring Basin was closer to the Greenland craton than to the Norwegian mainland, and a shale-out to the south and southeast of the deep-water deposits in the Vøring Basin excludes the Norwegian mainland as a source area for the turbidites. The shale-out is well-documented on seismic lines extending from the Nyk High into the Någrind Syncline (e.g. Fig. 7 below the Cam6 MFS marker).

The Greenland provenance is corroborated by information from heavy mineral assemblages and zircon dating for the Campanian in well 6607/5-2; these data indicate that North-East Greenland is the most likely provenance area for the Campanian sandstones in the Någrind Syncline (Morton and Grant, 1998). Zircon age data from the Traill Ø region in central East Greenland, which is the conjugate margin to the Vøring Basin, give the same age groups as those shown for the sandstones interpreted by Morton and Grant (1998) to be derived from North-East Greenland (Surlyk and Noe-Nygaard, 2001). Additional work in East Greenland has shown that large tracts of the East Greenland craton could have served as provenance for the deep-water sandstones in the Vøring Basin (Thrane, 2002; Kalsbeek et al., 2000). It is, however, difficult to identify the sediment source area from East Greenland and its shelf and influx routes/fairways to the Late Cretaceous basin-floor fans in the Vøring Basin, since the main part of the Greenland palaeo-shelf is covered by Paleogene basalt flows (Fig. 22A).

It is likely that fluvial systems were following the direction of the present-day fjords in the Wollaston Forland, Hold with Hope and Jameson Land areas. Thick successions of Devonian and Carboniferous continental sediments are preserved onshore Greenland today, and parts of these sand-dominated basins may have been eroded during the Late Cretaceous to provide the sand source for the Vøring deep-water deposits (Fig. 22). A more northerly provenance may also be envisaged, following an axial direction to the main fault blocks onshore Greenland close to the Shannon Ø. However, a possible provenance of sedimentary rocks has not been identified in the northern areas.

It is envisaged that rivers transported sediments from the Greenland mainland to a sand-rich shelf. During sea-level lowstand, the sand-rich sediments followed incised valleys across the shelf, and were redeposited by density flows down-slope into the Vøring Basin (Fig. 22). The transport corridors of sediments from the East Greenland shelf to the floor of the Vøring Basin can be indirectly identified from the mapped distribution of the deep-water fan systems and the likely structural locations of major sandstone fairways. The sediment entry points were probably controlled by the rift topography and the transfer zones in the Vøring Basin. Major fault lineaments commonly follow deep-sited weakness zones in the crust, and are expressed by breaks in rift fault continuity and shifts in rift shoulder orientations. Hence, the lineaments are suitable entry points for sediments crossing rift blocks from the Greenland mainland to the basin floor. The numerous minor lineaments also form breaks in the rift shoulder continuity, and hence contribute significantly to the total discharge of sediments to the basin floor.

The Upper Cretaceous deep-water sandstones show their thickest development in a basin-floor position in proximity to the Surt and Gleipne Lineaments, suggesting that these lineaments acted as major sediment fairways during the Late Cretaceous (Fig. 22C). Part of the thick basin-floor sheet sands in the Nyk–Vema area was probably transported through the Surt fairway, while the Gleipne Lineament was probably the main fairway for the density flows to the Gjallar area. These main lineaments were probably also the entry points for the most far-travelled flows, which deposited the sediments identified on seismic lines in the Vigrid syncline. The south-easternmost rift shoulder on the Gjallar Ridge blocked sediment entry across the Fenris Graben (Figs. 7, 14 and 19 at A), and a continuous seismic character suggests that sands entered at the Surt and Gleipne Lineaments, and passed into the Vigrid syncline (Fig. 20).

During the Campanian–Maastrichtian the sand distribution shifted in amount and time between the Nyk–Vema area and the Gjallar Ridge (Fig. 11). The lateral and vertical shift in sediment distribution may suggest that, in addition to the main lineaments, there are several sediment entry points into the basin, providing sediments to the basin floors of the Hel and Fenris Grabens along minor lineaments. A fill-spill mechanism is suggested as an additional potential sediment source to the basin-floor sheet sands from the Hel Graben across the Nyk High Fault Zone and into the thick basin-floor successions in the Någrind syncline.

The main reservoir unit in Well 6704/12-1 on the Gjallar Ridge is of Late Campanian–Maastrichtian age. The high seismic reflectivity of the reservoir interval is tied on seismic sections away from the well. The reflectivity suggests that the Upper Campanian–Maastrichtian density-flow sandstones are best developed in the Vigrid syncline (Springar SS on Fig. 20) and in the Gjallar Ridge fault blocks (Fig. 19 at B). The sands show a weaker seismic reflectivity in the Någrind Syncline (Fig. 17). They are also less well-developed in Wells 6607/5-2, 6707/10-1 and 6706/11-1 on the Utgard, Nyk and Vema structures, respectively, compared to Well 6704/12-1 on the Gjallar Ridge (Fig. 11, below the eroded Ma units). This suggests an improved Maastrichtian reservoir development towards the southwest, and a sand source in the areas between Hold With Hope and Jameson Land in East Greenland is likely.

During Campanian time, the Bivrost Lineament separated the deep-water Vøring Basin from a shallower shelf northeast of the lineament where large SW–NE trending ridges were exposed within the shallow shelf areas (Brekke et al., 2001). The Bivrost Lineament acted as a hinge zone, and sediments eroded from the ridges are likely to have been transported across the shelf areas down dip across the Bivrost slope towards the basin floor in the Vøring Basin. It is, however, questionable whether the ridges could contribute substantially to the coarse clastic successions that were deposited during the Campanian in the Vøring Basin. By analogy, the Nordland Ridge (Fig. 1) was eroded during tectonic episodes in the Late Cretaceous, but the coarse clastic deposits present on the neighbouring Dønna Terrace do not reach a thickness of more than 100 m in wells. The sandstones appear as single cycle reservoirs on seismic sections and cover only a limited area close to the ridge (Figs. 4 and 5). Erosion of the large ridges northeast of the Bivrost Lineament is thus considered of minor importance as a source for sediments to the basin-floor fans in the Vøring Basin.

Summary

The Upper Cretaceous basin-floor plays of the Vøring Basin are identified by seismic mapping and interpretation of distinct, high reflectivity seismic packages tied to wells drilled on the Nyk High, Utgard High, Vema Dome and Gjallar Ridge. Two main deep-water fan systems are recognised, comprising a Lower–Middle Campanian (Nise Formation) and an Upper Campanian–Maastrichtian (Springar Formation) system. Seismic character tied to wells shows the outline of the fan systems. The Lower–Middle Campanian system was developed in the Fenris and Hel Grabens, and was limited to the northeast by the Bivrost Lineament and to the southwest by the Gleipne Lineament. The deposits thicken basinward along the Surt Lineament and into the Nyk–Vema area, where they reach a maximum thickness of approximately 900 m. The fan system shales out towards the Utgard High and the Fles Fault zone.

Facies interpretation of cores shows that the Campanian–Maastrichtian deposits are dominated by stacked, massive, partly graded sandstones interpreted as deposited from hyper-concentrated to concentrated density flows (Facies C1 and C2) and to some extent also turbidity flows (Facies T1) forming sheet-like basin-floor fan systems. Characteristic, cross-stratified sandstones (Facies T2) are observed in one distinct interval in the Nyk well.

They are interpreted as concentrated flow to turbidity flow deposits modified by a wave field appearing at the ambient water-flow interface during deposition, as is commonly observed for channel spillover deposits. However, reworking by deep-storm waves cannot be completely ruled out. Heterolithic sandstones and mudstones (Facies H1) were deposited by turbidity currents in intra-channel and distal fan positions. Brecciated and slumped deposits (Facies S1) are also common in the cores.

During Campanian–Maastrichtian time, phases of relative uplift of the East Greenland mainland and adjacent shelf resulted in erosion and re-deposition of coarse clastic sediments by density flows down slope to the basin floors off the mainland and in the outer Vøring Basin. The sources were probably uplifted Palaeozoic and younger sandstones as well as contemporaneous shelf sands.

Changing depositional patterns along the Vøring Margin through Late Cretaceous time suggest that several sediment entry points were active, and sediment influx is believed to have followed incised valleys located along pre-existing lineaments. The Surt Lineament was a likely feeder valley for the stacked, massive basin-floor sandstones found in the Nyk–Vema area. The uniform nature of the sandstones and a lack of vertical trends suggest sand influx from a large, stable point source feeder valley. The thinner sandstone units of the proximal Gjallar and Naglfar areas may have been derived through smaller lineament fairways. Isopach maps and seismic character suggest that the Surt Lineament controlled the main feeder valley also for the deep-water sands that entered the distal parts of the Vigrid and Någrind synclines. The temporal evolution in the Vøring Basin illustrates the change from small and immature Lower Cretaceous deep-water clastic systems over extensive mid-Cretaceous single cycle fans to Campanian–Maastrichtian multi-cycle fan systems.

Acknowledgements

The authors would like to thank Vidar Friestad, Francis Mediavilla, Asgeir Bang, Steve Thomas, Gordon Blakely and Helge Kommedal for their contributions to discussions and figures of this paper. Thierry Jaquin contributed significantly to the sequence stratigraphic framework for the Vøring Basin wells. Special thanks to the referees Erik Lundin, Trond Lien and Stephen Lippard for their very thorough and constructive reviews of the paper. Permission to present seismic courtesy data is granted by TGS Nopec, WesternGeco and PL 122. Permission to present the study is granted by ExxonMobil.

References

Blystad, P., Brekke, H., Færseth, R.B., Larsen, B.T., Skogseid, J. and Tørudbakken, B., 1995. Structural elements of the Norwegian continental shelf. NPD Bull., 8: 1–45.

Bouma, A.H., 1962. Sedimentology of some flysch deposits: a graphic approach to facies interpretation. Elsevier, Amsterdam, 168 pp.

Bukovics, C. and Ziegler, P.A., 1985. Tectonic development of the Mid-Norway continental margin. Mar. Petrol. Geol., 2: 2–22.

Brekke, H., Dahlgren, S., Nyland, B. and Magnus C., 1999. The prospectivity of the Møre and Vøring Basins on the Norwegian Sea continental margin. In: A.J. Fleet and S.A.R. Boldy (Editors), Petroleum Geology of Northwest Europe. Proceedings of the 5th Conference, Geological Society, London, pp. 261–274.

Brekke, H. and Riis F., 1987. Tectonics and basin evolution of the Norwegian shelf between 620N and 720N. Norsk Geol. Tidsskr., 67: 295–322.

Brekke, H., Sjulstad, H.I., Magnus, C. and Williams, R.W., 2001. Sedimentary environments offshore Norway – an overview. In: O.J. Martinsen and T. Dreyer (Editors), Sedimentary Environments Offshore Norway — Palaeozoic to Recent. Norwegian Petroleum Society (NPF) Spec. Publ. 10, Elsevier, Amsterdam, pp. 7–37.

Dalland, A., Worsley, D. and Ofstad, K., 1988. A lithostratigraphic scheme for the Mesozoic and Cenozoic succession, offshore Mid- and Northern Norway. NPD Bull., 4: 1–65.

Dott, R.H. and Bourgeois, J., 1982. Hummocky stratification: significance of its variable bedding sequences. Geol. Soc. Am. Bull., 93: 663–680.

Doré, A.G. and Lundin, E.R., 1996. Cenozoic compressional structures of the NE Atlantic margin: Nature, origin and potential significance for hydrocarbon exploration. Petrol. Geosci., 2: 299–311.

Doré, A.G., Lundin, E.R., Fichler, C. and Olesen, O., 1997. Patterns of basement structure and reactivation along the NE Atlantic. J. Geol. Soc. London, 154: 85–92.

Doré, A.G., Lundin, E.R., Jensen, L.N., Birkeland, Ø., Eliassen, P.E. and Fichler, C., 1999. Principal Tectonic Events in the Evolution of the Northwest European Atlantic Margin. In: A.J. Fleet and S.A.R Boldy (Editors), Petroleum Geology of Northwest Europe. Proceedings of the 5th Conference, Geological Society, London, pp. 41–63.

Eldholm, O., Thiede, J. and Taylor, E., 1989. Evolution of the Vøring volcanic margin. In: O. Eldholm, J. Thiede, E. Taylor et al. (Editors), Proc. Ocean Drilling Program, Sci. Res., 104: 1033–1065.

Gabrielsen, R.H., Færseth, R.B., Hamar, G. and Rønnevik, H.C., 1984. Nomenclature of the main structural features on the Norwegian Continental Shelf north of the 62nd parallel. In: A.M. Spencer et al. (Editors), Petroleum Geology of the North European Margin. Proceedings of the North European Margin Symposium, Norwegian Petroleum Society, Trondheim, Graham and Trotman, London, pp. 41–60.

Gradstein, F.M., Agterberg, F.P., Ogg, J.G., Hardenbol, J., Van Veen, P., Thierry, J. and Huang, Z., 1994. A Mesozoic time scale. J. Geophys. Res. 99: 24051–24074.

Hardenbol, J., Thierry, J., Farley, M.B., Jacquin, T., Graciansky, P.C. de and Vail, P.R., 1998. Mesozoic and Cenozoic sequence chronostratigraphic framework of European basins. In: P.C. de Graciansky, J. Hardenbol, J. Thierry and P.R. Vail (Editors), Mesozoic and Cenozoic Sequence Stratigraphy of European Basins SEPM Spec. Publ., 60: 3–13.

Kalsbeek, F., Thrane, K., Nutman, A.P. and Jepsen, H.F., 2000. Late Mesoproterozoic metasedimentary and granitic rocks in the Kong Oscar Fjord region, East Greenland Caledonian fold belt: evidence for Grenvillian orogenesis? J. Geol. Soc., London, 157: 1215–1255.

Kittelsen, J.E., Olsen, R.R., Marten, R.F., Hansen, E.K. and Hollingsworth, R.R., 1999. The first deepwater well in Norway and its implications for the Upper Cretaceous play, Vøring Basin. In: A.J. Fleet and S.A.R. Boldy (Editors), Petroleum Geology of Northwest Europe. Proceedings of the 5th Conference, Geological Society, London, pp. 41–63.

Kneller, B.C., 1995. Beyond the turbidite paradigm: physical models for deposition of turbidites and their implications for reservoir prediction. In: D.J. Prosser and A. Hartley (Editors), Reservoir Characterization of Deep Marine Clastic Systems, Geol. Soc. London, Spec. Publ., 94: 31–49.

Kneller, B.C. and Branney, M.J., 1995. Sustained high-density turbidity currents and the deposition of thick massive sands. Sedimentology, 42: 607–616.

Lowe, D.R., 1982. Sediment gravity flows: II. Depositional models with special reference to the deposits of high-density turbidity currents. J. Sediment. Petrol., 52: 279–297.

Lundin, E.R. and Doré, A.G., 1997. A tectonic model for the Norwegian passive margin with implications for the NE Atlantic: Early Cretaceous to break-up. J. Geol. Soc., London, 154: 85–92.

Lundin, E.R. and Doré, A.G., 2002. Mid-Cenozoic post-breakup deformation in the 'passive' margins bordering the Norwegian-Greenland Sea. Mar. Petrol. Geol., 19: 79–93.

Morton, A.C. and Grant S., 1998. Cretaceous depositional systems in the Norwegian Sea: Heavy mineral constraints. AAPG Bull., 2: 274–290.

Mulder, T. and Alexander, J., 2001. The physical character of subaqueous sedimentary density flows and their deposits. Sedimentology, 48: 269–299.

Mutti, E., Tinterri, R., Biase, D., Fava, L., Mavilla, S. and Calabrese, L., 2000. Delta front associations of ancient flood-dominated fluvio-deltaic systems. Revista de la Sociedad Geológia España, 13: 165–190.

Osmundsen, P.T., Sommaruga, A., Skilbrei, J.R. and Olesen O., 2002. Deep structures of the Mid Norway rifted margin. Nor. J. Geol. (NGT), 82: 205–224.

Sanchez-Ferrer, F., James, S.D., Lak, B. and Evans, A.M., 1999. Techniques used in the exploration of turbidite reservoirs in a frontier setting — Helland Hansen licence, Vøring Basin, Offshore Mid Norway. In: A.J. Fleet and S.A.R. Boldy (Editors), Petroleum Geology of Northwest Europe. Proceedings of the 5th Conference, Geological Society, London, pp. 281–292.

Shanmugam, G., Lehtonen, L.R., Straume, T., Syvertsen, S.E., Hodgkinson, R.J. and Skibeli, M., 1994. Slump and debris-flow dominated upper slope facies in the Cretaceous of the Norwegian and Northern North Seas (61–67°N): Implications for sand distribution. AAPG Bull., 78: 910–937.

Skogseid, J., Pedersen, T., Eldholm, O. and Larsen, B.T., 1992a. Tectonism and magmatism during NE Atlantic continental break-up: the Vøring Margin. In: B.C. Storey, T. Alabaster and R.J. Pankhurst (Editors), Magmatism and the causes of continental break-up. Geol. Soc. Spec. Publ., 68: 305–320.

Skogseid, J., Pedersen, T. and Larsen, V.B., 1992b. Vøring Basin: Subsidence and tectonic evolution. In: R.M. Larsen, H. Brekke, B.T. Larsen and E. Talleraas (Editors), Structural and tectonic modelling and its application to petroleum geology. Norwegian Petroleum Society (NPF), Special Publication 1, pp. 55–82.

Surlyk, F., 1990. Timing, style and sedimentary evolution of Late Paleozoic — Mesozoic extensional basins of East Greenland. In: R.P.F. Hardman and J. Brooks (Editors), Tectonic Events Responsible for Britain's Oil and Gas Reserves. Geol. Soc. London, Spec. Publ., 55: 107–125.

Surlyk, F. and Noe-Nygaard, N., 2001. Cretaceous faulting and associated coarse-grained marine gravity flow sedimentation, Traill Ø, East Greenland. In: O.J. Martinsen and T. Dreyer (Editors), Sedimentary Environments Offshore Norway — Palaeozoic to Recent. Norwegian Petroleum Society (NPF) Spec. Publ. 10, Elsevier, Amsterdam, pp. 293–319.

Thrane, K., 2002. Relationships between Archaean and Paleoproterozoic crystalline basement complexes in the southern part of the East Greenland Caledonides: an ion microprobe study. Precambrian Res. 113: 19–42.

Ziegler, P.A., 1990. Geological Atlas of Western and Central Europe. 2nd ed., Shell Internationale Petroleum Maatschappij B.V., The Hague, 239 pp.

Late Permian to Triassic basin infill history and palaeogeography of the Mid-Norwegian shelf — East Greenland region

Reidar Müller, Johan Petter Nystuen, Frøydis Eide and Hege Lie

In the Late Permian to Triassic, the present Mid-Norwegian shelf and East Greenland represented an extensional basinal region, c. 400 km wide and 800 km long, composed of several sub-basins. This basinal region was the site of a highly varied sediment infill history during Late Permian to Late Triassic time, controlled by changes and variations in tectonic development, climate and eustasy. In the Late Permian minor movements along some intra-basinal faults resulted in the formation of several sub-basins. This tectonic event is defined as an *initial rift phase*. In the Mid-Norwegian shelf — East Greenland region the Permian–Triassic transitional interval is represented by a major shift in basin infill style, including erosion of carbonate margins and influx of silisiclastic sediments to marine sub-basins. This combination of processes is controlled by increased fault activity and fault block rotation, which was accompanied by a fall in the relative sea level. The Early Triassic (early Scythian) represents the *syn-rift phase* when the dominantly marine sediment infill pattern was controlled by continued fault-block rotation and tectonic activity along several structural lineaments. The basin infill in the late Scythian represents an overall shallowing up of the basinal region, and a marginal marine to continental depositional environment became established. The Middle Triassic *post-rift phase 1* is represented by a dominantly aggrading continental succession on the Mid-Norwegian shelf, probably with some short-lived marine transgressions. In East Greenland, this phase is characterised by a great variability of continental facies with a minor marine incursion. The establishment of a continental depositional environment is caused by a decrease in rate of accommodation relative to rate of sediment input, brought about by cessation in fault activity and reduced tectonic subsidence. However, the Vingleia Fault Zone was still active during deposition of the Middle Triassic succession. This indicates that certain structural elements continued to be tectonically active during the 'post-rift phase'. A second tectonic event influenced the basin infill of the lower part of the Upper Triassic succession in the *post-rift phase 2*. Thick evaporites formed in isolated marine sub-basins, was triggered by an arid climate, oscillation in the relative sea level, and likely, the establishment of a structural threshold to the Borealic open marine seaway. The upper part of the Upper Triassic succession of *post-rift phase 3* represents the establishment of a fluvio-lacustrine depositional environment. This facies shift was probably caused by reduced rate of subsidence, tectonic uplift in the hinterland and more humid conditions.

Introduction

Various aspects of the tectonic and sedimentological development of the Mid-Norwegian shelf in the Permo-Triassic are discussed in several papers (Hagevang and Rønnevik, 1986; Jackobsen and Van Veen, 1984; Bukovics et al., 1984; Brekke and Riis, 1987; Blystad et al., 1995; Grunnaleite and Gabrielsen, 1995; Roberts et al., 1999; Brekke et al., 1999; Brekke et al., 2001; Bugge et al., 2002). However, the basin infill and palaeogeography of the Permo-Triassic succession are still poorly understood on the Mid-Norwegian shelf, including the relation to the adjacent areas in the northern proto-Atlantic region, such as East Greenland. This is reflected by the different models and interpretations concerning the timing of tectonic events in various studies (Grunnaleite and Gabrielsen, 1995; their Table 1). In addition, the sedimentological development related to allogenic factors such as tectonism, eustasy and climate has received little attention (Roberts et al., 1999; Brekke et al., 2001), and no proper subdivison of the Triassic succession has been presented in earlier literature.

The studied area is an important link between the Triassic continental basin infill and palaeogeography in the North Sea (Nystuen et al., 1989; Steel, 1993; Frostick and Reid, 1992) and the north-western part of the British Isles (Steel, 1974; Swiecicki et al., 1995; Roberts et al., 1999), and the marine-dominated

Onshore–Offshore Relationships on the North Atlantic Margin
edited by B. Wandas et al. NPF Special Publication 12, pp. 165–189,

Table 1 Available wells from the Mid-Norwegian shelf used in this study with additional information. (* – not total thickness due to erosion of the top of the unit or where the base is not drilled.)

Well	Unit	Facies (See table 2)	Age	Top	Base	Total thickness	Cores
6611/9U-1	Unit Tr1 (Upper Turbidite Unit)	FA1	Early Triassic (Early Griesbachian)	366	559	193*	366–579
	Lower Turbidite Unit	FA1	Late Permian (Ufimian — ? Early Tartarian)	558	740	182	558–740
	Anhydrite Unit	FA5a	Late Permian (Ufimian — ? Early Tartarian)	740	756	16	740–756
	Shallow Marine Unit	FA2	Late Permian (Ufimian — ? Early Tartarian)	756	914 (TD)	158*	756–914
6611/9U-2	Unit Tr1	FA1	Early Triassic (Early Griesbachian)	366	639 (TD)	273*	366–639
6608/8-1	Tr2	FA2	Early Triassic (Dienerian)	1753	2221	468*	1833–1842
	Tr1	FA1 and FA2	Early Triassic (Griesbachian)	2221	2750	529	
	Carbonate Unit	FA6	Late Permian?	2750	3006 (TD)	256*	2766–2775
6609/7-1	Carbonate Unit	FA6	Late Permian	1876	1912	36*	
6507/6-1	Unit Tr5	FA4 and FA3b	Late Triassic (Norian-Rhaetian)	1306???	1754	448	
	Unit Tr4	FA5a	Late Triassic (Carnian)	1754	2020	266	
	Unit Tr3	FA3a	Middle Triassic (Anisian-Ladinian)	2020	3063	1043	3041–3045
	Unit Tr2	FA2	Early Triassic (Dienerian-Spathian)	3063	3769	706	
	Unit Tr1	FA1	Early Triassic (Griesbachian)	3769	4040 (TD)	271*	3995–4013
6610/7-2	Unit Tr5	FA3b and FA4	Late Triassic (Norian-Rhaetian)	1748	2651	903	
	Unit Tr4	FA5b	Late Triassic (Carnian)	2651	3451	800	
	Unit Tr3	FA2 and FA3a	Middle Triassic (Anisian-Ladinian)	3451	4215	764*	4180–4195
6510/2-1	Unit Tr5	FA3b	Late Triassic (Norian-Rhaetian)	2249	3083	834	
	Unit Tr4	FA5b and FA5a	Late Triassic (Carnian)	3083	3690	607	
	Unit Tr3	FA3a	Middle Triassic (Anisian-Ladinian)	3690	4700 (TD)	1010*	4082–4101
6507/12-2	Unit Tr5	FA4 and FA3b	Late Triassic (Norian-Rhaetian)	2607	3454	847	
	Unit Tr4	FA5b	Late Triassic (Carnian)	3454	4781	1327	
	Unit Tr3	FA3a	Middle Triassic (Anisian-Ladinian)	4781	5008 (TD)	227*	4795–4985
6608/11-1	Unit Tr5	FA4	Late Triassic (Nor. – E. Rhaet.)	1231	1620	389*	1372–1398
6507/12-1	Unit Tr5	FA4 and FA3b	Late Triassic (Norian-Rhaetian)	2757	3720 (TD)	963*	3708–3720

palaeogeography and basin infill in the Triassic of the Barents Sea-Svalbard region and the Sverdrup Basin (Mørk et al., 1989; Van Veen et al., 1992). The large-scale tectonic development of the proto-Atlantic rift system, with its multiple phases of rifting, is also of importance in the understanding of the initial break-up of Pangea. These problems include the tectonic style of rift-basin formation and controlling factors of the basin infill history.

The aim of this study of the Permo-Triassic in the northeastern part of the Proto-Atlantic region is therefore: (1) Suggestion of a new sedimentological subdivision of the Triassic succession on the Mid-Norwegian shelf, based on a detailed facies analysis of the available cores and well data, (2) Integration of seismic interpretation with facies analysis in order to discuss the timing of tectonic events and their influence on the sedimentary architecture, (3) Comparison of the tectono-sedimentary history of the Mid-Norwegian shelf and East Greenland for making an overall model of the basin infill history and palaeogeography of the region in the Permo-Triassic.

Data set

An extensive database of around 80 seismic surveys is applied in this study. Main attention has been given to the surveys, ST8707(NHR01), ST8804(NHR01), ST8808(NHR01), MB-84(NHR01), GMNR-94, SH9601 and VIWT-93. A few seismic lines are presented in this chapter to exemplify the interpretation of these surveys. Regional thickness maps of the Permo-Triassic based on these surveys are presented in non-published reports of former Saga Petroleum.

The well database represents the stratigraphically, the most important wells drilled in the Permo-Triassic succession (6507/6-1, 6507/12-1, 6507/12-2, 6608/8-1, 6611/9U-1, 6611/9U-2, 6608/11-1, 6510/2-1, 6510/2U-2, 6610/7-2) on the Mid-Norwegian shelf (Table 1, Fig. 1). These wells are correlated and the available cores described. Biostratigraphy and palynofacies are elaborated for the available well data set.

Biostratigraphic interpretations are based on palynological and micropalaeontological analyses performed by various consultant companies, scientific institutions and scientists. Samples were processed according to standard preparation methods for palynological and micropalaeontological material and analysed by a light microscope. All the results were merged and adjusted according to standard zonation schemes in order to give a homogeneous interpretation. Data quality varies among the wells and internally between intervals, dependent on sample type and on the quality of the

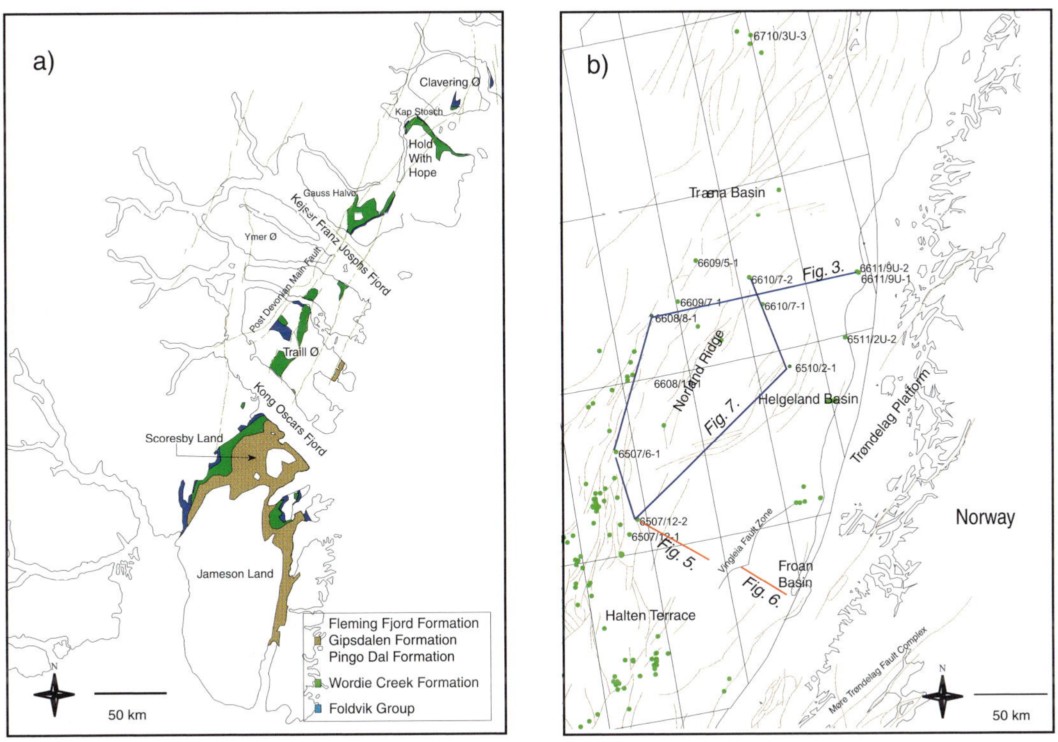

Fig. 1 (a) Upper Permian and Triassic outcrops are marked on map of East Greenland. (b) Locations of the applied wells and seismic lines are marked on the map of Mid-Norwegian shelf.

preserved palynomorphs. Most analyses are based on sidewall cores and the ditch-cutting samples. In addition, cores from the exploration wells are also analysed (Table 1). Shallow Drilling Project wells (IKU) are cored from top to their terminal depth. Thus, the analyses are exclusively from core material.

The palaeogeographic maps and basin infill models are made conceptual, based on the studied data set from the Mid-Norwegian shelf, supplied with data published on the palaeogeographic development of the Permo-Triassic succession in East Greenland (Clemmensen, 1980b; Surlyk, 1990; Seidler, 2000b; Stemmerik, 2000). The palaeogeographic reconstructions have been performed on a continental plate reconstruction at 170 Ma by Skogseid et al. (2000).

Stratigraphic subdivision

There is no consistent stratigraphic subdivision of the Permo-Triassic succession on the Mid-Norwegian shelf. The Upper Permian succession is subdivided into the 'Shallow Marine Sandstone Unit', 'Anhydrite Unit' and 'Lower Turbidite Unit', according to Bugge et al. (2002) (Fig. 2). The Upper Permian carbonates present in wells on the Nordland Ridge are termed 'Wegener Halvø Equivalent' by Bugge et al. (2002). In this chapter, the Upper Permian carbonates is entitled the 'Carbonate Unit'.

In the present study a new informal lithostratigraphic subdivision is introduced for the Triassic succession on the Mid-Norwegian shelf. The Triassic section has been divided into five stratigraphic units, Unit Tr1–Tr5 (Fig. 2). The principle of the subdivision is based on systematic trends in facies and biostratigraphy. The lowermost part of the Triassic succession, of early Scythian age (Griesbachian), is defined as Unit Tr1, despite the usage of 'Upper Turbidite Unit' by Bugge et al., 2002. The overlying successions are termed Unit Tr2 and Unit Tr3, dated to late Scythian (Dienerian–Spathian) and Middle Triassic (Anisian–Ladinian), respectively. The Upper Triassic evaporite units, named Lower Salt and Upper Salt by Hagevang and Rønnevik (1986), are termed Unit Tr4. The deposits stratigraphically above, of a Norian–Rhaetian age, are entitled Unit Tr5. In other studies, Unit Tr5 is termed 'Triassic Red Beds' and 'Triassic Grey Beds' (Hagevang and Rønnevik, 1986; Jacobsen and van Veen, 1984).

Seismic horizons

Important horizons, such as intra-Upper Permian (top Anhydrite Unit), base Trias (base Unit Tr1), intra-Triassic (base Unit Tr3), base Unit Tr4 (base Salt) and top Unit Tr4 (top Salt), are interpreted basin wide (Fig. 2). Seismic interpretation and further discussion regarding these reflectors are presented in the chapter.

Top Anhydrite Unit (intra-Upper Permian)

The top Anhydrite Unit (intra-Upper Permian) represents a reflector of high amplitude, as indicated by seismic sections adjacent to IKU-well 6611/9U-1 (Bugge et al., 2002). The reflector is interpreted to be present basin wide on the Trøndelag Platform (Bugge et al., 2002). However, due to limited resolution of seismic sections and poor well control in the deeper parts of the Trøndelag Platform, this reflector is difficult to interpret over the entire area studied.

Base Triassic (base Unit Tr1)

The base Triassic reflector is mainly interpreted on the Nordland Ridge, where it is a marked reflector of high amplitude. The reflector is penetrated in well 6608/8-1 where it rests on carbonates. However, tectonic overprinting by Late Jurassic faults makes the horizon difficult to follow in certain parts of the Nordland Ridge. In the Helgeland Basin, the interpretation of base Triassic is still more problematic. This is related to limited well control and lack of any lithological contrasts of the Permo-Triassic boundary, as documented in IKU-well 6611/9U-1. The top Anhydrite reflector is therefore applied as a correlative horizon instead of the base Triassic, in most parts of the Helgeland Basin and to the west of the Vingleia Fault Zone.

Intra-Triassic (base Unit Tr3)

This reflector is interpreted in the southern part of the Trøndelag Platform, where it drapes the underlying half-graben systems, both towards the eastern part of the Nordland Ridge and the western part of the Vingleia Fault Zone. The reflector is traced northwards into the Helgeland Basin, where it is identified on seismic sections, just below terminal depths of the wells 6610/7-2 and 6510/2-1. The reflector is generally present at a relatively constant level beneath the base Unit Tr4. The intra-Triassic reflector is only penetrated in well 6507/6-1, where it is interpreted to represent a shift from a

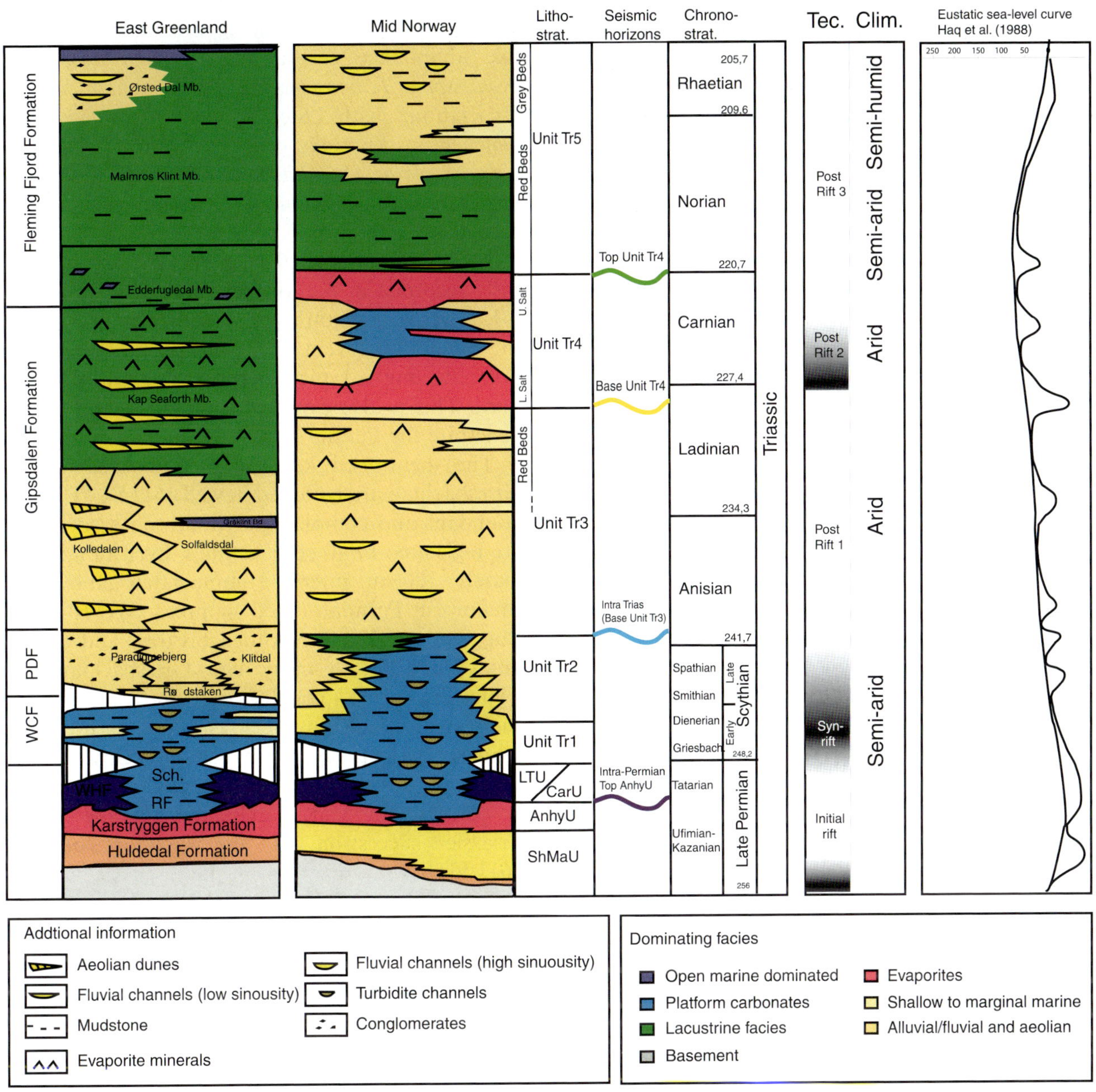

Fig. 2 Idealised chronostratigraphy illustrates the basin infill history of the Mid-Norwegian shelf and East Greenland. Chronostratigraphy in East Greenland is based on the work of Clemmensen (1980), Surlyk (1990), Seidler (2000b) and Stemmerik (2001). Chronostratigraphic time-scale is based on Gradstein and Ogg (1996). Chronostratigraphy (WCF: Wordie Creek Formation, PDF: Pingo Dal Formation, WHF: Wegener Halvø Formation, RF: Ravnefjeld Formation, Sch: Schuchert Dal Formation; LTU: Lower Turbidite Unit, CarU: Carbonate Unit, AnhyU: Anhydrite Unit, ShMaU: Shallow Marine Unit).

marine to a continental facies. However, there is no direct tie from well 6507/6-1 on the Nordland Ridge out into the Helgeland Basin. There are, therefore, some uncertainties regarding the identification of this horizon basin wide.

Base Unit Tr4 (base Salt)

The reflector base *Unit Tr4* (*base Salt*) represents the most important seismic horizon and is identified basin wide in most parts of the Trøndelag Platform area. This seismic horizon is characterised by strong amplitudes and relatively good well control (Table 1). The reflector provides helpful information about the age of the sediment infill and the deeper structures below base Unit Tr4 in both, the Helgeland and the Froan Basins.

Top Unit Tr4 (top Salt)

The contrast between the uppermost salt interval in Unit Tr4 and the mudstones and sandstones in

Unit Tr5 creates a reflector of relatively strong amplitude which is interpreted in parts of the Trøndelag Platform. The absence of a thick upper salt succession in Unit Tr4 in the northern part of the Helgeland Basin, on the Nordland Ridge and in the marginal areas has resulted in a subtler reflector in these parts of the basin (Table 1).

Intra-Åre reflector

This reflector horizon is characterised by a band of high amplitude reflectors and is relatively well defined in the whole area (Fig. 5). The high amplitudes are related to the presence of coal deposits.

Biostratigraphic dating and correlation

Permian

Carbonates encountered in the wells 6608/8-1, 6609/7-1 cannot be dated precisely, because no diagnostic fossils are recorded. A Late Permian age is suggested, based on the retrieval of bryozoan fragments in the ditch-cutting samples.

However, rich Late Permian fossil assemblages, mainly palynological, were recorded in IKU well 6611/9U-1. Palynological assemblages in the East Greenland deposits (Piasecki, 1984) correlate very well with those recognised in the Mid Norway shelf material. Abundant occurrences of taxa as *Vittatina spp.*, *V. sriata*, *V. costabilis* and *Protohaploxypinus samoilowitchii* in association with *Lueckisporites spp.*, *Florinites spp.*, *Raistrickia sp.* and *Unellium sp.* characterise this Upper Permian section in well 6611/9U-1. However, the correlation indicates that there might be a hiatus on top of the Permian at the site of the IKU wells, compared to East Greenland (Bugge et al., 2002). Consequently an Ufemian-early Tatarian age is suggested for this interval.

Triassic

Unit Tr1

The Griesbachian age for unit Tr1 is assumed to be, based mainly on palynological and micopalaeontological data. These data also provide information about the Permo-Triassic boundary. In well 6608/8-1, an incomplete lowermost Triassic is evident, since the most characteristic earliest Griesbachian floral elements are missing. Events like abundance of the marine taxa *Tymphanicysta spp.*, *T. stoschiana*, *Micrhystidium spp.* in association with the spores *Bharadwajispora labichensis*, *Kreuselisporites staplinii*, *Lundbladispora obsoleta*, *Maculatasporites sp.*, *Proprisporites pocockii* and *Stratiobieites richterii* can not be recognised.

Traces of some of the actual taxa are sporadically recorded in the lowermost part of well 6507/6-1. The association here is rather poor, lacking the abundance event of *Tymphanicysta spp.*, thus not reflecting a typical very earliest Griesbachian flora. This is also expected since the well does not penetrate the Permo-Triassic boundary.

The mentioned key events and taxa for the Griesbachian are recorded in well 6611/9U-1, where abundant and complete palynomorph associations are registered in all core samples, over an approximately 100 m interval, immediately above the shallowest Permian-dated sample. According to Bugge et al. (2002), recorded bivalves and ammonoids indicate an early late Griesbachian age. Following their conclusions, the absence of such macrofossils in the lowermost 43 m of the sequence might suggest an earliest Griesbachian dating of that part of Unit Tr1. If this assumption is correct, no hiatus exists within the Lower Triassic succession at the location of well 6611/9U-1.

Well 6611/9U-2 does not seem to penetrate the entire lowermost Griesbachian as the actual fossils, in particular *Tymphanicysta spp.*, are less frequently recorded than in 6611/9U-1. This observation is also in accordance with Bugge et al. (2002). They report a stratigraphic overlap of about 60 m between the two wells, approximately 100 m up in the Griesbachian, where 6611/9U-2 has its terminal depth.

Unit Tr2

Palynofloral assemblages within Unit Tr2, partly penetrated by 6507/6-1 and partly by 6608/8-1, are similar. Fossil abundance is, however, variable and very often sparse. A lot of the sidewall cores from 6507/6-1 were barren. On the other hand, cutting samples from 6608/8-1 were contaminated by cavings. Precise dating and correlation are therefore difficult.

Presence of common to abundant *Densoisporites nejburgii* in the upper part, and records of *D. playfordi* over the interval, indicate an age no younger than the latest Spathian for this unit.

Unit Tr3

The characteristic first downhole occurrence (FDO) of *Densosporites nejburgii* is registered within the lowermost part of the Unit Tr3 interval. The last downhole common occurrence (LDCO) of *Ovalipollis pseudoalatus* is another correlative event that can be followed through Tr3 in all wells (wells 6507/6-1, 6507/12-2, 6510/7-1, 6510/2-1). Generally, palynomorphs are rare within Tr3, probably a consequence of poor preservation conditions. However, sufficient palynomorphs are recorded to indicate that the unit was deposited over a time period lasting from the earliest part of the Anisian and into the Ladinian.

Unit Tr4

The first downhole abundant occurrence (FDAO) of *Triadispora spp.* can be traced in the upper part of unit Tr4 in the correlated wells (Fig. 7). This event also dates the unit as Carnian or older, but not older than late Ladinian. As a whole, the Tr4 is poor concerning palynomorphs, especially within the salt layers, which most often are barren.

Unit Tr5

Correlative events like first downhole occurrence (FDO) of *Triadispora spp.*, last downhole common occurrence (LDCO) of *Limbosporites lundbladii* and first downhole abundant occurrence (FDAO) of *Ovalipollis pseudoalatus* are registered in several exploration wells. Palynomorph assemblages are relatively rich in most of the samples analysed within this unit. Records of *Kyrtomisporites spp.*, *Carnisporites spiniger*, *Granuloperculatipollis rudis*, among others, signify a Norian–Rhaetian dating of Tr5.

Facies and facies associations

Facies are sorted into six facies associations. The subdivision is based on core descriptions, well logs, palynofacies and micropalaeontology. The facies associations are very generalised and are roughly subdivided due to the geographically scattered well data set and limited amount of cores. Additional information and summaries about the different facies associations are presented in Table 2.

Facies association 1 (*FA1*) is characterised by fining upward units with lower erosive contacts and upper laminated greyish mud- and siltstones. Sedimentary structures such as slumping, flame and water-escape structures are present in the sandstones. Marine palynomorphs (e.g. *Micrhystridium spp.* and *Veryhachium spp.*) and foraminiferal linings are registered regularly. The palynoflora are characterised by the dominance of allochthonous (pollen) versus autochthonous (spores) continental palynomorphs. FA1 is interpreted to represent submarine fan deposits. This facies assemblage comprises basin-floor fans deposited by turbidite currents.

Facies association 2 (*FA2*) is characterised by sandstone intervals with cross-bedding and parallel lamination with a thickness up to 10 m. The fine-grained intervals are characterised by bipolar ripples, fine lamination and unidirectional ripples. FA2 is defined by the scattered occurrence of brackish and marine algae. The dominance of an autochthonous relative to an allochthonous palynoflora implies relatively short distance to local vegetational areas. Facies association 2 represents marginal marine deposits that comprise different facies assemblages, such as shoreface sedimentation, lagoonal deposits, distal reaches of fan deltas, coastal plain and tidal flat/channels.

Facies association 3a (*FA3a*) is typified by blocky sandstone beds that alternates between 1 to 5 m, in thickness. These sandstone beds form stacked channel units with a maximum thickness of 10–15 m. The sandstone units are interbedded with brownish to red coloured mud- and siltstones. The mudstone units are characterised by extensive bioturbation, scattered root horizons, desiccation cracks and carbonate nodules. Occurrence of spores and pollen is limited, but the preserved palynoflora is dominated by local, continentally derived species. The sandstone units are interpreted to represent non-channelised deposits within ephemeral braided streams and sheet-flood deposits. The mudstone units are interpreted as former soil profiles or paleosols.

Facies association 3b (*FA3b*) is characterised by the occurrence of up to 20 m-thick fining upwards to the blocky sandstone units. The mudstone units are typified by greenish-grey to brown-red colour and coal fragments occur regularly. The palynofacies is characterised by the dominance of local, continentally derived spores and 'wet' vegetation types. Facies 3b is interpreted to represent high sinuousity channels deposited within an alluvial basin.

Table 2 Facies associations of the Permo-Triassic succession on the Mid-Norwegian shelf.

Facies association	Facies characteristics	Palynofacies/paleontology	Depositional environment
Facies association 1 FA1	Fining upward units. Slumping, flame and water escape structures. Limited bioturbation. Greyish colour of laminated mud- and siltstones.	Marine palynomorphs as algae/acritarcs, foraminiferal linings. Dominance of allochthonous versus autochthonous continental palynomorphs.	Submarine fan deposits
Facies association 2 FA2	Sandstone intervals with cross bedding, parallel lamination, bipolar ripples and ripples. Other sedimentary structures as convolute lamination, climbing ripples and flame structures.	Scattered occurrence of brackish and marine algae. High frequency of autochthonous relative to allochthonous continental palynomorphs.	Marginal marine deposits
Facies association 3a FA3a	Blocky sandstone beds alternating between 1 and 5 meters. Stacked channel units; max. thickness of 10–15 m. Mudstone units: brownish-red, extensive bioturbation, scattered root horizons, mud cracks and carbonate nodules.	Dominance of local, continentally derived taxa.	Non-channelized fluvial deposits
Facies association 3b FA3b	Thick fining upwards to blocky sandstone units (20 m). Greenish-grey to brown-red mudstone units. Coal fragments.	Dominance of continental, locally derived spores, marsh and peat-bog types.	Channelised fluvial deposits
Facies association 4 FA4	Thick greenish-grey to brownish-red mudstones. Scatttered occurrence of mud cracks, calcretes, root structures. Moderate bioturbation. Interbedded with isolated, blocky sandstone units of 2–5 m in thickness and small carbonate beds.	Continental palynomorphs dominate. Limited occurrence of marine algae.	Shallow, lacustrine basin/ playa
Facies association 5a FA5a	Anhydrite and carbonate beds (up to 5 m).	Marine algae. Allochthonous palynomorphs dominate.	Basin marginal evaporites
Facies association 5b FA5b	Thick massive and homogenous halite deposits	Marine algae.	Basin central evaporites
Facies Association 6 FA6	Massive and structureless dolomites and calcite beds. Drill breaks in well 6609/7-1 could be due to karstic cavities in the carbonate (Bugge et al., 2002).	Skeletal grainstone clasts, fragments of reef-building organisms, bryozoans, tabulate coral fragments and algal-coated foraminifera.	Platform carbonates

Facies association 4 (*FA4*) is mainly characterised by thick, greenish-grey mudstones interbedded with thin brownish to red mudstone beds. The mudstone beds have scattered occurrence of mud cracks, root structures and small carbonate nodules. The mudstones are interbedded with isolated and blocky sandstone units of 2–5 m in thickness and thin carbonate beds (5–10 cm). The mudstone beds contain mainly continental palynomorphs. The occurrence of marine algae is limited. Facies association 4 is interpreted to represent mainly shallow lacustrine floodbasin, which periodically was exposed and influenced by pedogenic processes.

Facies association 5a (*FA5a*) is typified by anhydrite and carbonate beds which interfingers with thin sandstone and mudstone beds. The thickness of an individual anhydrite bed is a maximum of 5 m, but they can form stacked units up to 20–30 m in thickness. The occurrence of marine algae emphasises the proximity to the marine environment. FA5a represents basin marginal evaporites, such as marginal marine and inland sabkha desposits.

Facies association 5b (*FA5b*) consists of thick massive and homogenous halite deposits with the presence of marine algae. The halite deposits can reach several hundred meters in thickness. FA5b is interpreted to represent basinal evaporites.

Facies Association 6 (*FA6*) is characterised by massive and mainly structureless dolomite and calcite units. The presence of skeletal grainstone clasts, fragments of reef-building organisms, algal-coated forams, tabulate coral fragments and bryozoans in the cutting samples indicates that carbonate deposition and reef development occurred (Bugge et al., 2002). Drill breaks experienced when drilling into these carbonates might suggest karstic cavities. The carbonates are interpreted to represent isolated and/or ramp carbonate platforms.

Sedimentology and structural style of the Mid-Norwegian shelf

Upper Permian (Ufimian–Kazanian to Tatarian)

The Carbonate Unit is interpreted to represent platform or ramp carbonates (FA6, Tables 1, 2). The Carbonate Unit is present in the wells 6609/7-1 and 6608/8-1 located on the Nordland Ridge, where the thickness is 36 and 256 m, respectively (Fig. 3, Table 1). In well 6609/7-1, the carbonates rest directly on a metamorphic basement.

In IKU well 6611/9U-1 370 m of marginal to submarine deposits, of a Late Permian age, are cored (Fig. 3). The lowermost succession, the 'Shallow Marine Sandstone Unit', is characterised by strongly bioturbated and massive sandstones. This sandstone unit rests directly on basement and is interpreted as upper shoreface deposits (Bugge et al., 2002) (Fig. 2). Above this sandstone unit, the 15 m thick 'Anhydrite Unit' was cored. The unit is interpreted as sabkha or reworked sabkha deposits (Bugge et al., 2002). The upper part of the Upper Permian succession (Lower Turbidite Unit) represents submarine fan deposits (FA1) (Table 1). Within these turbidite fan units, two organic-rich mudstone intervals are identified, which indicates deposition in a periodically anoxic environment (Bugge et al., 2002).

The Triassic

Unit Tr1 (Griesbachian)

The Permo-Triassic boundary is characterised by an abrupt facies shift on the present Nordland Ridge. Platform carbonate deposits are truncated by clastic sediments that are interpreted to represent a submarine fan/shallow marine environment (Fig. 3). The Permo-Triassic boundary represents, therefore, a considerable hiatus and a sequence boundary (Fig. 3). In IKU-well 6611/9U-1, located in the margins of the Helgeland Basin, there is no abrupt shift in facies and Upper Permian turbidite units are succeeded by Lower Triassic turbidite units (Table 1). However, a possible hiatus is registered between the Upper Permian and Lower Triassic in well 6611/9U-1, where the upper Tartarian is interpreted to be missing (Bugge et al., 2002).

Unit Tr1, the lowest part of the Triassic succession, is cored in six wells (Table 1) and comprises mainly the facies associations submarine fan deposits (FA1) and the marginal marine deposits (FA2) (Table 2) (Figs. 1 and 4). FA2 becomes more dominant in the upper part of Unit Tr1 and reflects a general, upward shallowing of the depositional environment (Fig. 3). This is observed in well 6608/8-1, located on the Nordland Ridge (Table 1, Fig. 3), which is the only well drilled completely through Unit Tr1. However, in well 6608/8-1, most of the lower Griesbachian is interpreted to be absent (see biostratigraphy), suggesting that non-deposition and/or erosion took place during the earliest Griesbachian time at the Nordland Ridge. Within the lower part of

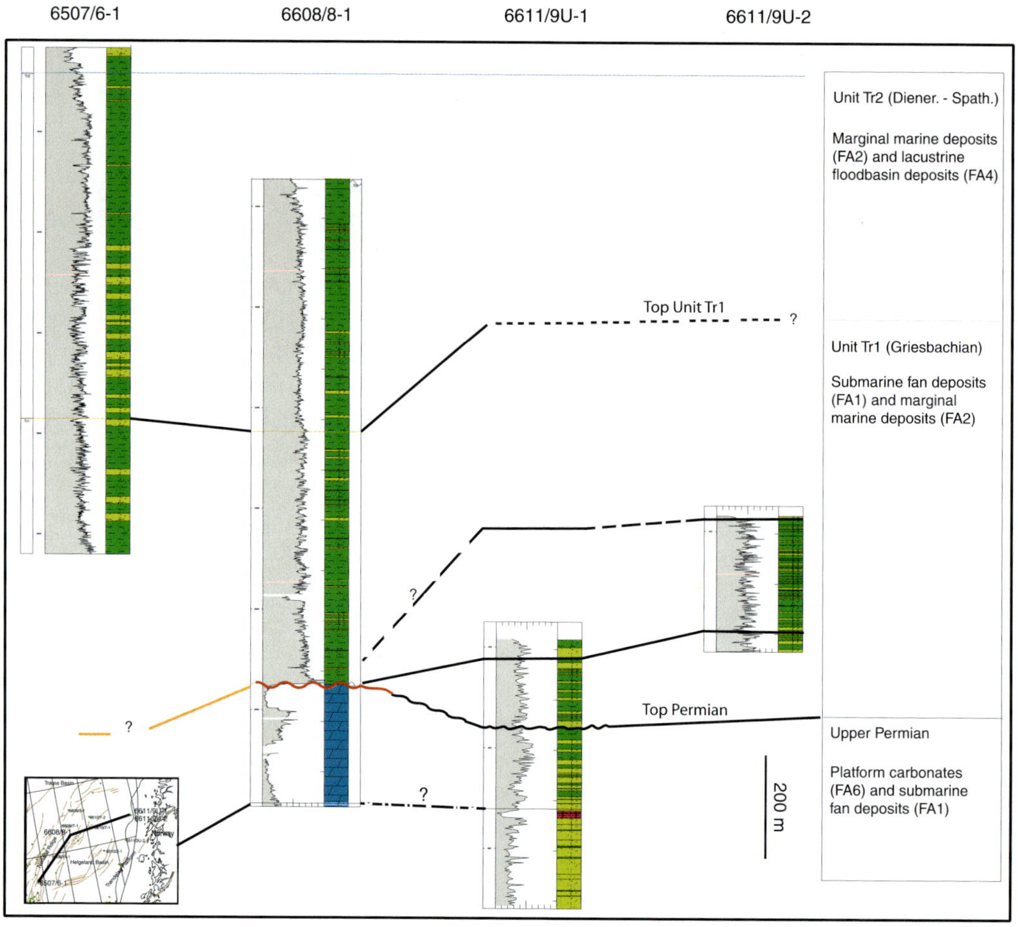

Fig. 3 Correlation panel of the Upper Permian and Lower Triassic succession on the Mid-Norwegian shelf. Notice the Permo-Triasssic unconformity in well 6608/8-1, located on the present Nordland Ridge. Adjacent to the Norwegian coast the boundary is more conformable, as indicated in 6611/9U-1. Thus, a hiatus is present on the Permo-Triassic boundary in both wells (see text). The boundary between Units Tr1 and Tr2 is transitional and difficult to detect on the well logs.

Unit Tr1 in the IKU wells, thin beds of eroded sabkha deposits are present (Bugge et al., 2002).

There is no well control of the Upper Permian and Lower Triassic succession in the Froan Basin and the central parts of the Helgeland Basin. The intra-Upper Permian and base Triassic seismic horizons are in these areas difficult to place and trace exactly. The estimated thickness of the Lower Triassic in areas with sufficient well control is, therefore, important, since it provides an indication of the position to the intra-Upper Permian and base Triassic reflectors. In addition, it provides information about the age of the sedimentary fill below the intra-Triassic and base Unit Tr4 reflectors in the Froan and Helgeland Basins. The total thickness of Units Tr1 and Tr2 is about 1250 m, when the thickness of these units in wells 6608/8-1 and 6507/6-1, both located on the Nordland Ridge, is combined (Table 1, Fig. 3). The absence of the early Griesbachian in well 6608/8-1 suggests non-deposition and/or erosion simultaneous with deposition in the Helgeland Basin (Fig. 3). The unconformity and hiatus in this well might imply a location on a structural high or the crestal areas of a half-graben (Fig. 3) (See correlation with East Greenland and Discussion). The total thickness of the Lower Triassic succession is therefore expected to be much higher in downdip areas of the half-grabens and in the basinal areas, in general. This is also indicated by an incompletely drilled 400 m thick Griesbachian succession in the IKU-wells, located off Norway, which reflects very high rates of sedimentation. The total thickness of the Lower Triassic succession on the Mid-Norwegian shelf is therefore assumed to reach up to at least 2000 m in the Helgeland Basin. The wedge-shaped sediment infill of the half-grabens, observed in the Helgeland and the Froan Basins, is therefore assumed to contain mainly Units Tr1 and Tr2 (Figs. 5 and 6).

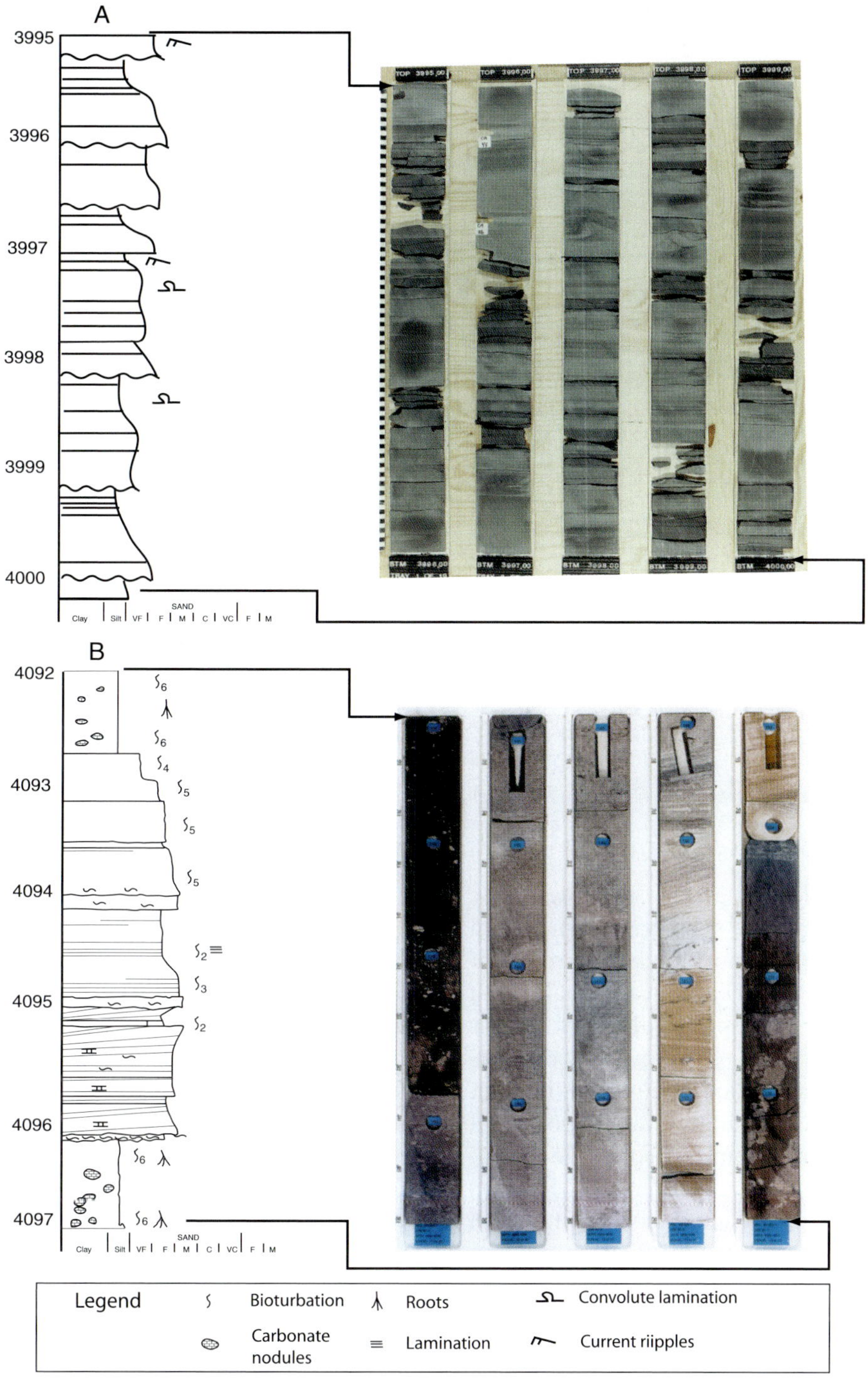

Fig. 4 (a) Beds with characteristic upward fining trends studied in cores from well 6507/6-1 show (FA1). (b) Cross-bedded and parallel-laminated channel sandstones studied in cores from well 6510/2-1. The sandstone beds are interbedded with bioturbated, brownish-red coloured mudstone intervals with carbonate nodules and root structures (Fa3a) (meters in mRKB).

These wedge-shaped depositional successions, present above the interpreted top Anydrite Unit, constitute the infill of large rotated fault blocks. Fault throw to these half-grabens is estimated to vary between 2000–3000 m along these faults. The apparently uniform thickness below this reflector may imply that an onset of fault growth occurred after the deposition of the Anhydrite Unit (Fig. 5).

Unit Tr2 (Dienerian–Spathian)

Unit Tr2 is poorly documented, and only two wells have been drilled in this unit (Fig. 3; Table 1). In well 6507/6-1, the thickness of Unit Tr2 is approximately 700 m and it is the only well which is completely drilled (Table 1). The transition from Unit Tr1 to Tr2 is characterised by a gradual shift

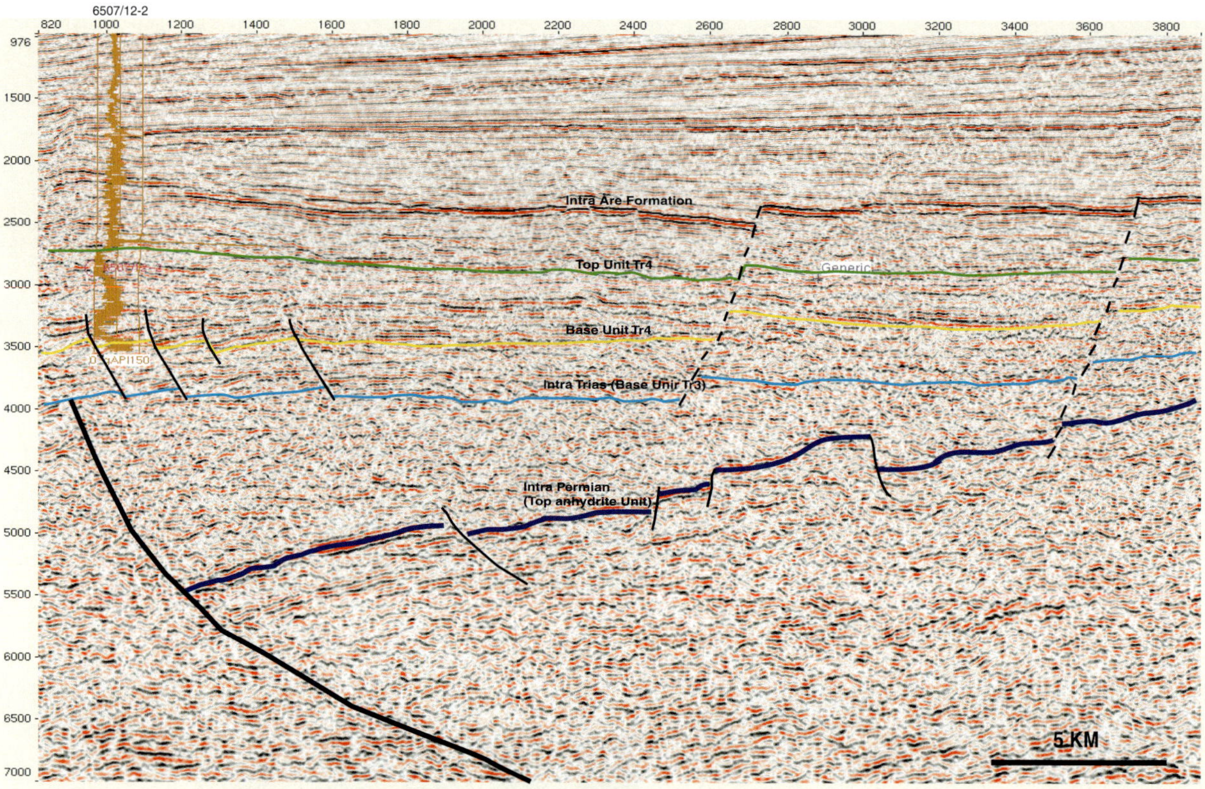

Fig. 5 Seismic line in the Helgeland Basin shows wedge-shaped sediment infill of mainly Early Triassic age, beneath the intra-Triassic reflector. The top Anhydrite Unit is tentatively. Thickness of Unit Tr4 increases towards west and is influenced by small-scale faults. Notice the relatively conform thickness of Tr3 and unit Tr5, in this part of the Helgeland Basin. (See location in Fig. 1).

in facies from a dominant submarine fan system to a marginal-marine depositional environment (FA2) (Tables 1 and 2). This upward shallowing of the basin during deposition of Unit Tr2 is indicated by: (1) Brownish-red-coloured sediments that occur more regularly and frequently in Unit Tr2, suggesting periodic oxidation and exposure of the depositional system. (2) Increased dominance of autochthonous terrestric palynomorphs (spores) upward through the succession which might reflect proximity to local vegetation. (3) Marine palynomorphs that occur less frequently in Unit Tr2 compared to Unit Tr1.

In well 6507/6-1, a 300 m-thick mudstone unit (FA4) was deposited in the upper part of Unit Tr2. This succession is characterised by variation between brownish-red and greyish colours. Deposition probably had taken place in a lacustrine and occasionally in a lagoonal sedimentary environment, with temporary exposure and oxidation of the deposits. Rare occurrences of marine algae indicate periodical marine influence.

Unit Tr3 (Anisian to Ladinian)

Unit Tr3 marks the transition from a dominantly marginal marine (FA2) to a continentally-dominated facies (FA3a) (Fig. 4b; Table 2). This unit is represented in five wells (Table 1), but is only completely drilled in 6507/6-1, where the thickness is c. 1100 m. Unit Tr3 is not penetrated completely in the wells 6610/7-2 and 6510/2-1, but the thicknesses are c. 760 m and 1010 m, respectively (Table 1) (Fig. 6). The base of Unit Tr3 is interpreted to correspond approximately to the intra-Triassic reflector, combined with additional information from biostratigraphic data.

Unit Tr3 is dominated by fine-grained, brownish-red sediments and exhibits an overall aggrading alluvial architecture (Fig. 6). In the uppermost part of Unit Tr3 the sediments are more greyish, and some scattered marine algae are present, which suggests that the environment became more influenced by marginal marine sedimentation (FA2) during deposition of the upper part of Unit Tr3. Facies between the different wells show limited variations. However, Unit Tr3 in well 6610/7-2 is characterised by the presence of marine algae through the whole succession. In addition, sedimentary facies in cores from this well show that the depositional system was dominated by marginal marine sedimentation (Table 2) (FA2). The location of well 6610/7-2, in the northern part of the Helgeland Basin, suggests that the marine

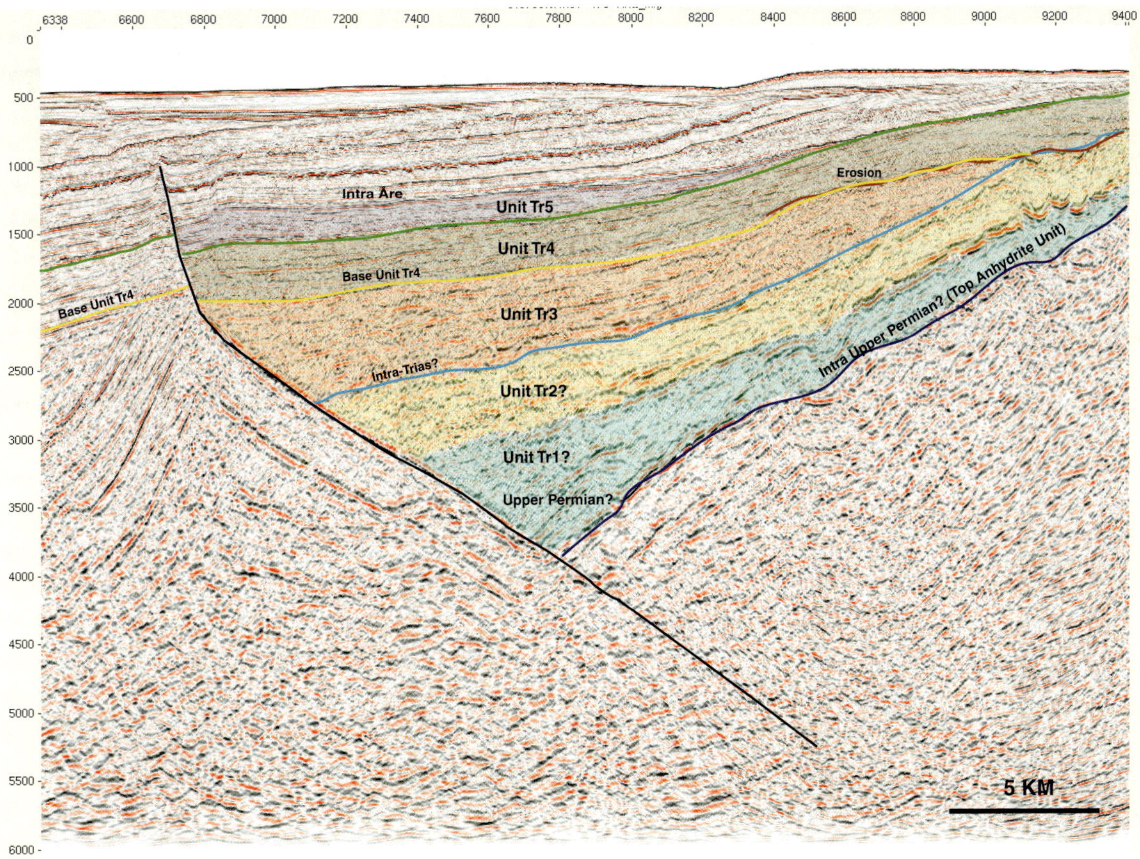

Fig. 6 Seismic line shows wedge-shaped Lower and Middle Triassic deposits beneath the base Unit Tr4 reflector in the Froan Basin. The different units are tentatively subdivided, based on the estimated thickness of the Triassic units from other parts of the Trøndelag Platform (See text and location in Fig. 1).

influence was more pronounced in the northern part of the Trøndelag Platform (Table 1).

An increase in the anhydrite content is recorded in Unit Tr3 compared to Unit Tr2 in well 6507/6-1. Anhydrite beds and traces of anhydrite are located in the fine-grained intervals of Unit Tr3 in all the wells, indicating that the alluvial flood-basin was influenced by a periodically evaporitic environment.

On the seismic sections, Unit Tr3 is characterised by limited thickness variations, especially in the Helgeland Basin (Fig. 5). However, sedimentary growth of Unit Tr3 towards the Vingleia Fault Zone is indicated by wedge-shaped deposits in the Froan Basin (Fig. 4). There are also identified surfaces internal within Unit Tr3 in the Froan Basin, which onlap on the intra-Trias reflector to the east in the basin.

Unit Tr4 (Carnian)

Unit Tr4 is completely penetrated in four wells, and the thickness varies from 260 m to 1330 m (Table 1, Fig. 7). The lower boundary of Unit Tr4 represents a distinct facies shift from a marginal marine/continental environment to a dominantly marine evaporite facies (FA5a, b) (Table 2).

Three evaporite successions are recorded in Unit Tr4. The *lowermost evaporite succession* is composed of thick halite bed sets and can reach up to 420 m in thickness, as documented by well 6507/12-2. These evaporite deposits represent basin central evaporites (FA5b) and identified over a large area of the Helgeland Basin. Thus, the thick basin central evaporites of the lowermost succession are correlated to thin (20–30 m), basin marginal evaporites (Fa5) on the present Nordland Ridge (Fig. 7). The *middle evaporite succession* is represented by relatively thin anhydrite rich units (10–40 m), present in wells 6610/7-2 and 6510/2-1, located in the northern part of the Helgeland Basin. The *uppermost evaporite succession* is an approximately 380 m-thick unit of halite deposits, representing basin central evaporites (FA5b). The thick halite deposits of the uppermost evaporite succession are correlated with significant-thinner basin-marginal evaporite deposits on the present Nordland Ridge (Fig. 7). This evaporite succession is not recorded in the northern part of the

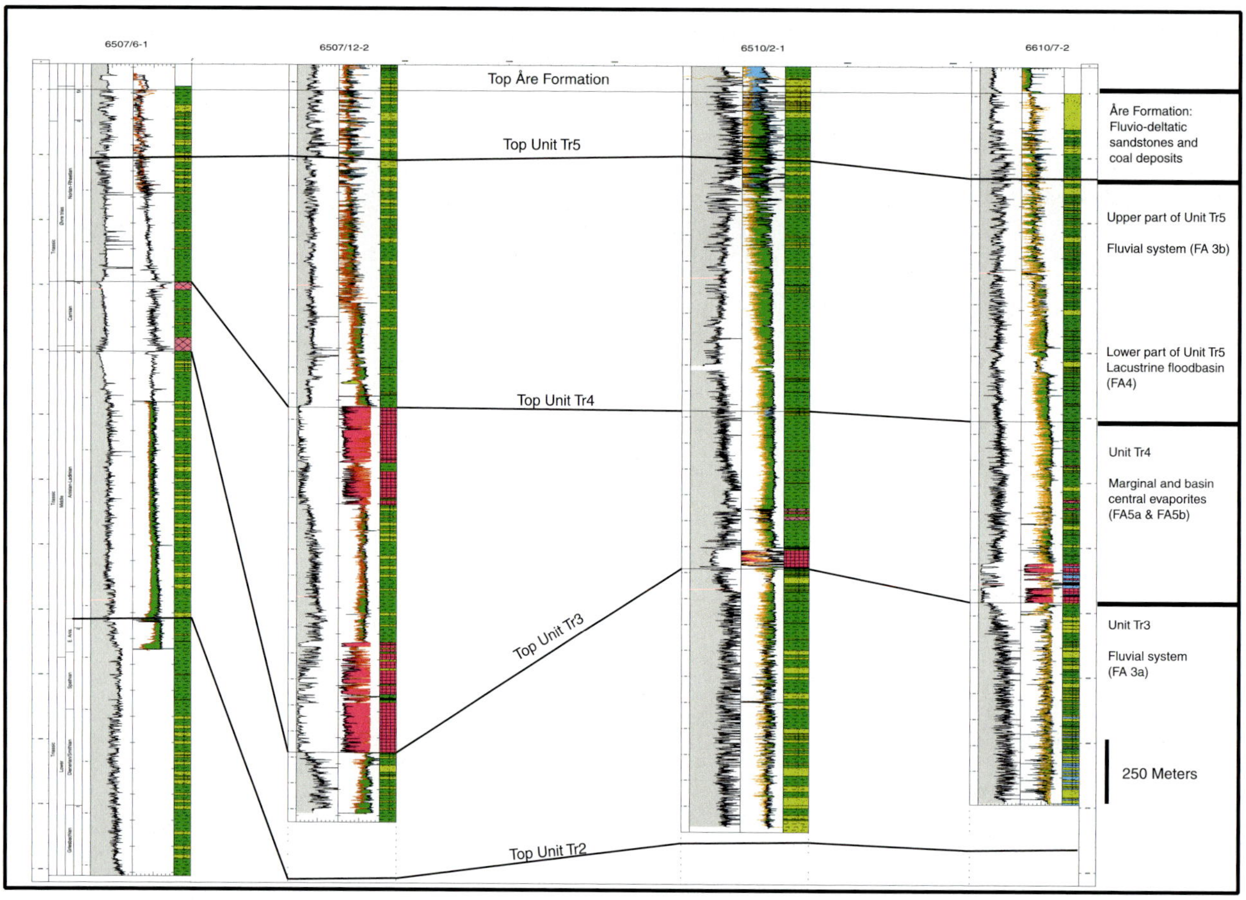

Fig. 7 Correlation panel of the Middle Triassic to Upper Triassic succession illustrates facies and thickness variation on the Mid-Norwegian shelf. Notice the increase in thickness and lateral facies variation between the wells, 6507/6-1 and 6507/12-2.

HelgelandBasin. The lateral extension of the succession indicates that the evaporites were restricted to the southern part of the Helgeland Basin, also identified on seismic sections in the Haltenbanken Area.

A marine origin of the halite beds is suggested by the presence of marine algae. The content of palynomorphs is generally low throughout Unit Tr4. This might reflect a stressed environment.

Small-scale block-rotated half-grabens and wedge-shaped sedimentary packages of Unit Tr4 are recognised in NNE–SSW trending sub-basins on seismic sections in various parts of the Helgeland Basin (Fig. 5). These basins follow the same trend as the Late Permian-Early Triassic basins, but they are located further to the west towards the present Nordland Ridge. The fault throw associated with these deposits is maximum 200–300 m (Fig. 6). Larger fault throws may exist from the Helgeland Basin to the Nordland Ridge, as suggested by the thickness variations, recorded in the wells (Fig. 7). However, Unit Tr4 shows no sedimentary growth towards the Vingleia Fault Zone and it generally thins towards the Norwegian Mainland (Fig. 6). In the eastern part of the Froan Basin, interpretation of base Unit Tr4 implies truncation and erosion of the underlying Unit Tr3.

Unit Tr5 (Norian–Rhaetian)

Unit Tr5 is represented in several wells, but in this study, there are six wells of importance (Table 1). The thickness of Unit Tr5 varies from 600 to 800 m in the wells (Fig. 7). The boundary between the underlying Units Tr4 and Tr5 is defined by the last occurrence of thin evaporite beds and the dominance of a continental palynofacies in Unit Tr5.

The lower part of Unit Tr5 was deposited in a shallow and extensive mud-dominated lacustrine basin (FA2) (Table 2). The thickness of the lacustrine succession exceeds 300 m and covers large parts of the present Mid-Norwegian shelf with a lateral extent of at least 150×200 km^2 (Fig. 7). The upper part of Unit Tr5 is characterised by thick sandstone units, interbedded with greenish-grey coloured mudstones and thin coal beds. These sediments are interpreted to be deposited within an extensive fluvial system dominated by moderate to high sinuous channels (FA3b). Unit Tr5, thus, represents an overall coarsening upward trend from a mudstone-rich lacustrine flood-basin to a sand-dominated fluvial environment.

Seismic sections show that the thickness of Unit Tr5 is relatively uniform, both in the Helgeland Basin and on the Nordland Ridge. No sedimentary growth towards intra-basinal lineaments is observed and this influenced the deposition of Unit Tr4 (Figs. 5, 6). Towards the Norwegian Mainland, a marked thinning of Unit Tr5 is observed (Fig. 6). The boundary between Unit Tr5 and the Early Jurassic Åre Formation is gradual and characterised by the occurrence of coal deposits, thicker channel sandstone units and the increasing number of greyish coloured mudstone beds (Fig. 7).

Correlation of the Permo-Triassic of East Greenland and the Mid-Norwegian shelf

The Permo-Triassic successions in East Greenland and the Mid-Norwegian shelf are compared and correlated to attain a regional model for the basin infill history during this period (Fig. 2). Facies analysis of outcrops from the Triassic succession in East Greenland has been undertaken in detail (e.g. Clemmensen 1980a,b; Seidler, 2000a,b), in contrast to what is possible from the well data on the Mid-Norwegian shelf. However, from the Mid-Norwegian shelf, an extensive seismic database provides important information about the structural framework and tectonic development of the region, whereas in outcrop it is more difficult to recognise large-scale basin infill architecture related to tectonic activity. Some precautions have to be considered when comparing the Mid-Norwegian shelf and East Greenland: (1) The data sets from the two sub-regions might represent different basinal positions as regarding proximity to hinterland regions and the regional basin axis. This might influence the type of facies' associations and their pattern of basinal distribution. (2) Biostratigraphy is particularly poor in the continental deposits of the Middle and Upper Triassic, thus, complicating the correlation.

Upper Permian

The Upper Permian succession on the Mid-Norwegian shelf and East Greenland consists of varied and mainly marine-dominated facies. The 'Shallow Marine Sandstone Unit', which represents the lowermost part of the Upper Permian succession on the Mid-Norwegian shelf, is correlated with the Huledal Formation on the East Greenland (Bugge et al., 2002). The Huledal Formation consists of poorly sorted, immature fluvial conglomerate beds and sandstone sheet deposits and rests on the Mid-Permian unconformity (Surlyk, 1990) (Fig. 2).

The 'Anhydrite Unit' is correlated to the Karstryggen Formation (Fig. 2), which is the most widespread Upper Permian unit in East Greenland (Bugge et al., 2002). It is therefore inferred that carbonate and evaporite deposits, similar to those in the Karstryggen Formation, covered large parts of the Trøndelag Platform. This correlation supports the assumption that the top Anhydrite Unit represents an important seismic reflector basin wide. The Karstryggen Formation is generally represented by limestone, with gypsum deposited under hypersaline conditions. These deposits rest either on the Huledal Formation or on the pre-Upper Permian rocks (Stemmerik, 1987).

The 'Lower Turbidite Unit' shows many similarities with the Ravnefjeld Formation. The Ravnefjeld Formation is a black, bituminous, laminated mudstone that occurs throughout the East Greenland basin (Surlyk et al., 1986). Ravnefjeld Foundation is a lateral equivalent to, and interfingers with, the reef-carbonates of Wegener Halvø Formation. The Wegener Halvø Formation occurs in an areally restricted depositional environment on structural highs or basin margins, infilling a karstic surface (Stemmerik et al., 1990; Stemmerik et al., 1993). The Upper Permian platform carbonates on the Nordland Ridge are correlated to Wegener Halvø Formation. The similarities with the Wegener Halvø Formation suggest that these carbonates also formed on structural highs, probably with clastic material trapped in the Helgeland Basin and in the local sub-basins. The upper part of the 'Lower Turbidite Unit' is correlated to the Schuchert Dal Formation in East Greenland. Schuchert Dal Formation was deposited within a submarine fan system and represents a shift in basin infill to a more siliciclastic-dominated system (Kreiner-Møller and Stemmerik, 2001).

The Upper Permian Foldvik Creek Group in East Greenland has a maximum thickness of 180 m (Stemmerik et al., 1993), whereas the thickness of the Upper Permian succession on the Mid-Norwegian shelf, recorded in the IKU-well 6611/9U-1, is c. 350 m. This may reflect the relative position of the sample points within the basin.

Lower Triassic

The Permian–Triassic boundary in East Greenland is reported to be both, unconformable and conformable. Variable and largely unknown amounts of the Upper Permian succession were eroded prior to the Griesbachian transgression (Stemmerik et al., 1997). Two Ammonite zones (*G. triviale*/*G. martini*) are missing in the Wegener Halvø sub-basins due to erosion or/and non-deposition (Seidler, 2000b). In other parts of East Greenland, these two ammonite zones represent a thickness up to c. 100 m (Seidler, 2000b). Numerous Lower Triassic incised submarine canyons were eroded down into the Upper Permian succession giving rise to reworked Permian sediments occurring high up in the Triassic succession (Birkenmajer, 1977, Seidler, 2000a, Wignall and Twitchett, 2002). The deeper part of the basins experienced a continuous marine sedimentation, as no hiatus between the Permian and the Triassic is observed (Piasecki, 1984). The same pattern in basin infill architecture is recorded on the Mid-Norwegian shelf. Submarine deposits of Early Triassic age rest unconformably on limestones of Late Permian age on the Nordland Ridge, whereas in the Helgeland Basin, the Permo-Triassic boundary is interpreted as more conformable.

The Lower Triassic Unit Tr1 on the Mid-Norwegian shelf is an equivalent to the Wordie Creek Formation in East Greenland (Fig. 2). Both these successions comprise submarine fan deposits, thus suggesting a genetic interrelationship regarding basinal control factors. The thickness of Wordie Creek Formation varies from c. 270 m in the marginal parts of the basin to 750 m basinward (Seidler, 2000a). The thickness variation depends on location in the sub-basins relative to syn-sedimentary faults (Seidler, 2000a).

In East Greenland, the boundary between the marine Wordie Creek Formation and the succeeding, 700 m-thick, continental Pingo Dal Formation (late Scythian) is marked by a major unconformity (Perch Nielsen et al., 1974; Clemmensen, 1980b; Seidler, 2000b). Coarse-grained alluvial fan successions (Klitdal Member and the Paradigme Bjerg Member) were formed along the N–S trending marginal fault zones. They pass from both sides of the graben, into the sandy floodplain deposits (Rødstaken Mb) with longitudinal drainage towards the north (Perch-Nielsen et al., 1974; Clemmensen, 1980b; Surlyk, 1990). Such a distinct change in facies during late Scythian has not been observed in the present data set from the Mid-Norwegian shelf. However, the basin was shallowing up from a submarine fan-setting (Unit Tr1) to a dominant marginal marine and alluvial setting (Unit Tr2). There is not observed conglomerate facies in Unit Tr2 yet, like in the Pingo Dal Formation.

The total thickness of the Lower Triassic in East Greenland is 1450 m, which is higher than the equivalent succession recorded from the wells located on the Nordland Ridge (1250 m). However, the estimated thickness of the Lower Triassic is expected to be significantly higher (up to at least 2000 m) in the basinal areas, such as the Helgeland and Froan Basins. The thick Lower Triassic succession indicates high rates of accumulation compared to that of the Upper Permian, and again supporting the creation of accommodation space by Early Triassic tectonic activity.

Middle Triassic

The Kolledalen, Solfaldsdal and the lower part of the Kap Seaforth Members of the Gipsdalen Formation are correlated to Unit Tr3 on the Mid-Norwegian shelf (Fig. 2). The Gipsdalen Formation is characterised by aeolian, fluvial and lacustrine sediments and has a maximum thickness of 350 m (Clemmensen, 1980b), indicating lower accumulation or/and accommodation rates compared with Unit Tr3 (1150 m), on the Mid-Norwegian shelf. The increase in abundance of anhydrite beds from Unit Tr2 to Unit Tr3 is similar to the lithological change that occurs between the Pingo Dal Formation and the Gipsdalen Formation in East Greenland.

The marine mudstones of the Gråklint Beds (up to 35 m thick) (?Anisian) on the East Greenland (Clemmensen, 1980b) are difficult to correlate with any specific transgressive event on the Mid-Norwegian shelf. However, the scattered occurrences of marine algae in intervals of Unit Tr3 indicate short and periodical marine incursions onto the alluvial plain. We, therefore, suggest that one of these trangressive events may be correlated with the marine Gråklint Beds.

Upper Triassic

Late Ladinian–Carnian

The upper parts of Kap Seaforth Member (Gipsdalen Formation) and Edderfugledal Member (Fleming Fjord Formation) in East Greenland are correlated with Unit Tr4 on the Mid-Norwegian shelf (Fig. 2). The Kap Seaforth Member is characterised by gypsum-bearing sandstone and mudstone. The sediments were deposited in continental sabkha environment with periodic aeolian influence (Clemmensen, 1978). The upper part of the Kap Seaforth Member is assumed to be deposited in the late Ladinan–Carnian and probably corresponds to the first evaporite bed in Unit Tr4. Edderfugledal Member was deposited in a shallow playa lake. The occurrence of halite pseudomorphs and stromatolite-bearing carbonate deposits (Clemmensen, 1978) resembles parts of Unit Tr4. However, the sedimentary environment during deposition of Unit Tr4 was mainly marine when compared to the fully continental sedimentary infill in the same period in East Greenland.

The thicknesses of the Kap Seaforth and the Edderfugledal Members are between 50 to 150 m and 30 to 70 m, respectively (Clemmensen et al., 1998), which is considerably lower than the estimated thickness of the corresponding interval (1350 m) on the Mid-Norwegian shelf (Table 1).

Norian–Rhaetian

The lower part of Unit Tr5 is correlated with the lacustrine and mud-rich Malmros Klint Member of the Fleming Fjord Formation in East Greenland (Fig. 2). The Malmros Klint Member is composed of cyclically bedded, intra-formational conglomerates, red siltstones and fine-grained sandstones and disrupted dolomitic sediments (palaeosols) (Clemmensen et al., 1998). The thick and laterally extensive lacustrine deposits, both on the Mid-Norwegian shelf and in East Greenland, indicate that an extensive lacustrine flood-basin covered most of the northeastern part of the Proto-Atlantic region.

The upper part of Unit Tr5, which is typified by an upward-increasing dominance of fluvial sediments, is correlated with the fluvial conglomerate units of the Bjergkronerne Beds, Ørsted Dal Member. However, a conglomerate facies is not recorded in Unit Tr5 in the present data set. Mud-rich intervals in the upper part of Unit Tr5 also indicate similarities with the lacustrine mudstones of Carlsberg Fjord Beds and the lacustrine carbonates and siliciclastic mudstones of the Tait Bjerg Beds (Clemmensen et al., 1998). Introduction of fluvial conglomerates in the upper part of the Fleming Fjord Formation (Clemmensen et al., 1998) might indicate a similar overall prograding trend as identified in Unit Tr5.

The thicknesses of the Malmros Klint and the Ørsted Dal Members vary between 50 to 150 m and 125 to 200 m, respectively (Clemmensen et al., 1998), which is less than the corresponding interval on the Mid-Norwegian shelf (ca. 800 m).

Discussion

Basin infill dynamics related to intra-basinal and extra-basinal tectonic development, climate and eustasy are discussed on a general basis in several studies (e.g. Prosser, 1993; Ravnås and Steel, 1998; Gawthorpe and Leeder, 2000). The temporal and the spatial influence of these factors on the sedimentary infill and palaeogeography for the Permo-Triassic succession in the northeastern part of the Proto-Atlantic region is discussed in this chapter. The Permo-Triassic is subdivided into five phases, an intitial rift, the syn-rift and the post-rift phases 1–3, each representing characteristic features concerning the basin infill dynamics (Fig. 8). All the post-rift phases are related to the same syn-rift event.

Initial rift phase (Late Permian)

A dominant marine depositional environment which characterizes the Late Permian in the northern Proto-Atlantic region is related to a general rise in the eustatic sea level (Figs. 8, 9) (Doré, 1992). The Upper Permian succession was deposited on a pronounced unconformable contact and marks the transition from a period of crustal extension in the Late Carboniferous/Early Permian to a period of subsidence governed mainly by thermal relaxation of the stretched and thinned crust in the Late Permian (Surlyk, 1990).

Some tectonic influence of the basin infill of the Upper Permian succession is inferred from East Greenland. The Wegener Halvø Formation and the associated Ravnefjeld Formation experienced some differential subsidence and fault activity, where the location of the reef build-ups was also influenced by the existing relief of the 'karstified' Karstryggen Formation (Seidler, 2000a; Stemmerik, 2001). In the latest Late Permian, a change in basin infill type, marked by the deposition of the Schuchert Dal Formation, was probably influenced by some block rotation and faulting, which is recorded on the Jameson Land (Kreiner-Møller and Stemmerik, 2001).

Indication of Late Permian tectonic activity is uncertain and difficult to asses on seismic sections from the Mid-Norwegian shelf. However, the interpreted intra-Upper Permian reflector (Top Anhydrite Unit) in the Helgeland Basin indicates

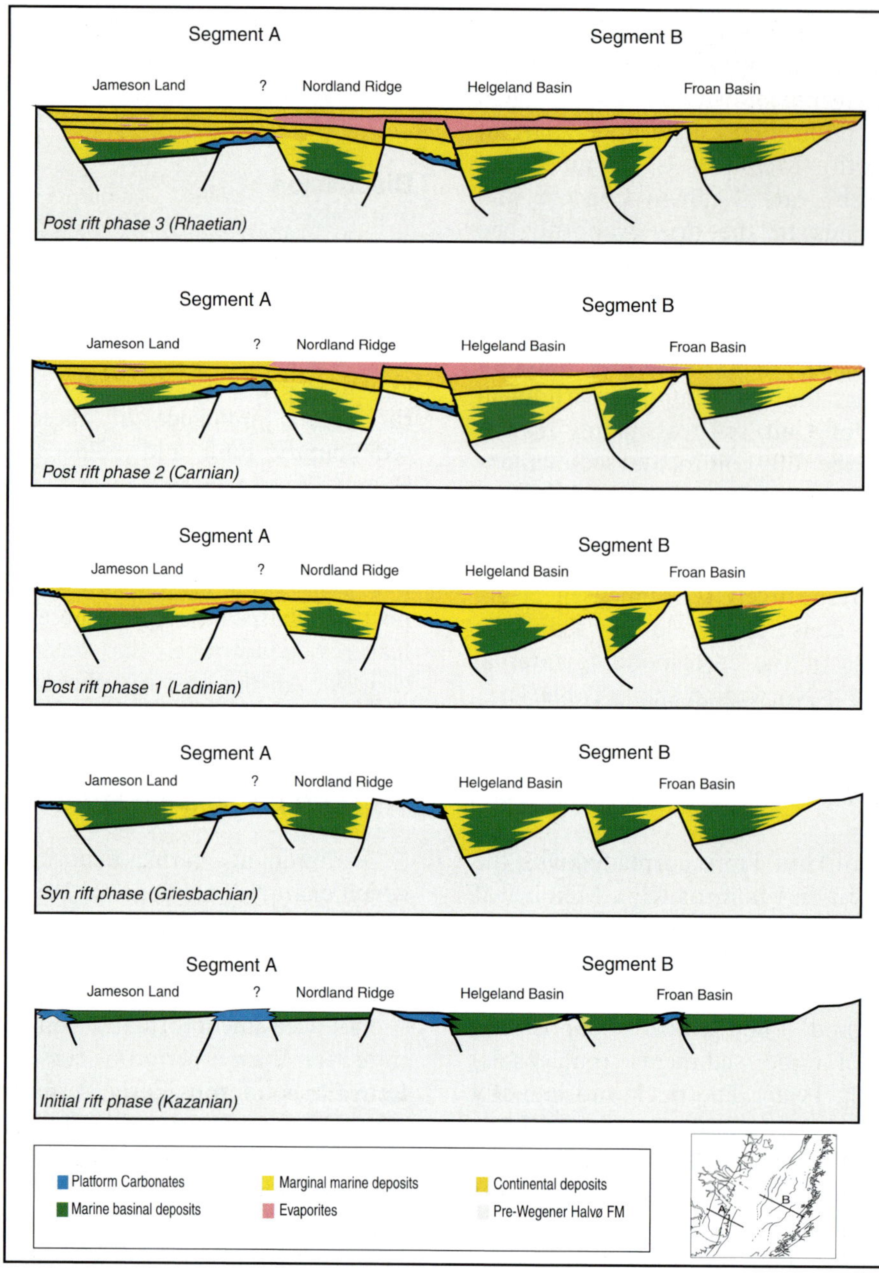

Fig. 8 Schematic and generalised figure showing the basin infill history of the Permo-Triassic of the northeastern part of the Proto-Atlantic region.

that the onset of the tectonic activity was in the Late Permian. The evidence of minor fault activity and block tilting is therefore interpreted to represent a period of initial rifting during the Late Permian on the Mid-Norwegian shelf — East Greenland region.

Syn-rift phase (Early Triassic)

The tectonic influence on the Triassic succession in the proto-Atlantic region has been discussed in several papers (Clemmensen, 1980b; Surlyk et al., 1981; Ziegler, 1982; Doré, 1992; Surlyk, 1990; Seidler, 2000a; Müller and Nystuen, 2002; Wignall and Twitchett, 2002). However, an Early Triassic tectonic phase is underestimated and even neglected in the literature from both the Mid-Norwegian shelf and East Greenland (Scholle et al., 1993; Blystad et al., 1995; Brekke et al., 1999). Several features indicate that rifting affected the basin infill in the Early Triassic. In the Helgeland and the Froan Basins, large wedge-shaped deposits of mainly Lower Triassic sediments suggest considerable syn-sedimentary faulting. This is also

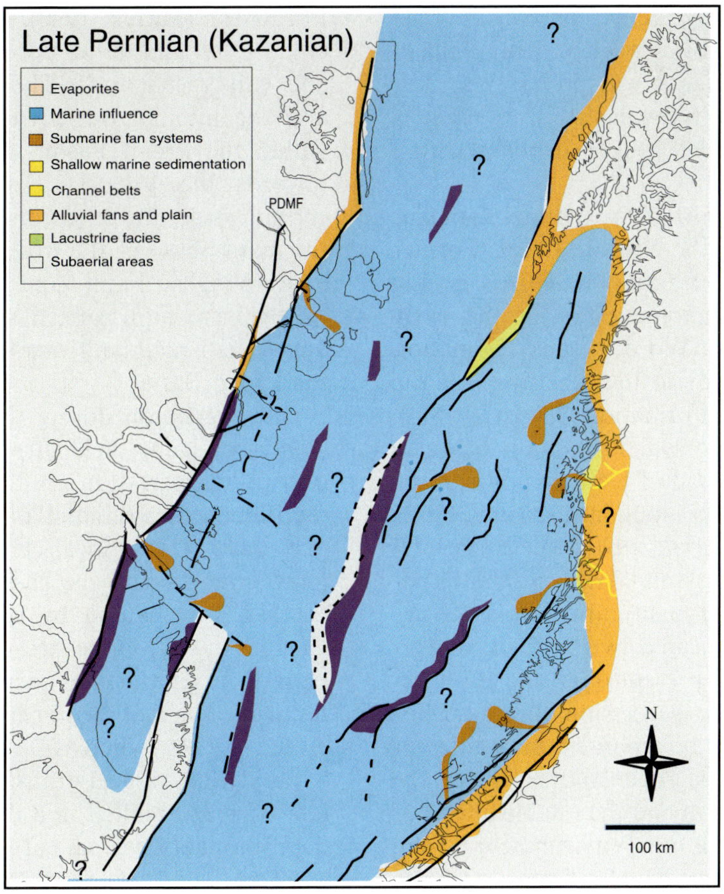

Fig. 9 Palaeogeographic map showing the location of Upper Permian platform carbonates and reefs. Submarine fan systems were located in the basinal areas.

recorded on East Greenland, where syn-sedimentary growth towards faults, and considerable and rapid thickness variations are observed in the Wordie Creek Formation of the Lower Triassic (Seidler, 2000a; Oftedal, 2002; Wignall and Twitchett, 2002). These structural and stratigraphic characteristics suggest that the Early Triassic was a period of profound extension in the northeastern part of the proto-Atlantic region.

Peak extension is mainly assigned to the Early Triassic, and not to the Late Permian. A thick Lower Triassic succession implies high rates of accommodation, probably related to rapid rates of subsidence, compared to the Upper Permian succession, both on the Mid-Norwegian shelf and on East Greenland. Thus, infilling of remnant topography created during active rifting commonly continues into the post-rift stage of thermal subsidence. The overall wedge-shaped geometry may not exclusively be, therefore, related to a period of active stretching, as suggested in most rift-basin models (cf. Prosser, 1993; Gawthorpe and Leeder, 2000). However, the Wordie Creek Formation, at Hold with Hope, was already filled to sea level already, in the lowermost part of the 750 m-thick succession (Oftedal, 2002). This suggests that most of the accommodation was created during deposition of the Wordie Creek Formation. Rifting of a similar age is also interpreted in adjacent areas, such as in the West Shetland area (Swiecicki et al., 1995), the northern North Sea (Steel, 1993) and the Barents Sea (Wood et al., 1989).

The formation of a distinct unconformity and hiatus between the Upper Permian and the Lower Triassic in parts of the basins can be related to uplift and erosion, due to continued and more intense block rotation in the Early Triassic. It might also be explained by an important fall in relative sea level which is recorded between the Permian and the Triassic in the Barents Sea region and adjacent areas (Mørk et al., 1989). Eustatic sea-level fluctuations are normally difficult to identify in structurally active basins (Prosser, 1993). The marked fall in relative sea level in the Early Triassic was followed by a rapid rise in the relative sea level in the Borealic Seaway (Mørk et al., 1989). This regional increase in the rise of the relative sea level, probably resulted in a mainly marine basin infill in the tectonically active basins in East Greenland and the Mid-Norwegian shelf. Palaeoclimate in the Early Triassic was probably arid to semi-arid in the Mid-Norwegian shelf area suggesting the presence of eroded sabkha deposits within lower part of Unit Tr1.

Seidler (2000b) reported a marked angular unconformity between the Wordie Creek Formation and the Pingo Dal Formation. This is mainly related to a second tectonic event in the early Scythian with erosion of Wordie Creek Formation and block rotation, prior to the deposition of the late Scythian Pingo Dal Formation (Seidler, 2000b). Deposition of the Pingo Dal Formation has also been attributed to a period of tectonic uplift in the hinterland and to activity along intra-basinal faults (Clemmensen, 1980b; Surlyk, 1990). The basin infill of the corresponding Unit Tr2 on the Mid-Norwegian shelf was probably also influenced by tectonic activity, as indicated by the thick wedge-shaped deposits. In the present data set, the basin infill of Lower Triassic succession on the Mid-Norwegian shelf cannot be separated into different tectonic events, as on East Greenland. However, the shallowing up of the basin and the establishment of a marginal marine depositional environment during deposition of Unit Tr2 probably reflects a decrease in rate of accommodation. This decrease may have been caused by reduced rate of subsidence due to ceasing tectonic activity. The period of deposition of Unit Tr2 may therefore represent a transitional phase from the late syn-rift to an early post-rift.

Post-rift phase 1 (early Anisian–Ladinian)

The palaeogeography and basin infill were characterised by change from a dominating marginal marine to a continental fluvial facies on the Mid-Norwegian shelf in the early Anisian (Figs. 8, 10). The establishment of a continental depositional environment is caused by a decrease in the rate of accommodation relative to rate of sediment input, brought about by cessation in fault activity and reduced tectonic subsidence. The cessation in tectonic activity is reflected by relative uniform lateral thickness of Unit Tr3, and that Unit Tr3 seems to drape the half-grabens with mainly Upper Permian and Lower Triassic sediment infill. In addition, accumulation rates are lower for the continental Unit Tr3, when compared with the over- and underlying units. The Middle Triassic represents a period of tectonic quiescence also in adjacent areas, such as East Greenland, during deposition of the corresponding Gipsdalen Formation (Surlyk, 1990), the North Sea (Nystuen et al., 1989) and in the Borealic domain (Jackobsen and van Veen, 1984). The wedge-shaped deposits and the internal onlap surfaces within Unit Tr3 within the Froan Basin, and sedimentary growth towards the Vingleia Fault Zone suggest that certain structural elements remained active after the main phase of the rifting ended. A tectonically active Vingleia Fault Zone has also an implication for the basin infill pattern, since sediments from the Norwegian mainland might have been trapped in the Froan Basin.

The regression during the early Anisian on the Mid-Norwegian shelf corresponds to a period of overall transgression, both in the Tethyan and the Borealic Sea (Gianolla and Jacquin, 1998). This may indicate that local changes in tectonism were more important to the basin infill than changes in the sea level caused by super-regional or global changes in intra-plate stress and/or eustasy. In this period, a continental depositional environment totally dominated the northern North Sea (Nystuen et al., 1989), whereas in the Borealic Seaway, a rapid and radical transgression starting in the late Anisian resulted in a dominantly open marine depositional environment in the Ladinian (Van Veen et al., 1992). The Mid-Norwegian shelf — East Greenland region is situated in an intermediate position between these two areas with an alluvial floodplain that was transgressed periodically. A regional peak transgression in the Ladinian is recorded, both in the Tethyan and the Boreale Sea domains (Gianolla and Jacquin, 1998). This event may correspond to the occurrence of short-lived marine incursions reflected by the Gråklint Beds in East Greenland and to the marginal marine deposits in uppermost part of Unit Tr3 on the Mid-Norwegian shelf. The Palaeoclimate was probably arid to semi-arid, reflected by the occurrence of red-coloured palaeosols with carbonate nodules, aeolian deposits and evaporites.

Post rift phase 2 (late Ladinian/early Carnian to late Carnian)

Establishment of a dominantly evaporitic depositional environment shows that a major shift in the basin infill history and palaeogeography occurred in the late Ladinian–early Carnian (Fig. 10). The base of Unit Tr4 represents the onset of a period of renewed tectonic activity. Wedge-shaped deposits and considerable lateral thickness variation indicate that faulting and fault block rotation influenced the basin infill.

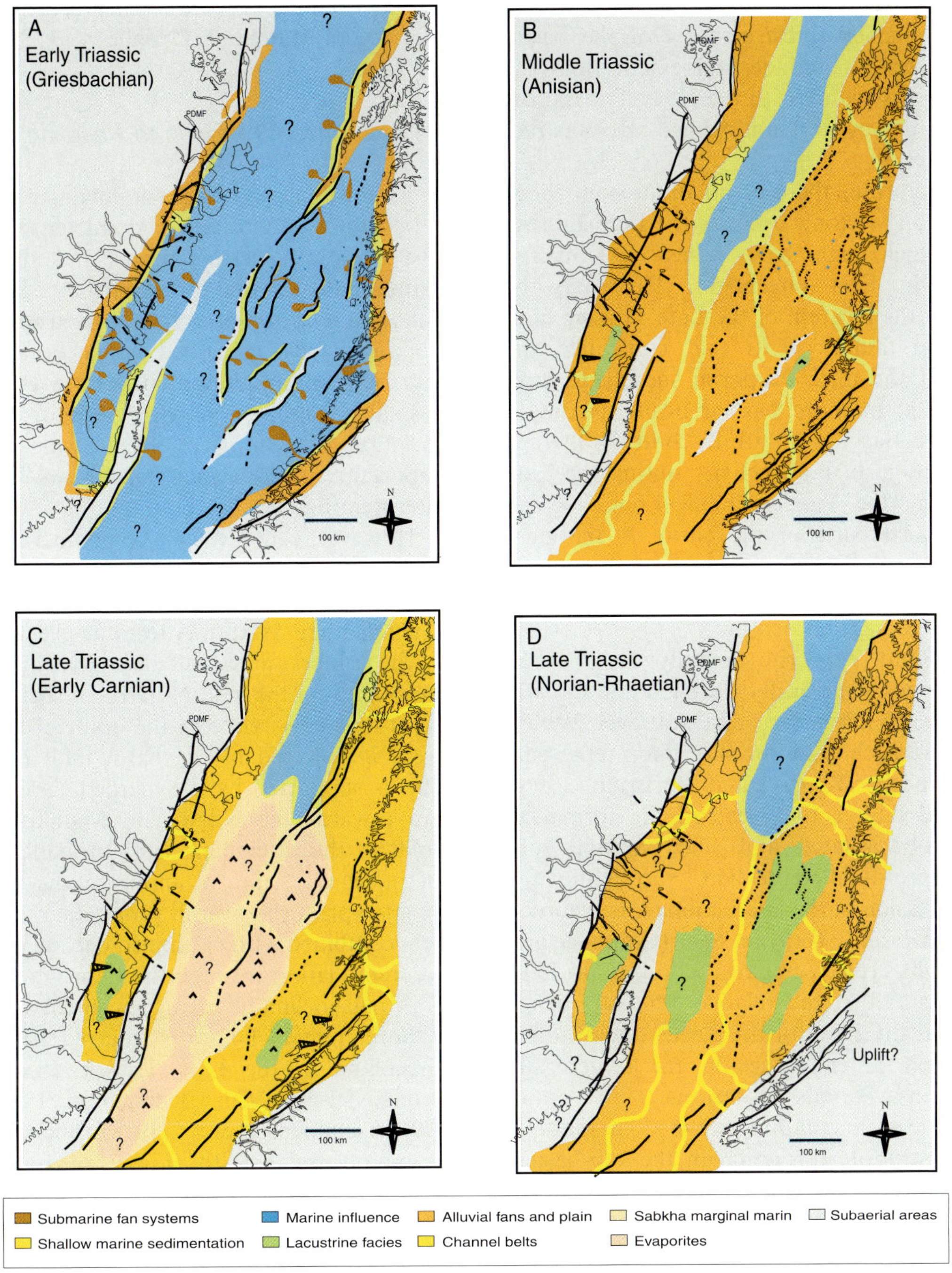

Fig. 10 (A) Palaeogeographic map showing the fault block rotated syn-rift system of the East Greenland-Mid-Norwegian region in the Early Triassic. The depositional environment was typified by mainly submarine fan systems and deep marine sedimentation. (B) The tectonic activity ceased in the Middle Triassic and a continental depositional system prograded into the basin. (C) A phase of restructuring and renewed tectonic activity in the early Carnian resulted in thick evaporite deposits formed within structurally silled and isolated basins. These basins were mainly located in the central parts of the rift-system, while the marginal parts were dominated by playas and aeolian depositions. (D) The latest Triassic (Norian-Rhatian) is characterised by the establishment of a fluvio-lacustrine depositional environment and a period of tectonic quiesence within the basin. The Norwegian Mainland probably experienced tectonic uplift during the basin infill of these sediments.

Evidence of Late Triassic fault activity is also recognised in adjacent areas, such as the Atlantic margin of the British Isles (Morton, 1993), Northern North Sea (Steel and Ryseth, 1990) and Northern Ireland (Ruffel and Shelton, 1999). In the eastern part of North America, the Late Triassic represents an important rifting event with the formation of the Newark Basin (Smoot, 1991). Veevers (1989) discussed, on a global scale, that the Middle/Late Triassic boundary marks the incipient dispersal of Pangea by the onset of continental rifting. This Late Triassic fault activity on the Mid-Norwegian shelf is probably related to a restructuring of the basin, but it is of less magnitude and extent than the fault activity associated with the Early Triassic rift event. Thus, this event is not defined as a separate syn-rift phase.

In the literature, there is no discussion about a corresponding tectonic event in the Late Triassic (Carnian) which might have influenced Unit Tr4 equivalents in East Greenland. This may be explained by the location of the Late Triassic fault systems. The fault systems that controlled the formation of the thick evaporite succession were mainly located in more basinal positions out in the Helgeland Basin (Fig. 10). In contrast, there is no observed sign of fault activity during the Late Triassic along the Vingleia Fault Zone, located in the marginal parts of the basin. A marginal position of the Triassic basins in East Greenland may therefore explain the limited Late Triassic fault activity along that margin.

Basin central evaporites mainly form when basins become structurally isolated or greatly restricted due to lowered eustatic sea level. Within marine basins, such evaporite succession represents lowstand deposits (Kendall and Harwood, 1996). The renewed fault activity caused the formation of structurally isolated and silled marine sub-basins. A profound regression, succeeding the peak transgression in the late Ladinian, is recorded in the Boreal and Tethyan Sea (Gianolla and Jacquin, 1998). This regressive period could have resulted in a fall in the relative sea level below the structural threshold of the basins, thus forcing the formation of evaporites in the structurally isolated basins. Periods of tectonic activity and eustatic sea-level fluctuations resulted in multiple periods of evaporite formation in the sub-basins. Thickness variations and lateral differences in the occurrence of salt intervals in Unit Tr4 are probably a result of different basin configuration and water exchange with the marine seaway to the north. The formation of these evaporites also indicates that the palaeoclimate, semi-arid to arid, prevailed in the northern part of the proto-Atlantic region.

The truncation of Unit Tr3 in the marginal areas towards the Norwegian Mainland probably suggests that a regional uplift of the Norwegian Mainland occurred, prior or time-equivalent, to the deposition of Unit Tr4. A general decrease in the thickness of Unit Tr4 towards the Norwegian Mainland also points to an uplift of the marginal areas. This may correspond to a Late Triassic–Middle Jurassic period of uplift and cooling recorded in Trondheimsfjord, Norwegian Mainland (Grønlie et al., 1994).

Post-rift phase 3 (Norian to Rhaetian)

Establishment of a dominantly continental depositional environment characterised by extensive shallow lacustrine basins and alluvial floodplains indicates that an important shift in basin infill and palaeogeography occurred in the Norian. This is partly explained by a decrease in the rate of accommodation related to ceasing intra-basinal tectonic activity. Limited thickness variation of Unit Tr5 basin wide and across important intra-basinal lineaments favour this interpretation.

However, the thick conglomerates in the Ørsted Dal Member in East Greenland are attributed to renewed tectonic activity in the hinterland (Clemmensen, 1980b). A tectonic event (e.g. uplift) in the hinterland might also have influenced the basin infill on the Mid-Norwegian shelf. A tectonic event that may relate to the uplift of the marginal areas is reflected in the basin infill of Unit Tr4. This is supported by the marked thinning of Unit Tr5 towards the Norwegian Mainland. Such an uplift of the catchment areas probably resulted in an increased sediment input to the basins. This might explain the overall prograding trend of Unit Tr5, combined with continued decline in the rate of accommodation through the Late Triassic.

A palaeoclimatic change from the semi-arid to a more humid climate is inferred from the East Greenland in the Late Triassic (Birkelund and Perch Nielsen, 1976; Clemmensen, 1980b; Surlyk, 1990; Clemmensen et al., 1998). This climatic shift is mainly related to a northward continental plate drift (Clemmensen et al., 1998). The lake deposits in the Fleming Fjord Formation also contain evidence of seasonal, orbital and long-term climatic changes, which influenced the sedimentary infill (Clemmensen et al., 1998). On the Mid-Norwegian shelf, there is evidence for a general increase in humidity of climate: (1) Colour change from reddish-brown to a more greenish-grey floodplain sediments, (2) Occurrence of wood debris and small coal beds in the upper part of Unit Tr5 and in the succeeding Åre Formation, (3) Disappearance of evaporite deposits in Unit Tr5.

The Middle and Late Triassic post-rift succession on the Mid-Norwegian shelf is considerably thicker than the equivalent succession in East Greenland. In addition, conglomerates are absent in the studied data set from the Mid-Norwegian shelf. These features probably reflect a more central position of the Mid-Norwegian shelf, relative to the regional basin axis compared to East Greenland. The East Greenland basins probably occupied a relatively marginal position in the rift-system. The basin infill during the Triassic in East Greenland was therefore more exposed to erosion, reworking and bypass, caused by lower rates of accommodation than the central basins. This difference in structural setting within the regional rift system might also explain the more profound marine influence on the depositional environment on the Mid-Norwegian shelf during the post-rift phases as compared to the dominating continental deposits of East Greenland basins.

Conclusions

- The Permo-Triassic basin infill history and structural development of the Mid-Norwegian shelf–East Greenland region can be subdivided into five different phases.
- The Upper Permian succession was deposited during the *initial phase of rifting*. This period was characterised by dominant marine basin infill, influenced by minor movement along intra-basinal faults.
- The main rifting event is assigned to the Early Triassic, representing the *syn-rift phase*. Major block rotation and activity along structural lineaments took place during basin infill of mainly submarine fan sediments. This event represents an important rifting episode during the pre-break up of the proto-Atlantic region.
- *Post-rift phase 1* was initiated in the Middle Triassic and represents a period of progradation of continental fluvial systems out in the basins. Tectonic activity along marginal faults, such as the Vingeleia Fault Zone, shows that certain structural elements can be tectonically active during the 'post-rift phase'.
- The sediment infill in *post-rift phase 2* in the late Ladinian/early Carnian was influenced by renewed tectonic activity. Thick evaporites were deposited within the structurally isolated and silled basins. The magnitude and extent of this tectonic event are of less importance, compared to the Early Triassic rift episode, and are mainly related to restructuring during the post-rift subsidence.
- *Post-rift phase 3* represents the termination of the Late Triassic tectonic event and the establishment of an extensive fluvio-lacustrine flood-basin. Progradation of the fluvial system out into the basin in the end of this phase is related to an uplift of the Norwegian Mainland and a cessation in tectonic activity within the basin. This period shows a gradual increase in the humidity of the palaeoclimate.
- In the Permo-Triassic the Mid-Norwegian shelf was located in a more basinal position within the regional rift system of the proto-Atlantic region compared to East-Greenland. Regional correlations of the sediment infill between the two regions must therefore consider the marginal depositional environment in East Greenland, more exposed to erosion, bypass and reworking than the Mid-Norwegian shelf.

Acknowledgments

We thank Norsk Hydro ASA and former Saga Petroleum ASA for financing this project. We are also indebted to Tom Bugge, Odd-Ragnar Heum, Snorre Olaussen, Bjørn Terje Oftedal, Lars Seidler, Ingar Throndsen and Terje Veum for their stimulating discussions, contribution and support. Reviewers, Robert Hunsdale and Jon Vold, are also thanked for their comments. Constructive comments of an earlier version of this manuscript were helpfully done by Bjørn Tore Larsen.

References

Birkelund, T. and Perch Nielsen, K., 1976. Late Palaeozoic–Mesozoic evolution of central East Greenland. In: A. Escher and W.S. Watt (Editors), Geology of Greenland, Geol. Surv. of Greenland, Copenhagen, pp. 304–339.

Birkenmajer, K., 1977. Erosional unconformity at the base of marine Lower Triassic at Wegener Ø, central East Greenland. Grønl. Geol. Under., Rapp., 85: 103–107.

Blystad, P., Brekke, H., Færseth, R., Larsen, B.T., Skogseid, J. and Tørudbakken, B., 1995. Structural elements of the Norwegian continental shelf, Part II: The Norwegian Sea Region: NPD, Bull., 8, 45 pp.

Brekke, H. and Riis, F., 1987. Tectonics and basin evolution of the Norwegian shelf between 62°N and 72°N. Norsk Geol. Tidsskr., 6: 295–322

Brekke, H., Dahlgren, S., Nyland, B. and Magnus, C., 1999. The prospectivity of the Vøring and Møre Basins on the Norwegian Sea continental margin. In: A.J. Fleet and S.A.R. Boldy (Editors), Petroleum Geology of Northwest Europe: Proceedings of the 5th Conference. Geological Society, London, pp. 261–274.

Brekke, H., Sjulstad, H.I., Magnus, C. and Williams, R.W., 2001. Sedimentary environments offshore Norway — an overview. In: O. Martinsen and Dreyer (Editors), Norwegian Petroleum Society (NPF) Special Publication 10, pp. 7–37.

Bugge, T., Ringås, J.E., Leith, D.A., Mangerud, G., Weiss, H.M. and Leith, T.L., 2002. Upper Permian as a new play model on the mid-Norwegian continental shelf; Investigated by shallow stratigraphic drilling. AAPG Bull., 86: 107–127.

Bukovics, C., Cartier, E.G., Shaw, N.D. and Ziegler, P.A., 1984. Structure and development of the mid-Norway Continental Margin. In: A.M. Spencer et al. (Editors), Petroleum geology of the north European margin, Norwegian Petroleum Society (NPF). Graham and Trotman, London, pp. 407–423.

Clemmensen, L.B., 1978. Alternating aeolian, sabkha and shallow-lake deposits from the middle Triassic Gipsdalen Formation, Scoresby Land, East Greenland. Paleogeogr., Paleoclimat., Paleoecol., 24: 111–135.

Clemmensen, L.B., 1980a. Triassic lithostratigraphhy of East Greenland between Scoresby Sund and Kejser Franz Josephs Fjord. Grønl. Geol. Unders., Bull., 139: 1–56.

Clemmensen, L.B., 1980b. Triassic rift sedimentation and palaeogeography of central East Greenland. Grønl. Geol. Unders., Bull., 136: 1–72.

Clemmensen, L., Kent, D.V. and Jenkins, F.A., Jr., 1998. A late Triassic lake system in East-Greenland: facies, depositional cycles and palaeoclimate. Paleogeogr., Paleoclimat., Paleoecol., 140: 135–159.

Doré, A.G., 1992. Synoptic palaeogeography of the Northeast Atlantic Seaway: late Permian to Cretaceous. In: J. Parnell (Editor), Basins on the Atlantic Seaboard: Petroleum Geology, Sedimentology and Basin Evolution. Geol. Soc., London, Spec. Publ., 62: 421–446.

Frostick, L.E and Reid, I., 1992. Tectonic and climate control of Triassic sedimentation in the Beryl Basin, northern North Sea. J. Geol. Soc., London, 149: 13–26.

Gawthorpe, R.L. and Leeder, M.R., 2000. Tectono-sedimentary evolution of active extensional basins. Basin Res., 12: 195–218.

Grønlie, A., Naeser, C.W., Naeser, N.D., Mitchell, J.G., Sturt, B.A. and Ineson, P.R., 1994. Fission-track and K-Ar dating of tectonic activity in a transect across the Møre-Trøndelag Fault Zone, central Norway. Norsk Geol. Tidsskr., 74: 24–34.

Gianolla, P. and Jacquin, T., 1998. Triassic sequence stratigraphic framework of western European Basins. Mesozoic and Cenozoic Sequence Stratigraphy of European Basins, SEPM Spec. Publ., 60: 643–650.

Gradstein, F.M. and Ogg, J.G., 1996. A Phanerozoic Time Scale. Episodes, 19 (1–2): 3–5.

Grunnaleite, I. and Gabrielsen, R.H., 1995. Structure of the Møre Basin, mid-Norway continental margin. Tectonophysics, 252: 221–251.

Hagevang, T. and Rønnevik, H., 1986. Basin development and hydrocarbon occurrence offshore Mid-Norway. In: M.T. Halbouty (Editor), Future Petroleum Provinces of the World: AAPG Mem., 40: 599–613.

Jacobsen, V.W. and Van Veen, P., 1984. The Triassic offshore Norway north of 62°N. In: A. M. Spencer et al. (Editor), Petroleum geology of the north European margin, Norwegian Petroleum Society (NPF), Graham and Trotman, London, pp. 317–327.

Kendall, A.C. and Harwood, G.M., 1996. Marine evaporites; aridshorelines and basins. In: H.G. Reading (Editor), Sedimentary Environments; processes, facies and stratigraphy, pp. 281–324.

Kreiner Møller, M. and Stemmerik, L. 2001. Upper Permian lowstand fans of the Bredehorn Member, Schuchert Dal Formation, East Greenland. . In: O.J. Martinsen and T. Dreyer (Editors), Sedimentary environments Offshore Norway Palaeozoic to Recent. Norwegian Petroleum Society (NPF) Special Publication 10, Elsevier, Amsterdam, pp. 51–65.

Morton, N., 1993. Potential reservoir and source rocks in relation to Upper Triassic to Middle Jurassic sequence stratigraphy, Atlantic margin basins of the British Isles. In: J.R. Parker (Editor), Petroleum Geology of Northwest Europe: Proceedings of the 4th Conference. Geological Society, London; pp. 285–297.

Müller, R. and Nystuen, J.P., 2002. Late Permian to Triassic basin infill history and paleogeography of the Mid-Norwegian shelf-East Greenland region. In: A. Hurst (Editor), Onshore-Offshore Relationships on the Nordic Atlantic Margin. NGF Abstracts and proceedings 2, 2002 of the Norwegian Petroleum Society (NPF) and Norwegian Geological Society (NGF) Conference, October 7–9 Trondheim, pp. 140–142.

Mørk, A., Embry, A.F. and Wolfgang, W., 1989. Triassic transgressive-regressive cycles in the Sverdrup Basin, Svalbard and the Barents shelf. In: J.D. Collinson (Editor), Correlation in Hydrocarbon Exploration, Norwegian Petroleum Society (NPF). Graham and Trotman, London, pp. 113–130.

Nystuen, J.P., Knarud, R. and Jorde, K., 1989. Correlation of Triassic and Jurassic sequences, Snorre Field and adjacent areas, northern North Sea. In: J.D. Collinson (Editor), Correlation in Hydrocarbon Exploration, Norwegian Petroleum Society (NPF), Graham and Trotman, London, pp. 273–289.

Oftedal, B.T., 2002. Tidlig triassisk tektono-sedimentær utvikling på Kap Stosch, Hold With Hope, Øst-Grønland. M.Sc. Degree. University of Oslo (In Norwegian). 144 pp.

Perch-Nielsen, K., Birkenmajer, K., Birkelund, T. and Aellen, M., 1974. Revision of Triassic stratigraphy of the Scoresby Land and Jameson Land region: East Greenland. Grønl. Geol. Unders., Bull., 109, 51 pp.

Piasecki, S., 1984. Preliminary palynostratigraphy of the Permian-lower Triassic sediments in Jameson Land and Scoresby Land, East Greenland. Bull. Geol. Soc. Den., 32: 139–144.

Prosser, S., 1993. Rift-related linked depositional systems and their seismic expression. In: G.D. William and A. Dobb (Editors), Tectonics and Seismic Sequence Stratigraphy, Geol. Soc., London, Spec. Publ., 71: 35–66.

Ravnås, R. and Steel, R., 1998. Architecture of marine rift successions. AAPG Bull., 82 (1): 110–146.

Roberts, D.G., Thompson, M., Mitchener, J., Hossack, J., Carmichael, S. and Bjørnseth H.-M., 1999. Palaeozoic to Tertiary rift and basin dynamics: mid-Norway to the Bay of Biscay- a new context for hydrocarbon prospectivity in the deep water frontier. In: A.J. Fleet and S.A.R. Boldy (Editors), Petroleum Geology of Northwest Europe: Proceedings of the 5th conference, Geological Society, London, pp. 7–40.

Ruffel, A. and Shelton, R., 1999. The control of sedimentary facies by climate during phases of crustal extension: examples from the Triassic of onshore and offshore England and Northern Ireland. J. Geol. Soc., London, 156: 779–789.

Scholle, P.A., Stemmerik, L., Ulmer-Scholle, D., Di Liegro, G. and Henk, F.H., 1993. Paleokarst-influenced depositional and diagenetic patterns in Upper Permian carbonates and evaporites, Karstryggen area, central East Greenland. Sedimentology, 40: 895–918.

Seidler, L., 2000a. Incised submarine canyons governing new evidence of Early Triassic rifting in East Greenland. Palaeogeogr., Paleoclimat., Palaeoecol., 161: 267–293.

Seidler, L., 2000b. Sedimentology and sequence stratigraphy of the Lower Triassic Wordie Creek Formation in northern Jameson Land, Scoresby Land and on Traill ø, East Greenland, Ph.D. Thesis, pp. 1–122.

Skogseid, J., Planke, S., Faleide, J.I., Pedersen, T., Eldholm, O. and Neverdal, F., 2000. NE Atlantic continental rifting and volcanic margin formation. In: A. Nøttvedt et al. (Editors), Dynamics of the Norwegian Margin. Geol. Soc., London, Spec. Publ., 167. 295–326.

Smoot, J.P., 1991. Sedimentary facies and depositional environment of early Mesozoic Newark Supergroup basins, eastern North America. Palaeogeogr., Paleoclimat., Palaeoecol., 84: 369–423.

Steel, R., 1974. New Red Sandstone floodplain and piedmont sedimentation in the Hebridean province, Scotland, Scotland. J. Sediment. Petrol., 44: 328–357.

Steel, R., 1993, Triassic-Jurassic megasequence stratigraphy in the Northern North Sea: rift to post-rift evolution. In: J.R. Parker (Editor), Petroleum Geology of NW Europe: Proceedings of the 4th Conference, Geological Society, London, pp. 299–315.

Steel, R. and Ryseth, A., 1990. The Triassic-early Jurassic succession in the Northern North Sea: megasequence stratigraphy and intra-Triassic tectonics. In: R.F.P. Hardmann and J. Brooks (Editors), Tectonic events responsible for Britain's oil and gas reserves. Proceedings Conference May 21–23 Bath, England, Geol. Soc., London, Spec. Publ., 55: 39–168.

Stemmerik, L., 1987. Cyclic carbonate and sulphate from the Upper Permian Karstryggen Formation, East Greenland. In: T.M. Peryt (Editor), Lecture Notes in Earth Sciences 10: Berlin Heidelberg, Springer Verlag, pp. 5–22.

Stemmerik, L., 2000. Late Paleozoic evolution of the North Atlantic margin of Pangea. Palaeogeogr., Paleoclimat., Palaeoecol., 161: 95–126.

Stemmerik, L., 2001. Sequence stratigraphy of a low productivity carbonate platform succession: the Upper Permian Wegener Halvø Formation, Karstryggen Area, East Greenland. Sedimentology, 48: 79–97.

Stemmerik, L., Christiansen, F.G., Piasecki, S., Jordt, B., Marcussen, C. and Nøhr-Hansen, N., 1993. Depositional history and petroleum geology of the Carboniferous to Cretaceous sediments in the northern part of East Greenland. In: T.O. Vorren et al. (Editors), Arctic Geology and Petroleum Potential, Norwegian Petroleum Society (NPF) Special Publication 2, Elsevier, Amsterdampp. 67–87.

Stemmerik, L., Scholle, P.A., Henk, F.H., Liegro, G.D., Mantovani, M. and Ulmer, D.S., 1990. Facies mapping and reservoir evolution of the Wegener Halvø Formation along the western margin of Jameson Land, East Greenland. Grønl. Geol. Unders., Rapp, 148: 105–108.

Stemmerik, L., Clausen, O.R., Korstgård, J., Larsen, M., Piasecki, S., Seidler, L., Surlyk, F. and Therkelsen, J., 1997. Petroleum geological investigations in East Greenland: project 'Resources of the sedimentary basins of North and East Greenland': Greenl. Geol. Surv., Bull., 176: 29–38.

Surlyk, F., 1990. Timing, style and sedimentary evolution of Late Palaeozoic-Mesozoic extensional basins of East Greenland. In: R.F.P. Hardman, J. Brooks (Editors), Tectonic events responsible for Britain's Oil and Gas Reserves. Geol. Soc., London, Spec. Publ., 55: 107–155.

Surlyk, F., Clemmensen, L.B. and Larsen, H.C., 1981. Post-Paleozoic evolution of the East Greenland continental margin. In: J.W. Kerr and A.J. Fergurson (Editors), Geology of the North Atlantic borderland. Can. Soc. Petrol. Geol. Mem., 7: 611–645.

Surlyk, F., Hurst, J.M., Piasecki, S., Rolle, F., Scholle, P.A., Stemmerik, L. and Thomsen, E., 1986. The Permian of the Western Margin of the Greenland Sea — A Future Exploration Target. In: M.T. Halbouty (Editor), Future Petroleum Provinces of the World: AAPG Mem., 40: 629–659.

Swiecicki, T., Wilcockson, P., Canham, A., Whelan, G. and Homann, H., 1995. Dating, correlation and stratigraphy of the Triassic sedimentation in the West Shetland area. Permian and Triassic rifting in Northwest Europe. Geol. Soc., London, Spec. Publ., 91: 57–85.

Van Veen, P.M., Skjold, L.J., Kristensen, S.E., Rasmussen, A., Gjelberg, J. and Stølan, T., 1992. Triassic sequence stratigraphy in the Barents Sea. In: T.O. Vorren et al. (Editors), Arctic Geology and Petroleum Potential, Norwegian Petroleum Society (NPF) Special Publication 2, Elsevier, Amsterdam, pp. 515–538.

Veevers, J.J., 1989. Middle/Late Triassic (230 ± 5 Ma) singularity in the stratigraphic and magmatic history of the Pangean heat anomaly. Geology, 17: 784–787.

Wignal, P.B. and Twitchett, R.J., 2002. Permian-Triassic sedimentology of Jameson Land, East Greenland: incised submarine channels in an anoxic basin. J. Geol. Soc., London, 159: 691–703.

Wood, R.J., Edrich, S.P. and Hutchinson, S.P., 1989. Influence of North Atlantic Tectonics on the Large-scale Uplift of the Stappen High and Loppa High, Western Barents Shelf. In: A.J. Tankard and H.R. Balkwill (Editors), Extensional Tectonics and Stratigraphy of the North Atlantic Margins. AAPG Mem., 46: 559–566.

Ziegler, P.A., 1982. Geological Atlas of Western and Central Europe, SEPM. Elsevier, Amsterdam.

Early Triassic syn-rift sedimentation at Hold with Hope, Northeast Greenland

Bjørn Terje Oftedal, Arild Andresen and Reidar Müller

The Early Triassic Wordie Creek Formation at Kap Stosch, Hold with Hope, represents sediments deposited during the rift climax stage in a Late Palaeozoic–Early Mesozoic rift episode in NE Greenland. The Wordie Creek Formation comprises over 750 m of clastic marine sediments deposited in less than four million years. The formation is dominated by mudstones. However, in the lower part, four coarse grained units have been identified from the Kap Stosch sections. These units are interpreted to represent southwards progradation of coarse clastic deltas filling the half-graben basin from the north. Foresets and cross-stratification measurement from the coarse clastic units demonstrate that the sediments were deposited axially into one or more N–S trending marine half-grabens. The intra-basinal faults controlling these half-grabens, on the basis of sedimentological data, are interpreted to have been active during deposition of the Wordie Creek Formation. The Kap Stosch area appears to be positioned on a large relay ramp linking two major N–S striking fault zones. Progradation of coarse clastic deltas into the half-grabens are suggested to be associated with contemporaneous movement along these two faults, resulting in rotation of the relay ramp. This led to increased uplift and erosion of the upper part of the ramp, and local subsidence and progradation of coarse clastic deltas in the lower parts of the ramp. The coarse clastic units show a back-stepping trend, related to a gradual increase in water depth in the half grabens. We link this to an overall transgression in the region controlled by rapid tectonic subsidence of the entire rift basin.

Introduction

Although rift sedimentation has been documented from East Greenland (Clemmensen, 1980; Surlyk, 1990), the Late Permian to Early Triassic time interval has often been regarded as a period of tectonic quiescence (Blystad et al., 1995; Brekke, 2000). However, lately, it has become more evident that the marine basin occupying the North Atlantic-Barents Sea Region in the Late Palaeozoic and Early Mesozoic was affected by significant tectonic activity (Seidler, 2000a, b; Tsikalas et al., 2001; Wignall and Twitchett, 2002; Müller et al., this volume). Detailed stratigraphic information which can document temporal and spatial changes in the basin infill dynamics in late Permian and early Triassic are relative sparse from the continental shelves offshore Norway and East Greenland. A different situation exists in Northeast Greenland where thick, well-exposed successions of Late Permian and Early Triassic marine sediments are relatively well-preserved in the coastal areas (Fig. 1). The proximity of these and other post-Devonian deposits in Northeast Greenland to the Mid-Norwegian shelf prior to ocean-floor spreading and opening of the Norwegian-Greenland Sea makes it a key area for deciphering the basin evolution of the continental shelf offshore Mid Norway. This contribution focuses on variations in the depositional environments of the Early Triassic Wordie Creek Formation at Hold with Hope (Figs. 1 and 3) and discusses the different factors influencing the Early Triassic basin infill.

Geological setting

The post-Caledonian/post-Devonian rocks in Northeast Greenland appear in series of down-to-the-east rotated fault blocks, with successively younger rocks being preserved and exposed towards the east (Fig. 1). Caledonian basement rocks are exposed in the uplifted footwall of some of these fault blocks implying significant block rotation. Much of the E–W extension recorded by

Onshore–Offshore Relationships on the North Atlantic Margin
edited by B. Wandas et al. NPF Special Publication 12, pp. 191–206,
Published by Elsevier B.V., Amsterdam

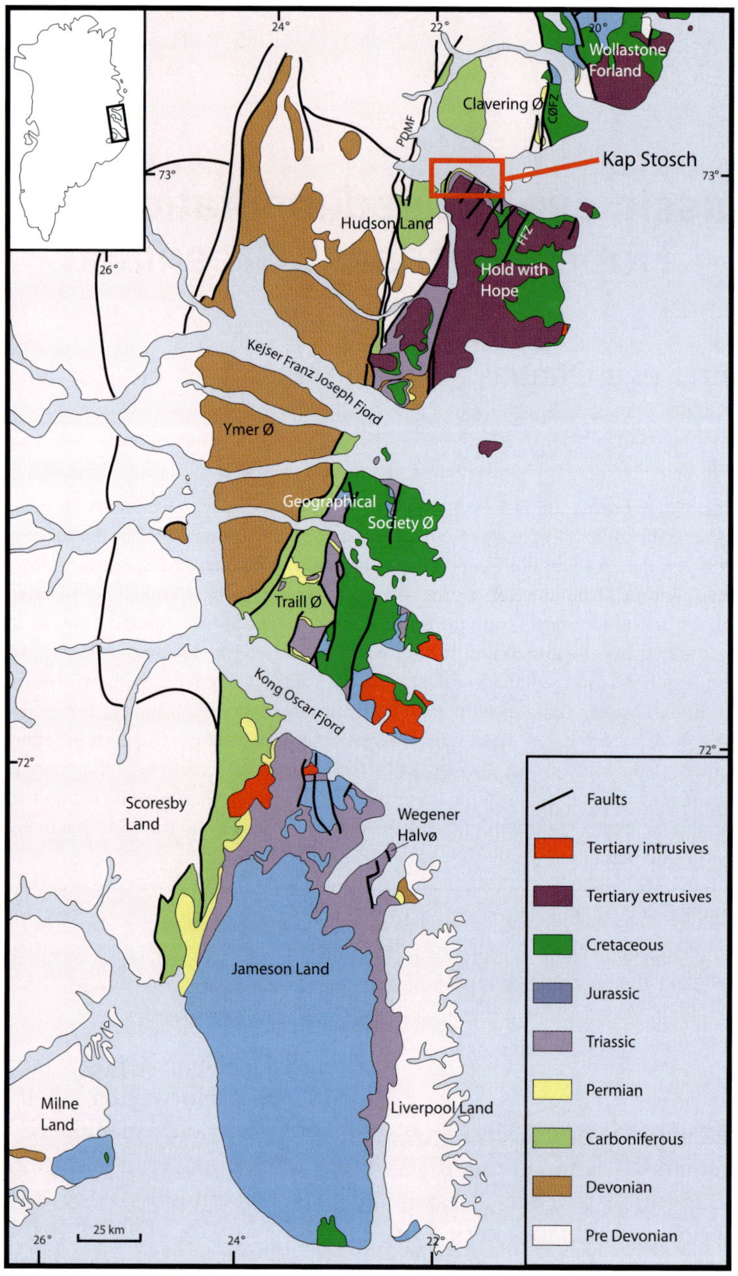

Fig. 1 Simplified geological map of Northeast Greenland (Modified from Stemmerik, 1997 and Hartz et al., 2002). The study area is marked with a red square.

these rotated fault blocks is Palaeogene in age, as plateau basalts (c. 55 Ma) capping many of the mountains are displaced by several down-to-the-east, N–S striking, extensional faults with displacement in the order of several hundred meters (Price et al., 1997). Dramatic facies variations and unconformities within some of the rotated fault blocks also document phases of earlier tectonic activity in the Mid-Jurassic–Early Cretaceous and Late Cretaceous–Palaeogene (Kelly et al., 1998; Hartz et al., 2002).

The coastal areas of Northeast Greenland are cut by several regional N–S trending fault zones. Many of these fault zones appear to have originated as Devonian strike-slip faults, indicated by their rectilinear geometry over tens to hundreds of kilometers (Andresen, 2002). These fault zones appear to be arranged in an en echelon right stepping pattern (Hartz et al., 2002). The oldest post-Caledonian deposits occur just east of one of these N–S trending fault zones, commonly referred to as the 'Post Devonian Main Fault'

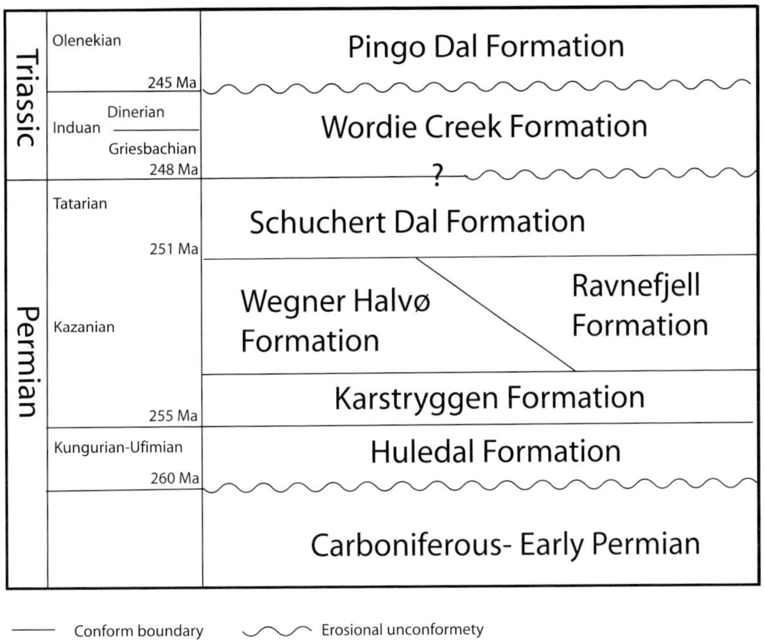

Fig. 2 Simplistic stratigraphic Table of the Upper Permian and Early Triassic formations in Northeast Greenland, based on Stemmerik, 2000 and Seidler, 2000b.

(PDMF) (Vischer, 1943) (Fig. 1). A thick succession (>1000 m) of westward-dipping Carboniferous and Early Permian continental deposits (Vigran et al., 1999; Stemmerik et al., 1993) was deposited in large half-grabens bounded by this fault zone. An angular unconformity between these continental deposits and the overlying Upper Permian conglomerates and marine deposits (Surlyk, 1990) is indicative of renewed tectonic activity in the Late Permian. Other N–S trending faults were also reactivated, resulting in the formation of a large Late Permian basin, subdivided into a series of smaller sub-basins. The lowermost Upper Permian deposits are dominated by red conglomerates indicative of continental conditions. However, during Late Permian, the sea transgressed the region from the north (Stemmerik, 2001) resulting in the deposition of marine evaporites, carbonate reefs, organic shales and turbidites (Stemmerik, 1993). The Upper Permian deposits, which show evidence of significant fluctuations in relative sea level (Kreiner-Møller and Stemmerik, 2001; Stemmerik, 2001), are grouped together in the Foldvik Creek Group (Fig. 2). These relative sea-level fluctuations are interpreted to be caused by eustatic sea-level changes associated with Late Permian glaciations (Scholle et. al., 1993), although it has been argued that local tectonic activity also was an important factor (Kreiner-Møller and Stemmerik, 2001; Stemmerik, 2001). The Foldvik Creek Group shows indications of syn-tectonic deposition (Surlyk, 1990; Kreiner-Møller and Stemmerik, 2001), but depositional rates in the Late Permian were likely relatively low (Stemmerik, 2001).

The Permian-Triassic boundary has traditionally been interpreted to represent a significant subaerial erosional hiatus in large parts of the basin (Birkenmayer, 1977; Seidler, 2000b; Stemmerik et al., 2001). Wignall and Twitchett (2002), however, suggested that sedimentation was continuous without a hiatus across the boundary. Major incisions recorded in Permian deposits (Seidler, 2000a) were interpreted to represent features caused by submarine gravity flows (Wignall and Twitchett, 2002) in the Earliest Triassic. In Jameson Land (Fig. 1) the lowermost marine Triassic Wordie Creek Formation is unconformably overlain by the continental Pingo Dal Formation of Late Schytian age (Seidler, 2000b).

The Early Triassic Wordie Creek Formation, in the Hold with Hope area, comprises clastic marine sediments up to 750 m thick. Abundant ammonite fossils have previously been used to subdivide the formation into seven ammonite zones (Spath, 1935; Grasmück and Trumpy, 1969). The Wordie Creek Formation is dominated by silty mudstones which locally contain fish fossils. The biostratigraphical data indicate that the Wordie Creek Formation was deposited in less than four million years (Griesbachian to Dienerian) (Seidler, 2000b).

In the lower part of the Wordie Creek Formation (lowermost 400 m), four coarse grained units have been identified. The units are, in ascending order, referred to as coarse grained units (CGU) 1–4 in the description and discussion given in this

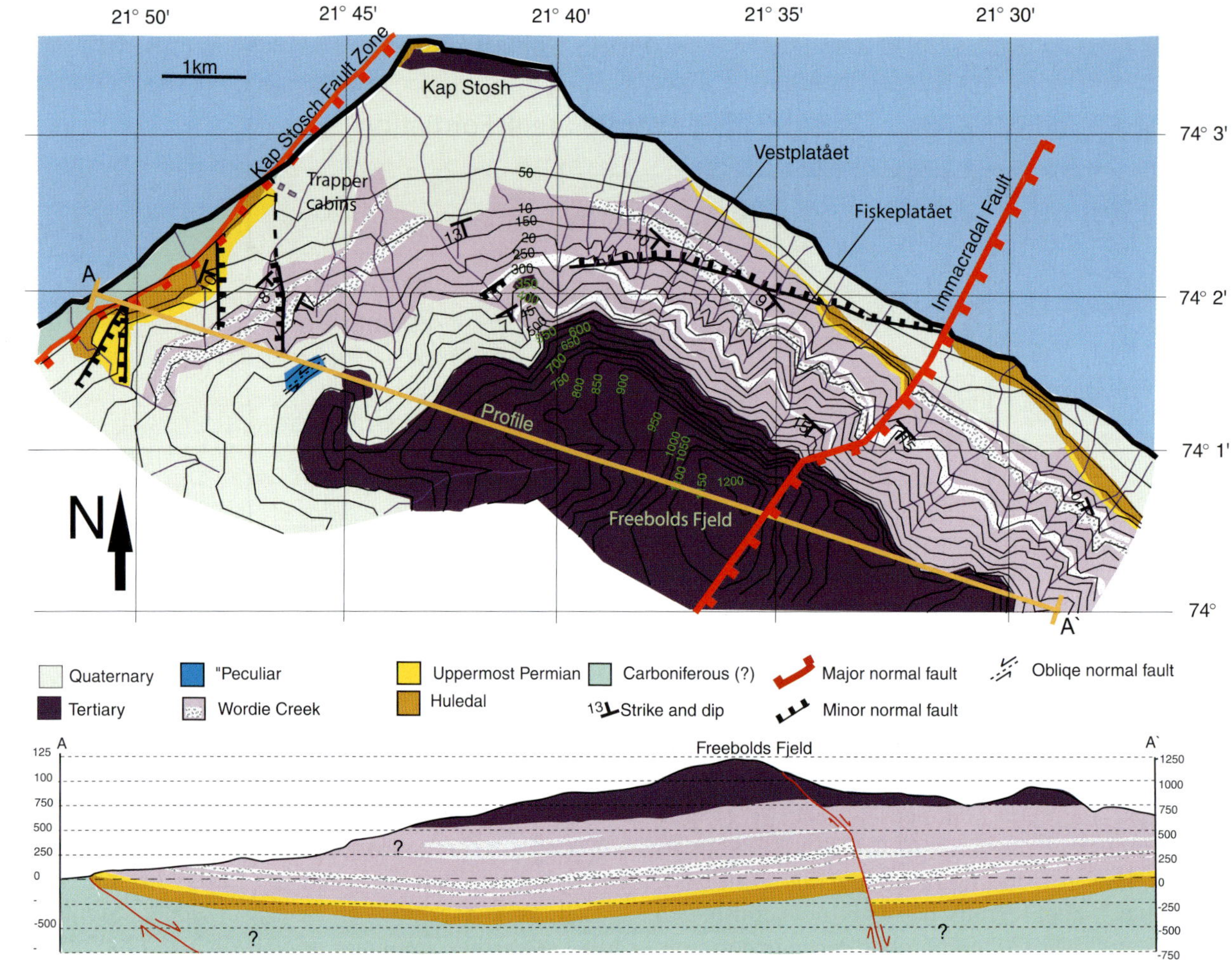

Fig. 3 Geological map and profile of investigated area at Kap Stosch.

chapter. The two lowermost units (CGU 1 and 2) are conglomeratic whereas CGU 3 and CGU 4 are sandstone-dominated. The mudstone-dominated intervals between the coarse-grained units are referred to as fine-grained unit 1 (FGU 1) (lowermost 100 m of the formation), fine-grained unit 2 (FGU 2) (between CGU 2 and CGU 3), fine-grained unit 3 (FGU 3) (from CGU 4 to approximately 650 m) fine-grained unit 4 (FGU 4) (650–750 m). Figure 5 shows a simplified log through the entire Wordie Creek Formation at Kap Stosch.

The Permian and Triassic rocks at Kap Stosch, the north-westernmost part of Hold with Hope, are located approximately 20 km east of Post Devonian Main Fault and 25 km west of the Clavering Ø Fault. The Kap Stosch area appears to be down-dropped relative to the Carboniferous deposits on Clavering Ø to the North as well as to the Carboniferous deposits on Hudson Land to the South. Neither the location nor the age of formation of the faults controlling the inferred down-dropped block with the Triassic deposits at Kap Stosch, are known, due to the fjords surrounding Kap Stosch and the glacial/alluvial cover in the low-lying areas. Several workers (Clemmensen, 1980; Seidler, 2000a) have suggested that NW–SE trending cross-faults parallel to present fjords were active during the Triassic. These faults may have created significant topography, explaining the large lateral thickness variations within Triassic basins (Seidler, 2000a). An intra-basinal fault, striking WNW–ESE and dipping steeply towards north and offsetting the Triassic Wordie Creek Formation c. 60 m, is located along the north-eastern shore of Kap Stosch (Fig. 3). This fault may be an antithetic fault to a 'graben-bounding' fault separating the Triassic sediments at Kap Stosch from the Caledonian basement gneisses and the overlying Carboniferous rocks at Clavering Ø. However, we found no thickness variations in thickness or changes in sedimentary facies across the fault which would indicate Early Triassic activity.

Peacock et al. (2000) associated the inferred Triassic palaeotopography with a large relay ramp

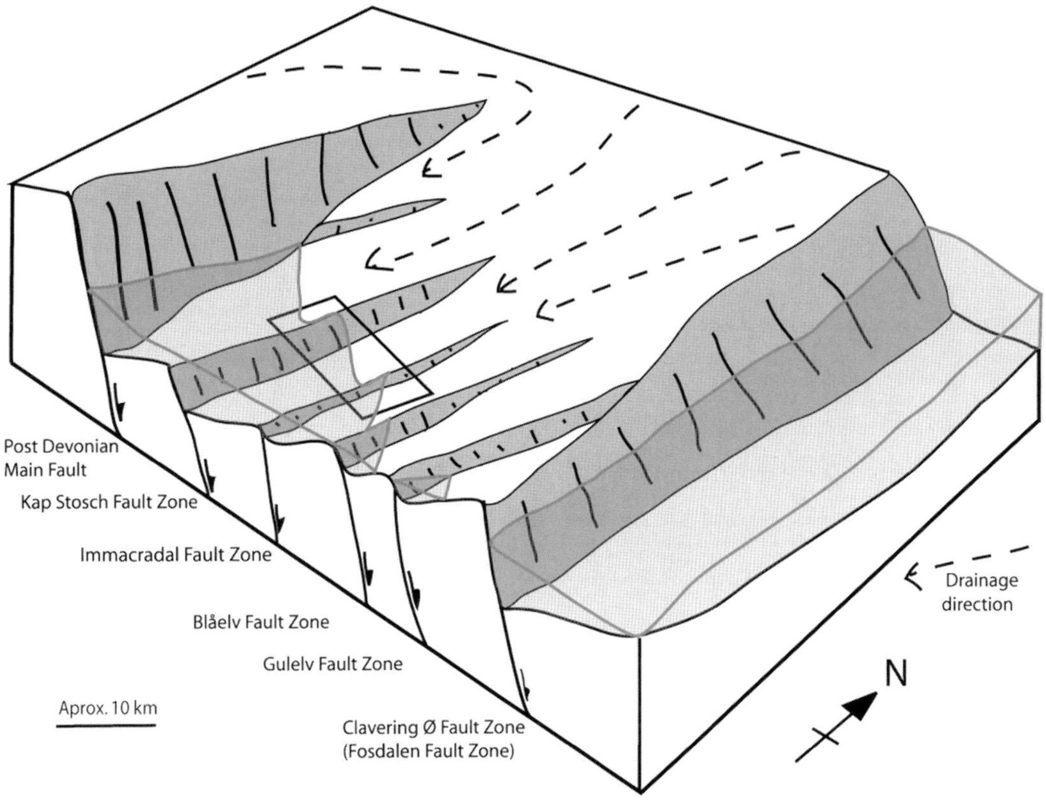

Fig. 4 Simplified model of the structural setting for the Early Triassic basin at Hold with Hope/Clavering Ø, based on Peacock et al., 2000. Possible active E–W orientated faults are not shown in this model. Square marks the approximate position of the investigated area. Note the close proximity to the hinterland in the North.

connecting the N–striking Post Devonian Main Fault with the Clavering Ø Fault Zone and Hochstetters Foreland Fault as part of a right-stepping en échelon fault system. This relay ramp resulted in uplift of the Clavering Ø area relative to the Hold with Hope area (Peacock et al., 2000). A series of NNE-striking extensional faults on Hold with Hope are interpreted as intra-basinal faults (Peacock et al., 2000). These intra-basinal faults may represent connecting faults (hard links) across the ramp (Fig. 4) (Peacock et al., 2000).

Two intra-basinal, down-to-the-east-southeast extensional faults are identified in the investigated area; the Immacradal Fault (Nielsen, 1935) and the Kap Stosch Fault (Vischer, 1943) (Fig. 3). The movement history on the Immacradal Fault (total displacement c. 250) is composite and includes both syn-depositional Triassic displacement (c. 150 m), based on thickness variations in the Wordie Creek Formation, as well as post-Palaeocene displacement, indicated by up to 100 m offset of the c. 50–54 Ma old basalts in the region. The Kap Stosch Fault (Vischer, 1943), or more correctly, the Kap Stosch Fault Zone is composed of a series of sub-parallel NNE-striking faults (Oftedal, 2002), that separate continental Carboniferous deposits in the footwall from Upper Permian and Lower Triassic sediments in the hanging wall. Throw across this fault zone is unknown, as suitable marker beds for correlation between the footwall and the hanging wall are missing. The Immacradal and the Kap Stosch Faults are considered to be two intra-basinal faults within a half-graben in which the Wordie Creek Formation was deposited.

Facies association and depositional environments

To decipher lateral and temporal variations in the depositional environment of the Wordie Creek Formation at Hold with Hope, sedimentary logs in key areas were established (Oftedal, 2002). We have identified thirteen different facies which are presented in Table 1. These thirteen facies were grouped into six facies associations, observed at several stratigraphic levels within the c. 750 m thick section at Kap Stosch. A summary of the 13 facies, identified on the basis of the various sedimentary facies, is given in Table 1.

Table 1

Facies	Description	Interpretation
1: Silty shales	Grey, purple, grey-green plane parallel laminated mudstones, commonly good lateral continuity, fish and ammonite fossils. (The shales are not studied in great detail)	Mainly suspension fallout and low concentration turbidity currents (Wignall and Twitchett, 2002).
2: HCS and SCS Sandstones	Swaley — and hummocky cross stratified sandstone (SCS and HCS respectively), 20 cm to 1 m thick beds (commonly apx. 0.5 m), very fine to medium sand, little or non grain size gradient within a bed.	Wave reworked sediments deposited below normal wave base (HCS) or right above normal wave base (SCS) (Harms et al., 1985).
3: Upward graded sandstones	20 cm to 1 m (commonly apx. 0.5 m) thick upward graded sandstone beds, sharp but rarely erosional lower boundaries, rip-up clast, thin basal lag occur.	Turbidity current deposits (Lowe, 1982; Stow and Johansson, 2000).
4: Massive sandstones	Sandstones with little or no grain size grading, bed s > 1 meter thick, normally medium sand, rip-up clast, water escape structures common, weak stratification occur, sharp but seldom erosive lower boundaries, stacked beds are typical.	Sandy debris flow or turbidites (Stow and Johansson, 2000).
5: Plane parallel stratified sandstone	Plane parallel stratified sandstones, 20 cm to 1 m thick beds, little or non grain size grading, sharp lower boundary, thin basal lag with small (<2) clasts may occur, long lateral continuity, little thickness variation.	Sediments deposited from submarine turbidity currents /debris flows, by wave reworking of sand or by fluvial processes.
6: Matrix supported conglomerates	Matrix-supported conglomerates, clasts varying from 1 cm to a meter in diameter, random orientated clasts, imbricated clasts are not common, matrix of yellow medium to coarse grained sand, usually erosive lower boundaries (locally hard to identify).	Deposited from cohesive debris flows (Nemec, 1990) generated on a coarse clastic fan delta.
7: Normal graded conglomerates	Distribution normal-graded conglomerates, bed thickness up to a meter, grain size varies from coarse gravel to medium sand, plane parallel and cross stratification are common in upper parts of the beds, trough cross bedding, erosive lower boundaries.	Deposited from turbulent gravity driven flows (Falk and Dorsey, 1998; Nemec, 1990) on a steep delta front.
8: Inverse graded conglomerate	Distribution inverse graded conglomerate beds, bed thickness up to two meter, grain size varies from medium sand to coarse gravel, trough cross bedded, sharp and often erosive lower boundaries.	Deposited from debris falls (Nemec, 1990) on steep delta front.
9: Cross stratified sandstone	Cross stratified sandstones, beds up to 0.5 m thick, grain size varies from fine to coarse sand, no grain size grading, both planar and tangential cross stratification, commonly sharp but seldom erosive lower boundaries.	Deposited from shallow marine dune migration associated with tide or wave induced currents, or from different types of fluvial banks.
10: Sandstones with thin conglomerate lags	Well sorted sandstone successions with thin gravel sized conglomerate lags, sandstones are typically medium grained, cross-stratification occur in sandstone beds, clasts are well rounded, each conglomerate lag is seldom more than 1 clast (1–3 cm) thick, conglomerate beds are commonly sub–parallel but can form trough and low angle cross bedding.	Sediments that is reworked or strongly influenced by strong wave energy.
11: Sandstone with vortex ripples	10 cm to 50 cm sandstone beds with symmetrical, relatively straight crested vortex ripples.	Wave reworked sand, deposited in shallow water.
12: Clast supported conglomerate	Clast supported conglomerates beds 0.5–5 m thick, clasts range from 3–10 cm, longest clast axis oriented parallel to bedding, imbrications rare, beds can be distribution normal or inverse or distribution graded or a combination of both, erosive lower boundary (locally hard to identify).	Sediments deposited from, either from fluvial low angle longitudinal bars (Gustavson, 1974) or debris flows (Hwang and Chough, 1990).
13: Redbeds	Brownish-red coloured mudstones, thickness ranging from 5 cm to 1 m.	Subaerial exposed sediments deposited on coastal plain (Seidler, 2000a).

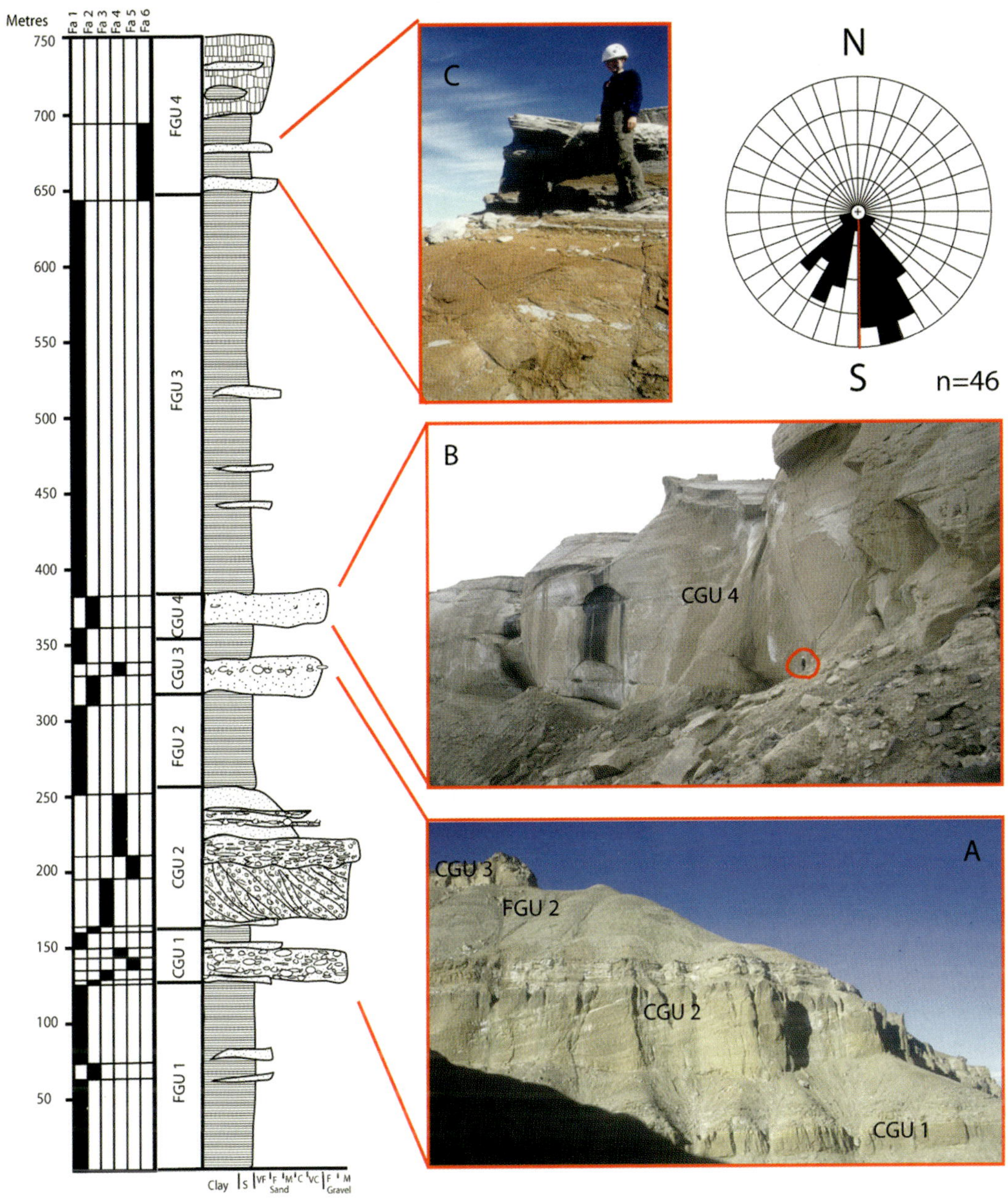

Fig. 5 Schematic log of the Wordie Creek Formation. The formation has large lateral variations and the log is therefore a simplification. Lithostratigraphic units and facies associations are shown to the left of the log. The stacking pattern of facies associations displays the interpreted deepest facies to the left. The pictures are linked to the associated stratigraphic interval at the schematic log. The rose diagram displays the palaeocurrent measured on foresets and cross-stratified beds in the lower part (300 m) of the Wordie Creek Formation. The total numbers of measurement (n) are 46. Red line represents mean transport direction. (A) The picture shows the lowermost three coarse grained units at the Fiskeplatået location. Outcrop of part of the Wordie Creek Formation at Fiskeplatået, approximately 5 km SE of Kap Stosch (Fig. 3). Thickness of section is approximately 150 m. Coarse-grained units 1–3 can be observed in this picture. Foresets in the middle unit (CGU2) exceed 30 m in thickness. (B) Massive sandstones in coarse grained unit 4. Outcrop at Vestplatået, 3 km east of Kap Stosch (Fig. 3). The succession is interpreted as prodelta lobe/basin-floor fan deposits and reaches a maximal thickness over 50 m (person for scale). (C) Redbeds in the uppermost part of the Wordie Creek Formation (or Pingo Dal Formation?) at Kap Stosch. These redbeds are interpreted to have been deposited in coastal plain environment.

Facies association 1: Marine mudstones

Description

Facies association 1 (Fa1) is composed of thick (> 10 m) intervals, dominated by silty mudrocks (facies 1). These mudrocks are commonly well-laminated and the laminae have large lateral continuity. The mudrocks can be interbedded with fine-grained, relatively thin-bedded sandstone beds (facies 2, 3 and 5) (Table 1). Facies association 1 is dominant in the lower, middle, upper and the uppermost shale intervals (Fig. 5).

Interpretation

Fa1 deposits are interpreted to be deposited from suspension fall-out and from low-density turbidity currents, as these are defined by Stow and Johansson (2000). The small grain size within the laminated mudstones indicates deposition in relatively low-energy conditions. These kinds of environments can be observed at shallow water depth. However, the thick shale intervals and the appearance of hummocky cross-stratified beds indicate that a distal and deep-marine setting (usually deeper than normal wave base) is dominant.

Facies association 2: Submarine fans/prodelta lobe deposits

Description

Facies association 2 (Fa2) is composed of massive- (facies 4), upward fining- (facies 3) and plane-parallel stratified sandstone beds (facies 5). The sandstone beds can reach thicknesses up to 10 m and can be interbedded with up to 1 m thick matrix-supported conglomerate beds (facies 6). Fa2 successions reach thicknesses up to 60 m, but show large lateral thickness variations. The lower boundary of Fa2 successions are often poorly exposed, but are conformable on localities we have investigated. Where Fa2 successions overlie Fa1 successions, the underlying shales are often deformed vertically but still show extensive lateral continuity. Facies association 2 dominates CGU 3 and CGU 4 (Fig. 5), but is also observed in the lower part of CGU 1 and CGU 2 (especially in the distal part of these units).

Interpretation

Fa2 is interpreted to represent deposition from sand-rich turbidity currents or debris flows. The thick accumulations of well-sorted sandstone suggest deposition in submarine fans or lobes. Proximity to coarse clastic deltas is supported by the appearance of thick (>1 m) matrix-supported conglomerate beds at some locations. Since the lower boundaries of a Fa2 package are conform, the thickness variations are most likely attributed to the tendency of accumulation of gravity driven flows in the deeper parts of the basin (Stow and Johansson, 2000), and not to erosion. Vertical deformation of shales is interpreted to be caused by differential loading. Even though the investigated outcrops show no sign of major erosion preceding deposition of a Fa2 succession, this does not mean that large-scale erosional features can be excluded. Nevertheless, Fa2 successions are interpreted as delta toe-lobed deposits accumulating in front of a fan delta.

Facies association 3: Delta front deposits

Description

Facies association 3 (Fa3) consists of 3–30 m thick successions of upward-coarsening (facies 8) and upward-fining (facies 7) conglomerate beds. The beds of these successions are inclined (c. 15–25°) compared with the underlying and overlying strata. Debris-fall deposits (facies 8) tend to dominate in clast-rich successions, whereas gravity-flow deposits (facies 7) dominate in the more sand-rich distal succsessions. Fa3 deposits are commonly through cross-bedded with a distinct erosion surface at the base of each bed. The beds normally exhibit a concave geometry. However, the beds often form packages where the dip-angle of each bed increases upwards in the package. These packages are cut by a planar erosion surface, often traceable through the whole Fa3 succession. Facies association 3 occurs in both CGUs 1 and 2. However, the thickest succession of Fa3 (c. 30 m) is recorded in CGU 2.

Interpretation

Fa3 was deposited as foresets at the delta front of a Gilbert-type delta indicated by the up to 30 m-thick cross-bedded succession. The sediments are interpreted to be deposited from debris falls (Nemec, 1990), cohesionless debris flows (Nemec, 1990) or turbulent mud-poor gravity flows (Falk and Dorsey, 1998). The through-shaped erosional base of each bed indicates that the gravity driven flow/falls are commonly deposited in erosional chutes on the delta front. The planar erosion surfaces are interpreted to be caused by major avalanches bypassing the entire delta front.

Facies association 4: Shallow marine sandstones

Description

Facies association 4 (Fa4) consists of thin (<1 m) sub-horizontal sandstone beds (facies 3, 5, 9, 11) and sandstone interbedded with thin gravel-sized conglomerate lags (facies 10). Up to 1 m thick matrix-supported conglomerate beds (facies 6), interpreted as sand-rich cohesive debris-flow deposits, are locally associated with facies association 4. Sandstone beds with symmetrical wave ripples (facies 11), thin gravel-sized conglomerate lags (facies 10) and low-angle cross-stratification (facies 9) indicating reworking from waves or tidal currents, are commonly found in lower parts of a succession with Fa4. Upper parts of Fa4 successions are often dominated by upward graded (facies 3) indicating deposition from turbidity currents. Facies association 4 commonly overlies delta front deposits (Fa3) and/or fluvial delta top deposits (Fa5). Facies association 4 is found in CGU 1, CGU 2 and more sporadically in CGU 3.

Interpretation

Symmetrical wave ripples, gravel-sized lags and low-angle cross-stratification, often seen in lower part of a Fa4 succession, indicate deposition in relatively shallow water. The general trend in Fa4 successions is that the upper parts are dominated by turbidites, implying sedimentation in a slightly greater water depth. This increasing water depth is linked to deposition during transgression. Facies association 4 deposits normally overlie delta front deposits or fluvial delta-top deposits, which indicate that Fa4 normally represents shallow-marine delta top deposits, associated with transgression.

Facies association 5: Fluvial delta top deposits

Description

Facies association 5 (Fa5) consists mainly of clast-supported conglomerates (facies 12) interbedded with upward-graded conglomerates (facies 7) and plane-parallel sandstones (facies 5). Fa5 varies from 2 to 10 m in thickness, and commonly overlies foreset beds (Fa3) with an erosional base, and underlies shallow-marine sandstones (Fa4). Fa 5 is recognised in CGU 1 and 2.

Interpretation

The clast-supported conglomerate beds are interpreted to channel infill deposits, formed as low-angle longitudinal bars (cf. Gustavson, 1974) or by debris flows (cf. Hwang and Chough, 1990), implying deposition above or just below sea level. Fa5 successions commonly overlie delta-front deposits (Fa3) indicating deposition in a delta top environment.

Facies association 6: Coastal plain deposits

Description

Facies association 6 consists of redbeds (facies 13) (Fig. 5C) greyish plane- and cross-stratified sandstones (facies 5 and 9) and laminated mudrocks (facies 1). Each redbed can be up to 2 m thick.

Interpretation

The redbeds indicate periods of subaerial exposure during deposition of Fa6. The co-existence of redbeds with stratified sandstone beds and mudstone suggest deposition in a coastal plain/lagunal environment. Occurrence of Fa4 is restricted to fine-grained unit 4 (FGU 4) (Fig. 5).

Palaeocurrent direction

Description

Palaeocurrent directions based on orientation of foresets and cross-stratification in CGU 1–4 in the Wordie Creek Formation show a transport direction towards south (Fig. 5). Limited palaeocurrent data from cross-stratified sandstone beds in fine grained unit 4 indicate transport towards the southeast.

Interpretation

A southward transport direction is slightly oblique to the strike of the intra-basinal faults, as well as the PDMF (Figs. 1 and 4), and indicates sediment transport parallel to the axis of the inferred half-grabens. The slight change in palaeocurrent direction may indicate a different drainage pattern during deposition of the uppermost part of the formation. However, the limited amount of measurements alone cannot support a change in drainage direction.

Provenance

Description

Pebble- and cobble-sized clasts in CGU 1–4 can be grouped into three categories; (1) well-rounded quartzite clasts (commonly $>80\%$), (2) well-rounded clasts of granite and gneissic granite (generally $<20\%$), and (3) angular to sub-angular clasts of grey limestone ($<10\%$). The carbonate clasts are restricted to CGU 1–3, and appear to be more abundant and angular in the vicinity of the intra-basinal Kap Stosch Fault Zone. In this area, limestone blocks up of 1 m in size are identified. Although no dating of the limestone clasts has been done, they were most likely derived from the Upper Permian Wegener Halvø or Karstryggen Formations (Fig. 2).

Interpretation

Presence of angular carbonate clasts mixed with well-rounded basement clasts points toward two different source areas, a feature further discussed in conjunction with our tectono-sedimentary model at the end of the chapter.

Interpretation of depositional environments for the Early Triassic Wordie Creek Formation

Facies association stacking pattern and T–R sequences

Even though the different units within the Wordie Creek Formation show significant lateral thickness variations in the studied area, a simplified log of the formation has been established (Figs. 5 and 6).

Fine grained unit 1 (FGU 1) (Figs. 5 and 6), comprising of 100–120 m of marine mudstone (Fa1), is indicative of deposition in a low-energy deep-water environment. However, the appearance of sandstone beds with hummocky cross-stratification indicates that the water depth was less than the storm-wave base in periods. FGU 1 is overlain by the two lowermost coarse clastic units (CGU 1 and CGU 2), signalling a basinward shift in facies. These units represent extensive conglomeratic deltas that prograded far into the basin. The lowermost unit (CGU 1) is interpreted to be associated with a fan-delta complex (cf. Nemec and Steel, 1988). The thickness of CGU 1, commonly 15 m, indicates that the fan delta prograded into a basin with shallow water depth, most likely less than 15 m. CGU 2 represents deposits typical of a Gilbert-type delta (Nemec and Steel, 1988) with foreset beds (Fa3) exceeding 30 m (Fig. 5A). The basin depth is therefore estimated to have been at least 30 m implying that a relative increase in water depth occurred in the basin between the deposition of CGU 1 and CGU 2. The erosive contact between the foreset and the topset beds in a Gilbert-type delta suggests that the delta was fluvial and not wave-dominated (Colella, 1988), proposing deposition on a relative low-energy coastline (Oftedal, 2002). This is consistent with small half-graben sub-basins.

Deposition of fine-grained unit 2 (FGU 2) (Figs. 5 and 6) indicates a period of marine transgression. The overlying CGU 3 and CGU 4, composed of mainly massive sandstone, are interpreted to be deposited in pro-delta lobes (Fa2) in front of the large fan deltas. The fact that neither CGU 3 nor CGU 4 contains fluvial sediments (Fa5), indicates that these fan deltas were unable to prograde as far into the basin as the fan deltas that deposited CGUs 1 and 2. The upper two coarse clastic units, therefore, represent a landward shift in facies compared to the two underlying coarse clastic units. This landward shift in facies is probably a result of increased water depth in the basin, and suggests back-stepping of the coarse clastic deltas. The overlying thick succession of well-laminated mudstones in fine-grained unit 3 (Figs. 5 and 6) indicates deposition in a deep-marine environment over a substantial period of time. This implies that the water depth continued to increase after deposition of coarse-grained unit CGU 4. The fine grained unit 4 (FGU 4) (Figs. 5 and 6). contains several redbeds and some sandstone beds of possibly fluvial or shoreface origin (Fa6). These successions are most likely deposited in a coastal plain setting, indicating a significant basinward shift in facies in the period.

As seen from the variation in depositional environments described, the basin at Kap Stosch displays significant fluctuations in relative sea level during deposition of Wordie Creek Formation. The facies association stacking pattern, shown in Figure 6, reveals five minor fourth-order T–R sequences, in the terminology of Embry (1993). It is important to stress that these fourth-order T–R sequences can most likely be observed only locally due to fault block rotation. The basin was influenced by a gradual increase in relative sea level during deposition of the lower 650 m of the Wordie Creek Formation. This is indicated by back-stepping of the four coarse clastic units and the overall shift in depositional environment from deltaic to deep marine. The lower T–R sequences (1–4) and the transgressive system tract of the fifth T–R sequence (Fig. 6) are interpreted to represent the transgressive system tract of a composite third-order T–R sequence. However, the lower and the upper sequence boundaries of this composite sequence are uncertain. The basinward shift in facies from the upper to the uppermost shale interval, in the fifth T–R sequence, indicates a period of significant regression during the deposition of the uppermost part of the Wordie Creek Formation

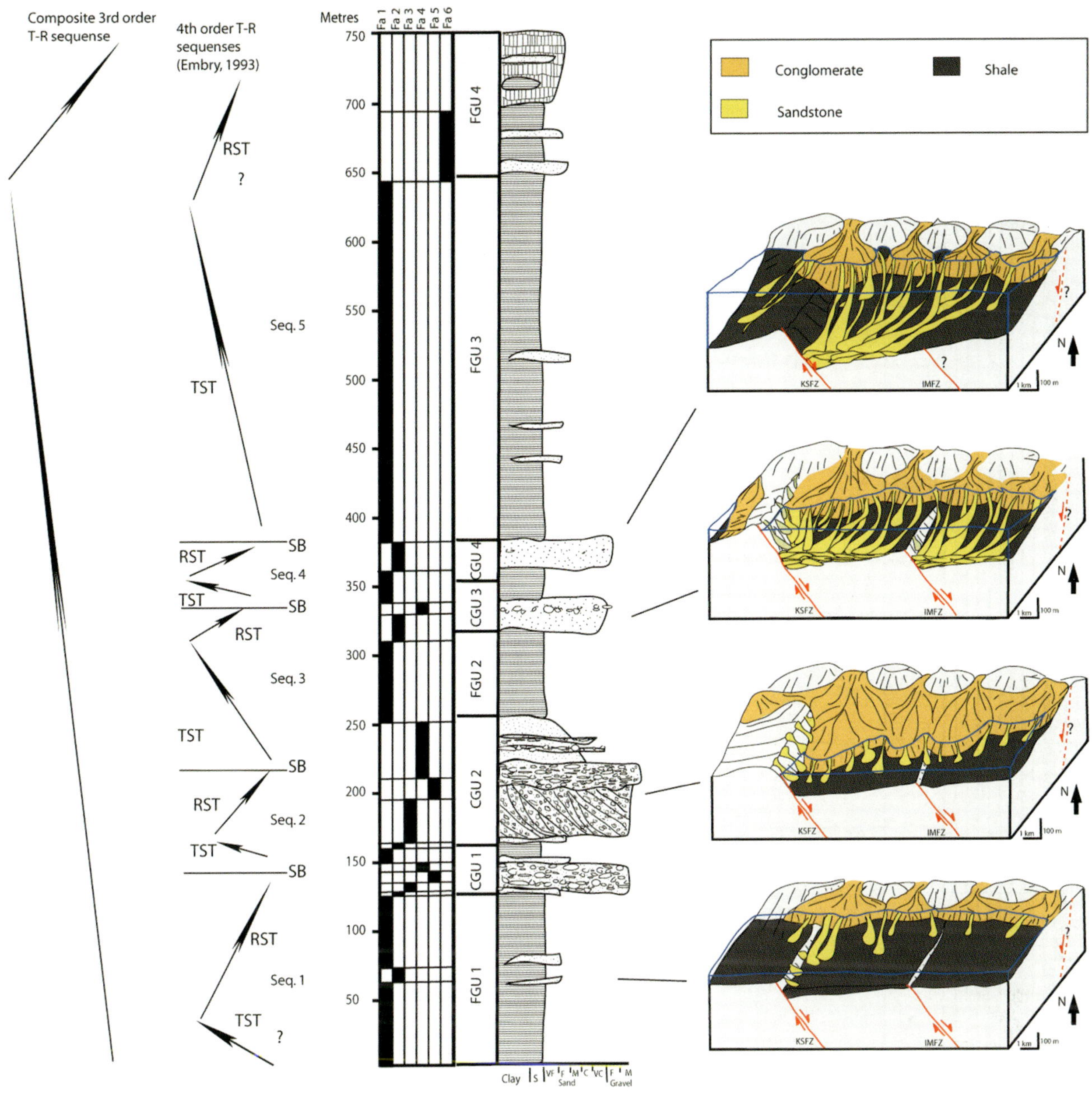

Fig. 6 Schematic log of the Wordie Creek Formation with interpreted facies association. The most basinward and landward facies associations are placed to the left- and the right-hand side in the column respectively. The arrows at the left of the log displays interpreted transgressive–regressive (T–R) sequences, as described by Embry (1993). The formation has been divided into 5 fourth-order sequences (cf. Embry, 1993). These local fourth-order sequences are interpreted to be a part of composite third-order sequence, with uncertain lower and upper boundaries, shown to the right in the figure. On the right-hand side of the figure, depositional models of the stratigraphic intervals of the Wordie Creek Formation are linked to the schematic log.

(Dinerian age). This uppermost shale unit is, therefore, interpreted to represent the regressive system tract of the composite T–R sequence.

Lateral variations

Both CGU 1 and CGU 2 are laterally extensive and can actually be traced to poorly studied outcrops to the south of the study area (Fig. 3). The units show a relatively minor decrease in thickness towards the inferred distal southern part of the basin. However, these units show significant lateral changes in facies. For instance, the height of foreset beds in CGU 2 decreases distally, while the fluvial delta top is actually thicker in some distal localities (Oftedal, 2002). Little or no thickness variation in this coarse-grained unit is identified across intra-basinal faults.

The two upper coarse clastic units (CGU 3 and 4) are not as laterally extensive as the two lower ones.

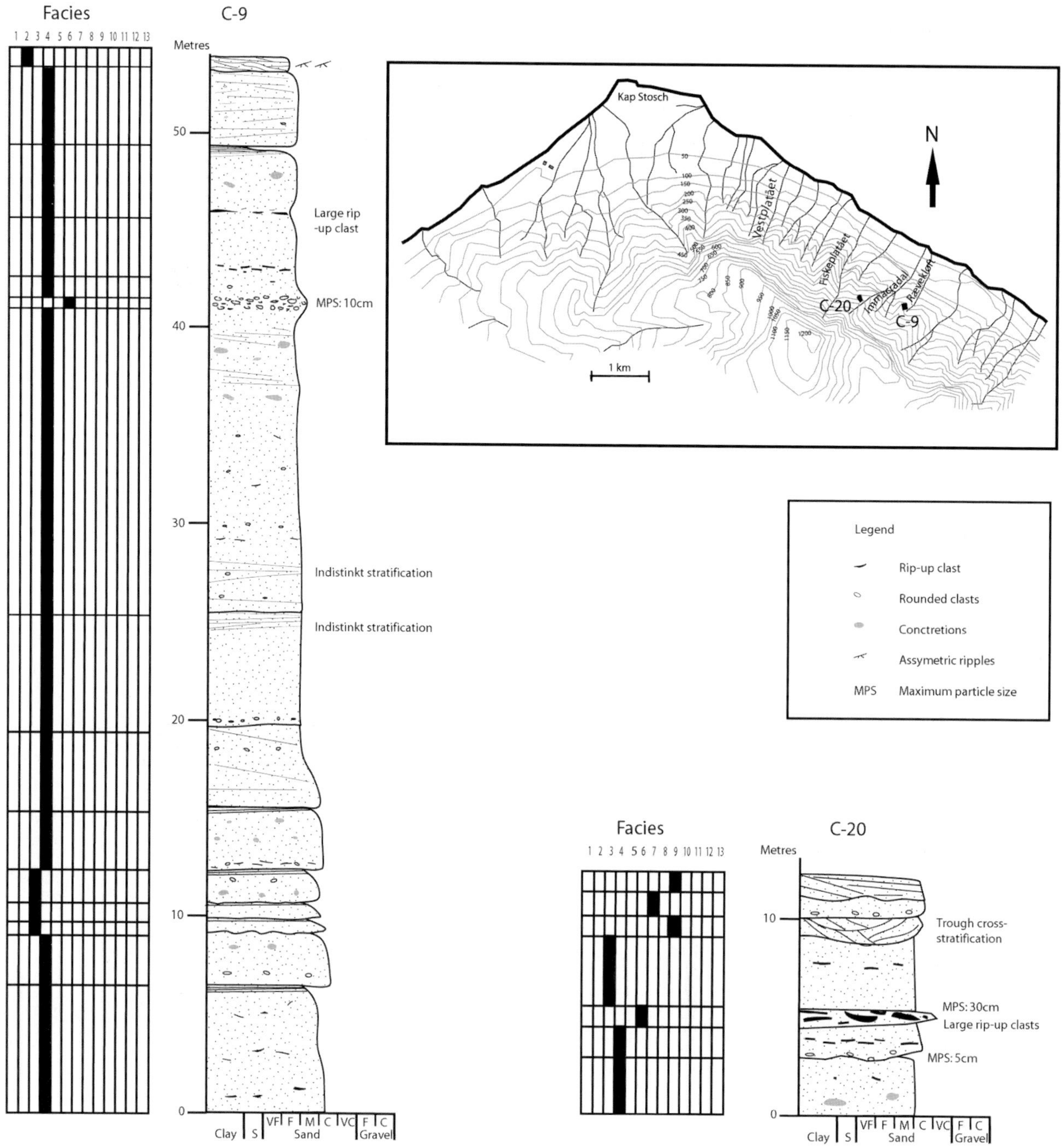

Fig. 7 Compressed and simplified logs of coarse grained unit 3 on both sides of the Immacradal Fault Zone. The unit shows considerable thickness variations, implying syn-depositional movement of the fault zone.

Locally, CGUs 3 and 4 show dramatic variations in thickness with maximum thickness up 60 m, adjacent to intra-basinal faults. This is particularly well-demonstrated across the Immacradal Fault Zone, where CGU 3 changes in thickness from 11 m in the footwall to 60 m in the hanging wall (Fig. 7). CGU 3 also displays distinct wedge-shape geometry, with the greatest thickness adjacent to the intra-basinal faults. A wedge-shaped geometry is also observed for CGU 4 in the hanging wall, adjacent to the Kap Stosch Fault Zone.

Implication from provenance

The coarse-grained units in the Wordie Creek Formation are dominated by large amounts of well-rounded basement clasts, implying relatively long sediment transport or several transport events. Carboniferous/Devonian molasse deposits or a basement source area are therefore probable for these clasts. A different source area and transport history has to be called upon to explain the angular limestone clasts observed in the three lowermost clastic units. As the largest and most angular clasts are found in proximity of the Kap Stosch Fault Zone, we consider the uplifted footwall to an intra-basinal fault as the most likely source area.

In CGU 4, carbonate clasts are rare, even in localities close to Kap Stosch Fault Zone. One possible explanation for this could be a low relief across Kap Stosch Fault Zone during deposition of CGU 4. However, the increase in thickness of CGU 4 (see lateral variations) towards the Kap Stosch Fault Zone suggest, otherwise. The lack of carbonate clasts might, therefore, be related to the drowning of the footwall of this intra-basinal fault.

Discussion

Early Triassic tectonism

A significant change in sedimentation rate appears to have taken place in the Hold with Hope area when the thickness of the Late Permian and Early Triassic successions are compared. The Upper Permian succession, with a maximum thickness in the Clavering Ø area of 200 m (Stemmerik, 1993), is deposited during a time-interval of approximately 10 million years (Wignall and Twitchett, 2002). Even though this succession might have major erosional hiati (Stemmerik, 1993), the depositional rate is low, when compared with the Lower Triassic successions. The Wordie Creek Formation at Hold with Hope is at least 750 m thick, probably deposited in less than four million years. An increase in sedimentation rate of this magnitude is most likely the result of increased tectonic activity and local subsidence in the region. The low sedimentation rate in the Late Permian may have caused a partly starved basin, implying relict accommodation space in the Early Triassic. However, our observations indicate that the Wordie Creek Formation deposits filled the basin up to sea level in its lower parts (CGU 1 and 2, see Fig. 6), indicating that a possible sediment delay was limited. The basin continued to deepen after deposition of these coarse clastic units, showing that most of the accommodation space was created during (and not earlier) deposition of the formation. The rift climax stage is, therefore, interpreted to be in the Lower Triassic and not in the Upper Permian successions. Accordingly, the Late Permian is interpreted to represent the early stage of the rift episode.

Several other features also support a syn-tectonic depositional history for part of the Wordie Creek Formation. One feature is the identification of a Gilbert-type delta in the lower part of the formation (CGU 2) indicating that the Early Triassic basin at Kap Stosch was bounded by relatively steep basin margins, typical for rift basins (cf. Ravnås and Steel, 1998). The geometry of these deposits furthermore indicates deposition in a basin with low-wave energy, suggesting a confined basin. The wedge-shaped geometry of CGUs 3 and 4 in the proximity of intra-basinal faults, as well as significant variation in thickness of CGU 3 across Immacradal Fault Zone, indicate syn-depositional tectonic activity. These thickness variations are primarily controlled by the tendency for axial gravity-driven submarine flows to accumulate in the deeper parts of rift basins, often located on the down-dropped hanging wall controlled by an intra-basinal fault, as suggested in rift models by Gawthorpe and Leeder (2000).

The appearance of carbonate clasts indicates an intra-basinal catchment area linked to the exhumed footwall of Kap Stosch Fault Zone. Elevation and erosion of the footwall can be explained by fault-block rotation, due to significant tectonic activity in the Early Triassic. In rift sub-basins, fault-block rotation will result in the rise and fall of local sea level, depending on the position relative to fulcrum (Ravnås and Steel, 1998). As a result, Upper Permian carbonates were elevated, exhumed and eroded west of the intra-basinal Kap Stosch Fault Zone. The carbonate clasts were then transported into the basin by small transverse fans dominated by gravity falls (Fig. 8). These transverse fans were often toe-cut by the axial depositional system and the carbonate clast were transported further out in the half-graben basin.

Basin infill history

The fourth-order sequences in the Wordie Creek Formation (Fig. 6) at Kap Stosch are interpreted to record local tectonic events in the region. Movement on the basin-bounding fault zones (PDMF and Clavering Ø Fault Zone, see figure 4) may have resulted in rotation of large-scale relay ramp. The related increase in relief resulted in increased erosion rate causing axial progradation of coarse clastic deltas into the 'Kap Stosch sub-basin'. Rotation of the relay ramp was likely

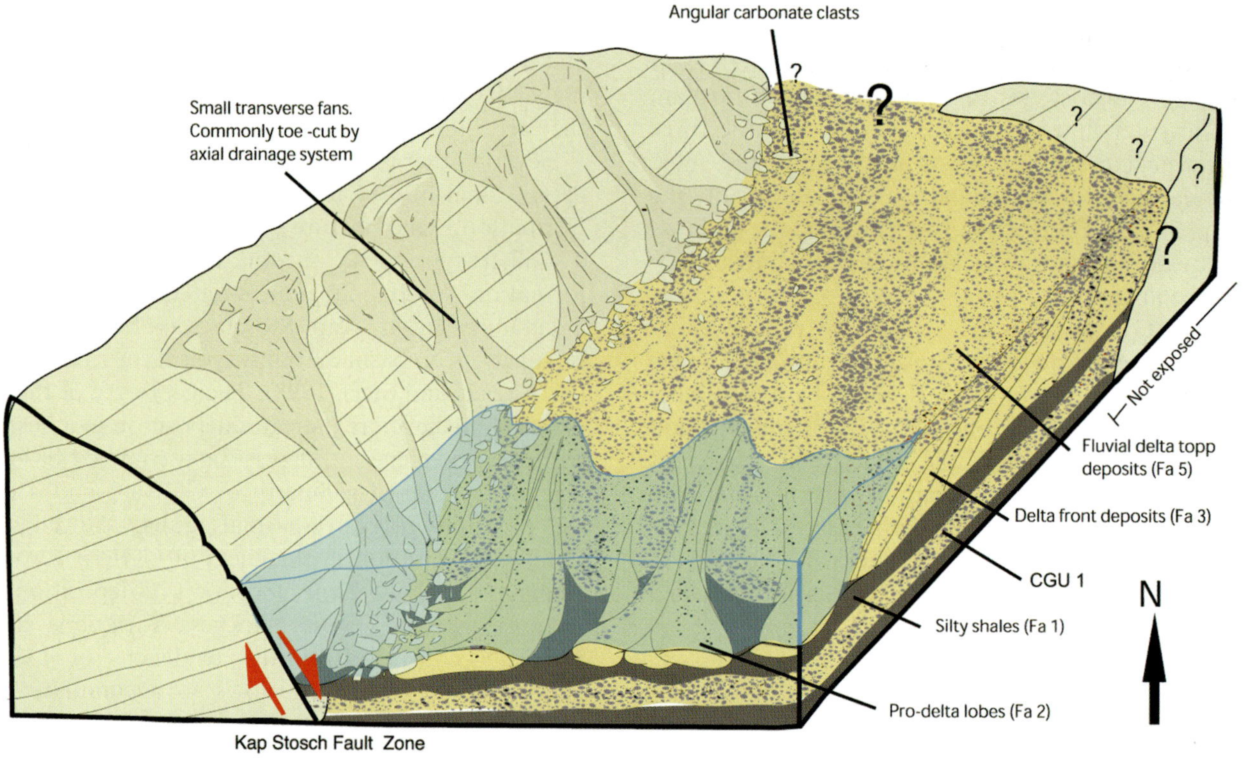

Fig. 8 Depositional model of coarse grained unit 2 (CGU 2) in proximity to Kap Stosch Fault Zone. The footwall of the fault zone was elevated above sea level as a result of fault block rotation, which resulted in erosion of the underlying Permian limestones.

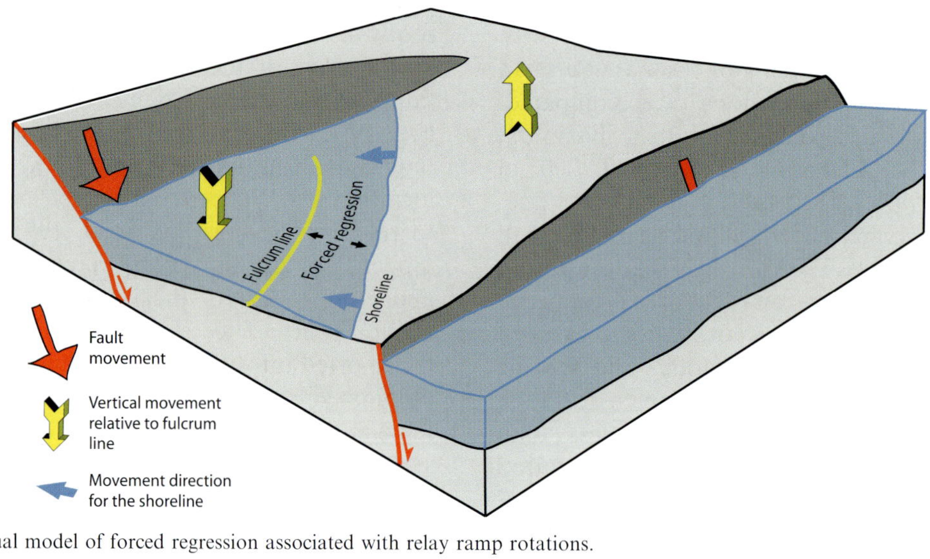

Fig. 9 Conceptual model of forced regression associated with relay ramp rotations.

associated with forced regression of the shoreline, if it was located above the fulcrum line on the relay ramp (Fig. 9). This is based on the assumption that deltas on a rotating relay ramp will behave in a similar way to those on a rotating fault block, as described by Ravnås and Steel (1998). Delta progradation associated with forced regression can explain the relatively pronounced delta progradations seen in the two lowermost coarse clastic units. Forced regression during deposition can also explain the relative uniform thickness of these units by syn-progradational erosion of proximal parts of the deltas. Syn-progradational erosion can also have caused the thickness variations seen in the fluvial delta top deposits (Fa 5).

We propose that the fourth-order sequences represent part of a composite higher-order sequence (probably third-order sequence of Embry, 1993, linked to long-term variation in the tectonic subsidence rate controlling the rift episode. The overall landward shift in facies seen in all of the Wordie Creek Formation, except for fine-grained unit 4, implies a transgressive trend often identified in rift climax stages (Prosser, 1993; Ravnås and Steel, 1998). This transgressive trend is supported by drowning of the footwall associated with Kap Stosch Fault Zone after deposition of CGU 3. Drowning of intra-basinal highs is also a common feature during the rift climax stage (Prosser, 1993). The overall increase in water depth indicates that delta progradations were back-stepping, onlapping the gradually more-inclined relay ramp. The regressive trend interpreted for fine-grained unit 4 (containing redbeds) indicates infilling of the Early Triassic marine sub-basin at Kap Stosch. This regressive trend can also be observed in the Triassic sub-basin, farther south (Seidler, pers. comm. 2003). This overall regression is probably caused by a period of low tectonic activity with the immediate post-rift stage (Prosser, 1993).

Eustatic sea-level fluctuations are difficult to identify in tectonically active marine rift basins (Ravnås and Steel, 1998) and probably had limited effect on the Early Triassic depositional system due to the relatively steep margins of the rifted basin. Steep basin margins in this period are supported by the existence of Gilbert-type delta deposits and the thick conglomeratic intervals within the Wordie Creek Formation. Thus, the Upper Permian succession is highly influenced by eustatic sea-level variation due to lower basin relief and large eustatic sea-level fluctuations (Scholle et al., 1993) in this period.

Conclusion

The Early Triassic successions in East Greenland are interpreted to represent the rift climax stage deposits associated with a Late Palaeozoic–Early Mesozoic rift episode. The best indications of a rift climax stage in the Wordie Creek Formation are: (1) high sedimentation rate; (2) erosion of underlying Upper Permian carbonates located in the uplifted footwall of intra-basinal faults; (3) lateral thickness variations across intra-basinal faults; (4) wedge-shaped geometry of coarse clastic units (increasing thickness towards faults); (5) Gilbert-type delta deposits indicative of relatively steep basin margins; (6) overall deepening of the basin and drowning of intra-basinal footwalls with time, due to tectonic subsidence.

Acknowledgements

The authors would like to thank Wojtec Nemec, Lars Seidler, Ebbe Hartz, Caroline Pickles, editor Johan Petter Nystuen, reviewers and others, who have contributed and commented upon this work. This field-based study would not have been possible without the financial suport from The Norwegian Polar Institute (Arctic Fellowship to BTO), Nansenfondet (BTO) and Norsk Hydro (AA and BTO), an invaluable logistical support by the Danish Polar Center field crew in Mestersvig.

References

Andresen, A., 2002. Late Caledonian transpressional/transtensional tectonics in the North Atlantic Region and its influence on Late Paleozoic Basin evolution. In: A. Hurst (Editor), Onshore-Offshore Relationships on the Nordic Atlantic Margin. NGF Abstracts and proceedings 2, 2002 of the Norwegian Petroleum Society (NPF) and Norwegian Geological Society (NGF) Conference, October 7–9 Trondheim, pp. 9–11.

Birkenmayer, K., 1977. Erosional unconformity at the base of marine Lower Triassic at Wegner Ø, Central East Greenland. Bull. Geol. Soc. Den., 25: 107–116.

Blystad, P.H., Brekke, R.B., Færseth, R., Larsen, B.T., Skogseid, J. and Tørudbakken, B., 1995. Structural elements of the Norwegian continental shelf, Part II: The Norwegian Sea Region: NPD, Bull., 8, 45 pp.

Brekke, H., 2000. The tectonic evolution of the Norwegian Sea Continental Margin with emphasis on the Vøring and Møre Basins. In: A. Nøttvedt et al. (Editors) Dynamics of the Norwegian Margin. Geol. Soc., London, Spec. Publ., 167: 327–378.

Clemmensen, L., 1980. Triassic rift sedimentation and palaeogeography of Central East Greenland. Grønl. Geol. Unders., Bull., 136: 11–72.

Colella, A., 1988. Pliocene–Holocene fan deltas and braid deltas in the Crati Basin, southern Italy: a consequence of varying tectonic conditions. In: W. Nemec and R.J. Steel (Editors), Fan Deltas: Sedimentology and Tectonic Settings, Blackie and Son, pp. 50–74.

Embry, A., 1993. Transgressive–regressive (T–R) sequence analysis of the Jurassic succession of the Sverdrup Basin, Canadian Artic Archipelago. Can. J. Earth Sci., 30: 301–320.

Falk, P.D and Dorsey, R.J., 1998. Rapid development of high-density turbidity currents in marine Gilbert-type fan deltas, Loreto Basin, Baja California Sur, Mexico. Sedimentology, 45: 331–349.

Gawthorpe, R.L. and Leeder, M.R., 2000. Tectono-sedimentary evolution of active extensional basins. Basin Res., 12: 195–218.

Grasmück, K. and Trumpy, R., 1969. Triassic stratigraphy and general geology of the country around Fleming Fjord (East Greenland). Meddelelser fra Grønland, 168 (2): 5–74.

Gustavson, T.C., 1974. Sedimentation on gravel outwash fans, Malaspina Glacier Foreland, Alaska. J. Sediment. Petrol., 44 (2): 374–389.

Harms, P.H., 1985. Sedimentology of saline density current deposits in the Delaware Mountain Group, Delaware Basin. In: Permian carbonate/ clastic sedimentology, Guadalupe

Mountains; analogs for shelf and basin reservoirs. Soc. Econ. Paleontol. Miner., 17 pp.

Hartz, E., Eide, E.A., Andresen, A., Midbøe, P., Hodges, K.V. and Kristiansen, S.N., 2002. Ar-40/Ar-39 geochronology and structural analysis: Basin evolution and detrial feedback mechanisms, Hold with Hope region, East Greenland. Nor. J. Geol. (NGT), 82: 341–358.

Hwang, I.G. and Chough. S.K., 1990. The Miocene Chunbuk Formation, southeastern Korea; marine Gilbert-type fan-delta system. Int. Assoc. Sedimentol., Spec. Publ., 10: 235–254.

Kelly, S.R.A., Whitham, A.G., Koraini, A.M. and Price, S.P., 1998. Lithostratigraphy of the Cretaceous (Barremian-Santonian) Hold with Hope Group, NE Greenland. J. Geol. Soc., London, 155: 993–1008.

Kreiner-Møller, M. and Stemmerik, L., 2001. Upper Permian lowstand fans of the Bredehorn Member, Schuchert Dal Formation, East Greenland. In: O. Martinsen and Dreyer (Editors), Norwegian Petroleum Society (NPF) Special Publication 10, pp. 51–65.

Lowe, D.R., 1982. Sediment gravety flows: II. Depositional models with spesial reference to the deposit of high-density turbidity currents. J. Sediment. Petrol., 52: 279–297.

Müller, R., Nystuen, J.P., Eide, F. and Lie, H., 2004. Late Permian to Triassic basin infill history and palaeogography of the mid-Norwegian shelf — East Greenland region. In: B.T.G. Wandås, E.A. Eide, F. Gradstein and J.P. Nystuen (Editors), Onshore-Offshore relationships on the North Atlantic Margin. Norwegian Petroleum Society (NPF), Special Publication 12. Elsevier, Amsterdam.

Nemec, W., 1990. Aspects of sediment movement on steep delta slopes. In: Coarse grained deltas. Int. Assoc. Sedimentol., Spec. Publ., 10: 29–73.

Nielsen, E., 1935. The Permian and Eotriassic vertebra bearing beds at Godthaab Gulf (East Greenland). Meddelelser om Grønland, 98 (1): 1–111.

Nemec, W. and Steel, R.J., 1988. What is a fan delta and how do we recognize it? In: W. Nemec and R.J. Steel. (Editors). Fan Deltas : Sedimentology and tectonic setting. Blackie and Son, pp. 3–13.

Oftedal, B.T., 2002: Tidlig triassisk tectono-sedimentær utvikling på Kap Stosch, Hold with Hope, Nordøst Grønland. Cand. Scient thesis, University of Oslo (In Norwegian). 144 pp.

Peacock, D.C.P., Price, S.P, Whitham, A.G. and Pickles, C.S., 2000. The World's biggest relay ramp: Hold with Hope, NE Greenland. J. Struct. Geol., 22: 843–850.

Price, S., Brodie, J., Whitham, A. and Kent, R., 1997. Mid-Tertiary rifting and magmatism in the Traill Ø region, East- Greenland. J. Geol. Soc., London, 154: 419–434.

Prosser, S., 1993. Rift–related linked depositional systems and their seismic expresspions. In: G.D. Williams and A. Dobb. (Editors), Tectonics and Seismic Sequence Stratigraphy. Geol. Soc., London, Spec. Publ., 71: 35–66.

Ravnås, R. and Steel, R.J., 1998. Architecture of marine rift successions. Am Assoc. Pet. Geo., Bull., 82 (1): 110–146.

Scholle, P.A., Stemmerik, L., Ulmer, D., Di-Liegro, G. and Henk, F.H., 1993. Palaeokarst-influenced depositional and diagenetic patterns in Upper Permian carbonates and evaporites, Karstryggen area, central East Greenland. Sedimentology, 40 (5): 895–918.

Seidler, L., 2000a. Sedimentology and sequence stratigraphy of the Lower Triassic Wordie Creek Formation in northern Jameson Land, Scoresby Land and on Traill Ø, East Greenland. Ph.D. thesis, University of Copenhagen.

Seidler, L., 2000b. Incised submarine canyons governing new evidence of Early Triassic rifting in East Greenland. PALAEO, 161: 267–293.

Spath, L.F., 1935. Additions to the Eo-Triassic invertebrate fauna of East Greenland. Meddelelser om Grønland, 98 (2): 1–115.

Stemmerik, L., 2001. Sequence stratigraphy of a low productivity carbonate platform succession: the Upper Permian Wegner Halvø formation, Karstryggen Area, East Greenland. Sedimentology, 48: 79–97.

Stemmerik, L., Bendix-Almgreen, S.E. and Piasecki, S., 2001. The Permian-Triassic boundary in central East Greenland: past and present views. Bull. Geol. Soc. Den., 48: 159–167.

Stemmerik, L., Christiansen, F.G., Piasecki, S. Jordt, B., Marcussen, C. and Nøhr-Hansen, H., 1993. Depositional history and petroleum geology of the Carboniferous to Cretaceous sediments in the northern part of East Greenland. In: T.O. Voren, E. Bergsager, Ø.A. Dahl- Stamnes, E. Holter, B. Johansen E. Lie and T.B. Lund (Editors), Artic geology and petroleum potential. Norwegian Petroleum Society (NPF) Special Publication 2, pp. 67–87.

Stow, D.A.V. and Johansson, M., 2000. Deep Water massive sands: nature, origin and hydrocarbon implications. Mar. Petrol. Geol., 17: 145–174.

Surlyk, F., 1990. Timing, style and sedimentary evolution of Late Paleozoic- Mesozoic extensional basins of East Greenland. In Hardman, R.F.P. and Brooks, J. (Editors) Tectonic Events Responsible for Britain's Oil and Gas Reserves. Geol. Soc., London, Spec. Publ., 55: 107–125.

Tsikalas, F., Faleide, J.I. and Eldholm, O., 2001. Lateral variation in tectono-magmatic style along the Lofoten-Vesterålen volcanic margin off Norway. Mar. Petrol. Geol., 18: 807–832.

Vischer, A., 1943. Die postdevonische Tektonik von Ostgønland zwichen 74 und 75 N. Br., Kuhn Ø, Wollaston Forland, Clavering Ø und angrenzende Gebiete. Meddelelser om Grønland, 133 (7): 195 pp.

Vigran, J.O., Stemmerik, L. and Piasecki, S., 1999. Stratigraphy and depositional evolution of the uppermost Devonian-Carboniferous (Tournaisian-Westphalian) non-marine deposits in North-east Greenland. Palynology, 23: 115–152.

Wignall, P.B and Twitchett, R.J., 2002. Permian-Triassic sedimentology of Jameson Land, East Greenland: incised submarine channels in an anoxic basin. J. Geol. Soc., London, 159: 691–703.

Controls on depositional history and architecture of the Oligocene–Miocene succession, northern North Sea Basin

Yngve Rundberg and Tor Eidvin

The tectonostratigraphic framework of the Oligocene–Miocene succession in the northern North Sea Basin (58–62°N) is closely linked to the large-scale structural evolution of the NW European passive margin. Fairly contemporaneous with the structural doming on the Mid-Norwegian margin uplift activity also affected the Shetland Platform and southern Fennoscandia, including the sedimentary basin of the northern North Sea. This uplift caused a gradual shallowing-upward trend of the northern North Sea Basin, which culminated in severe submarine and possibly also subaerial erosion during middle Miocene, creating a northward increasing stratigraphic break (20 million years in northernmost North Sea), which is visible as a distinct seismic unconformity. Uplift of the East Shetland Platform caused three major phases of sand influx to the basin (1) an early Oligocene phase, resulting in deposition of gravity flow sands in the northern Viking Graben (Statfjord–Tampen area); (2) an early Miocene phase, resulting in deposition of turbiditic sands (Skade Formation) in southern Viking Graben; and (3) a late Miocene–early Pliocene phase, resulting in deposition of shelfal sands (Utsira Formation). During the latter phase, the northern North Sea Basin formed a relatively shallow marine, shelfal strait between deeper marine settings to the north and south. The Utsira Formation sands accumulated in this narrow strait in a high-energy, possibly tidal-current controlled regime.

This chapter also presents an improved lithostratigraphic and chronostratigraphic subdivision of the Oligocene–Miocene including redefinitions of the Skade and Utsira formations. The Oligocene–Miocene succession in the northern North Sea has been subdivided into two megasequences, separated by a seismically distinct unconformity (mid-Miocene break). The age diagnostic *Bolboforma* assemblages, known from ODP/DSDP boreholes in the North Atlantic and on the Vøring Plateau, have aided in correlation between wells and have been important in resolving the basin history.

Introduction

In this chapter, we present our latest understanding of the depositional history of the Oligocene–Miocene succession in northern North Sea. The main object has been to view the depositional history of this area in a larger-scale tectonic perspective. Focus has been on the depositional architecture, stratigraphical outline and the coarse clastic input to the basin, with special emphasis on the Utsira Formation. We also present an improved chronology of the Oligocene–Miocene, with more precise age constraints of the Skade and Utsira formations.

The study area embraces the northern North Sea between 58 and 62°N. Some results of the work carried out in the Møre and Faeroe–Shetland Basins are also presented. The erosive mid-Miocene surface forming the top of the Hordaland Group and its continuation to the south has been crucial to our work, and subdivides the strata described in this chapter into two distinct megasequences. Critical features that are preserved in the southern part of the basin are applicable to the interpretation of strata, farther north.

This work synthesizes several data sets and methods. The interpretations are results of detailed biostratigraphic works; integrated with $^{87}Sr/^{86}Sr$ isotope stratigraphy, seismic and wireline log studies. A suite of regional 2-D lines from an extensive database has been interpreted. Six selected lines are presented in this chapter. An extensive well database has also been available for study. The well data include gamma ray, resistivity and sonic logs; in some wells the density and neutron logs were also available. Interpreted log data from

Onshore–Offshore Relationships on the North Atlantic Margin
edited by B. Wandas et al. NPF Special Publication 12, pp. 207–239,

17 wells located along or close to the selected lines are shown in this chapter.

Biostratigraphic data were obtained from ditch-cuttings, sidewall cores and conventional cores from 17 selected wells. Main results from seven wells in the southern Viking Graben (wells 15/9-A-11, 15/9-A-23, 15/9-13, 15/12-3, 16/1-2, 16/1-4 and 24/12-1) are presented in two tables and also are shown in the log correlation diagrams. Detailed fossil descriptions from these wells and results from strontium isotope stratigraphy will be reported in a paper, to be submitted shortly. Results from well 15/12-3 are previously also presented in Eidvin et al. (1999) but samples from this well have been reanalysed and reinterpreted in the current study. Biostratigraphic data and results from strontium isotope analyses from 10 wells in northernmost North Sea have previously been presented in Eidvin and Rundberg (2001).

Much of the background of this work is based on the comprehensive study of the Cainozoic stratigraphy and basin evolution of Norwegian northern North Sea Basin, carried out by Rundberg (1989). Earlier works dealing with the litho- and seismic stratigraphy of the Oligocene–Miocene in the Norwegian Northern North Sea have been presented by Isaksen and Tonstad (1989), Rundberg (1989), Rundberg and Smalley (1989), Galloway et al. (1993), Gregersen et al. (1997), Jordt et al. (1995), Martinsen et al. (1999), Galloway (2002).

Oligocene–Miocene palaeogeography and palaeotectonic evolution

The sea-floor spreading history of the Norwegian and Greenland Seas and its effects on the NW European passive margin have been addressed by a number of earlier studies (e.g. Talwani and Eldholm, 1977; Vogt et al., 1981; Eldholm et al., 1990; Doré and Lundin, 1996; Vågnes et al., 1998; Brekke, 2000; Lundin and Doré, 2002 and Mosar et al., 2002). At end Eocene times, major plate reorganisations initiated a compressive structural regime, which had dramatic effects on the geohistory of the NW European margin. In the Norwegian Sea (at about anomaly 13; 35 Ma), the plate movements were characterised by a 30° counterclockwise rotation (Lundin and Doré, 2002) and a westward jump in sea-floor spreading axis to the south of the Jan Mayen Fracture Zone (JMFZ; Fig. 1) which led to the formation of the Jan Mayen microcontinent (Fig. 1). Sea-floor spreading also commenced in the Greenland Sea, with shearing affecting the Spitsbergen Margin and gradually during Oligocene establishing a seaway link to the Arctic Sea (Fig. 1).

The structural activity of the margin was heavily controlled by compressional strain, developed as a response to movements along major fracture zones. Between Lofoten and the Faeroe Islands, three large fracture zones (Bivrost, Jan Mayen and Erlend; Fig. 1), broadly subdivide the margin into three compartments, each of which has undergone a different structural post-Eocene evolution. (1) The margin between the Jan Mayen and Bivrost Fracture Zones comprises the Vøring Basin and Vøring Marginal High. This area has undergone a complex structural evolution involving the growth of large domes (Helland Hansen and Modgunn Arches, Naglfar and Vema Domes; Figs. 1, 2) and also inversion structures (e.g. Fles Fault Complex). Southward, along the Helland Hansen structural trend, the Ormen Lange dome developed at the transition to the Møre Basin. (2) The area between the Jan Mayen and the Erlend Fracture Zones, comprising the Møre Basin and Møre Marginal High, underwent a totally different evolution. It is characterised by overall subsidence and the almost absence of compressional structures, apart from a slight uplift of the Møre Marginal High. These two contrasting segments of the passive margin could thus reflect a lower-plate origin for the Møre Basin and an upper-plate origin for the Vøring Basin (Rundberg, 1989 p. 235; Mosar et al., 2002). (3) The margin to the south of the Erlend Fracture Zone was affected by severe compressional folding which developed the strongly anticlinal Fugløy Ridge and the narrow Faeroe–Shetland Basin. Further to the south, a complex structural evolution took place involving formation of the Wyville Thompson, Ymir and Munkegrunnur Ridges (Boldreel and Andersen, 1993 and 1994). More distant from the ocean-continent boundary, and sub-parallel to the Fugløy Ridge, structural uplift also affected the Shetland Platform and southern Fennoscandia during Oligo-Miocene times, as shown in Figures 1 and 2.

Most of the anticlinal structures or domes along the margin show a multi-phase growth history with important phases in (1) Middle Eocene to Early Oligocene and (2) the Miocene (Doré and Lundin, 1996; Brekke, 2000; Lundin and Doré, 2002).

As a result of the mid-Cainozoic plate reorganisations, an important change in palaeogeography took place during Oligocene–Miocene. The semi-enclosed basin of the Norwegian and North Seas that existed during late Palaeocene–Eocene

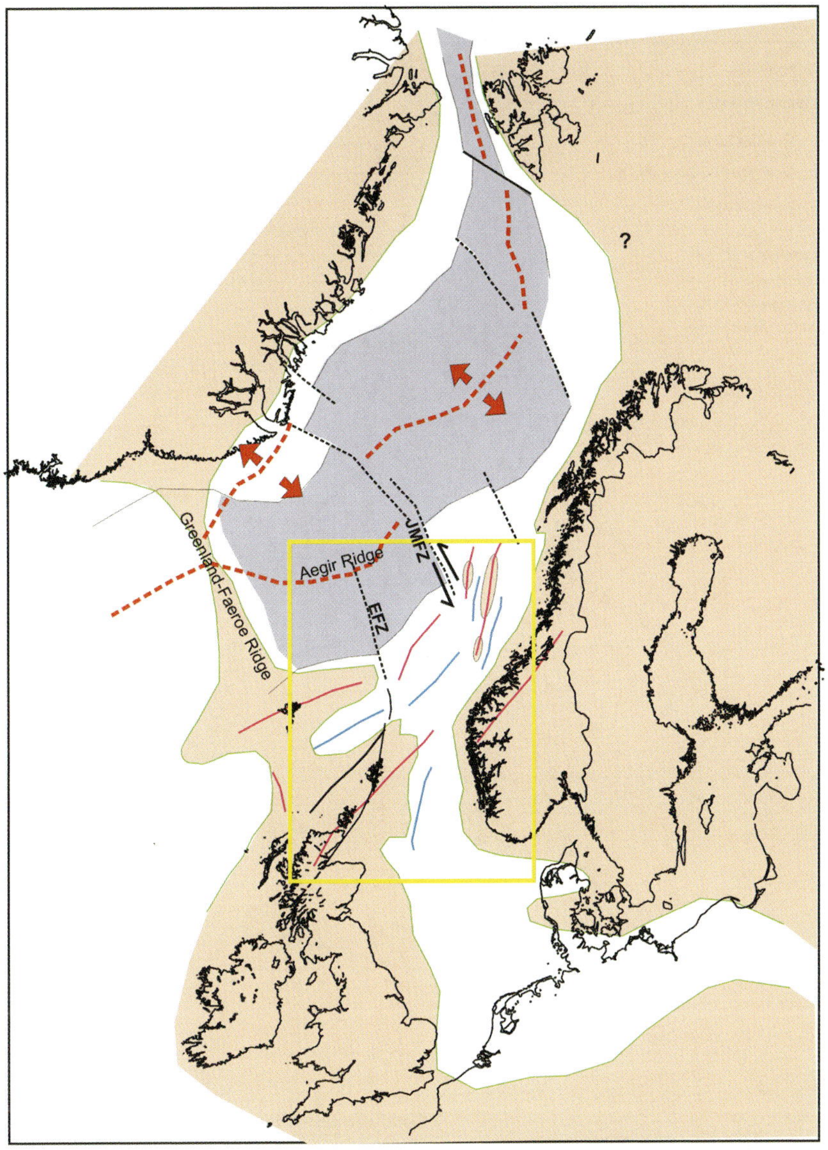

Fig. 1 Late Oligocene palaeogeography of NW Europe illustrating some key elements: (a) westward jump in spreading axis to the south of the Jan Mayen Fracture Zone (JMFZ), extinction of Aegir spreading axis and formation of the Jan Mayen microcontinent; (b) opening of a seaway to the Arctic Sea; (c) the mainly subaerially exposed Greenland–Faeroe Ridge and (d) the structurally affected margin between Faeroe and Lofoten Islands (modified from Rundberg, 1989). Red lines illustrate anticlinal areas or areas affected by uplift; blue lines illustrate synclinal areas. Yellow rectangle shows location of Fig. 2. It is uncertain whether the entire Barents Sea was exposed during this time, as indicated here.

(e.g. Rundberg, 1989) gradually changed during Oligocene into a more open basin with seaway connections to the Arctic Sea to the northeast (Fig. 1). The Greenland–Scotland Ridge acted as a barrier to the Atlantic, although surface water connections probably existed, via the Faeroe–Shetland Channel and the Denmark Strait (Eldholm and Thiede, 1980). Its subsidence history and the connection to the Atlantic have been discussed by a number of authors (e.g. Wold, 1994; Wright and Miller, 1996).

In the study area of the northern North Sea, the Eocene–Oligocene boundary probably represents one of the most important breaks within the Cainozoic. Rundberg (1989) suggested that the distinct changes in lithostratigraphy, mineralogy and biostratigraphy observed at this boundary are thought to be controlled by the large drop in global temperature at terminal Eocene time (Kennett, 1982; Ruddiman, 2000; Zachos et al., 2001), coupled with an increased oceanic circulation pattern and the rise of the passive margin at the intercept with the North Sea (Figs. 2 and 3). The latter was probably a result of the compressional regime, which also affected the northernmost North Sea.

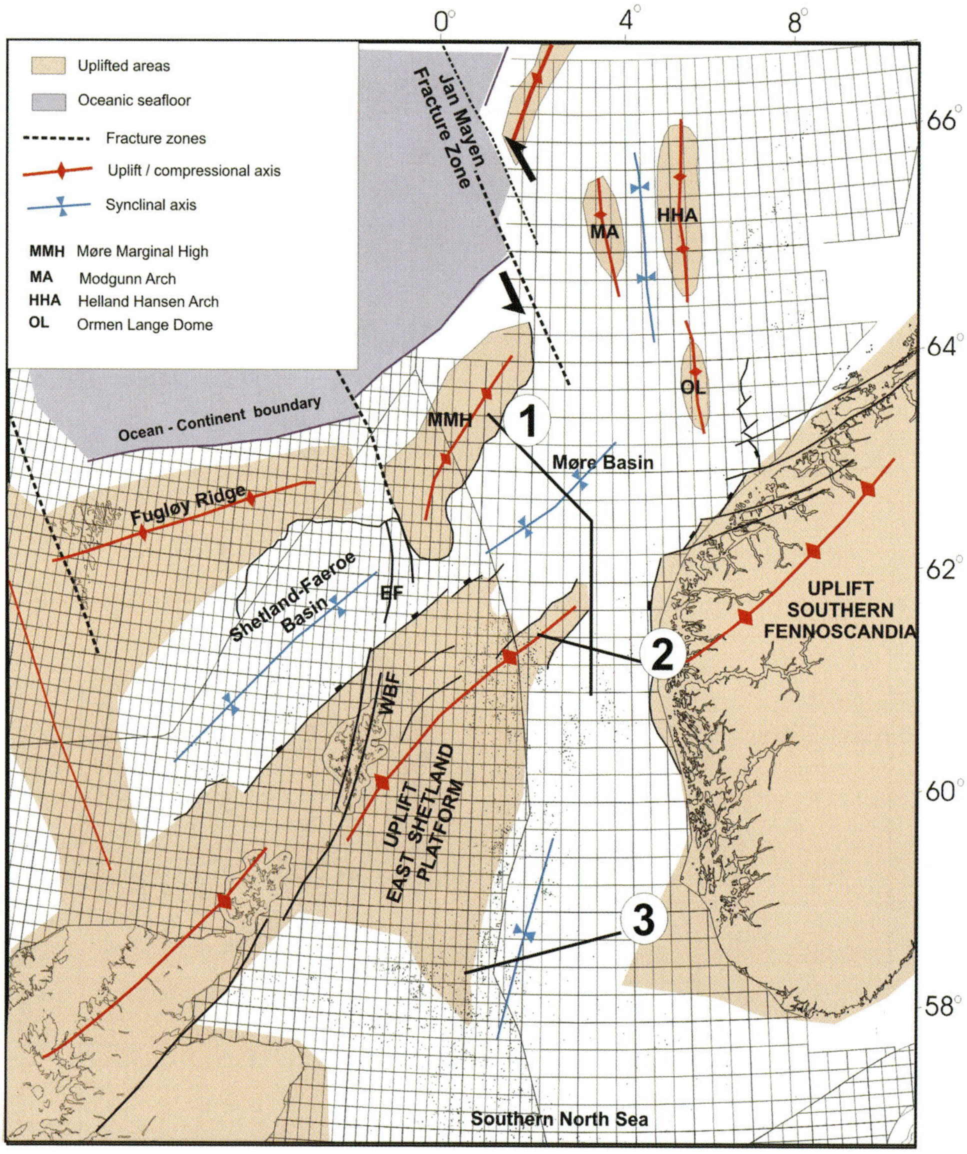

Fig. 2 Structural map of Oligocene–Miocene showing main areas affected by uplift, large structural domes (red lines). Note location of Erlend Fracture Zone and Erlend transfer Fault (EF) and its southern link with the Walls Boundary Fault (WBF) on the Shetland Platform. See insertion of map in Fig. 1 (yellow rectangle) for larger scale structural setting. Line 1, 2 and 3 refer to lines shown in Figs. 3, 4a and 4b, respectively.

During the Oligocene–Early Miocene, the northern North Sea progressively shallowed, and by the end of the period fairly shallow-marine conditions were established in the northernmost North Sea. The sediments are mainly fine-grained in nature and contain an abundance of diatoms and sponge spicules. The climatic conditions were still relatively warm and resulted in intense, locally lateritic weathering on the continents (Buchardt, 1978; Rundberg, 1989). Late Miocene paleogeographic reconstruction and sedimentary evolution of the northern sea are presented in a later section of this chapter (see Figs. 17, 18).

Oligocene–Miocene stratigraphy

Present day depositional architecture

A schematic north–south profile (Fig. 3) illustrates some key features of the present day architecture of the northern North Sea and Møre Basins. In the northern North Sea, a distinct northwards thinning of both Oligocene and Miocene strata is illustrated. A mirror image of this architecture is presented for the Møre Basin. Severe erosion is indicated at base Miocene in northernmost North Sea. This architecture

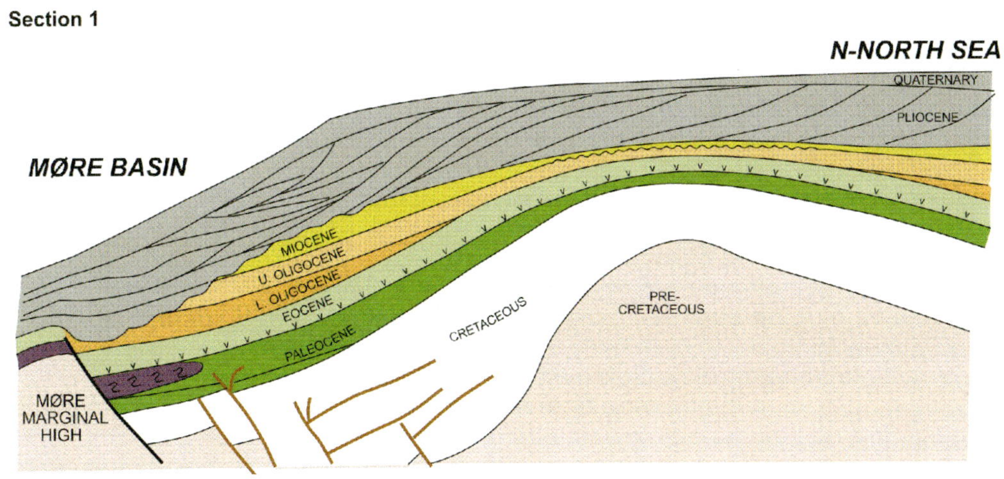

Fig. 3 Schematic illustration of the present day structural and sedimentary architecture of the Cainozoic along a N–S transect from the Møre Marginal High to the Viking Graben to the south. (Location of line shown in Fig. 2). Note the Oligocene–Miocene thinning of strata towards the Tampen Spur crest of the northernmost North Sea.

demonstrates that uplift has affected the northernmost North Sea during Oligocene–Miocene.

Two interpreted transects across the northern and southern Viking Grabens are presented in Figs. 4a and 4b, respectively. As seen, there are significant differences in the stratigraphical architectures of northern and southern Viking Grabens. The largest difference is linked to the stratal relationships just above and beneath the mid-Miocene unconformity. To the south (Fig. 4b), a very distinct infilling of sediments can be observed in the middle of the basin, above the mid-Miocene unconformity. To the north (Fig. 4a), strata below the mid-Miocene unconformity that are present to the south are either absent or heavily affected by erosion. Furthermore, the eastern flank of the basin has been heavily affected by Late Pliocene uplift, and a very characteristic thick, clinoformal system can be observed.

Lithostratigraphy

The post-Eocene lithostratigraphy of the North Sea is poorly subdivided in the Norwegian sector. Deegan and Scull (1977) subdivided Eocene to Lower Miocene strata into the Hordaland Group and the early Miocene to Recent into the Nordland Group. The only formation defined by these authors within the post-Eocene was the sandy Utsira Formation at the base of the Nordland Group. Isaksen and Tonstad (1989) adopted this nomenclature, and also recognised two sandy formations in the Oligocene part of the Hordaland Group, which they termed, Skade and Vade Formations, present in the Viking Graben and central North Sea, respectively. In his work on the Tertiary sediments of the Norwegian North Sea (60–62°N), Rundberg (1989) subdivided the Hordaland and Nordland Groups into four lithostratigraphic associations and nine lithostratigraphic units. This subdivision was based on a detailed study of sediments from eleven wells along two transects in the eastern part of the northern North Sea.

In the UK part of the basin, Knox and Holloway (1992) established the Westray Group as a new lithostratigraphic unit. It formed the upper of the two groups which they had introduced to replace the Hordaland Group. They introduced the Lark Formation for the distal, mudstone-dominated facies of the Westray Group and used the Skade Formation for glauconitic sandstones and siltstones of shelf-facies. Recently, Fyfe et al. (2002) published an updated lithostratigraphy of the central and northern North Sea, based on Isaksen and Tonstad (1989), Knox and Holloway (1992) and recent dating by Eidvin et al. (1999) and Eidvin and Rundberg (2001).

We present in this work a revised lithostratigraphic and chronostratigraphic subdivision of the Oligocene–Miocene of the Norwegian northern North Sea. This subdivision is illustrated in Figure 5. The Skade Formation is in this work assigned to the Early Miocene, and not to Oligocene, as defined earlier by Isaksen and Tonstad (1989) and a more precise age assignment is given to the Utsira Formation. As shown in Fig. 5, a large hiatus separates the Hordaland and Nordland Groups, which is widespread in the Norwegian continental shelf and has been termed

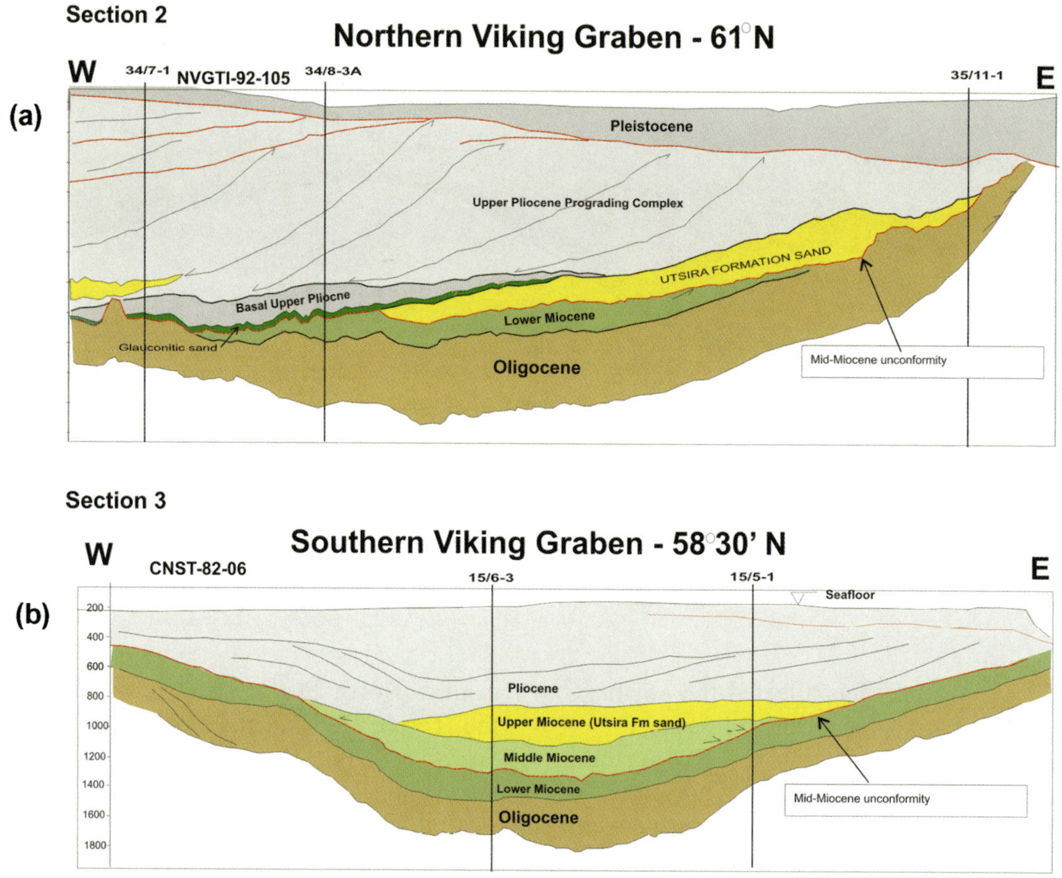

Fig. 4 (a) East–west transect of the northern North Sea at about 61°N illustrating main sequences and sedimentary architecture of the post-Eocene strata. Note the mid-Miocene unconformity (red line) and the seismic truncation of the Lower Miocene. See Fig. 2 for location of the line. (Modified from Eidvin and Rundberg, 2001). (b) East–west transect of the southern Viking Graben at about 58°30′N illustrating main sequences and sedimentary architecture of the post Eocene. Note the mid-Miocene unconformity and the infilling of the Middle Miocene sequence (this sequence is absent to the north). See Fig. 2 for location of the line.

the mid-Miocene hiatus by many workers. In the northern North Sea, it increases dramatically northwards, from almost no time-gap in the southern Viking Graben to about 15–20 million years, in the northernmost North Sea. The succession below the hiatus comprises sediments of Oligocene and early Miocene age. The Lower Oligocene sands are unnamed in the Norwegian sector.

Of particular notice is the recognition of an inconsistency in the definition of Skade and Utsira Formations. As presented in Figure 6, there is an overlap in definitions of the Utsira and Skade Formations in which the Skade Formation, as defined in type well 24/12-1, correlates to the lower part of the Utsira Formation as defined in type well 16/1-1. Using the type well boundaries for the Utsira, it would embrace sediments of almost the entire Miocene epoch and would also include the topmost part of the Hordaland Group. This is obviously in conflict with the definition and common usage of the Utsira Formation, being related in time and place to the Nordland Group.

Problems in seismic subdivision and biostratigraphic dating

The Oligocene–Miocene succession of the northern North Sea has been described seismically by many workers (e.g. Rundberg, 1989; Rundberg and Smalley, 1989; Galloway et al., 1993; Jordt et al., 1995; Gregersen et al., 1997; Martinsen et al., 1999). Still, however, there are many problems with a detailed sequential understanding and mapping of these strata. The reasons for this are threefold.

Firstly, much of the Oligocene–Miocene strata display a poor seismic resolution due chaotic seismic reflection pattern in large parts of the basin. The chaotic reflections can be observed particularly within the Viking and the Sogn Grabens and within parts of the southern Møre Basin. Much of the chaotic pattern probably represents highly deformed strata resulting from severe gas leakage from the Jurassic source rocks. A discussion of the origin of the chaotic facies has recently been presented by Løseth et al. (2003).

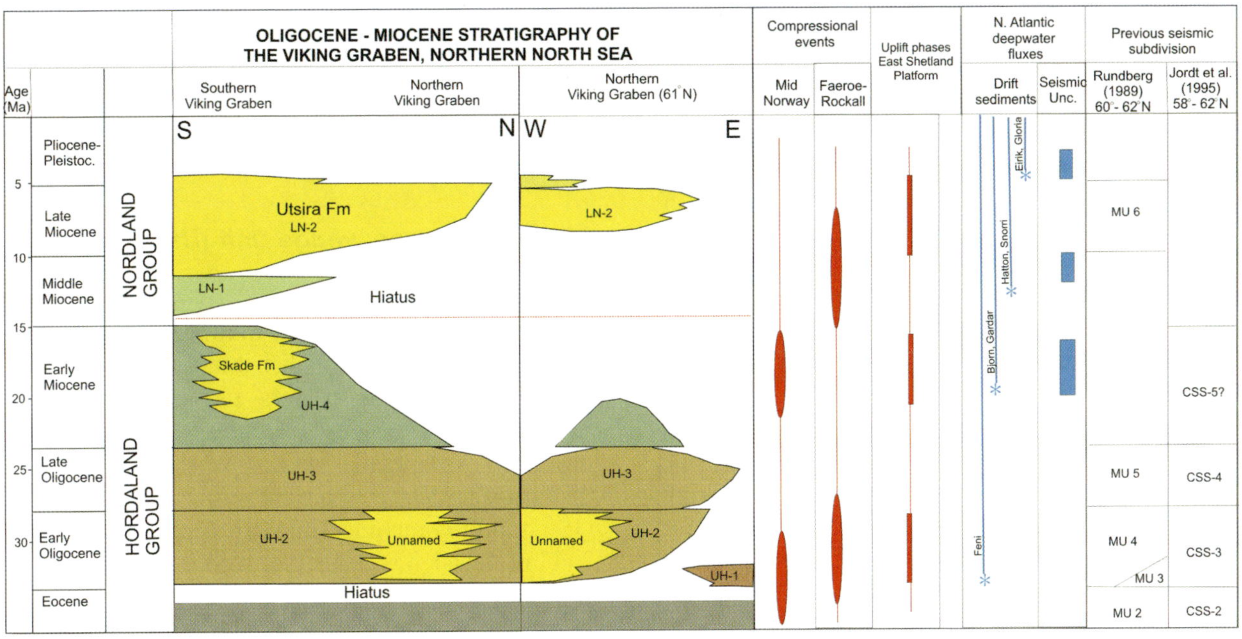

Fig. 5 Oligocene–Miocene stratigraphy of the Viking Graben, northern North Sea. Note the Middle Miocene sequence (new) and the chronology of the Skade Formation sands (new dating by Eidvin et al. 2002). North Atlantic deep-water fluxes showing periods of major drift accumulation to the south of the Greenland–Scotland Ridge (after Wold, 1994) and timing of seismic unconformities (after Wright and Miller, 1996). Compressional events of the Northwest European Atlantic margin after Lundin and Doré, 2002.

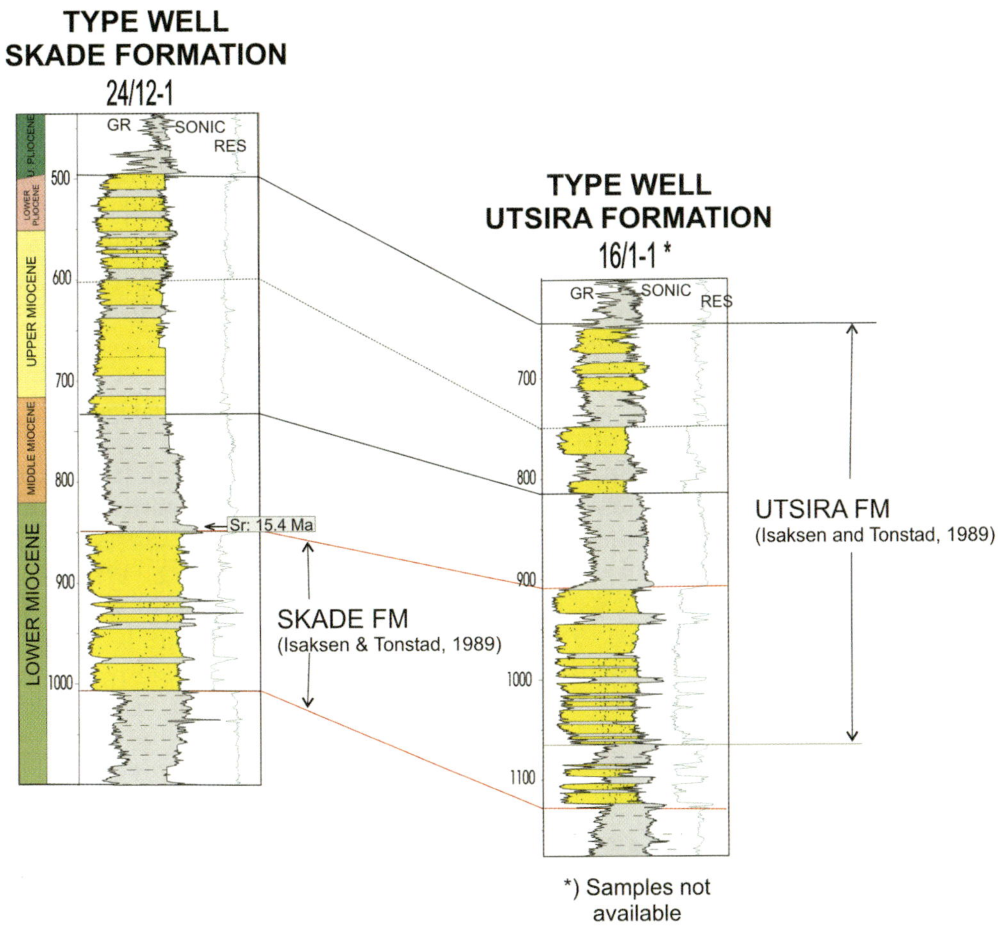

Fig. 6 Log correlation between well 24/12-1 (type well of the Skade Formation) and well 16/1-1 (type well of the Utsira Formation) illustrating conflict in previous definition of the Utsira Formation (see text for details). Left panel in well 24/12-1 shows results of biostratigraphic dating. Sr = Sr isotopic dating.

Secondly, there are often problems with the biostratigraphic dating of the Upper Oligocene and Lower Miocene sections, and the top Oligocene is a difficult boundary to assign using micropalaeontology. This is often seen in well completion logs from the northern North Sea. In many cases, the top Oligocene varies from well to well, and even between closely spaced wells. This is probably due to the fact that different biostratigraphic workers and consultants use different index fossils for the Oligocene/Miocene boundary, and due to severe reworking of fossil tests.

Thirdly, an inconsistency in the definition of Skade and Utsira Formations (see Fig. 6) has apparently caused much conflict for seismic workers in this part of the basin. Gregersen et al. (1997) for example, included Lower Miocene strata in their mapping of the Utsira in southern Viking Graben (e.g. their Fig. 6). This error probably results from the definition of the Utsira Formation in its type well 16/1-1. In their mapping of the Lower Miocene of the northern North Sea, Jordt et al. (1995) erroneously included included Upper Miocene strata (Utsira sands) in Unit CSS-5 (their Figs. 9 and 15).

Depositional systems and seismic units

The Oligocene–Miocene succession of the northern North Sea can broadly be subdivided into two megasequences. The base of each megacycle is marked by an unconformity or a regional hiatus. The lower megasequence comprises the Lower Oligocene to Lower Miocene succession, or the upper part of the Hordaland Group. It consists of a compounded system of sequences, which are best defined towards the margins of the basin. In general, much of the Oligocene strata display a chaotic seismic reflection pattern, which make it difficult to map regionally. Locally, however, it is possible to map the Oligocene in a more detailed manner. The upper megasequence comprises the Middle Miocene to Lower Pliocene succession, or the lower part of the Nordland Group. This megasequence has been subdivided into two seismic units. All of these units will briefly be described here.

Lower Oligocene–Lower Miocene megasequence

The Oligocene to Lower Miocene succession (upper part of Hordaland Group) has been subdivided into four seismic units (UH-1–UH-4; Fig. 5): (1) A wedge-shaped seismic unit (UH-1) confined to the eastern part of the basin, assigned to lowermost Oligocene; (2) a Lower Oligocene unit (UH-2) derived mainly from the west; (3) an Upper Oligocene unit (UH-3), and (4) a Lower Miocene unit (UH-4).

Lower Oligocene wedge unit (UH-1)

This unit occurs parallel to the Norwegian coastline and can easily be mapped between ~60° and 61°30′N. The areal distribution of this unit is shown in Fig. 7a. It was termed map unit 3 (lithologic unit B3) by Rundberg (1989) and was described in detail in his work. It was not distinguished as a separate unit by Jordt et al. (1995), and is included in their sequence CSS-3.

Unit UH-1 has a distinct wedge-shaped geometry with a thickness in excess of 200 m, and rapidly pinches out basinwards, as illustrated in seismic sections (Figs. 8a, 8b, 9). Seismically, it displays a characteristic low-amplitude, occasionally transparent reflection pattern, which can be obviously related to a uniform lithology as expressed by well log data. Lithologically, the unit consists of a very uniform series of noncalcareous, dark brownish claystones, which become slightly coarser upwards. In well 31/3-1 (Fig. 8b), the clays grade upward into glauconitic, sandy siltstones. Rundberg and Smalley (1989) reported Early Oligocene Sr isotope ages of samples from well 31/3-1. Such ages have later been confirmed in the same well by Eidvin and Rundberg (2001) using the same dating methodology and biostratigraphical correlation, and by Sejrup et al. (1995) in a borehole, cored nearby.

Lower Oligocene unit (UH-2)

The Lower Oligocene unit has been distinguished seismically in the northern part of the basin (between 60–61°N). It corresponds to seismic map unit 4 of Rundberg (1989) and to seismic sequence CSS-3 of Jordt et al. (1995). The outline of the Lower Oligocene unit is shown in Fig. 7a. It overlies Eocene strata and pinches out eastward, close to the western limit of the Lower Oligocene clastic wedge (UH-1; Figs. 8, 9).

Between 60–61°N the top of the unit is defined by a moderate to high-amplitude, semi-continuous seismic reflector in the eastern part of the basin (Figs. 8, 9). This reflector corresponds to an abrupt downward increase in sonic and particularly density levels, as illustrated in Figure 10. The top of unit UH-2 corresponds to a diagenetic horizon characterised by the transition from opal-A to

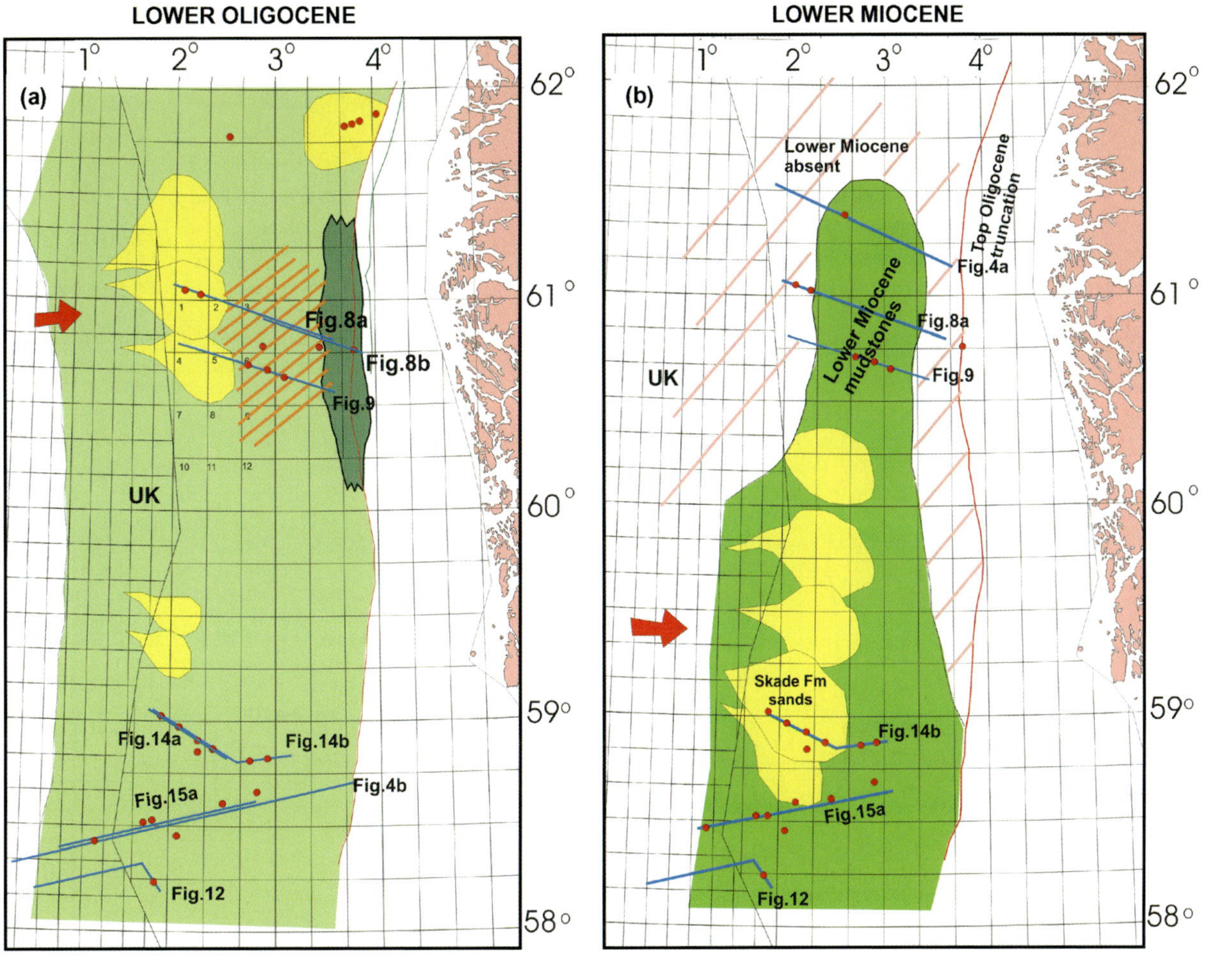

Fig. 7 (a) Distribution of Lower Oligocene sediments in northern North Sea; seismic units UH-1 (dark green) and UH-2 (light green). Approximate outlines of sands derived from west (Statfjord–Tampen area) and sands derived from east (Agat area). Hatched area shows approximate outline of siliceous-rich sediments (in the form of opal-CT). (b) Distribution of Lower Miocene sediments (seismic unit UH-4) in northern North Sea with approximate outline of the sandy Skade Formation. Hatched area shows erosion of Lower Miocene. (c) Distribution of Middle Miocene sediments (seismic unit LN-1) in northern North Sea. The northward extent is uncertain. Hatched area shows not deposited Middle Miocene strata. (d) Distribution of Utsira Formation sands (seismic unit LN-2) in northern North Sea. Yellow area shows outline of lower part (main Utsira sands, predominantly ?Late Miocene age), orange area shows outline of upper part of the Utsira Formation (Early Pliocene age). Green area to the north shows outline of thin glauconitic member extending beyond the main sand.

opal-CT rich mudstones (Fig. 11). Locally, at the top of the unit, it defines a flat seismic event cutting inclined reflectors. This siliceous-rich mudstone lithology is thought to be present locally within the northern North Sea, particularly between 60–61°N (see hatched area Fig. 7a), as interpreted from the seismic and wireline log data.

Westward, the Oligocene strata are severely affected by seismic disturbance, and it is difficult to map the top of the unit. It is probably best defined in the northern part, at about 60°45′–61°N, as illustrated in the two seismic sections, shown in Figs. 8a and 9. Along both of these profiles, the seismic reflector defining the top of the high-density zone can, with some degree of certainty, be correlated to discontinuous, high-amplitude seismic events further to the west. These events define the top of a thick sandy interval, which is penetrated in two wells (34/10-17 and 34/10-23) along the seismic profile, shown in Figure 8a. The sands make up a gross thickness of about 400 m in block 34/10 (Fig. 8c). They are clearly turbiditic in origin, and their areal extent is shown in Figure 7a. The Lower Oligocene sands are unnamed in the Norwegian sector.

Lower Oligocene sands are also present in the Agat area (block 35/3; Fig. 7a), as described by Rundberg (1989) and Rundberg and Smalley (1989). These sands (termed, subunit 3 in Rundberg, 1989) are distinguishable from the underlying sands (subunits 1 and 2) by their relatively high-content of glauconite, shell debris and lignites and by common calcite-cemented sandstone horizons. Rundberg (1989) interpreted subunits 1 and 2 to

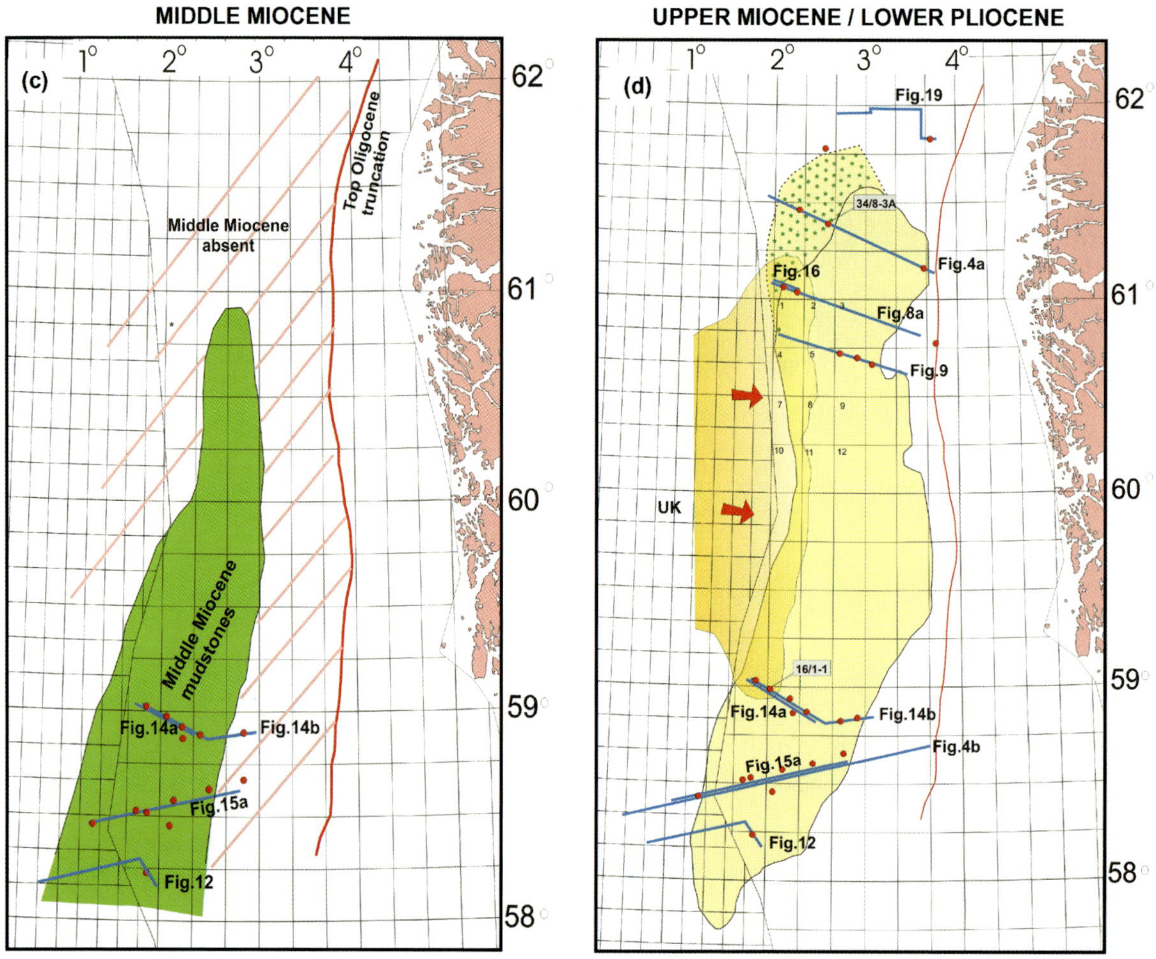

Fig. 7 Continued.

represent gravity-flow sands and interpreted a dramatic shallowing to take place with the incoming of subunit 3.

Upper Oligocene unit (UH-3)

The top of the Oligocene is commonly difficult to pick biostratigraphically in wells in the basin centre, as described earlier. Also seismically, it may be difficult to distinguish from the conformably overlying Lower Miocene sediments. In addition, a chaotic seismic reflection pattern causes mapping problems over large parts of the basin. Earlier workers in the northern North Sea have therefore grouped all sediments of the upper part of the Hordaland Group in one compound unit (e.g. Map unit 5 of Rundberg, 1989; CSS-4 of Jordt et al., 1995).

The outline of the Upper Oligocene unit (UH3) is not presented here, but is largely similar to that of the Lower Oligocene unit UH2 (Fig. 7a). Toward the eastern margin (between 60–61°N), it clearly onlaps the underlying wedge-shaped unit UH-1 (Figs. 8, 9). To the north of 60°N, the top of the unit becomes eroded at both margins, as illustrated in Fig. 4a, and is here unconformably overlain by Upper Miocene and Pliocene sediments.

We have defined the top of the Oligocene by detailed biostratigraphic investigations in wells 15/9-13, 15/12-3, 16/1-4 and 24/12-1 in the southern Viking Graben. The Miocene–Oligocene boundary is mainly based on the last appearance datum (LAD) of Diatom sp. 3 (King, 1983) (Tab. 2). In wells 15/12-3 and 16/1-4, the biostratigraphic interpretations are confirmed by Sr isotope stratigraphy (see Figs. 12 and 13). Similarly, in the Tampen area in the northern North Sea, we defined the Miocene–Oligocene boundary in wells 34/8-1 and 34/8-3A by biostratigraphic correlation and Sr isotope stratigraphy (Eidvin and Rundberg, 2001).

In well 15/12-3 in southern Viking Graben, the top of the Oligocene corresponds to a low-amplitude seismic event (Fig. 12) which can be traced with relatively good precision to the south of about 58°30′N. In the Tampen area, the top of

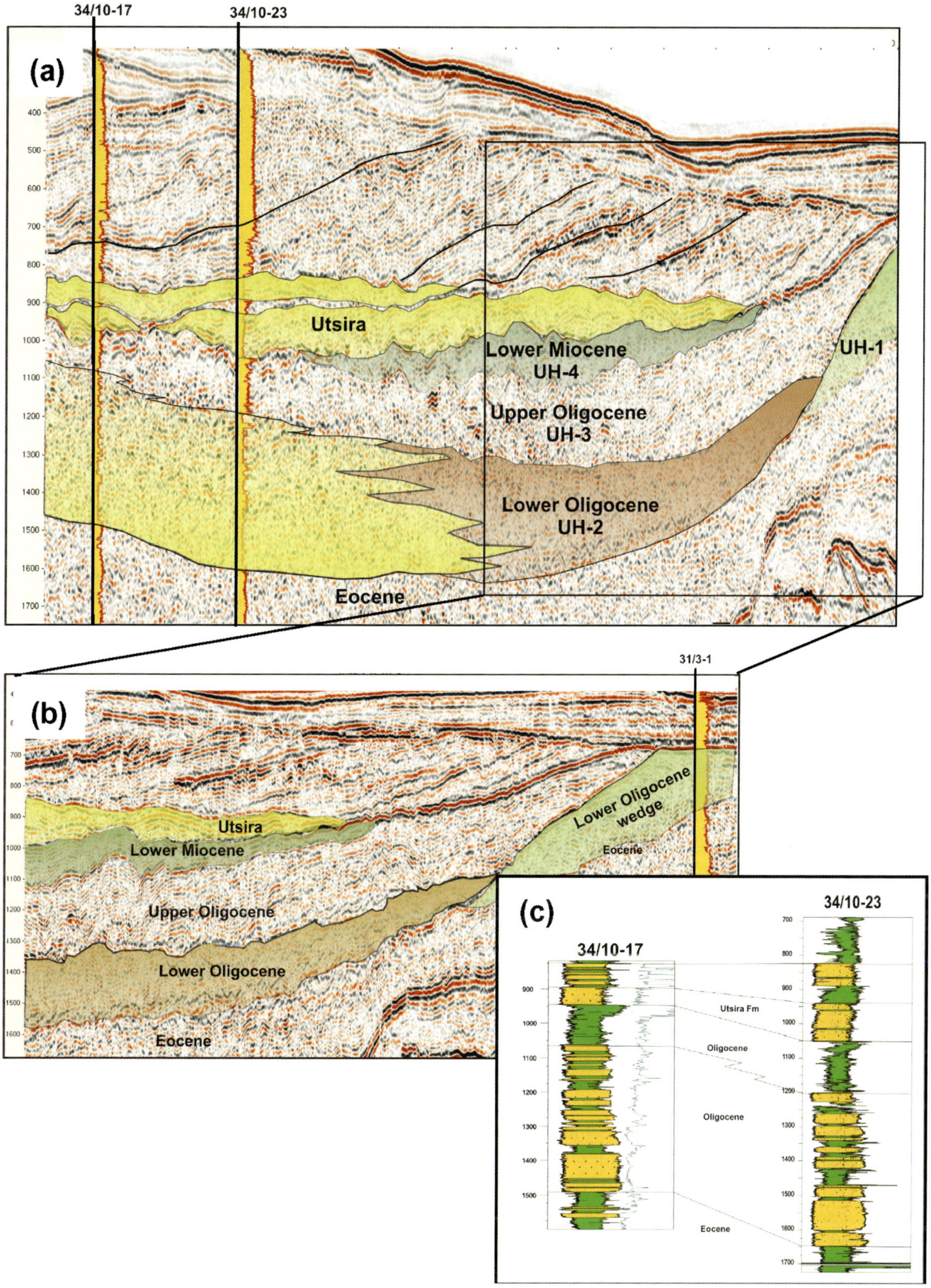

Fig. 8 (a) Seismic section (line NVGTI-92-208) across northern Viking Graben through wells 34/10-17 and 23 showing subdivision of Oligocene–Miocene strata at about 61°N. Note wedge-shaped Lower Oligocene seismic unit (UH-1) to the east; sand-rich wedge of sediments (seismic unit UH-2) pinching out to the east; lower Miocene strata (seismic unit UH-4) preserved in the middle of the basin. Location of line shown in Figs. 7a, 7b and 7d. (b) Seismic section (eastern part of NVGT-92-208) through well 31/3-1 illustrating stratal relationship between lower Oligocene units. Location of line shown in Fig. 7a. (c) Log correlation between wells 34/10-17 and 34/10-23 along seismic section shown in Fig. 8a.

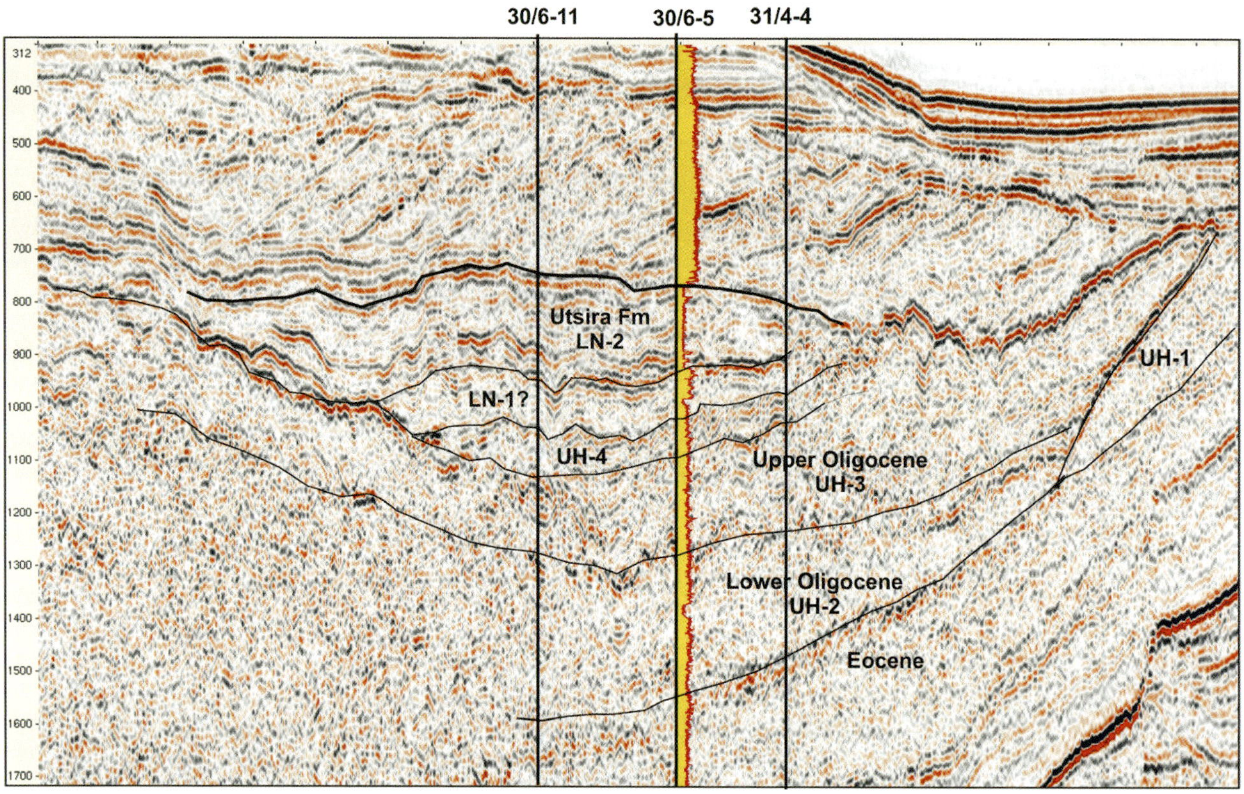

Fig. 9 Seismic section through wells 30/6-11, 30/6-5 and 30/6-4 illustrating seismic subdivision of the Oligocene–Miocene succession. Note main Utsira Formation sands (lower part) pinching out in westward direction and high-amplitude reflectors to the west (upper part of Utsira Formation) denoting influx of sands from the East Shetland Platform. Location of line shown in Figs. 7a–d.

the Oligocene also closely corresponds to a seismic reflector (well 34/8-3A), which allows a precise mapping in the northern part of the basin. Between 59–61°N, we have not executed new, biostratigraphical investigations, since the top of the Oligocene may be difficult to pick seismically. The mapping is also complicated by a chaotic seismic reflection pattern, which affects the Hordaland Group, over much of the central basin area.

Lithologically, the Upper Oligocene unit comprises dominantly; these are, mudstones however, only scattered with thin sands. Some sands are however noted in wells of block 30/2 and 30/3. In the northern part of the basin, the unit coarsens upward to silty sands and siltstones at the top (e.g. well 34/2-2, described by Rundberg, 1989). The siltstones are typically rich in sponge spicules and glauconite. In the Agat area, close to the eastern margin of the basin, the unit is represented by a progradational system with fine-grained, glauconitic sands at the top. The planktonic fossil assemblage is dominated by pyritised diatoms and radiolaria. Calcareous foraminifera dominate a moderately rich benthic fauna in most wells, but agglutinated forms are common in some areas (e.g. wells 15/12-3 and 15/9-13).

On wireline logs, the unit displays a slightly serrated, but otherwise stable, low gamma-log profile in wells to the north of 60°N (e.g. wells presented in Figs. 10 and 11). In the southern Viking Graben, the topmost part of the Oligocene shows an upward change to higher gamma-ray levels (e.g. well 15/12-3, Fig. 12). In the Statfjord area (block 34/10) for example, the very top of the Oligocene section displays characteristic high velocity and resistivity log values (Fig. 8c).

Lower Miocene unit (UH-4)

This unit comprises the topmost part of the Hordaland Group. The outline of the unit is shown in Fig. 7b. In the southern Viking Graben, it conformably overlies Oligocene strata (Fig. 4b). It is overlain by Middle Miocene sediments in the centre of the basin and Pliocene sediments at the margins. To the north of 60°N, the Lower Miocene unit is only present in the central basin and absent at the margins to the west and east (Figs. 4a, 7b). On seismic sections, the top of the unit can be defined by erosional truncation, onlap or downlap reflection terminations, as schematically illustrated in Figures 4a and 4b. In the

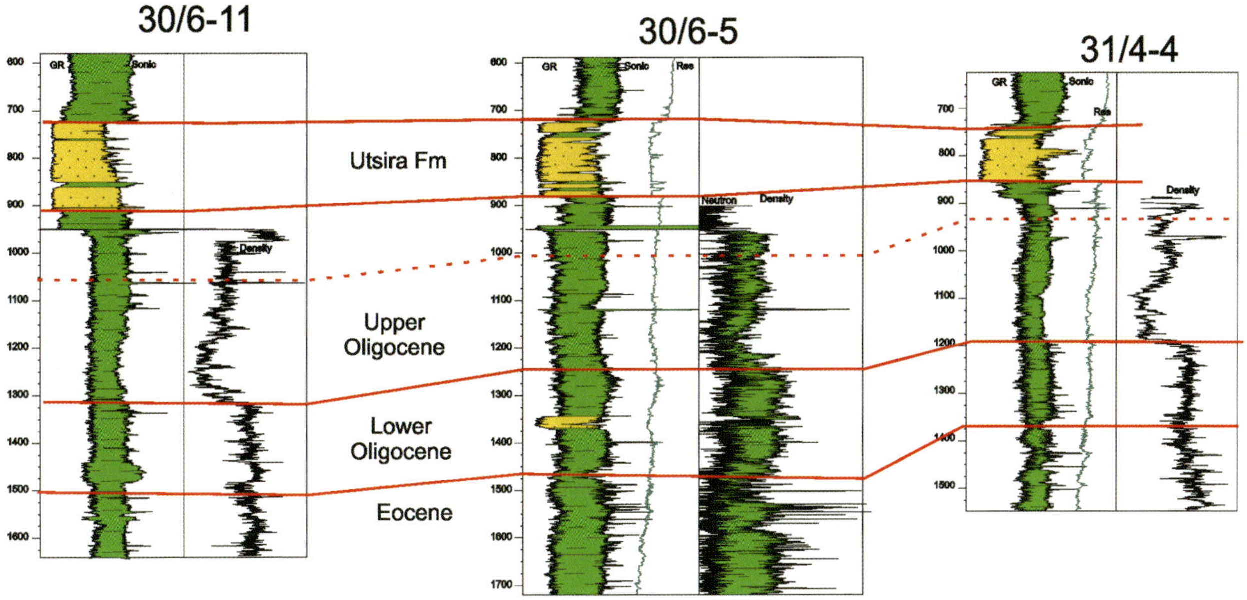

Fig. 10 Log correlation between wells 30/6-11, 30/6-5 and 31/4-4 along seismic section shown in Fig. 9 illustrating very abrupt downward density increase which marks the top of the lower Oligocene unit (UH-2). This boundary can be seen as a discontinuous, high-amplitude seismic response in Fig. 9.

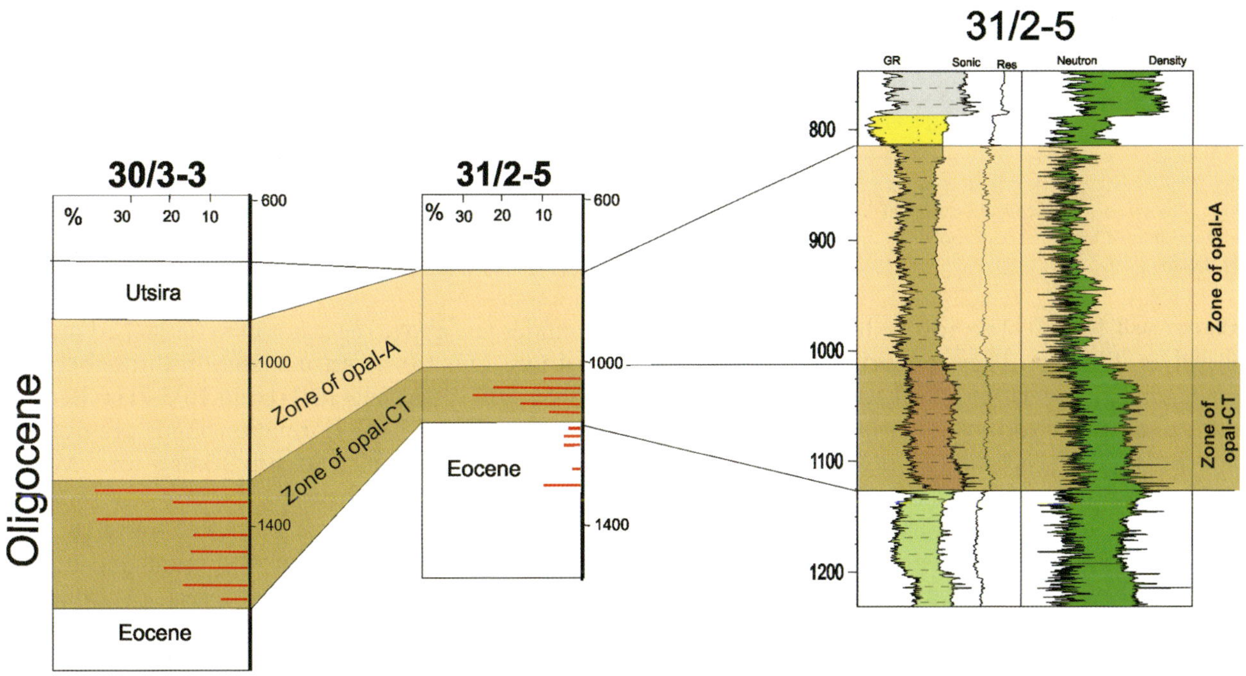

Fig. 11 Distribution of siliceous sediments (opal-CT) in wells 30/3-3 and 31/2-5. Note abrupt increase in opal-CT at the Eocene–Oligocene boundary. Top of opal-CT zone corresponds to abrupt downward density increase in well 31/2-5. This marks the top of seismic unit UH-2 (see Fig. 9). Modified from Rundberg (1989).

northernmost North Sea, between 61°30′ and 62°N, the unit has been completely eroded. This erosional period, termed mid-Miocene erosional event, is further dealt with in this chapter.

Unit UH-4 corresponds to the upper part of map unit 5 of Rundberg (1989) and to CSS-5 of Jordt et al. (1995). In the latter work, however, there are conflicts in the interpretation of the Miocene strata within the northern North Sea, in which Upper Miocene Utsira sands have been mistakenly included in CSS-5 (their Figs. 3, 9).

Lower Miocene strata have recently been described in the Tampen area by Eidvin and Rundberg (2001), and comprise mud-prone lithologies. In large parts of the Viking Graben, a sandy section makes up a great proportion of the Lower Miocene unit. These sands are referred to as the Skade Formation and reach a gross thickness up to

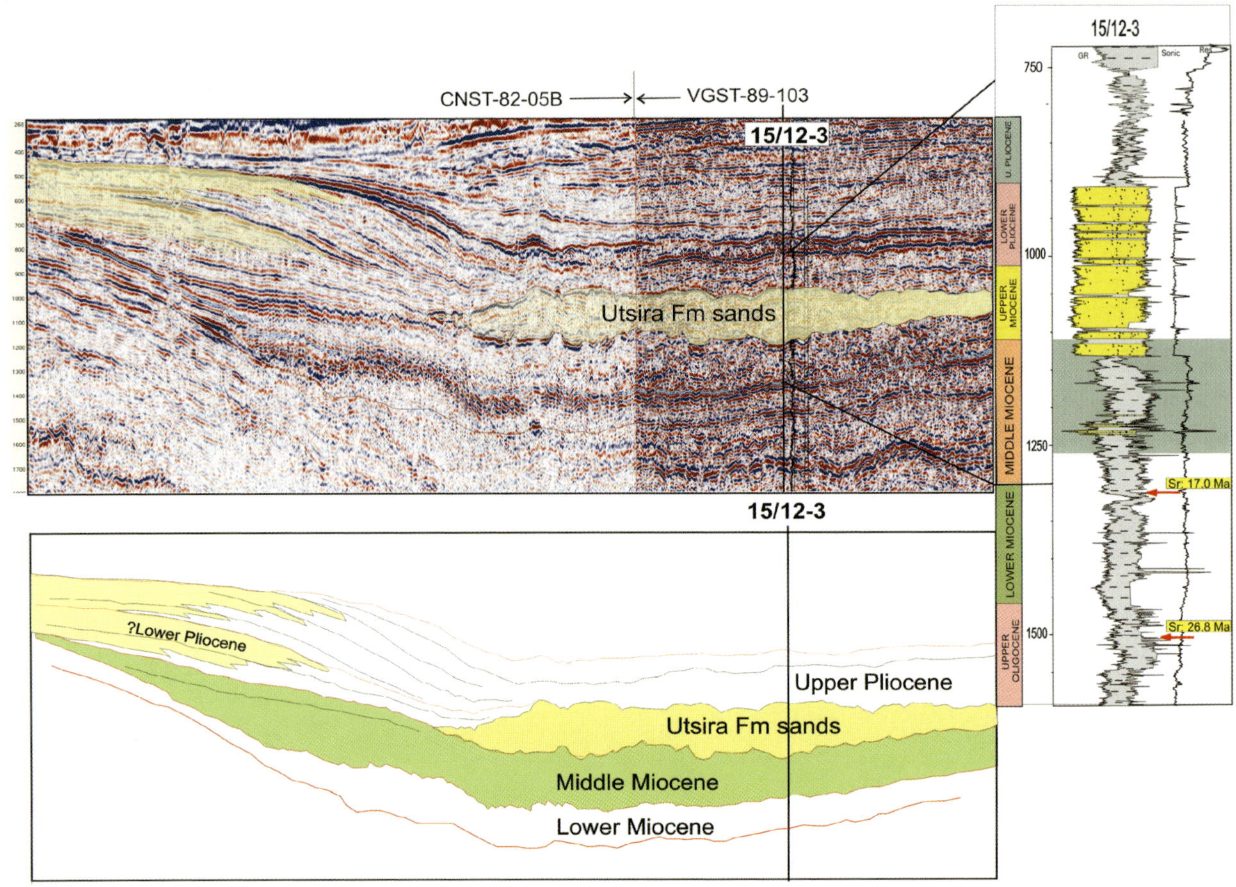

Fig. 12 Seismic profile across southern Viking Graben through wells 15/12-3 (well panel inserted) showing stratigraphical relationships of Miocene sediments. Location of line shown in Figs. 7a–d. Left panel of 15/12-3 shows results of biostratigraphic dating. Green-coloured part of log panel corresponds to the occurrence of the Middle Miocene *Bolboforma badenensis* and *B. reticulata* assemblages. See text for details.

300 m (well 16/1-4). The areal extent of the sands is shown in Fig. 7b. They comprise a sequence of amalgamated sands in alternation with thinner mudstones. Detailed biostratigraphic investigations of well 24/12-1 (type well of the Skade Formation) show that the sands are Early Miocene in age (Tables 1 and 2; Fig. 13) and not Late Oligocene, as suggested by Isaksen and Tonstad (1989).

A log correlation diagram between key wells in blocks 16/1 and 24/12 is presented in Fig. 13. As can be seen from this figure, the top of the Skade sands corresponds closely to the top of the Lower Miocene in wells with good biostratigraphic control. In well 24/12-1 (type well of the Skade Formation), the Early Miocene *Uvigerina tenuipustulata–Asterigerina guerichi staeschei* benthic foraminiferal assemblage and the Early Miocene *Globorotalia zealandica–Globigerina ciperoensis* planktonic foraminiferal assemblage have been identified for the intervals 820–1020 and 840–1040 m, respectively (Tables 1 and 2), both embracing the strata above and below the Skade Formation sands. The biostratigraphic data are supported by Sr isotope stratigraphy, which yield ages of about 15 Ma for samples just above the sands in wells 24/12-1 and 16/1-4 (see Fig. 13). A seismic section through all the three wells (plus wells 16/1-1 and 16/3-2) is shown in Fig. 14a. The Skade Formation sands are overlain by mudstones which clearly onlap the Middle Miocene surface to the east.

Further to the south, in blocks 15/6 and 16/7, the Lower Miocene unit has a maximum thickness of about 250 m. In Fig. 15a, is shown a seismic line through seven wells, close to the southern pinchout of the Skade sands (see Fig. 7b). Here, the Lower Miocene deposits comprise a stacked series of upward-coarsening subunits (up to 50 m thick), particularly well observed in wells 15/6-3 and 15/6-5. The top of the unit is taken at a very distinct high radioactive marker (defined in 15/12-3; Fig. 12), which can be identified in a number of wells in the southern Viking Graben (Fig. 15b). This marker defines the transition between the Hordaland and Nordland Groups, in this part of the basin.

The Lower Miocene section contains a rich planktonic assemblage including foraminifera, diatoms and radiolaria. Calcareous foraminifera

Table 1 Benthic Foraminiferal assemblages in southern Viking Graben wells.

Benthic Foraminiferal assemblages	Age interpretation	15/9-A-11	15/9-A-23	15/9-13	15/12-3	16/1-2	16/1-4	24/12-1
Elphidium excavatum–Haynesina orbiculare	Early to Middle Pliocene				200–380 m			
Elphidium excavatum–Cassidulina teretis	Early Pleistocene				380–600 m		357.5–480.5 m	
Cibicides grossus	Late Pliocene to Early Pleistocene	912. 4 m (one sample)			600–900 m	710–740 m	480.5–760 m	480–500 m
Cibicidoides pachyderma	Late Pliocene	912.8–913.1 m						
Monspeliensina Pseudotepida	Early Pliocene					740–750 m		500–520 m
Uvigerina venusta saxonica	Late Miocene to Early Pliocene		1080 m (one sample)		900–1110 m	750–870 m	760–770 m	520–720 m
Uvigerina pygmea langeri–Uvigerina pygmea langenfeldensis	Middle Miocene			1110–1160 m		870–880 m	860–912.5 m	720–820 m
Uvigerina pygmea langenfeldensis	Middle Miocene				1110–1250 m			
Asterigerina guerichi staeschei	Middle Miocene			1160–1190 m	1250–1300 m			
Uvigerina tenuipustulata–Asterigerina guerichi staeschei	Early Miocene						912–1090 m	820–1020 m
Uvigerina tenuipustulata	Early Miocene			1190–1320 m	1300–1340 m			
Plectofrondicularia seminuda	Early Miocene			1320–1480 m	1340–1460 m			1020–1090 m
Spirosigmoilinella compressa	Latest Late Oligocene to Early Miocene						1090–1190 m	
Turrilina alsatica	Late Oligocene to Earliest Miocene			1480–1550 m			1190–1260 m	1090–1240 m
Annectina biedai–Turrilina alsatica	Late Oligocene				1460–1520 m			
Rotaliatina bulimoides	Early Oligocene						1260–1400.5 m	

Table 2 Planktonic fossil assemblages in southern Viking Graben wells.

Planktonic fossil assemblages	Age interpretation	15/9-13	15/9-A-11	15/9-A-23	15/12-3	16/1-2	16/1-4	24/12-1
Neogloboquadrina pachyderma (dextral)	Late Pliocene				790–840 m		650–670 m	
Upper *Neogloboquadrina atlantica* (dextral)	Late Pliocene		913.1 m (one sample)		840–850 m		670–720 m	
Globigerina bulloides	Early to Late Pliocene				850–860 m			480–510 m
Neogloboquadrina atlantica (sinistral)	Early to Late Pliocene				860–940 m	710–740 m	720–763.5 m	
Globorotalia puncticulata	Early Pliocene			1080 m (one sample)	940–1010 m	740–780 m	763.5–770 m	510–550 m
Lower *Neogloboquadrina atlantica* (dextral)	Late Miocene					780–870 m		550–700 m
Neogloboquadrina atlantica (dextral)–*Neogloboquadrina acostaensis*	Late Miocene				1010–1110 m			
Bolboforma fragori	Late Miocene							700–720 m
Bolboforma badenensis	Middle Miocene	1110–1140 m				870–880 m		720–790 m
Bolboforma badenensis–*Bolboforma reticulata*	Middle Miocene				1110–1260 m		860–912.5 m	
Bolboforma reticulata	Middle Miocene	1140–1160 m						790–840 m
Globigerina praebulloides	Middle Miocene				1260–1300 m			
Globigerina praebulloides–*Globigerinoides quadrilobatus triloba*	Early- Middle Miocene	1160–1200 m						
Globorotalia zealandica–*Globigerina ciperoensis*	Early Miocene	1200–1310 m			1300–1340 m		912.5–950 m	840–1040 m
Diatom sp.4	Early Miocene	1310–1480 m			1340–1460 m		1030–1180 m	1040–1130 m
Diatom sp.3	Early Oligocene to Late Oligocene	1480–1550 m			1340–1460 m		1180–1400.5 m	1130–1240 m

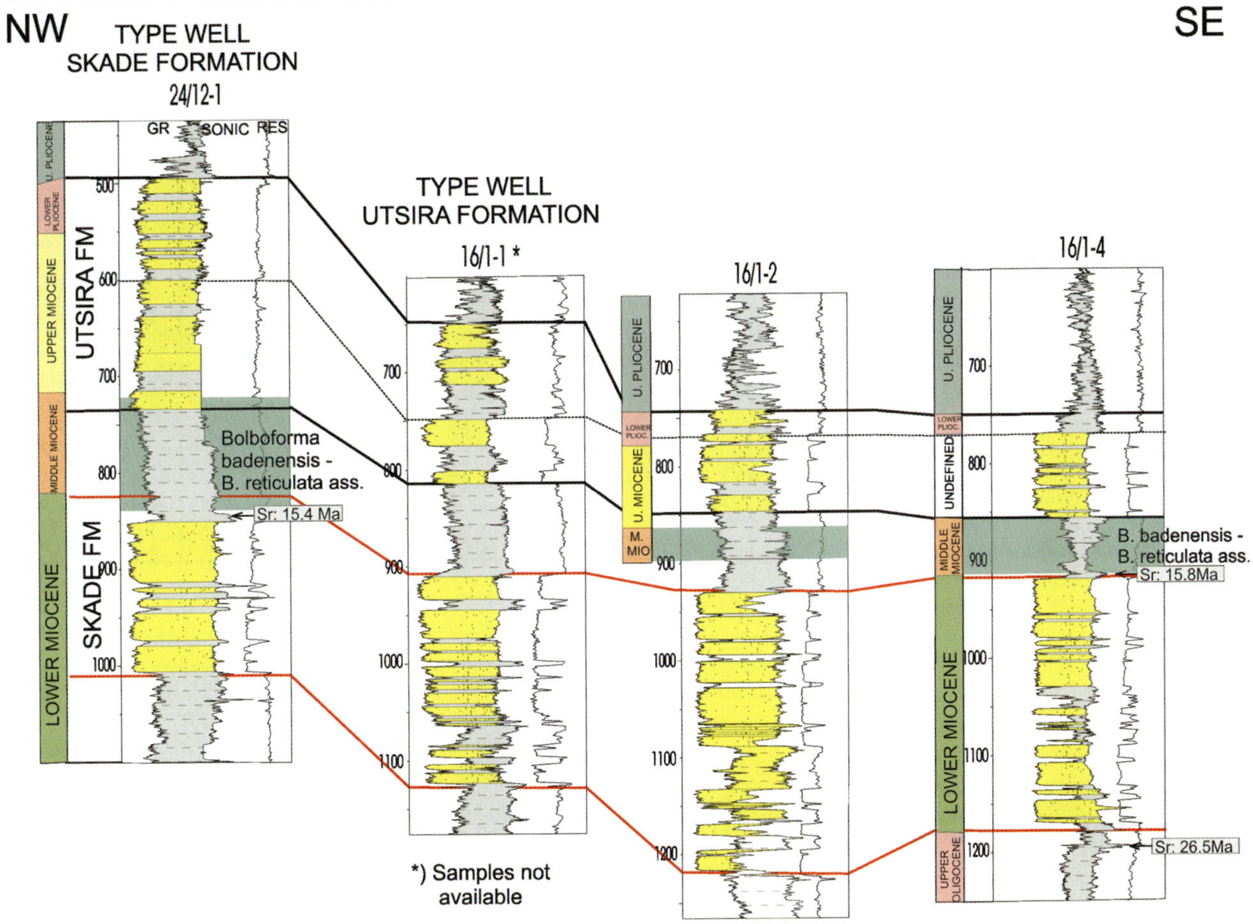

Fig. 13 Log correlation between wells 24/12-1, 16/1-1, 16/1-2 and 16/1-4 showing Skade Formation sands within the Lower Miocene section overlain by Middle Miocene mudstones characterised by the diagnostic *Bolboforma badenensis* and *B. reticulata* assemblages (interval marked in green) which again are overlain by Utsira Formation sands. Note lower main sands of the Utsira Formation yielding Late Miocene age and upper part yielding Early Pliocene age. Note also thick sands of the Utsira in well 24/12-1. See text for details. Location of wells shown in Figs. 7a–d.

dominate a moderately rich to sparse benthic fauna. Locally, agglutinating forms are numerous (e.g. 16/1-4 in the well).

Lower Nordland megasequence (Middle Miocene–Lower Pliocene)

The lower Nordland megasequence has been subdivided into two seismic units; (1) a Middle Miocene unit of dominantly mudstones at the base, overlain by (2) an Upper Miocene–Lower Pliocene unit (Utsira Formation) comprising dominantly thick, blocky sands.

Middle Miocene unit (LN-1)

Detailed biostratigraphic investigations of key wells in the southern Viking Graben have proved the existence of a distinct Middle Miocene unit in the northern North Sea. Sediments of this age have been identified in wells 24/12-1, 16/1-2 and 16/1-4 in southern Viking Graben (Tables 1 and 2). The biostratigraphic dating is summarised in the log correlation diagram (Fig. 13) and is also presented in Tables 1 and 2. The mudstone sequence overlying the Skade sands contains the diagnostic planktonic microfossil *Bolboforma badenensis* and *B. reticulata* assemblages. These assemblages that are known from the ODP/DSDP deep sea boreholes in the North Atlantic and the Vøring Plateau (Spiegler and Müller, 1992; Müller and Spiegler, 1993), suggest an age of approximately 14–12 Ma for this depositional unit. The presence of the *Bolboforma* assemblages is also indicated on the seismic section (Fig. 14a).

Farther to the south we have also identified the same *Bolboforma* assemblages in well 15/9-13 and 15/12-3 (Fig. 12). On the GR log from well 15/12-3, a very distinct high-radioactive marker occurs close to the boundary between the Lower and Middle Miocene strata. As can be seen from Figure 12, this

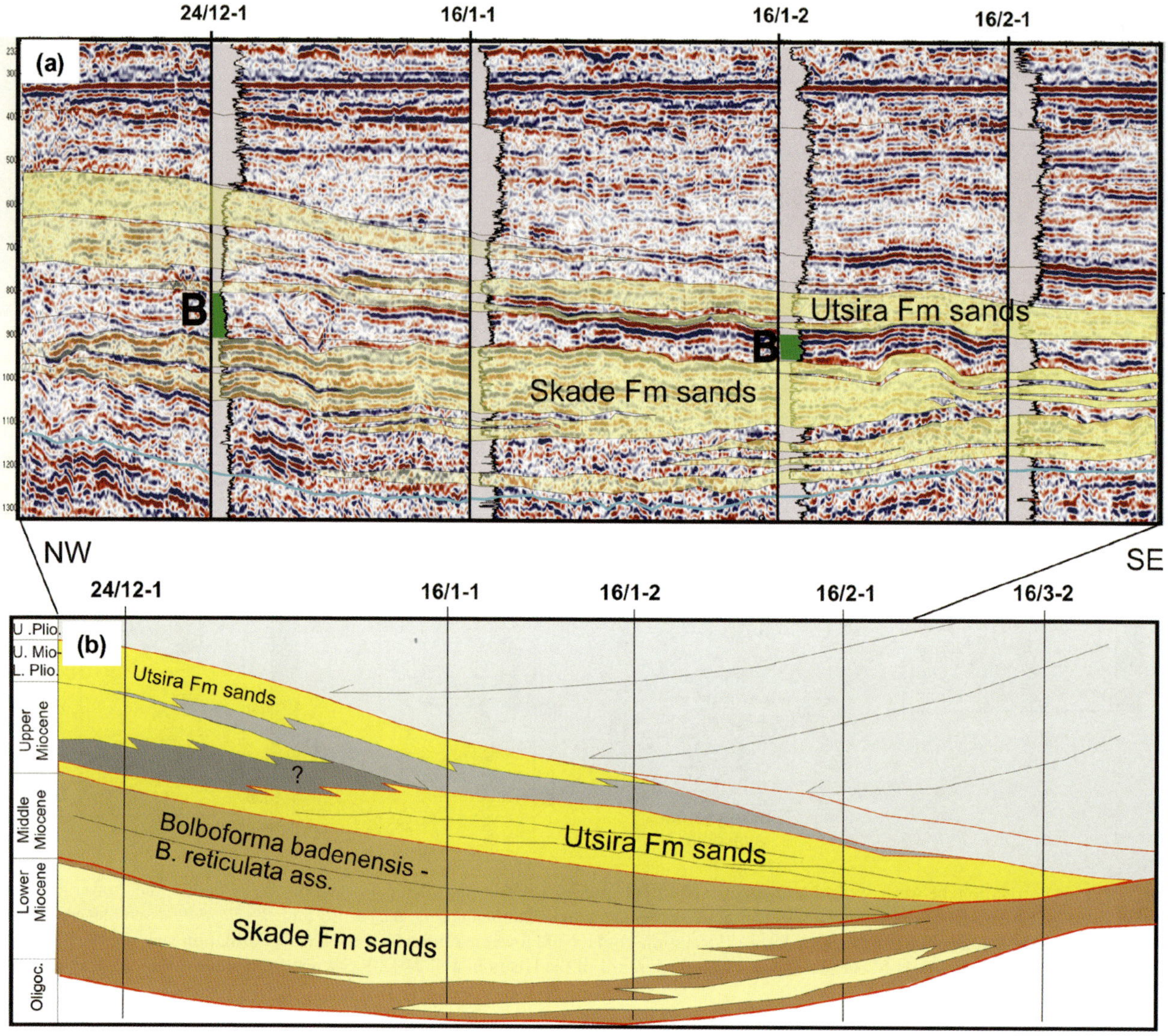

Fig. 14 (a) Seismic line through wells 24/12-1, 16/1-1, 16/1-2 and 16/2-1 across southern Viking Graben showing Skade and Utsira formations. (Well correlation between three of the wells is shown in Fig. 13). Green-coloured part of GR-logs denotes Middle Miocene *Bolboforma badenensis* and *B. reticulate* assemblages. (b) Interpretation of seismic line shown in Fig. 14a (extended towards east through well 16/3-2) illustrating Middle Miocene strata onlapping the top Lower Miocene surface (mid-Miocene unconformity) to the east, and interpretation of Utsira Formation (see text for details). Location of line shown in Figs. 7a–d.

marker can be tied closely to a high-amplitude seismic reflector, which to the east can be interpreted as a major sequence boundary. This high-amplitude reflector is also prominent on the seismic section presented in Fig. 15a. It ties very well to a high GR log marker in wells along this transect, as illustrated in the log correlation diagram (Fig. 15b). This GR marker also serves as a key for the definition of the base Middle Miocene sequence boundary in other wells that are located in the centre of the basin.

The Middle Miocene unit forms a basin infilling sequence which onlaps the underlying Lower Miocene. This is well-illustrated in seismic sections (Figs. 4b and 15a). It clearly postdates the mid-Miocene unconformity, thus forming the basal part of the Nordland Group. It comprises dominantly mudstones with only sparse thin sands present in some wells. The unit attains a maximum thickness of about 250 m in well 15/6-5 (Fig. 15b).

The northern extent of this sequence is difficult to map seismically due to chaotic reflections, but is thought to be present in the centre of the basin along the Viking Graben, as presented in Fig. 7c.

The seismic section (Fig. 15a) illustrates very clearly the sequential relationships and the geometry of the three Miocene units, and represents a key line to the understanding of the stratigraphic framework of the northern North Sea. This line is also schematically presented in its

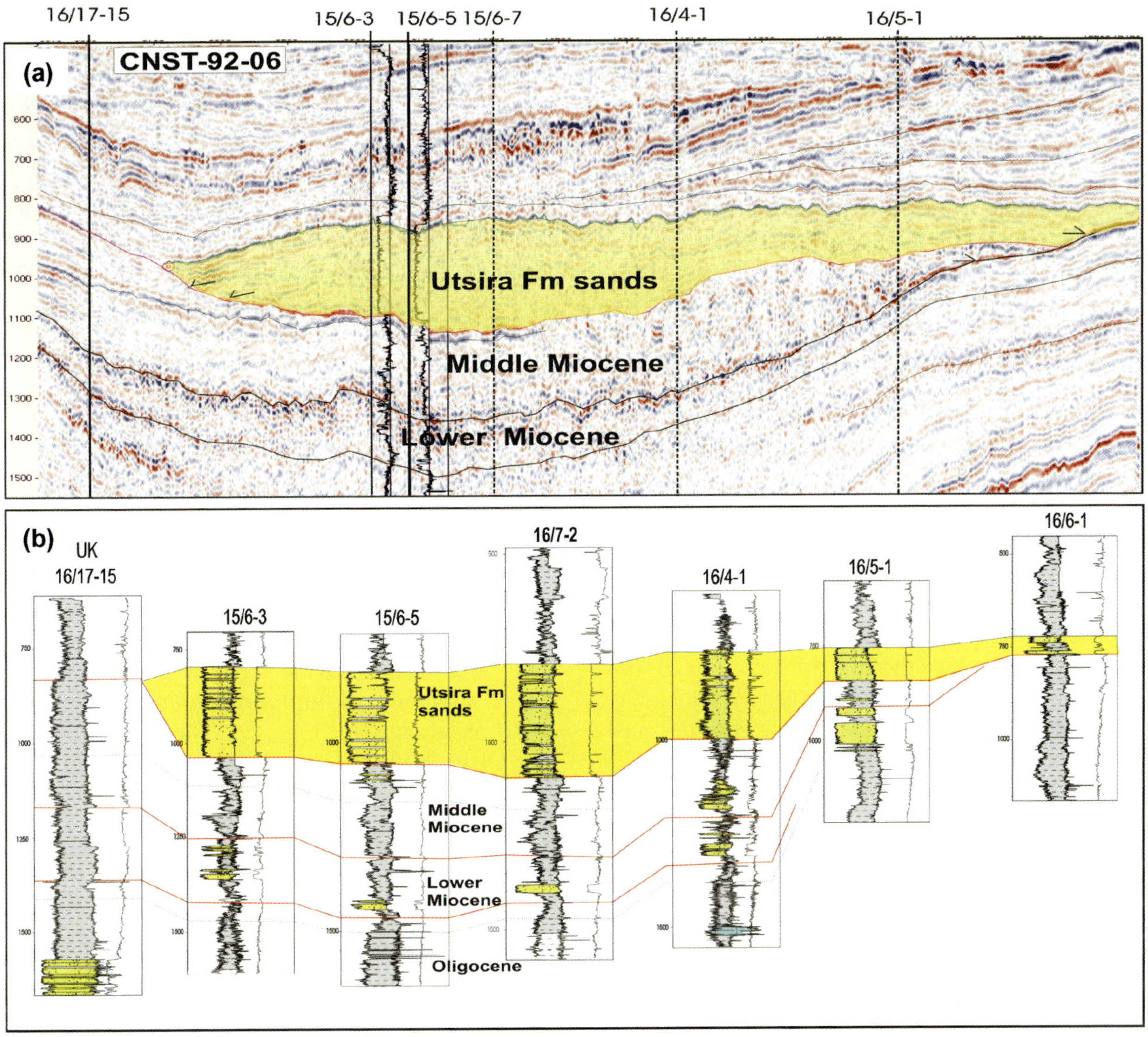

Fig. 15 (a) Seismic section (CNST-92-06) across southern Viking Graben through wells 16/17-15, 15/6-3 and 15/6-5 and close to wells 15/6-7, 16/4-1 and 16/5-1 showing Utsira Formation sands as a thick infilling unit in the basin centre. Note also the infilling nature of the Middle Miocene seismic unit above the mid-Miocene unconformity (see text for details). Location of line is shown in Figs. 7a–d. (b) Log correlation between wells shown in Fig. 15a (well 16/6-7 has been substituted with well 16/7-2). Note the overall blocky GR log profile of the Utsira Formation sands. Skade Formation sands penetrated in well 16/5-1. See text for details.

full length in Fig. 4b, illustrating the regional sequential architecture across the southern Viking Graben.

Upper Miocene–Lower Pliocene unit (LN-2)

The Upper Miocene–Lower Pliocene unit comprises the Utsira Formation, encompassing a huge sandy system with subordinate intercalated thin mudstones. The sands overlie Lower Miocene and Oligocene strata to the north and Middle Miocene strata to the south (Figs. 4a and 4b).

The outline of the sandy Utsira system is shown in Fig. 7d. It forms an elongated sandbody about 450 km long and 90 km wide, mainly deposited in the centre of the northern North Sea basin. To the north, in the Tampen area, the Utsira Formation is represented by a thin glauconitic member overlying Oligocene strata and deposited close to the Miocene–Pliocene transition (Eidvin and Rundberg, 2001). This member is thought to cap the main Utsira Formation sands in the northern part of the basin. Similar glauconitic sands have been observed at the same position in many wells in the Viking Graben to the south (Rundberg, 1989).

The thickness of the sands reaches about 250–300 m at maximum. Previous reported thickness estimates of 400–600 m (Gregersen et al., 1997) are probably incorrect. We believe that these estimates result from an error in the definition of the Utsira and Skade Formations (Fig. 6).

The Utsira Formation displays a very complex depositional architecture, as recently described by Galloway (2002). We present the Utsira Formation along four transects in the northern North Sea; two from the southern Viking Graben and two from the northern Viking Graben.

The southernmost transect (Fig. 15a) shows the Utsira Formation in the southern Viking Graben, at about 58°30′N. In this part of the basin, the Utsira is developed as a giant mounded sand system, pinching out in both the eastward and the westward directions. A log correlation diagram (Fig. 15b) through seven wells along the seismic profile (Fig. 15a) shows a very characteristic, blocky GR log profile of the Utsira sands with only scattered thin mudstone intervals. A maximum thickness of about 300 m is recorded in well 16/7-2. The top of the sands is marked by an abrupt increase in the GR log values. As can be interpreted from Fig. 15a, the base of the sands clearly erodes into the underlying sequence. This may explain differences in the GR log profiles just beneath the sands (Fig. 15b), which in some wells define coarsening-upward and in other wells fining-upward motifs.

At about 59°N, a different development of the Utsira Formation appears. Along this transect (Fig. 14a; Utsira type well 16/1-1) the Utsira Formation sands are distinctly thinner (blocky sands, 25–100 m thick). Based on seismic and log data (Fig. 13), there is no sign of erosion into the underlying mudstone sequence. The Utsira Formation can here be subdivided into two subunits: (1) a lower subunit characterised by dominant blocky sands, forming the main sandbody (thickest developed to the east); and (2) an upper subunit displaying a clear coarsening-upward trend (see Fig. 13). This subunit clearly thins in the eastward direction and probably represents a progradational system which downlaps the underlying blocky sands. This is also illustrated in Fig. 14b, which shows a more basin-wide interpretation of the Utsira Formation sands, along this transect (depositional model of the Utsira Formation dealt with below). The westward extension of the main Utsira Formation sandbody (lower subunit) is not clearly resolved along this transect. It is thought to pinch out close to well 24/12-1. This boundary is difficult to map precisely between 59° and 61°N.

In the northern Viking Graben, at about 60°45′N, the Utsira Formation forms a large mounded sandbody with thickness of almost 200 m in the basin centre (Fig. 9). Along this transect, it is penetrated by wells 30/6-5 and 30/6-11 and 31/4-4 (Fig. 10). The sands display a blocky GR response with only subordinate thin mudstone intervals. The top of the sands is well-expressed by a relatively continuous moderate to high-amplitude reflector. To the west, the main sandbody apparently thins out. Notably in the western part of the transect, a number of high-amplitude reflectors apparently overlie (downlapping) the main Utsira Formation sandbody in a retrogradational pattern. There are no wells penetrating these reflectors in blocks 30/4 and 30/5, but they probably reflect a very sandy system shed from the East Shetland Platform. This sandy system is thought to represent a northern equivalent to Utsira Formation subunit 2, as defined in the transect at about 59°N (Figs. 13, 14a).

Further to the north, at about 61°N, a similar development of the Utsira Formation can be observed (Fig. 8a). Along this transect, the Utsira Formation sands are penetrated by wells 30/10-15 and 30/10-23, close the western pinchout of the main sandbody. A log correlation between these wells is shown in Fig. 8c. The upper part of the Utsira Formation probably represents westerly-derived strata equivalent to those observed in Fig. 9. This upper subunit of the Utsira Formation is further dealt with in the discussion part of this chapter.

Rundberg (1989) reported that the Utsira Formation sands in wells from the northern Viking Graben (blocks 30/3, 30/6) are extremely well-sorted, mainly coarse-grained and texturally mature (subordinate feldspar content). They are also rich in shell debris and glauconite. Intercalated siltstones from the lower part of the sands were rich in glauconite and sponge spicules.

The age of the Utsira Formation sands has been discussed previously by several authors (Rundberg and Smalley, 1989; Goll and Skarbø, 1990; Smalley and Rundberg, 1990; Eidvin and Rundberg, 2001). Eidvin and Rundberg (2001) concluded that the main deposition of this sand to took place between 12 and 5 Ma (based on biostratigraphic correlations and Sr isotopic dating). New investigations of the dinoflagellate flora (Piasecki et al., 2002) and the foraminiferal fauna (this work) in a cored section in well 15/9-A-23 (Tables 1 and 2) indicate an age as young as ca 4.5 Ma, for the upper part of the Utsira.

A moderately rich planktonic fossil assemblage of mainly foraminifera is recorded throughout the Utsira Formation. *Bolboforma* are also recorded in parts of the section, but many of these are probably reworked. Calcareous foraminifera dominate a rich benthic fauna. Mollusc and mollusc fragments are also common.

Discussion

The data presented in this study permit a reconstruction of the sedimentary and tectonic history of the northern North Sea, whose general stratigraphic framework is schematically shown along two transects across the basin in Figs. 4a and 4b. We propose in this chapter that the stratigraphy and depositional history of the northern North Sea was heavily controlled by the major compressional tectonic regime that affected the Atlantic margin of northwest Europe during Oligocene–Miocene times. The northern transect (Fig. 4a) clearly demonstrates that the northern domains of the North Sea have been affected by severe erosion at mid-Miocene time. The southern transect (Fig. 4b) illustrates that a major drop in relative sea-level took place close to the Early-Middle Miocene transition. This drop in sea level probably resulted in subaerial exposure of much of the northern North Sea, particularly the flanks of the basin.

The Eocene–Oligocene transition: controls of lithostratigraphic changes

Globally, the Eocene–Oligocene boundary marks the last major transition from greenhouse to icehouse condition (Zachos, 2001; Ivany et al., 2003). It is marked by a drop in bottom water temperatures of about 4–5°C, based on oxygen isotope signatures from deep-sea core microfossils. This transition coincides with the first build up of significant ice on Antarctica and appears to have been the onset of deep-sea thermohaline circulation, with an overall increase in ocean fertility (Zachos, 2001). Recent work has demonstrated a large global mass extinction at the E/O-boundary (Ivany et al., 2000), and remarkable faunal changes from the late Eocene to the early Oligocene (Ogasawara, 2002). Sequences from areas of the Antarctic margin show an increase in biogenic sedimentation, usually biogenic silica (diatoms) during the Early Oligocene, suggesting significant cool-water upwelling (Kennett and Barker, 1988; Shipboard Scientific Party, 2000). In the western part of the equatorial Atlantic, Mikkelsen and Barron (1997) reported a distinct increase in biogenic silica accumulation during the Early Oligocene, which they related to the global cooling.

As stated earlier, the Eocene–Oligocene boundary in the northern North Sea is probably one of the most important breaks within the Cainozoic. This break is documented in the work of Gradstein and Bäckström (1996) and also described by Martinsen et al. (1999). The boundary is characterised by the following (Rundberg, 1989): (1) It marks an end to the extremely fine-grained, greenish, smectite-rich depositional regime that had dominated since the Danian, (2) It marks the onset of a clastic regime characterised by brownish, progressively coarser mudstones; (3) It marks the re-appearance of calcareous benthic foraminifera in the study area and a strong increase in microfossil siliceous sedimentation, which includes the marked incoming of opal-CT (Fig. 11). The very marked lithostratigraphic changes observed at the Eocene–Oligocene boundary in the northern North Sea are also observed in sediments exposed in Denmark (Heilmann-Clausen, 1985) and in wells offshore Mid Norway (Haltenbanken and Vøring Basins).

The Eocene–Oligocene boundary in the northern North Sea records many similar characteristics that are described from the boundary in ODP sites of the world oceans. For example, Rundberg (1989) reported a conspicuous change in clay mineralogy characterised by a decrease in smectite at the expense of illite and kaolinite, similar to what has been described from the boundary in the Southern Hemisphere (e.g. Shipboard Scientific Party, 2002). In the northern North Sea, the clay mineral changes are accompanied by other distinctive changes of the mudrocks, such as coarser grain size and a change in dominantly olive-green to yellowish-brown mudrock colouration. Rundberg (1989) suggested that the clay mineral changes at this boundary were primarily controlled by the abrupt climatic deterioration and, for some wells, also controlled by uplift activity of the Norwegian margin. In NE Belgium, de Man et al. (2003a), based on a study on stable oxygen isotopes reported a main temperature drop just above the Eocene–Oligocene boundary. It is also interesting that de Man et al. (2003b), based on biofacies analysis of the shallow marine Rupelian section in Belgium and Germany, could suggest relatively cold-water conditions for the Rupelian and a return to warm-water conditions at the Rupelian–Chattian boundary, the latter coinciding with the widespread Chattian warming event (e.g. Zachos, 2001). Warm and humid Chattian conditions were also suggested by Rundberg (1989) from mineralogical data, in particular the occurrence of gibbsite, boehmite and the traces of goethite peloids in sands from Block 35/3 (Agat area).

We agree that the onset of siliceous-rich sedimentation in the northern North Sea was most likely related to the dramatic decrease in global temperature at the Eocene–Oligocene boundary, as suggested by Rundberg (1989). He further suggested that the siliceous sedimentation was related to more oceanic circulation in response to seafloor spreading between Greenland and Spitsbergen (Fig. 1) and upwelling conditions that developed by uplift of the northernmost North Sea, along the Tampen Spur extension. Thyberg et al. (1999) also pointed to upwelling conditions in their work on the diatomaceous Oligocene deposits of the northern North Sea, but suggested that such conditions developed by a northward-oriented marine current system. It is evident that the opening of the Fram Strait during Early Oligocene permitted at least surface-water exchange with the Arctic Sea, but was the abyssal circulation of the Norwegian Sea affected? Kaminski and Austin (1999), based on a study of deep-water agglutinated foraminifers at ODP Site 985, suggested that the Oligocene deep water of the Norwegian Sea was poorly oxygenated, advocating no particular change in the bottom-water circulation pattern. These workers suggested that the Norwegian Sea was strongly dysaerobic, in contrast to the more ventilated environment south of the Greenland–Scotland Ridge. If this view is correct, the siliceous sedimentation of the northern North Sea cannot be linked to the increased deep-water circulation. It should, however, be noted that other workers have indicated intensification of deep-water circulation at the Eocene/Oligocene transition; for example, Eldrett (2004), in order to explain the Upper Eocene hiatus, observed in many ODP holes (and also reported from the northern North Sea), and Hull (1995) in a study of radiolarians from ODP Leg 151.

It is, however, likely that the end Eocene global cooling, which caused the start of the glaciations of Antarctica, must have led to a change in global wind systems. Obviously, such an increase in wind energy must have caused an increase in surface-water circulation. Thus, if there was no particular change in deep-water circulation in the Norwegian Sea, as suggested by Kaminski and Austin (1999), the distinct increase in siliceous sedimentation in the northern North Sea (see Fig. 11) could also be controlled by local increase in productivity of diatoms in response to wind-induced upwelling.

Influence of Oligocene–Miocene compressional tectonics on the sedimentary and structural evolution of the northern North Sea

We argue in this study that the compressive tectonic episode that affected the NW Atlantic margin also influenced the Shetland Platform, the northern North Sea Basin and southern Fennoscandia, as well (as shown in Fig. 2). The structural uplift of these areas corresponds broadly with the large-scale plate movements on the Atlantic margin (see Fig. 5). Such uplift may be difficult to envisage from seismic data alone, but is suggested from a series of observations and data, which make this statement more than mere guesswork:

(1) Two huge sandy depositional systems, with gross thickness in excess of 300 m, were sourced from the East Shetland Platform during the Early Oligocene and Early Miocene. These sands are most likely linked to uplift and erosion of the Platform (Fig. 5);
(2) The southern shift in depocentre for these sandy systems (from northern to southern Viking Graben) could be a result of change in basin physiography, in response to uplift of the northernmost North Sea (Tampen Spur area);
(3) The strong influx of sediments from southern Fennoscandia during early Oligocene time (wedge of mudstones between 60–61°N which northward passes into thick sands) indicates contemporaneous uplift of the eastern hinterland;
(4) The upward coarsening pattern in Oligocene mudstones for wells in northernmost North Sea indicates shallowing of the basin;
(5) The upward change from deeper marine, gravity flow to shallow marine facies in Oligocene sands, block 35/3 (Agat area) indicates rapid shallowing along the margin;
(6) The isopach of Oligocene strata between 60–62°N shows northward thinning (Rundberg, 1989);
(7) The distinct mid-Miocene unconformity and the northward increase in erosional hiatus (Fig. 5) are probably linked to uplift of the basin.

The mid-Miocene unconformity separates the Hordaland and Nordland Groups (Fig. 5). It is well-expressed from the stratal geometries of southern and northern Viking Grabens, as presented in Figs. 4a and 4b. In the southern Viking Graben, a very large fall in the relative sea level can be inferred from the architecture of the Lower and Middle Miocene seismic units (Fig. 4b). In the northern Viking Graben, the Lower Miocene strata are truncated at both margins, whereas Middle Miocene deposits are totally absent (Fig. 4a). As seen from Fig. 5, the mid-Miocene erosional unconformity records a hiatus in the order of 15–20 million years, and even more, in areas where the Utsira Formation is absent. This erosion period corresponds broadly with the last compressional phase activity along the mid-Norwegian margin (Lundin and Doré, 2002) and in the Faeroe–Rockall region (Boldreel and Andersen, 1993). Mid-Miocene unconformities are also a widespread phenomenon on the Norwegian Continental shelf (Eidvin et al., 2000; Gradstein and Bäckström, 1996, Brekke, 2000).

Sedimentary response of uplift

As discussed in the previous sections, three major sandy systems were deposited in the Norwegian North Sea, during the Oligocene and Miocene. These systems were largely sourced from the East Shetland Platform, as shown in Figs. 7a, 7b and 7d. Coarse clastic influx from the east were also recorded, but to a much lesser extent.

Lower Oligocene sandy system

The first major sandy influx to the northern North Sea Basin was deposited in the Statfjord–Tampen area during the Early Oligocene (Fig. 7a) and in our opinion represents marine gravity-flow facies. The thick pile of Oligocene sands (gross thickness of about 400 m, Fig. 8c) suggests that significant tectonic uplift of the Shetland Platform has taken place.

It is tempting to relate this uplift to compressive strain along the Erlend Transfer Zone separating the Møre and Shetland-Faeroe Basins (Figs. 1, 2). As can be seen in Fig. 2, movements along this transfer fault could be taken up along its southward extension, the Walls Boundary Fault. Erosional products from the uplifted areas of the East Shetland Platform have then been shed eastwards by river systems (~61°N), focussing delta progradation and gravity flow transport towards the Statfjord–Tampen area of the northern North Sea. This sandy system is illustrated by clear wedging of the Lower Oligocene strata, particularly well-observed along the seismic sections shown in Figures 8a and 9.

We suggest that the abrupt incoming of this depositional system marks the first signal of the uplift that affected the East Shetland area during Late Eocene–Early Oligocene time. It is broadly contemporaneous with a tectonic pulse that caused domal growth along the margin (Boldreel and Andersen, 1994; Doré and Lundin, 1996, Brekke, 2000; Lundin and Doré, 2002; see Fig. 5). We suggest that the sands are linked to this tectonic event, rather than to global eustasy.

Sands were also derived from an eastern source area during the Early Oligocene as recorded in block 30/3 (Agat area, Fig. 7a). In this part of the basin, thick sands of turbiditic origin are capped by sands of shallow marine facies (Rundberg, 1989). These sands are contemporaneous with sands of the Statfjord area and most likely represent Early Oligocene uplift of the Agat hinterland to the east. Further south towards the Troll area, a distinct wedge of organic-rich mudstones occurs parallel to the western coast of Norway (Fig. 7a). This wedge (seismic unit UH-1), termed map unit 3 of Rundberg (1989), was suggested to been an erosional product from the uplift of southern Fennoscandia at this time.

Lower Miocene sandy system

The second phase of large sand input took place during Early Miocene. Sands belonging to the Skade Formation were deposited mainly in the southern Viking Graben, as shown in Fig. 7b. They represent a southern shift in coarse clastic influx to the basin, relative to Oligocene time, and are also clearly turbiditic in origin. The sands pinch out to the east and were sourced from the East Shetland Platform. This sandy system has a magnitude in the same order as the Lower Oligocene system, i.e. a maximum gross thickness in excess of 300 m. According to our dating, the sands were deposited during Early Miocene, between 20 and 15 Ma We infer that they are a result of new tectonic event of uplift of the East Shetland Platform, possibly associated with a renewed compressional tectonic phase along the northwest European margin.

Lundin and Doré (2002) reported a compressional phase during Early Miocene along the Mid-Norwegian margin, which affected a number of domes between the Jan Mayen and Bivrost Fracture zones. This phase corresponds favourably with the deposition of the Skade Formation sands, as shown in Figure 5. Boldreel and Andersen (1994) suggested that a compressional phase affected the Faeroe–Rockall area during Middle or Late Miocene times. As indicated by these authors, there is a lack of a detailed well control in the Neogene section, which, consequently, leads to less precise timing for the last compressional phase. If the timing is correct, this phase post-dates deposition of the Skade Formation sands but coincides better with deposition of the Utsira Formation sands (Fig. 5).

Upper Miocene–Lower Pliocene sandy systems

The third phase of large sand input took place during Late Miocene–Early Pliocene time. This sand input, comprising the huge Utsira Formation, was the largest sandy influx to the North Sea Basin during the Cainozoic. It is dealt with in more detail later.

The mid-Miocene unconformity and the creation of a shallow seaway

A question, which arises, is whether the North Sea could have been subaerially exposed during this Middle Miocene erosion period, as suggested by several authors (Jordt et al., 1995; Rundberg et al., 1995; Martinsen et al., 1999). In the wells examined, no clear subaerial signature of sediments immediately below the unconformity has been observed. There are, however, a number of features that indicate a shallowing of the basin during late Oligocene–Miocene:

(1) The overall architecture of the depositional sequences, in particular the change from Early Miocene to Middle Miocene deposition, clearly signifies a very dramatic relative sea-level fall which must have caused a rapid decrease in water depths;
(2) The uppermost preserved Oligocene strata in wells on the Tampen Spur (block 34/2) consist of glauconitic, spicule-rich siltstones (Rundberg, 1989) which most likely represent a marine shelf setting. Such sediments are also typical of the Lower Miocene in wells, farther to the south (Blocks 30/6 and 30/3);
(3) The top Oligocene surface reveals a strong character of subaerial erosion (incisional features) in much of quadrant 35, particularly in areas to the northeast of Utsira Formation sand extent (see Fig. 7d);
(4) At the base of the Utsira Formation, there are several erosional features (Gregersen et al., 1997; Galloway, 2002), which have been interpreted as tidal channel scours (Galloway, 2002);
(5) Shell beds within the Utsira Formation sands point to periods with shallow-water conditions.

The long channel feature extending from offshore Sognefjorden (block 35/8) to about 62°N (block 34/3) incising deeply into Oligocene strata has previously been interpreted to indicate subaerial erosion (Rundberg et al., 1995; Gregersen, 1998; Martinsen et al., 1999). Work carried out by Eidvin and Rundberg (2001), however, shows that this incision is Late Pliocene in age. It cannot, therefore, be linked to the formation of the mid-Miocene unconformity.

We consider that it is more likely that a major part of the mid-Miocene unconformity is a result of submarine erosion, which is a very common phenomenon. Such erosion has been reported from work in the North Atlantic waters to the south of the Greenland–Scotland Ridge (e.g. Wold, 1994; Wright and Miller, 1996). Three major erosional phases, that can be seismically identified as distinct unconformities (age estimates, shown in Fig. 5), were linked to the development of strong deep-water currents. These unconformities correspond with the development of prominent drift sediments (Bjorn and Gardar Drifts, Hatton and Snorri Drifts, Eirik and Gloria Drifts). The accumulation of the drift sediments and their link to deep-water fluxes and Greenland–Scotland Ridge overflow have been thoroughly discussed by Wold (1994) and Wright and Miller (1996). Stoker et al. (2002) interpreted the latest Oligocene/Early Miocene unconformity on the Faeroe–Shetland Channel as a result of major change in oceanic regime, linked to the establishment of deep-water exchange between the Arctic and North Atlantic oceans.

In accordance with these workers we suggest that the mid-Miocene unconformity could have developed as a result of increased marine circulation and vigorous current erosion, which affected the uplifted northernmost North Sea. The abrupt Middle Miocene climatic cooling at about 14 Ma (Zachos, 2001) may also have intensified oceanic circulation systems at this time. Another aspect which may be important in this discussion is related to the change in North Sea Basin physiography that developed during Middle and Late Miocene, in which the northernmost North Sea gradually became shallower due to the tectonic uplift. The resultant basin physiography of the North Sea with a shallow threshold to the north may be ideal for the formation of strong tidal current regimes. Such currents may therefore have swept across the Viking Strait and caused vigorous erosion of the uplifted sea floor. The tidal effect would probably increase as the strait became shallower. The fact that the hiatus is largest to the north and decreases in the southward direction, is in accordance with this model.

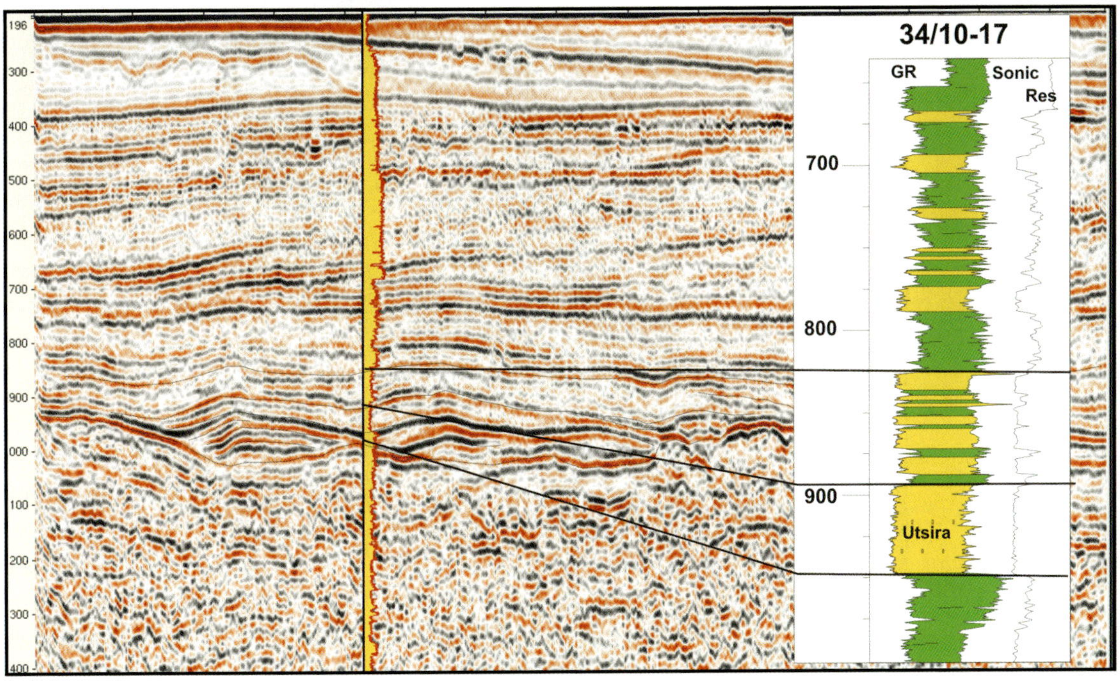

Fig. 16 Seismic section through well 34/10-17 (detail of line shown in Fig. 8a) showing erosive base of the Utsira Formation and infill with distinct westward migrating stacking pattern. The lower part of the Utsira Formation sands forms a mounded geometry whereas the upper part is more sheet-like.

The mid-Miocene (or top Hordaland) surface is locally very irregular forming diapirs, large ridges and troughs, as illustrated in Figs. 8a and 9. Some of these structures could have been formed as a result of instability and pore-pressure release of the underlying mudrocks in response to the sudden sea-level fall at mid-Miocene time. A number of incisional features are also observed at this surface below the Utsira Formation. One example is illustrated, close to well 34/10-17 in Fig. 16. This erosion may have developed as tidal channel scouring during late Miocene, marking the end of the erosional phase and the onset of a new clastic depositional regime.

Timing of the mid-Miocene tectonic and erosional events

Based on our biostratigraphic and seismic work in the southern Viking Graben, we are now able to give more precise age constraints for the tectonic and erosional events. We suggest that the first response of the Miocene tectonic activity was felt by the incoming Skade sands. This sandy system was deposited during Early Miocene and abruptly ceased at about 15 Ma, as shown by Sr isotopic dating of carbonate foraminiferal tests from the top of the Skade Formation (15.5 and 15.1 Ma in wells 24/12-1 and 16/1-4, respectively). Such ages are supported by biostratigraphic data from the overlying mudstones, containing the diagnostic *Bolboforma badenensis* and *B. reticulata* assemblages. These assemblages suggest deposition at about 14–12 Ma years. Deposition of the Skade sands was followed by a large relative sea-level fall, which was probably associated with the end of the Mid-Norwegian compression event. We infer that the erosional activity in the northern North Sea started at this time, beginning first in the Tampen Spur area to the north and then successively working in a southward direction.

Late Miocene palaeogeography and deposition of the Utsira Formation

As discussed earlier, we infer that the large-scale tectonic movements have exerted control on the basin evolution and depositional history of the northern North Sea. As a response to the structural uplift, the northern North Sea gradually changed into a shallow shelfal seaway between deeper waters in the Møre Basin to the north and central and southern North Sea to the south. The palaeogeographic reconstruction is illustrated in Figure 17.

When compared with the early Oligocene map (Fig. 1a), three interesting aspects can be observed: (1) Much of the Greenland–Faeroe-Scotland Ridge has subsided below sea level; (2) deep-water link is established to the Arctic Ocean, via the Fram Strait and to the Atlantic Ocean, via the

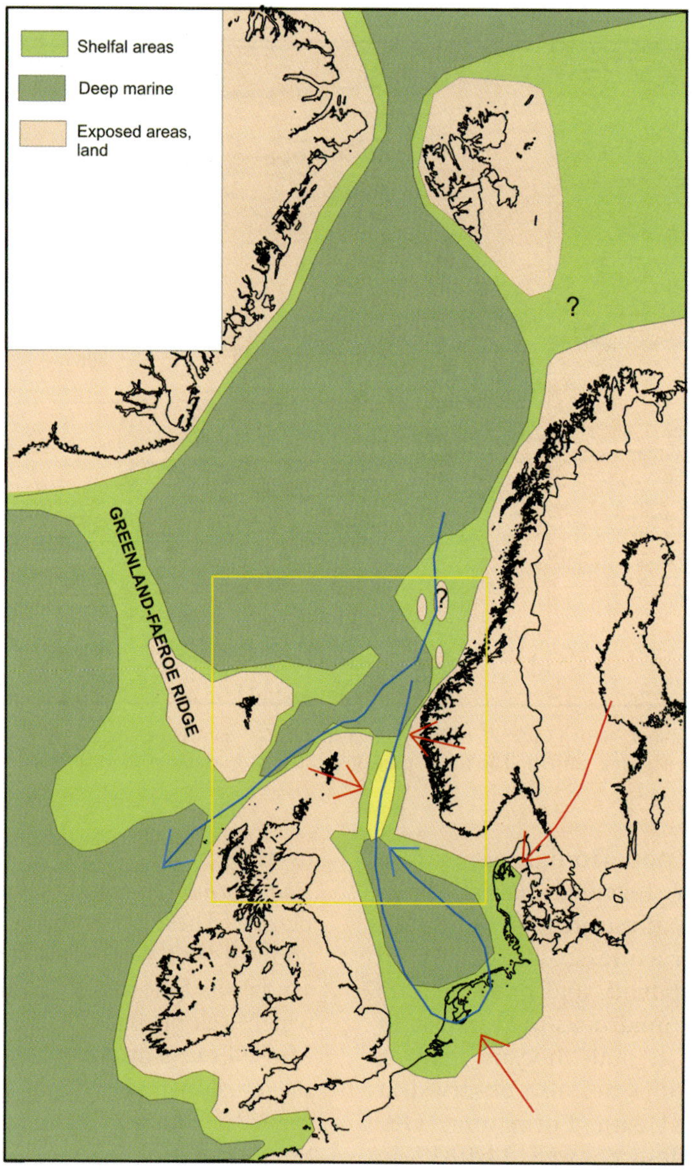

Fig. 17 Late Miocene palaeogeography of northwest Europe. The North Sea forms a semi-enclosed sea with deeper waters in central and southern North Sea and a narrow shallow strait in northern North Sea. The Greenland–Faeroe Ridge submerged during the Middle Miocene. Large amounts of coarse clastics were shed into the basin from the East Shetland Platform (northern North Sea), Fennoscandia (northern North Sea and Denmark) and the Alpine Region (Netherlands). Yellow rectangle shows location of map in Fig. 18.

Faeroe–Shetland Channel; and (3) the North Sea has changed into a semi-enclosed sea with a narrow shelfal passage to open marine waters to the north.

The Miocene submergence and the subsidence history of the Greenland–Scotland Ridge has been reported by many authors (e.g. Eldholm and Thiede, 1980; Wold, 1994; Wright and Miller, 1996). As mentioned earlier, the Faeroe–Shetland Channel probably operated as a conduit for deep-water passage from early Miocene time (Stoker et al., 2002). This gateway was characterised by vigorous deep-water circulation, which created severe erosion and subsequent deposition of drift sediments in the Rockall Trough.

The North Sea Basin progressively developed into a more or less landlocked or silled basin, with deeper waters in the central and southern North Sea and shallower waters in the northernmost North Sea. This situation is illustrated in the Late Miocene paleogeographic reconstruction (Fig. 17). The central North Sea formed a broader and deeper basin, which underwent subsidence during the entire Miocene. The shallow gateway of the northern North Sea or Viking Strait (term proposed by Galloway, 2002) persisted for a long period, probably from about 15 to 5 Ma. As mentioned above, an important consequence for this type of basin physiography is that it is ideal for the formation of large tidal cycles, which would create high-energy currents across the Viking Strait. It is under such conditions that the Utsira Formation sands probably were deposited.

Depositional model of the Utsira Formation sands

Several authors have discussed the depositional model of the Utsira Formation sands. In his work on the sedimentary history and basin evolution of the northern North Sea, Rundberg (1989) inferred that the northern North Sea formed a shallow marine passage between deeper waters of the North Sea to the south and the Møre Basin to the north during deposition of the Utsira Formation, and that the Utsira Formation sands were deposited under fairly uniform high energy conditions in a relatively stable tectonic setting. Based on a number of factors, such as the palaeogeographic setting, the elongate morphology of the total sand body, the overall slow accumulation rate, the uniform log patterns and the extreme maturity of the sands, he pointed to a model where strong, shallow marine and possibly tidal currents flowed between the deeper seas to the north and south. These currents probably caused sands to accumulate in parallel, N–S linear ridges, which subsequently became moulded and amalgamated into elongate sheet sands, as the shelf periodically became shallower due to local? tectonic activity and/or slight fluctuations in sea level.

Galloway et al. (1993), in a study of the Cainozoic sediments from the northern North Sea, interpreted the Utsira Formation to consist of a sand-dominated contourite drift system throughout the Viking Trough, and that the sands were derived almost entirely from the Scandinavian Platform.

Gregersen et al. (1997) questioned Rundberg's (1989) depositional model of the Utsira Formation and suggested a turbiditic origin, largely because of its typical blocky GR log profile coupled with seismic mounded and bi-directional onlap. These authors also interpreted the sands as representing stacked lowstand fan deposits. A problem with their interpretation, however, is the fact that they, by mistake, have included sands of the Skade Formation into the Utsira Formation. Martinsen et al. (1999), on the other hand, favoured a shallow-marine origin within a high-energy current setting.

Recently, Galloway (2002) has presented a thorough discussion of the Utsira Formation sands. His new model involves sediment input through a prograding platform (Shetland), coast-to-shelf bypass, and regional basin-centred transport within an elongate seaway characterised by (1) high-energy marine regime, (2) very low rates of sediment supply, (3) high sediment reworking and (4) regional along-strike sediment transport. He suggested that the deposition was concentrated in a southern and northern shoal system, which could be explained by a combined current system, probably tidal in origin, characterised by inflow into the North Sea along the western margin and outflow along the eastern side of the strait.

We agree in much of Galloways (2002) model, which largely follows Rundberg's (1989) interpretations, although in much more detail. Galloway included time-equivalent sands, which are present over large areas in UK waters to the Utsira Formation. Earlier, these sands have been termed Hutton sands, and their relationship to the Utsira Formation sands in Norwegian waters, has been discussed by Gregersen et al. (1997). Galloway (2002) interpreted these sands to represent an eastward prograding strandplain, which provided a source of sediment to the Viking Strait (Fig. 7d).

Our work in the southern Viking Graben demonstrates that such a relationship probably exists. This can be seen in Figure 14b, which represents an interpretation of the Utsira Formation between key wells at about 59°N. In the westernmost well (24/12-1), the Utsira Formation is distinctly thicker than seen in wells to the east. By reference to Galloway's depositional model of the Utsira Formation (Fig. 10), well 24/12-1 would be located in the easternmost part of the Shetland strandplain. It represents, therefore, a different facies than the sediments penetrated in wells to the east. Consequently, this may explain the sudden change in thickness between wells 16/1-1 and 24/12-1.

The biostratigraphic dating for these wells also yields important accounts to the depositional history. The results of the dating (Fig. 13; Tables 1 and 2) indicate that the upper prograding subunit of the Utsira is Early Pliocene in age whereas the lower subunit is Late Miocene in age. Farther to the south, in well 15/12-3 (Fig. 12), where the Utsira Formation is developed as a 300 m-thick, uniform sand throughout, we have recorded Late Miocene age for the lower half of the Utsira Formation and Early Pliocene age for the upper half. In a short core from the upper part of the Utsira Formation sands in well 15/9-A-23 (Sleipner area), Piasecki et al. (2002) reported Early Pliocene dinoflagellate cysts. We have analysed the same core and have confirmed the Early Pliocene age using foraminiferal correlation (Tables 1 and 2) and Sr isotopic analyses. These findings suggest that a prominent part of Utsira Formation deposition in the southern Viking Graben took place during early Pliocene.

The western source of sediments is evident from seismic data (Fig. 9). Supply from the east is more difficult to assess from seismic data as much of the Neogene section has been eroded. Remnants of easterly derived clastics are, however, preserved locally, e.g. in block 35/11 (penetrated in well 35/11-1). Farther to the north, off the Møre Margin at about 62°N, a very pronounced Neogene progradational system comprising nine depositional sequences (N1–N9) is preserved (Fig. 18). The lower part of this system is penetrated in well 35/3-1. It consists of spicule-rich, glauconitic fine-grained sands of possibly Early Miocene age (Rundberg, unpublished data), which are overlain by coarse, yellowish, marginal marine sands (lithologic unit D2 of Rundberg, 1989).

The relationship between this progradational system and the Utsira Formation is not clear, but may be interpreted as follows. As seen from Figure 18, a very pronounced basinward shift in depocentre (or forced regression) took place at the end of sequence N4. If this progradational system represents continuous Miocene deposition, as outlined in the Wheeler diagram of Figure 18, the forced regression would start at about Middle Miocene time, estimated very close to 15 Ma. Such an age fits into our understanding of the structural evolution of the northern North Sea and, as noted earlier, would correspond to the timing of the large relative sea-level fall, based on data from the southern Viking Graben. The Utsira Formation may thus be equivalent to the upper part of this prograding system, most likely sequences N5–N7. The major incision to the west of sequence N7 represents the northernmost part of an incisional feature extending from off Sognefjorden to about 62°N (Rundberg et al., 1995). Eidvin and Rundberg (2001) assigned a late Pliocene age of this incision, based on data from well 34/2-2, penetrating strata to the west of the incision in Fig. 18.

Noteworthy also is that at maximum progradation of this Neogene system, during latest Miocene/earliest Pliocene, the shoreline was probably as far east as block 35/1. This also indicates a narrowing of Viking Strait during deposition of the Utsira Formation sands.

The overall depositional model and palaeogeographic setting of the giant sand system of the Utsira Formation is reconstructed for the Late Miocene time interval (Fig. 18). The sands were deposited within a shelfal seaway, probably by high-energy marine current systems. These currents also had strong erosive capacity and created large scouring features, which can be observed particularly to the north of 59°30′N, as exemplified in Figure 16. Water depths probably fluctuated between fairly deep shelf to shallow-marine conditions. The planktonic and benthic foraminifera indicate that intermediate (shelfal) water depths periodically must have existed. The abundance of glauconite suggests longer periods with non-deposition, probably related to transgressive phases and relatively deep shelf settings. The concentration of molluscs in parts of the Utsira Formation suggests that faunal colonization periodically was established, most likely, within shallow-marine waters. The sands were deposited largely in a southern and northern shoal system (Galloway, 2002), and their accumulation was interrupted by frequent periods with non-deposition followed by vigorous reworking and amalgamation.

We had argued earlier (Eidvin and Rundberg, 2001) that the Utsira Formation in some wells is directly overlain by Upper Pliocene sands that are difficult to distinguish from the Utsira Formation sands. Some of these sands are genetically different and represent turbiditic sands belonging to the huge Upper Pliocene prograding complex. The westerly-derived sands, however, which represent the Lower Pliocene sand system are genetically related to the underlying main Utsira Formation sand system, and should, therefore, be included in the proper Utsira Formation.

One of the key issues regarding the model of Utsira Formation sand deposition is the lack of ancient analogues. In much of the published literature, shelf sands only make up 10–30 m sequences with a clear coarsening-upward grain size profile (e.g. Johnson and Baldwin, 1996). Rundberg (1989) therefore proposed that the Utsira Formation sands were atypical of shelf settings, and pointed out that a unique depositional setting was required for the deposition of the sands. Galloway (2002) discussed many sedimentologists reluctance to interpret thick (30–100 m) aggradational sand units as products of shelf transport and deposition. As pointed out in his work, some modern shelves (in particular, the Mallaca Strait) have many similarities in geographic setting, sediment composition and sand morphologies as the Utsira Formation.

It appears, however, that ancient analogues to the Utsira sand system do exist. The Middle Jurassic Garn Formation sandstones off Mid Norway comprise an aggradational system of

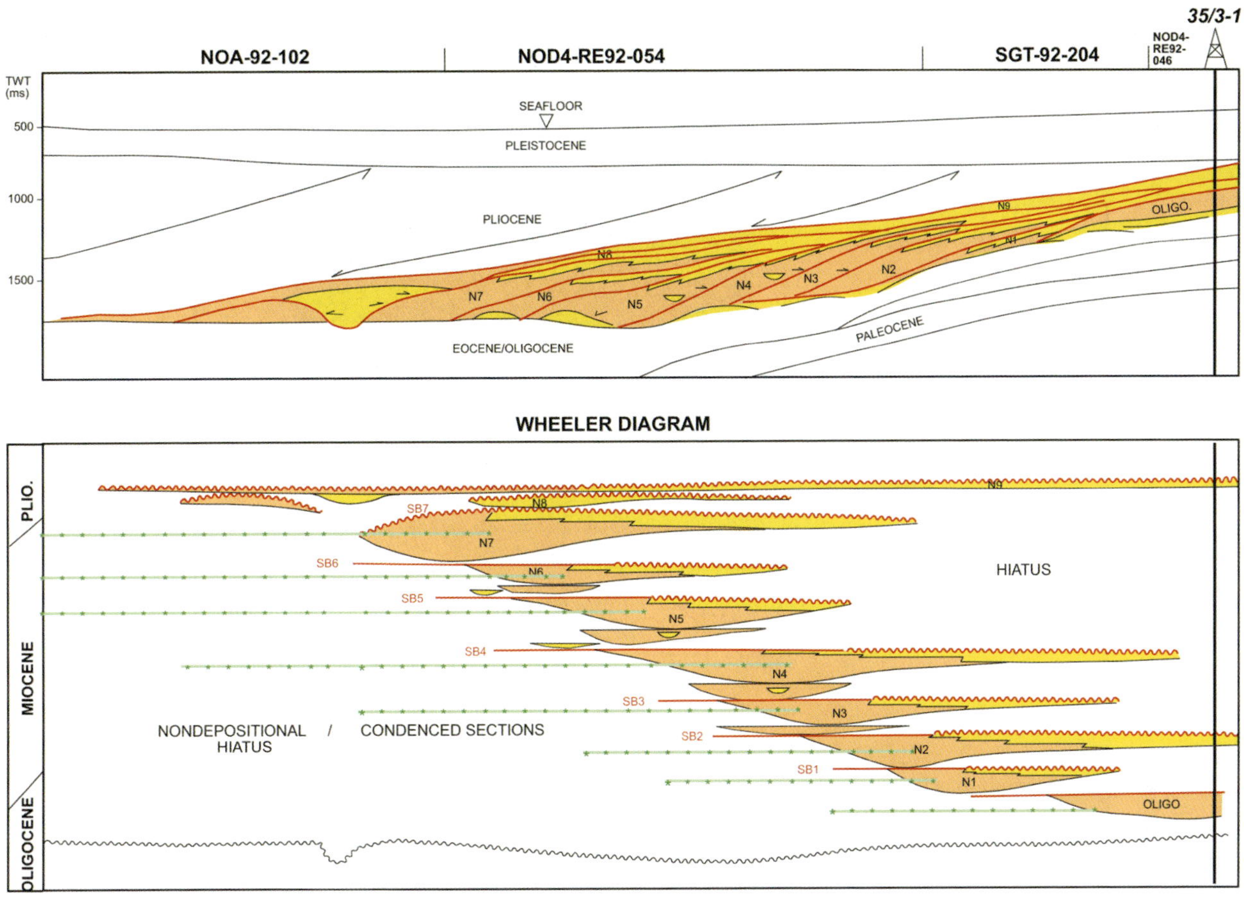

Fig. 18 (Above) Interpreted seismic composite section close to 62°N showing Neogene progradation from Møre Margin to the east into the southern Møre Basin, with tie to Agat well 35/3-1. This line shows the preservation of an almost complete post-Oligocene progradational system, which is absent further to the south due to uplift and erosion. Note stratigraphical position and dating of the large channel with incision into Oligocene strata. Location of line shown in Fig. 7d. (Below) Wheeler diagram of the composite section. Green lines illustrate glauconitic concentration at maximum flooding surfaces within each of the sequences.

blocky sands virtually without interbedding, reaching thickness in the order of 100–200 m. This sand has been variably interpreted as a fluvial to shelfal deposit, but its depositional environment remains relatively poorly constrained. In parts of the basin it is, however, interpreted as compounded tidal sandwaves, deposited in a high-energy current regime (Gjelberg et al., 1987; Corfield et al., 2001).

Conclusions

An improved seismic-lithostratigraphic subdivision of post-Eocene to Lower Pliocene strata has been established for the Norwegian North Sea between 58 and 62°N. This succession has been subdivided into a lower megasequence (Upper Hordaland Group) and an upper megasequence (Lower Nordland Group) separated by a major hiatus.

The stratigraphic framework has been revised by assigning the Skade Formation to the Early Miocene (previously Late Oligocene) and by introducing a new mudstone unit of Middle Miocene age at the base of the Nordland Group. The base of the Utsira Formation has been redefined in its type well (16/1-1; Isaksen and Tonstad, 1989), as the previous definition obviously overlapped with the upper part of the Skade Formation. Based upon diagnostic *Bolboforma* assemblages in the underlying mudstone, we conclude that the base of the Utsira Formation is not older than 12 Ma.

We suggest that the major 'mid-Tertiary' compressional tectonic regime that affected the passive margin of northwest Europe had a major impact on the stratigraphic and depositional evolution of the northern North Sea Basin.

On a larger scale, the Early Oligocene plate reorganisations caused seaway link to the Arctic Ocean, introducing a change in oceanic and palaeogeographic conditions. In the northern North Sea Basin, these changes coupled with the global climatic deterioration resulted in (1) a marked lithostratigraphic change at the Eocene–Oligocene boundary; and (2) an increased ocean

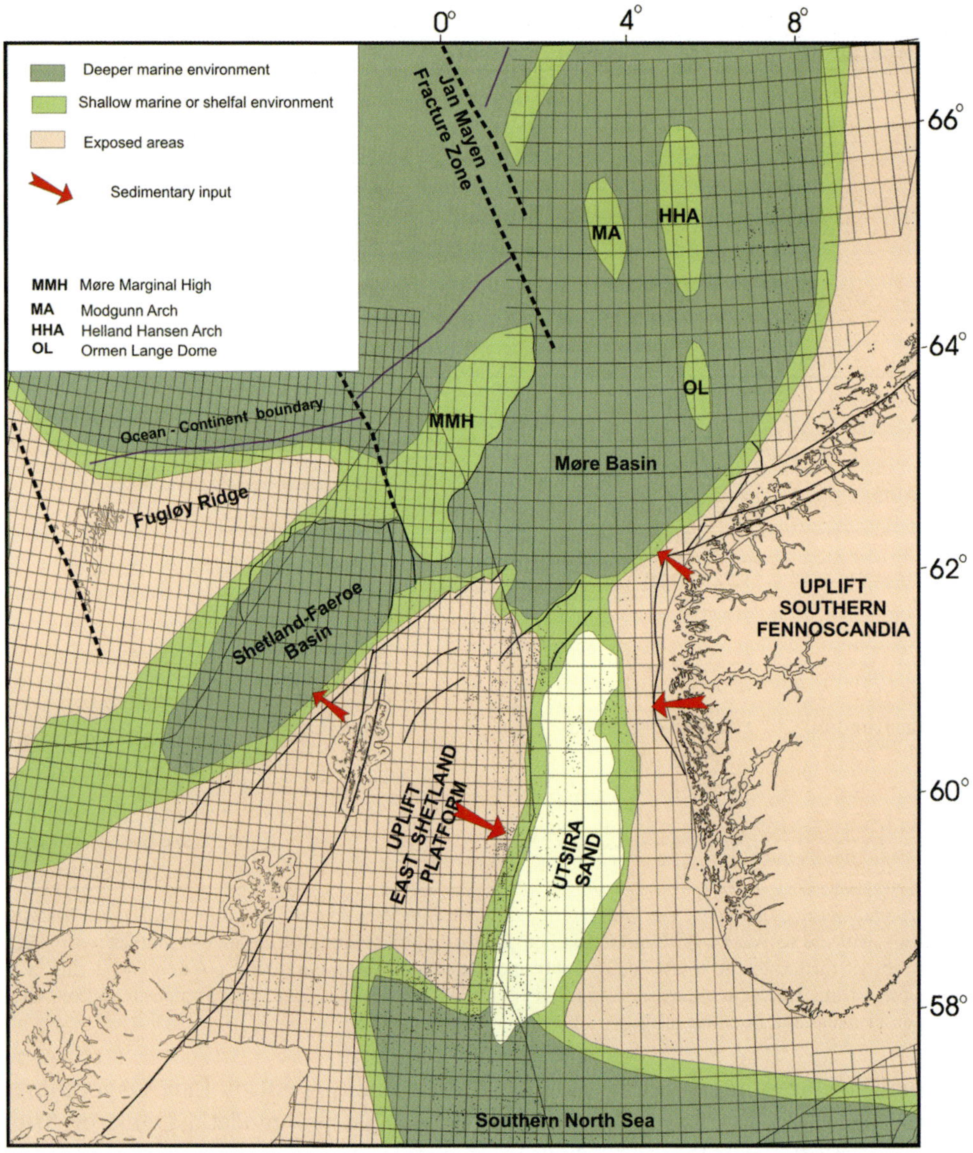

Fig. 19 Late Miocene palaeogeographic map of the northern North Sea and the Faeroe–Norwegian margin showing deposition of the Utsira sandbody in a narrow, shelfal strait within the Viking Graben of northern North Sea. The sands were deposited through a long time span (7–8 million years), and their accumulation was interrupted by frequent periods of non-deposition, followed by vigorous reworking and amalgamation. High-energy, (?tidal) marine currents probably operated across the narrow strait during the Late Miocene.

fertility, probably associated with upwelling conditions, leading to the marked increase in silica-rich deposition during Early Oligocene times.

The Oligocene–Miocene structural activity heavily affected the Shetland Platform and northernmost North Sea and was also felt in southern Fennoscandia (Fig. 5). We suggest that the East Shetland Platform underwent three significant phases of post-Eocene uplift, which resulted in deposition of major sandy systems into the northern North Sea Basin. The first sandy system was deposited during Early Oligocene (33–28 million years), the second input, during Early Miocene (20–15 million years) and third input, during Late Miocene/earliest Pliocene (12–4.5 million years). The first and the second sandy systems were deposited as turbiditic sands in the northern and southern Viking Graben, respectively, whereas the third sandy system was deposited in shallower marine settings as a consequence of uplift of the marine basin.

The progressive uplift of the northernmost North Sea during Oligocene–Miocene coupled with increased marine circulation led to vigorous current erosion and possibly also subaerial erosion in northernmost North Sea resulting in the formation of the mid-Miocene stratigraphic break. This erosional break is of 20 m.y. duration at maximum in northernmost North Sea to almost zero in southern Viking Graben. We suggest that this erosional event was associated with peak compression, beginning first, in the Tampen Spur area to the north and then successively working in the southward direction. Based on faunal and isotopic dating of samples from the southern Viking Graben, an age of about 15 Ma can be assigned for the transition between the Hordaland and Nordland Groups, at minimum stratigraphic break. Such an age also yields a timing of the uplift that affected the northern North Sea Basin, comparing closely to age estimates for the growth history of anticlinal structures and domes, along the passive margin.

During the Late Miocene to Early Pliocene, the northern North Sea formed a narrow seaway between deeper waters in the Møre Basin to the north and central North Sea to the south. This narrow strait received large amounts of coarse clastics from the East Shetland Platform and probably, also from southern Fennoscandia (although evidence for the latter is lacking due to later uplift and erosion of the Norwegian basin margin). The sands accumulated in a high-energy current regime, probably, first as elongated shelfal sand ridges which subsequently became moulded and amalgamated, and then as elongated sheet sands, as the shelf periodically became shallower due to local tectonic activity and/or fluctuations in sea level.

Acknowledgements

We thank Felix Gradstein and Morten Smelror for their critical reviews, which increased the quality of this chapter. We also thank Norsk Hydro ASA and Norwegian Petroleum Directorate for funding the project and giving us the permission to publish. The authors acknowledge TGS-Nopec for permission to publish seismic lines.

References

Boldreel, L.A. and Andersen, M.S., 1993. Late Palaeocene to Miocene compression in the Faeroe-Rockall area. In: J.R. Parker (Editor), Petroleum Geology of Northwest Europe. Proceedings of the 4th Conference. Geological Society, London, pp. 1025–1034.

Boldreel, L.A. and Andersen, M.S., 1994. Tertiary development of the Faeroe-Rockall Plateau based on reflection data. Bull. Geol. Soc. Den., 41: 162–180.

Brekke, T., 2000. The tectonic evolution of the Norwegian Sea continental margin with emphasis on the Vøring and Møre Basins. Geol. Soc., London, Spec. Publ., 167: 327–378.

Buchardt, B., 1978. Oxygen isotope palaeotemperatures from the Tertiary period in the North Sea area. Nature, 275: 121–123.

Buchardt, B., 1981. Tertiary deposits of the Norwegian-Greenland Sea region (Svalbard, Northeast and East Greenland, Iceland, the Faeroe islands and the Norwegian-Greenland Sea) and their correlation to NW Europe. Can. Soc. Petr. Geol., Mem., 7: 585–610.

Corfield, S., Sharp, I., Häger, K.-O., Dreyer, T. and Underhill, J., 2001. An integrated study of the Garn and Melke formations (Middle to Upper Jurassic) of the Smørbukk area, Halten terrace, mid-Norway. In: Martinsen, O. and Dreyer, T. Sedimentary environments offshore Norway — Palaeozoic to Recent. Norwegian Petroleum Society (NPF), Special Publication 10, Elsevier, Amsterdam, pp. 199–210.

Deegan, C.E. and Scull, B. (Editors), 1977. A proposed standard lithostratigraphic nomenclature for the central and northern North Sea. Institute for Geological Sciences, London, Rep., 77 (25).

DeMan, E., Ivany, L. and Vanderberghe, N., 2003a. Stable isotope record of the Eocene-Oligocene transition in NE-Belgium. Symposium on the Paleogene Preparing for Modern Life and Climate, Leumen, Belgium (abs.)

DeMan, E., Van Simaeys, S., Vanderberghe, N. and Steurbaut, E., 2003b. The Rupelian/Chattian boundary in the North Sea Basin: New micropaleontological and geochemical data from the stratotype areas. Symposium on the Paleogene Preparing for Modern Life and Climate, Leumen, Belgium (abs.)

Doré, A.G. and Lundin, E.R., 1996. Cenozoic compressional structures on the NE Atlantic margin: Nature, origin and potential significance for hydrocarbon exploration. Petrol. Geosci., 2: 299–311.

Eidvin, T. and Rundberg, Y., 2001. Late Cainozoic stratigraphy of the Tampen area (Snorre and Visund fields) in the northern North Sea, with emphasis on the chronology of early Neogene sands. Norsk Geol. Tidsskr., 81: 119–160.

Eidvin, T., Jansen, E., Rundberg, Y., Brekke, H. and Grogan, P., 2000. The Upper Cainozoic of the Norwegian continental shelf correlated with the deep sea record of the Norwegian Sea and North Atlantic. Mar. Petrol. Geol., 17: 579–600.

Eidvin, T., Riis, F. and Rundberg, Y., 1999. Upper Cainozoic stratigraphy in the central North Sea (Ekofisk and Sleipner fields). Norsk Geol. Tidsskr., 79: 98–129.

Eidvin, T., Rundberg, Y. and Smelror, M., 2002. Revised chronology of Neogene sands (Utsira and Skade formations) in the central North Sea. In: A. Hurst (Editor), Onshore-Offshore Relationships on the Nordic Atlantic Margin. NGF Abstracts and proceedings of the Norwegian Petroleum Society (NPF) and Norwegian Geological Society (NGF) Conference, 7–9th Oct. Trondheim, pp. 51–53.

Eldholm, O. and Thiede, J., 1980. Cenozoic continental separation between Europe and Greenland. Palaeogeogr., Palaeoclimat., Palaeoecol., 30: 243–259.

Eldholm, O., Skogseid, J., Sundvor, E. and Myhre, A.M., 1990. The Norwegian-Greenland Sea. In: Grantz, A., Johnson, L. and Sweeney, J.F. (Editors), The Geology of North America (Vol. L): The Arctic Ocean Region. Geol. Soc. Am., 351–364.

Eldrett, J.S., Harding, I.C., Firth, J.V. and Roberts, A.P., 2004. Magnetostratigraphic calibration of Eocene-Oligocene dinoflagellate cyst biostratigraphy from the Norwegian-Greenland Sea. Marine Geology, 204 (1–2), 91–127.

Fyfe, J.A., Gregersen, U., Jordt, H., Rundberg, Y., Eidvin, T., Evans, D., Stewart, D., Hovland, M. and Andresen, P., 2003. Chapter 16: Oligocene to Holocene. In: D. Evans, C. Graham, A. Armour and P. Bathurst (Editors), The Millennium Atlas: Petroleum geology of the Central and Northern North Sea. Geological Society, London, pp. 279–287.

Galloway, W.E., Garber, J.L., Liu, X., and Sloan, B.J., 1993. Sequence stratigraphic and depositional framework of the

Cenozoic fill, Central and Northern North Sea Basin. In: J.R. Parker (Editor), Petroleum Geology of Northwestern Europe: Proceedings of the 4th Conference. Geological Society, London, pp. 33–43.
Galloway, W.E., 2002. Paleogeographic setting and depositional architecture of a sand-dominated shelf depositional system, Miocene Utsira Formation, North Sea Basin. J. Sediment. Res., 72 (4): 477–490.
Gjelberg, J., Dreyer, T., Høie, A., Tjelland, T. and Lilleng, T., 1987. Late Triassic to Mid-Jurassic sandbody development on the Barents and Mid-Norwegian shelf. In: J. Brooks and K. Glennie (Editors), Petroleum Geology of North West Europe, Graham and Trotman, London, pp. 1105–1129.
Goll, R.M. and Skarbø, O., 1990. High resolution dating of Cenozoic sediments from the northern North Sea using 87Sr/86Sr isotope stratigraphy. Discussion. AAPG Bull., Discussion, 74: 1283–1286.
Gradstein, F. and Bäckström, S., 1996. Cainozoic Biostratigraphy and Paleobathymetry, northern North Sea and Haltenbanken. Norsk Geol. Tidsskr., 76: 3–32.
Gregersen, U., Michelsen, O. and Sørensen, J.C., 1997. Stratigraphy and facies distribution of the Utsira Formation and the Pliocene sequences in the northern North Sea. Mar. Petrol. Geol., 14 (7/8): 893–914.
Heilmann-Clausen, C., Nielsen, O.B. and Gersner, F., 1985. Lithostratigraphy and depositional environments in upper Paleocene and Eocene of Denmark. Bull. Geol. Soc. Den., 33: 287–323.
Hull, D.M., 1996. Paleogeography and biostratigraphy of Paleogene radiolarians from the Norwegian-Greenland Sea. In: Thiede, J., Myhre, A.M., Firth, J.V., Johnson, G.L., and Ruddiman, W.F. (Editors). Proc. ODP. Sci. Results, 151: College Station, TX (Ocean Drilling Program).
Isaksen, D. and Tonstad, K., 1989. A revised Cretaceous and Tertiary lithostratigraphic nomenclature for the Norwegian North Sea. NPD, Bull., 5: 59 pp.
Ivany, L.C., Patterson, W.P., and Lohmann, K.C., 2000. Cooler winters as a possible cause of mass extinctions at the Eocene-Oligocene boundary. Nature, 407: 887–890.
Ivany, L.C., Nesbitt, E.A., and Prothero, D.R., 2003. The marine Eocene-Oligocene transition: a synthesis. In: Prothero, D.R., Ivany, L.C., and Nesbitt, E.A. (Editors). From Greenhouse to Icehouse: The Marine Marine-Oligocene Transition. Columbia University Press, New York, pp. 522–534.
Johnson, H.D. and Baldwin, C.T., 1996. Shallow siliciclastic seas. In: H.G. Reading (Editor), Sedimentary environments, Processes, Facies and Stratigraphy. Oxford, Blackwell Science Ltd, pp. 232–280.
Jordt, H., Faleide, J.I., Bjørlykke, K. and Ibrahim, M.T., 1995. Cenozoic sequence stratigraphy of the central and northern North Sea Basin: tectonic development, sediment distribution and provenance areas. Mar. Petrol. Geol., 12: 845–879.
Kaminski, M. and Austin, W.E.N., 1999. Oligocene deep-water agglutinated foraminifers at Site 985, Norwegian Basin, southern Norwegian Sea. In: M.E. Raymo, E. Jansen, P. Blum and T.D. Herbert (Editors), Proceedings of the Ocean Drilling Program, Scientific Results, 162: 169–177.
Kennett, J.P., 1982. Marine Geology. Prentice-Hall, Inc., New Jersey, 813 pp.
Kennett, J.P. and Barker, P.F., 1990. Latest Cretaceous to Cenozoic climate and oceanographic developments in the Weddell Sea, Antarctica: an ocean-drilling perspective. In: Barker, P.F. and Kennett, J.P. et al. (Editors). Proc. ODP Sci. Results, 113: College Station, TX (Ocean Drilling Program), 937–960.
King, C., 1983. Cenozoic micropaleontological biostratigraphy of the North Sea. Institute for Geological Sciences, Rep., 82: 40 pp.
Kitchell, J.A. and Clark, D.L., 1982. Late Cretaceous-Paleogene paleogeography and paleocirculation: evidence of North Polar upwelling. Palaeogeogr., Palaeoclimat., Palaeoecol., 40: 135–165.
Knox, R.W.O'B. and Holloway, S., 1992. Paleogene of the central and northern North Sea. In: R.W.O'B. Knox and W.G. Cordey (Editors), Lithostratigraphic nomenclature of the UK North Sea. Nottingham, British Geological Survey.
Lundin, E. and Doré, A.G., 2002. Mid-Cenozoic post-breakup deformation in the 'passive' margins bordering the Norwegian-Greenland Sea. Mar. Petrol. Geol., 19: 79–93.
Løseth, H., Wensaas, L., Arntsen, B. and Hovland, M., 2003. Gas and fluid injection triggering shallow mud mobilization in the Hordaland Group, North Sea. In: P. van Rensbergen, R.R. Hillis, A.J. Maltman and C.K. Morley (Editors), Subsurface sediment mobilization. Geol. Soc., London, Spec. Publ., 216: 139–157.
Martinsen, O.J., Bøen, F., Charnock, M.A., Mangerud, G. and Nøttvedt, A., 1999. Cenozoic development of the Norwegian margin 60–64°N; sequences and sedimentary response to variable basin physiography and tectonic setting. In: A.J. Fleet and S.A.R. Boldy (Editors), Petroleum Geology of Northwest Europe: Proceedings of the 5th Conference, Geological Society, London, pp. 293–304.
Mikkelsen, N. and Barron, J.A., 1997. Early Oligocene diatoms on the Ceara Rise and the Cenozoic evolution of biogenic silica accumulation in the low-latitude Atlantic. In: N.J. Shackleton, W.B. Curry, C. Richter and T.C. Bralower (Editors), Proceedings of the Ocean Drilling Program, Scientific Results, 154.
Mosar, J., Lewis, G. and Torsvik, T.H., 2002. North Atlantic sea-floor spreading rates: implications for Tertiary development of inversion structures of the Norwegian-Greenland Sea. J. Geol. Soc., London, 159: 503–515.
Müller, C. and Spiegler, D., 1993. Revision of the Late/Middle Miocene boundary on the Voering Plateau (ODP Leg 104). Newsletter on Stratigraphy, 28 (2/3): 171–178.
Ogasawara, K., 2002. Responses of Japanese Cenozoic molluscus to Pacific gateway events. Revista Mexicana de Ciencias Geológicas, 19: 206–214.
Piasecki, S., Gregersen, U. and Johannessen, P., 2002. Lower Pliocene dinoflaggelate cysts from cored Utsira Formation in the Viking Graben, northern North Sea. Mar. Petrol. Geol., 19: 55–67.
Ruddiman, W.F., 2000. Earth's Climate, Past and Future. W.H. Freeman and Company, New York, 465 pp.
Rundberg, Y., 1989. Tertiary sedimentary history and basin evolution of the Norwegian North Sea between 60–62°N — An integrated approach. PhD thesis Univ Trondheim, (Reprinted 1991: NTH, Geol. Inst., Report Series, 25, 292 pp).
Rundberg, Y. and Smalley, P.C., 1989: High resolution dating of Cenozoic sediments from the northern North Sea using 87Sr/86Sr isotope stratigraphy. AAPG Bull., 73: 298–308.
Rundberg, Y., Olaussen, S. and Gradstein, F., 1995: Incision of Oligocene strata; evidence for northern North Sea Miocene uplift and key to the formation of the Utsira sands. Geonytt, 22: 62.
Sejrup, H.P., Haflidason, H., Løvlie, R., Bratten, Å., Tjøstheim, G., Forsberg, C.F. and Ellingsen, K.L., 1995. Quaternary of the Norwegian channel: glaciation history and palaeooceanography. Norsk Geol. Tidsskr., 75: 65–87.
Shipboard Scientific Party, 2000: Leg 189 Preliminary Report; The Tasmanian Seaway between Australia and Antarctica — Paleoclimate and Paleoceanography. ODP Prelim. Rpt, 89 (Online). Available from World Wide Web: (http://www-odp.tamu.edu/publications/prelim/189_prel/189PREL.PDF)
Smalley, P.C. and Rundberg, Y., 1990. High-resolution, dating of Cenozoic Sediments from northern North Sea using $^{87}Sr/^{86}Sr$ Stratigraphy. AAPG Bulletin, 74. Reply to Discussion, 74: 1287–1290.

Shipboard Scientific Party, 2002: Leg 199 Preliminary Report; Paleogene Equatorial Transect. ODP Prelim. Rpt, 99 (Online). Available from World Wide Web: (http://www-odp.tamu.edu/publications/prelim/199_prel/199PREL.PDF)

Spiegler, D. and Müller, C., 1992. Correlation of Bolboforma zonation and nannoplankton stratigraphy in the Neogene of the North Atlantic: DSDP sites 12-116, 49-408, 81-555 and 94-608. Mar. Micropaleontol., 20: 45–58.

Stoker, M.S., 1997. Mid-to late Cenozoic sedimentation on the continental margin off NW Britain. J. Geol. Soc., London, 154: 509–515.

Stoker, M.S., Nielsen, T., van Weering, T.C.E. and Kuijpers, A., 2002. Towards an understanding of the Neogene framework of the NE Atlantic margin between Ireland and the Faeroe Islands. Mar. Geol., 188: 233–248.

Talwani, M. and Eldholm, O., 1977. Evolution of the Norwegian-Greenland Sea. Geol. Soc. Am. Bulletin, 88: 969–999.

Thyberg, B.I., Stabell, B., Faleide, J.I. and Bjørlykke, K., 1999. Upper Oligocene diatomaceous deposits in the northern North Sea — silica diagenesis and paleogeographic implications. Norsk Geol. Tidsskr., 79: 3–18.

Vogt, P.R., Perry, R.K., Feden, R.H., Fleming, H.S. and Cherkiz, N.Z., 1981. The Greenland-Norwegian Sea and the Iceland environment; geology and geophysics. In: A.E.M. Nairn, M. Churkin and F.G. Stehli (Editors), The ocean basins and margins, 5, The Arctic Ocean. Plenum Press, N.Y., pp. 493–598.

Vågnes, E., Gabrielsen, R.H. and Haremo, P., 1998. Late Cretaceous-Cenozoic intraplate contractional deformation at the Norwegian continental shelf: timing, magnitude and regional implications. Tectonophysics, 300: 29–46.

Wold, C.N., 1994. Cenozoic sediment accumulation on drifts in the northern North Atlantic. Paleoceanography, 9: 917–941.

Wright, J.D. and Miller, K.G., 1996. Control of North-Atlantic deep-water circulation by the Greenland-Scotland Ridge. Paleoceanography, 11: 157–170.

Zachos, J., Pagani, M., Sloan, L., Thomas, E. and Billups, K., 2001. Trends, Rhytms, and aberrations in global climate 65 Ma to present. Science, 292: 686–693.

3D Cretaceous to Cenozoic palaeobathymetry of the northern North Sea

Stein Tore Wien and Tomas Kjennerud

3D Palaeobathymetry maps have been constructed for 14 time-steps for Cretaceous and Cenozoic times in the northern North Sea (58–62°N). The restorations have been carried out by integration of relevant information from seismic sequence geometries, zero or shallow water depth indicators, compaction, micropalaeontological interpretations from wells, and isostasy. The high spatial resolution results show that the basin was greatly segmented and locally deep in the early Cretaceous post-rift phase, while it was broadening in the early late Cretaceous. Although during most of the Cretaceous period the underlying Jurassic rift structures controlled the individual depocenters, at the K–T boundary all traces of late Jurassic rift-structures had been levelled out. During the Palaeogene period, the region was characterised by episodic uplift of both western and eastern flanks. Water depths increased dramatically in the basin centre in the latest Paleocene, while shallowing induced by tectonic uplift began in the Oligocene. Subsequently, the late Miocene deep was filled in the late Pliocene to early Quaternary.

Introduction

Palaeobathymetry is a crucial parameter in almost all basin modelling applications and in the prediction of sedimentary facies variations, particularly in deep marine environments. In most basin modelling studies, palaeobathymetry as an input has been either ignored or treated in a very simplified manner. Some recent studies have, however emphasised the use of palaeo-water depth in 2D tectonic modelling (Gabrielsen et al., 2001), 2D depositional modelling (Øvrebø et al., 2001) and 3D secondary hydrocarbon migration modelling (Kjennerud and Sylta, 2001). These studies have emphasised on increasing the relief control in a sedimentary basin as it evolves over geological time.

The types of approaches available in determining palaeobathymetry in presently submerged extensional basins where field observations are not possible, can be grouped into two: (1) well data (e.g. micropalaeontology and sedimentology), and (2) 2D/3D seismic interpretations and modelling. Palaeobathymetry estimates from well cores and cuttings show a good temporal but low spatial resolution, while the introduction of seismic data offers a shift from the 1D well estimates to 2D/3D control, leading to a drastic increase of the spatial resolution. Based on the Kjennerud et al. (2001) structural restoration method along seismic profiles, a map-based method for 3D structural restoration was developed by Kjennerud and Sylta (2001). By utilising these two approaches, we combine in this paper estimates from both the well data and the seismic reflection data and present a complete 3D restoration of the Cretaceous and Cenozoic bathymetry in the northern North Sea, 58–62°N (Fig. 1). The results were integrated with the micropalaeontological water-depth estimates of Gillmore et al. (2001). Our analysis thus enables more unified reconstructions than earlier studies.

3D Palaeobathymetric restoration

The 3D structural restoration method of Kjennerud and Sylta (2001) has been implemented as a separated module in the hydrocarbon migration software Semi (Sylta, 1993) for reconstructing palaeobathymetry. This software was used in the present work. We estimate 3D palaeobathymetry by combining relevant information from depositional geometries, shallow or zero water depth indicators, and micropalaeontological interpretation from

Onshore–Offshore Relationships on the North Atlantic Margin
edited by B. Wandas et al. NPF Special Publication 12, pp. 241–253,

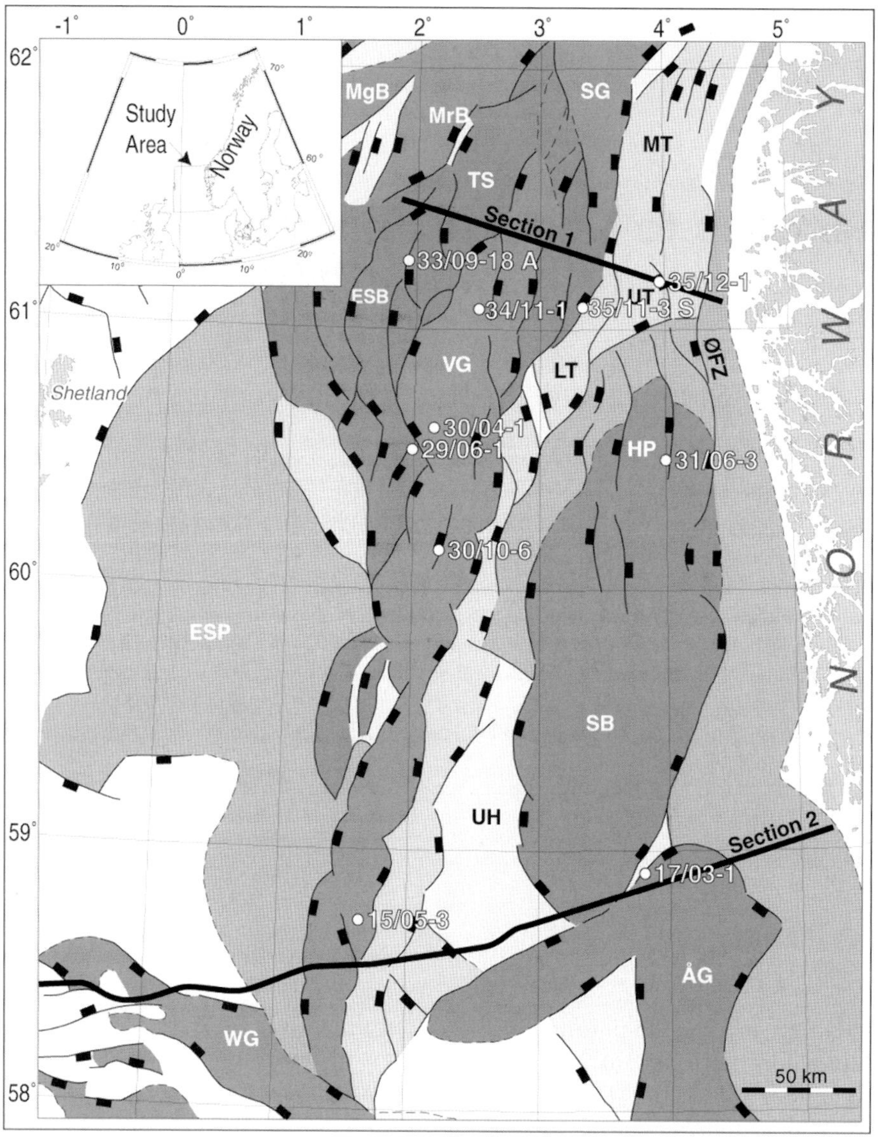

Fig. 1 The northern North Sea, late Jurassic structure map. The map includes the wells used for the micropalaeontological integration, together with the position of the seismic lines Section 1 and Section 2. ESB: East Shetland Basin, ESP: East Shetland Platform, HP: Horda Platform, LT: Lomre Terrace, MgB: Magnus Basin, MrB: Marulk Basin, MT: Måløy Terraces, SB: Stord Basin, SG: Sogn Graben, TS: Tampen Spur, UH: Utsira High, UT: Uer Terrace, VG: Viking Graben, WG: Witch Ground Graben, ØFZ: Øygarden Fault Zone, ÅG: Åsta Graben.

cores and cuttings. In addition, decompaction using the Sclater and Christie (1980) empirical compaction curves from the North Sea, and flexural or Airy isostasy are accounted for. Our approach uses a present day 3D depth converted geological model as a basis and restores the basin geometries backwards in time. For each restoration step decompaction is performed, correcting the palaeo-geometries of underlying carriers and source rocks.

Depositional geometries

The initial geometric restoration uses the type of sequence geometries observed in the seismic data. These are either based on whether the sequence is passively filling into a relief or actively prograding into it. These techniques were initially systematised by Kjennerud et al. (2001).

Deep marine infill

The post-rift stage is generally characterised by deep marine infill into a relief created during rifting, if the sediment supply is low (Kjennerud et al., 2001). The Cretaceous post-rift sequence in the northern North Sea fits this description (Gabrielsen et al., 2001; Kjennerud et al., 2001). In this setting, there is a direct correlation between the maximum sediment thickness and the water depth. Thus, the maximum bathymetry is reflected by the maximum sediment thickness in the basin centre. The decompacted thickness of the infilling unit is used to create

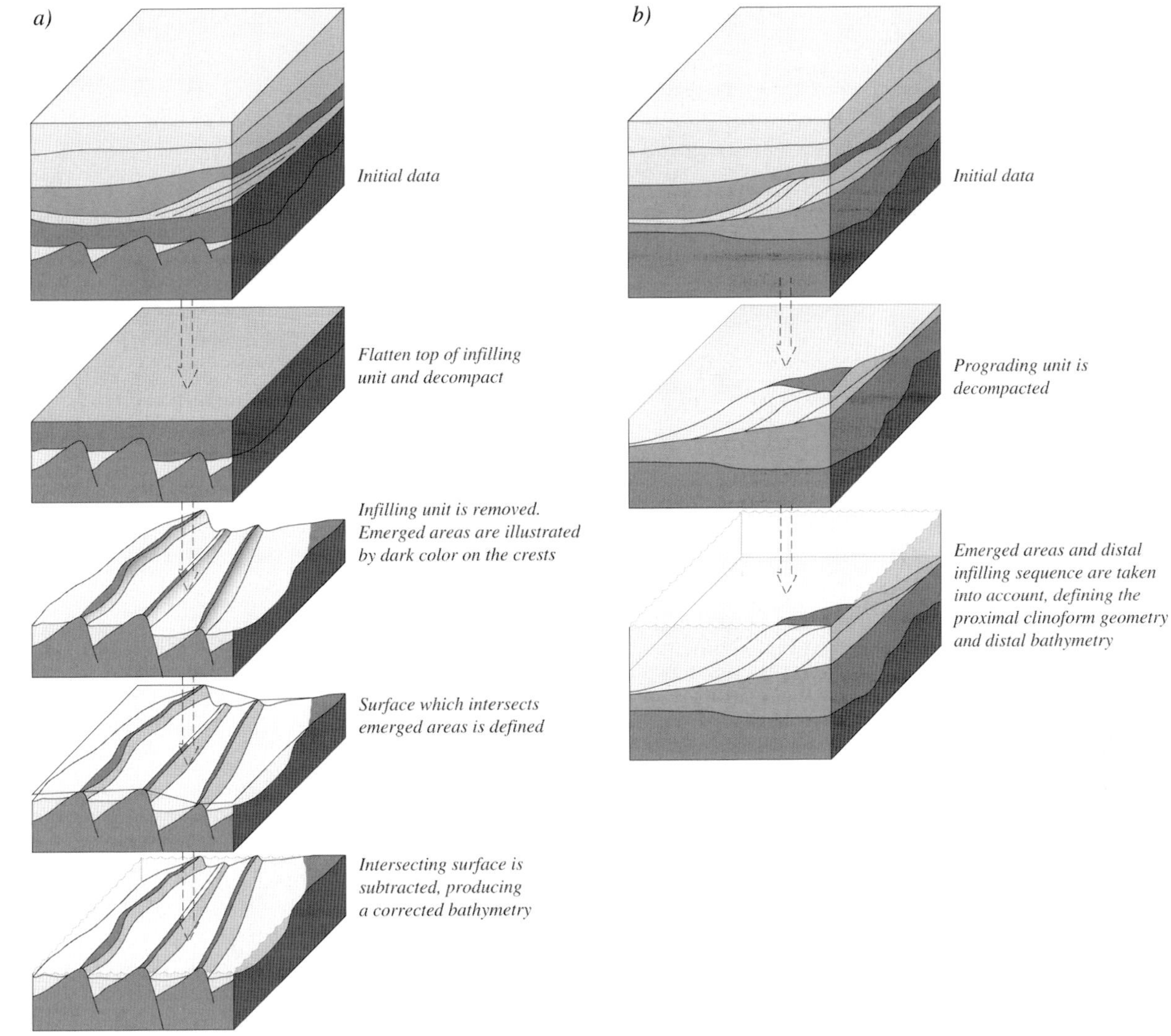

Fig. 2 Stepwise scheme showing bathymetric restoration using (a) deep marine infilling and (b) prograding sequence approach.

a relative relief, which is further constrained by calibration points from biostratigraphical, sedimentological and/or seismic sequence stratigraphical methods. The approach restores the bathymetry at the base of each infilling unit. The step-by-step deep marine infill restoration is shown in Fig. 2a.

Prograding sequences

Prograding sequences generally show an inverse correlation between maximum thickness and water depth, as the thickest part records the most shallow areas and vice versa. The proximal geometry is determined by the decompacted clinoform geometry, while the distal part may either be shaped according to the relief it was filling into, or by using the geometry of the overlying sequence. The step-by-step prograding sequences restoration is shown in Fig. 2b.

Micropalaeontology

Micropalaeontological depth estimates can easily be integrated in the 3D palaeobathymetric reconstructions in Semi, however, care should be taken in this process. Kjennerud and Gillmore (2003) has demonstrated that for prograding sequences, the geometrical estimates have proved to be much more precise than micropalaeontological estimates and should therefore be given priority in the integration. The opposite relation has been demonstrated for infilling sequences, where micropalaeontological estimates is crucial in constraining the restoration. A schematic 2D illustration for integrating micropalaeontological interpretations of palaeo-water depth with geometrical restorations is shown in Fig. 3. The structural restoration (Fig. 3a) is here constrained by the depth estimates from the wells (Fig. 3b). It is important to note that this will

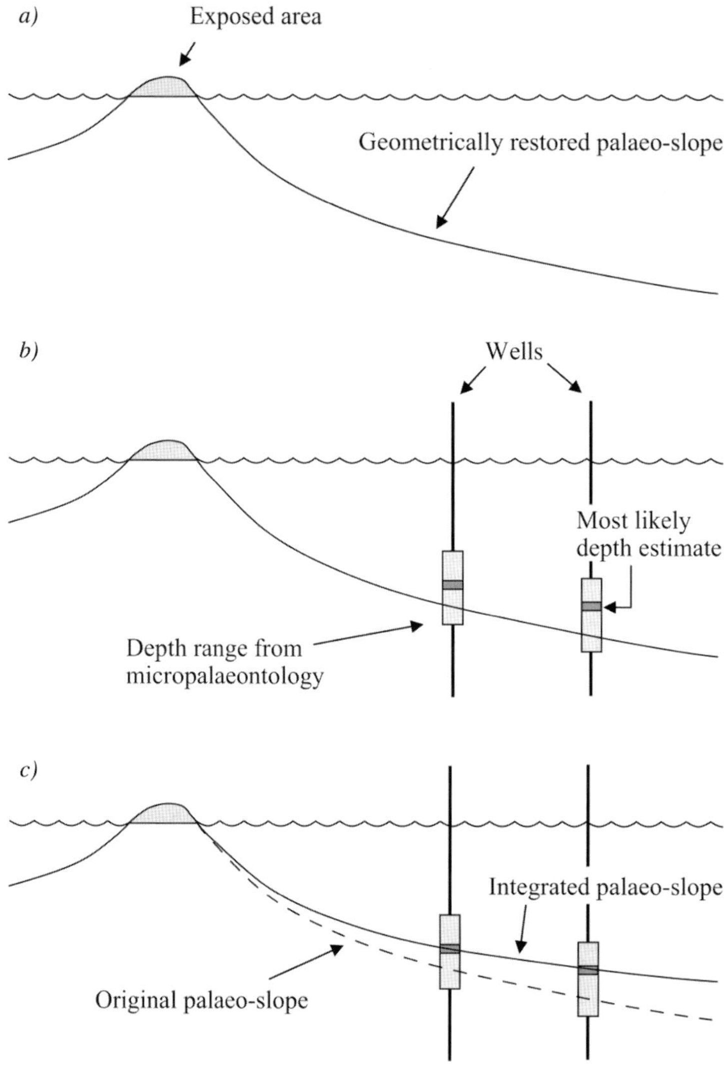

Fig. 3 Idealised illustration of integration between geometrical restoration and micropalaeontological water depth restoration. (a) Initial geometrical restoration, constrained by exposed area. (b) Depth ranges from micropalaeontological interpretations, including most likely estimates. (c) Geometrical restoration is constrained by most likely micropalaeontological depth estimates.

mainly change the basin amplitude, not the morphological features (Fig. 3c).

Isostasy

Both 1D Airy and 3D flexural isostasy can be applied with the restorations in Semi. The Airy isostatic correction implies zero strength of the lithosphere, while a given strength of the lithosphere is assumed using flexural correction. Only Airy isostasy has been applied in the present restorations.

Cretaceous to Cenozoic Palaeobathymetry

Following the late Silurian–early Devonian Caledonian Orogeny, two main extensional phases occurred in the northern North Sea, during the late Permian–early Triassic (Badley et al., 1988; Færseth, 1996; Gabrielsen et al., 1990; Steel and Ryseth, 1990) and latest Bajocian-earliest Bathonian–late Ryazanian (Badley et al., 1988; Gabrielsen et al., 1990; Rattey and Hayward, 1993). Prior to the Jurassic rift episode, the doming of the central North Sea led to the deposition of the northward prograding Brent Delta (e.g. Underhill and Partington, 1993).

Restoration of 14 time-steps have been performed; 7 Cretaceous and 7 Cenozoic reconstructions. The 3D geological model employed was based on a seismic database covering the northern North Sea consisting of most regional seismic lines. The seismic isopachs for each sequence are documented in Kyrkjebø (1999) and Kjennerud (2001). Depth conversion was performed using well information and earlier depth conversion of deep crustal transects (Faleide et al., 1999; Christiansson et al., 2000). The 10 wells shown in Fig. 1, on which micropalaeontological analysis of palaeo-water depth was performed, were used in the integration. These interpretations are documented in Gillmore et al. (2001) and Kyrkjebø et al. (2001). Gabrielsen et al. (2001) and Kjennerud et al. (2001) divided the Cretaceous stratigraphy into six seismostratigraphic units, K1–K6, while Jordt et al. (1995) divided the Cenozoic into ten Cenozoic Seismic Sequences, CSS-1–CSS-10. A correlation between the seismo-stratigraphic sequences and the lithostratigraphy of Isaksen and Tonstad (1989) is shown in Fig. 4, and the Cretaceous and Cenozoic sequences are illustrated by the seismic sections given in Fig. 5.

Cretaceous

Deep marine conditions prevailed along the basin axis during the Cretaceous post-rift stage. The Lower Cretaceous is represented by the K1 and K2 units, while the K3–K6 units corresponds to the Upper Cretaceous (Fig. 4). Due to the dominant passive infilling nature of the Cretaceous, the infilling restoration approach was used for all the reconstructions.

The Cretaceous restorations were constrained by using the zero and near zero depth areas described by Gabrielsen et al. (2001); Kjennerud et al. (2001) and Kyrkjebø et al. (2001). It should be emphasised that these areas are particularly significant in confining the basin topography.

The **late Ryazanian** bathymetric restoration (Fig. 6a) shows a segmented basin, where small depocenters were separated by transfer zones. The Viking and Sogn Grabens were completely separated by the exposed Utsira High and Lomre Terrace. The East Shetland Platform was also a prominent area of zero water depth. Maximum bathymetric values were observed along the rift-axis, with values exceeding 1200 m in both the Sogn Graben and Marulk Basin. Large water depths were also found in the Stord Basin and Åsta Graben, with values ranging from 400–1200 m. The very restricted nature of this restoration is due to incomplete seismostratigraphic framework in the East Shetland Basin. The K1–K2 sequences correspond to the incipient post-rift stage of Gabrielsen et al. (2001). During this stage, the Jurassic rift topography exerted a strong influence on the sediment distribution, with transfer zones separating local sub-basins along the rift axis.

A widening and shallowing of the basin is evident on the **earliest Cenomanian** restoration (Fig. 6b). The shallowing is in particular evident

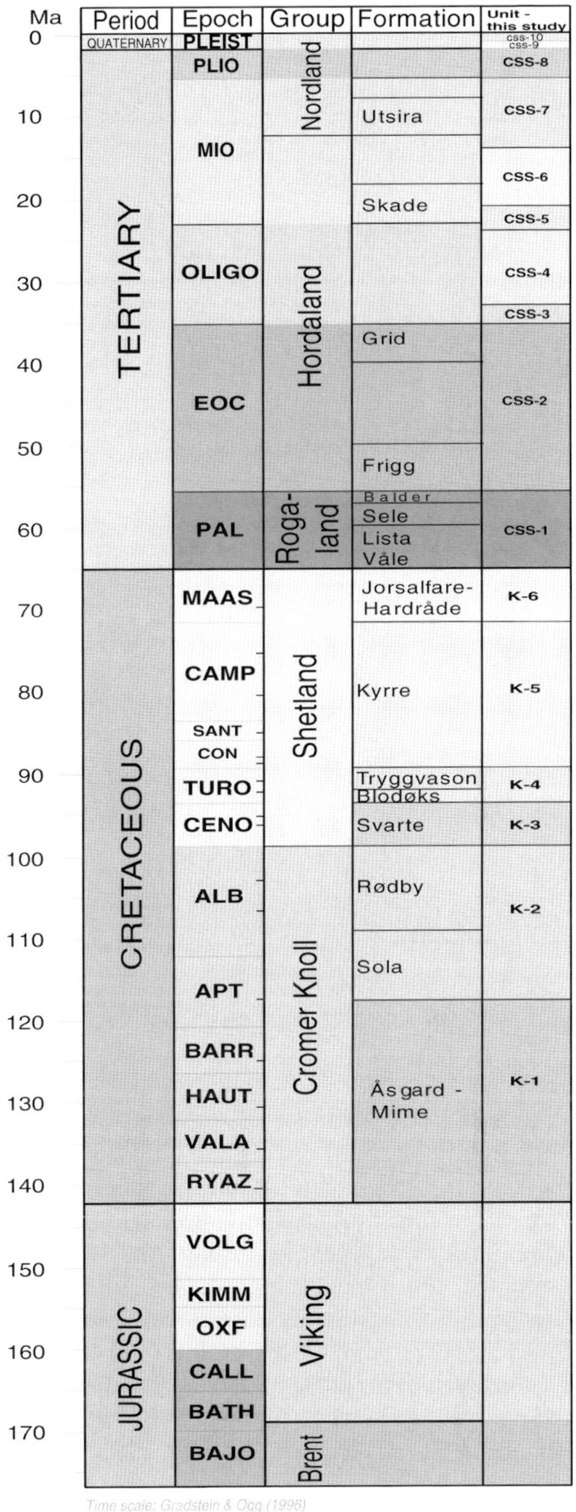

Fig. 4 Geological timescale with correlated lithostratigraphy and seismic stratigraphy, modified from Kyrkjebø et al., 2001.

in the Stord Basin, where maximum depths reached 300 m. The major depocenters were situated in the Magnus and Marulk Basins reaching approximately 1000 m. Along the Viking Graben two less pronounced deep water areas occurred, with maximum depths of 700 m. The decreasing water

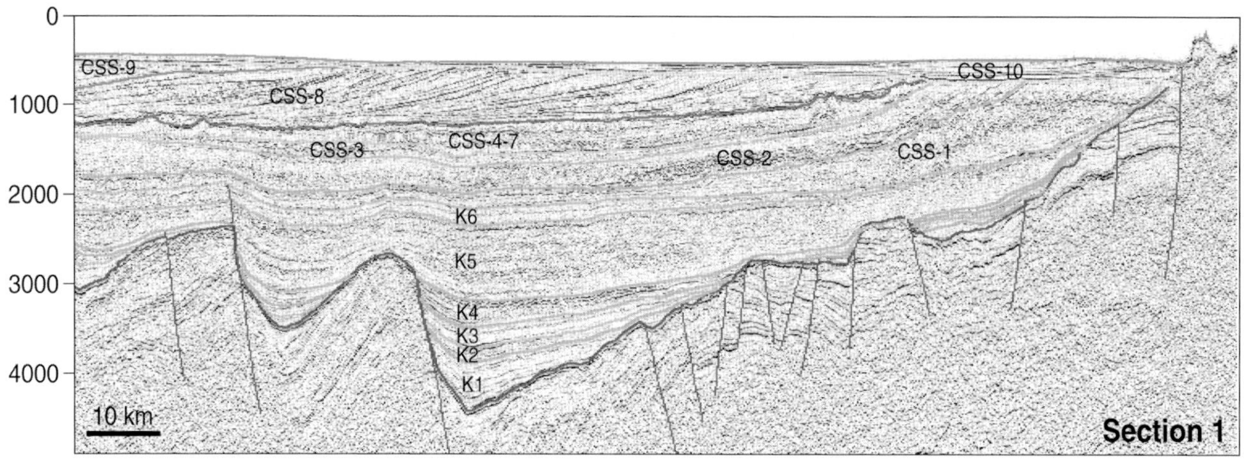

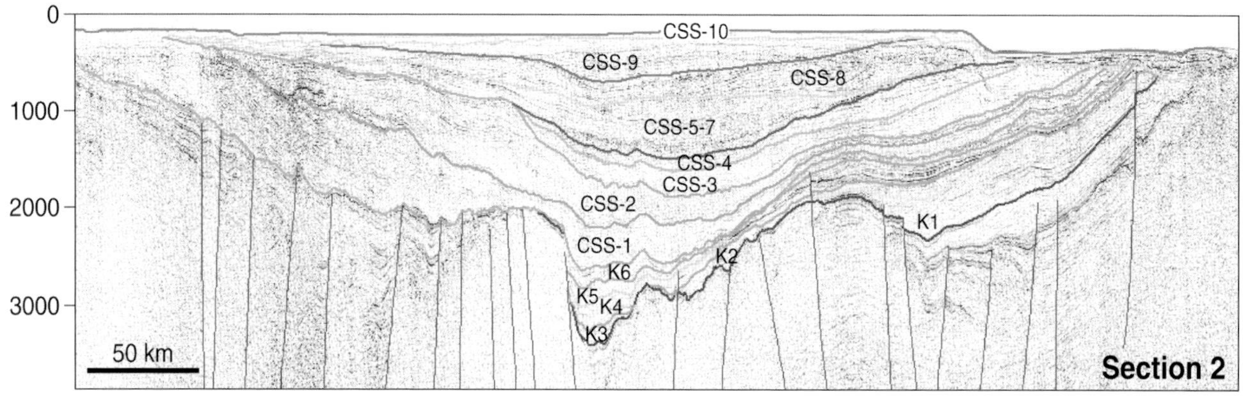

Fig. 5 The Cretaceous and Cenozoic seismostratigraphy illustrated by seismic lines Section 1 and Section 2, trending W/NW–E/SE and W/SW–E/NE respectively. Section 2 exceeds the study area westwards by approximately 50 km. The location of the seismic lines is showed in Fig. 1.

depth trend continued during the Cretaceous with an exception occurring in the **late Turonian** (Fig. 6c) when a prominent deepening event took place. Three major depocenters along the rift axis dominated the topography, along with a deep water area in the Magnus Basin area. The K3–K4 sequences correspond to the middle post-rift stage of Gabrielsen et al. (2001).

Renewed shallowing occurred in the **latest Campanian** (Fig. 6d), where a 400–500 m deep trench along the western flank was the main bathymetric feature. In the **latest Maastrichtian** (Fig. 7a) no prominent depocenters were evident, and water depths ranged from 200–300 m for the entire basin. The K5–K6 sequences correspond to the mature post-rift stage of Gabrielsen et al. (2001).

During the Cretaceous period, the basin widened significantly; from being strictly limited to the Jurassic rift relief in early Cretaceous, it covered the entire study area at the latest Maastrichtian. The Cretaceous gradual basin widening is illustrated by both Sections 1 and 2 (Fig. 5). A partial merging of the Viking Graben and the Stord Basin occurred during the early Aptian, although complete amalgamation was not evident until the latest Campanian. Major overprinting of the rift relief occurred first during the latest Campanian. At this time, the major depocenter in the northern part was situated on the East Shetland Platform, west of the Viking Graben. This is illustrated by Section 1 (Fig. 5), where the K6 sequence shows maximum thickness west of the Viking Graben. The Jurassic rift relief was totally diminished in the late Maastrichtian.

Together with the widening and shallowing of the basin, a change in the subsidence and deposition pattern took place. As illustrated, the earliest Cretaceous was dominated by small isolated basins, while a long wavelength deposition characterised the late Cretaceous (Fig. 6). This was due to the combined effect of increasing stiffness in the

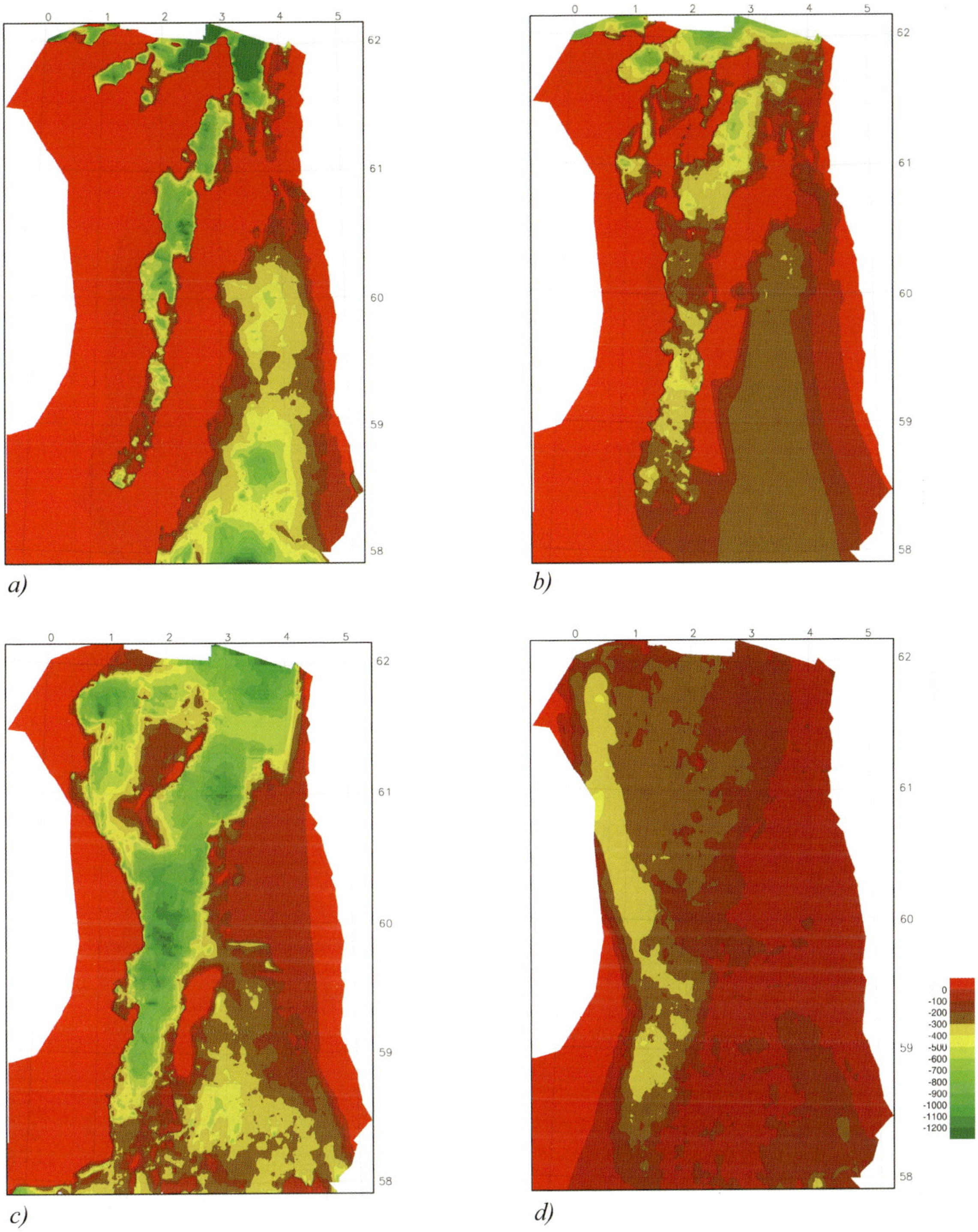

Fig. 6 Palaeo-water depth reconstructions of the (a) late Ryazanian, (b) earliest Cenomanian, (c) latest Turonian and (d) latest Campanian.

crust, differential compaction and burial of the relief (Gabrielsen et al., 2001).

Cenozoic

While the Cretaceous was dominated by passive infill of the Jurassic rift relief, the Cenozoic was characterised by episodic flank uplift (Jordt et al., 1995; Faleide et al., 2002). The basin fill was mainly sourced from the East Shetland Platform and southern Norway. During the latest Paleocene, a fall in the eustatic sea level combined with uplift of the East Shetland Platform led to increased sediment influx from the west. The platform remained the main sediment source until the Miocene, when shallowing led to exposure in the early Miocene. The Pliocene was dominated by uplift of southern Norway, leading to increased sediment input from the east.

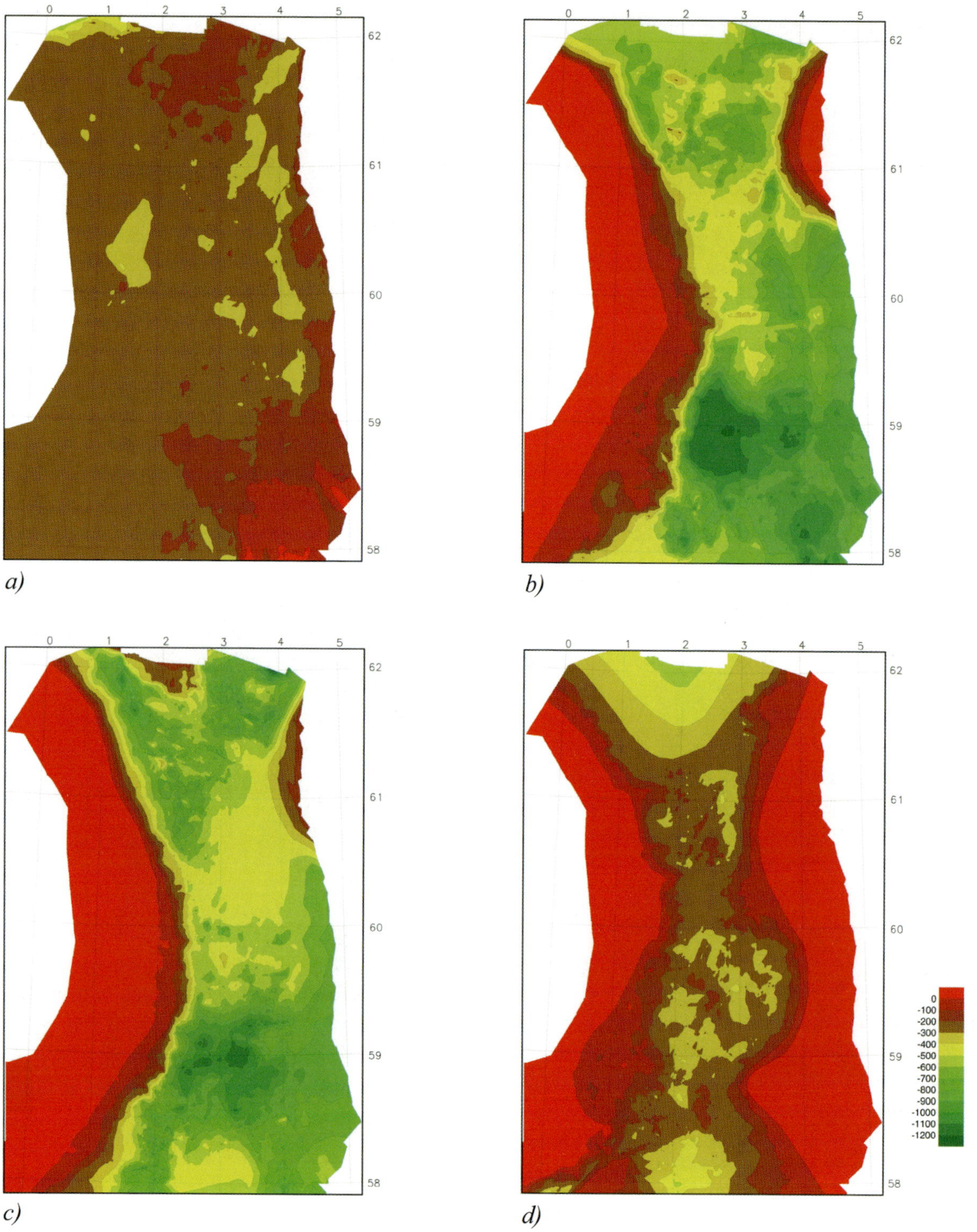

Fig. 7 Palaeo-water depth reconstructions of the (a) latest Maastrichtian, (b) latest Paleocene, (c) latest Eocene and (d) early Oligocene.

In the Pleistocene, glaciations eroded the underlying tilted Tertiary sediments. This resulted in a regional angular unconformity, which the middle–late Quaternary sediments were deposited upon.

Due to the more diverse nature of the Cenozoic sequences, both the infilling and prograding approaches were used in the Cenozoic reconstructions.

The **latest Paleocene** and **latest Eocene** were characterised by major outbuilding from the western flank (Figs. 7b and 7c). In addition, a westward prograding system is observed in the north along the eastern flank on the latest Paleocene reconstruction. The eastward latest Paleocene and the latest Eocene outbuildings are showed by both Sections 1 and 2 (Fig. 5). Increased sediment influx from the east is evident on northern Section 1, whereas the southern Section 2 shows no thickening along the eastern flank. Water depths increased drastically at the transition to the Cenozoic period, and exceeded 1000 m in the latest Eocene. Further, the latest Eocene restoration shows that sediments entered the basin from NW at this time.

By **early Oligocene**, water depths had decreased significantly (Fig. 7d). This was most evident along the eastern flank, where near-zero or zero water-depth is recorded on the bathymetric restoration. A N–S trending depocenter south of 60°N dominated the basin topography, with maximum water depths reaching 500–600 m. Shallow water depths prevailed throughout the Oligocene.

The Miocene bathymetry is poorly constrained, because of low stratigraphic control. Several wells show a hiatus during the Miocene (e.g. Gradstein and Bäckström, 1996), and it has been greatly debated whether this was caused by sub-aerial erosion or due to non-deposition in marine conditions.

A prominent deepening dominated **late Miocene**. The restoration (Fig. 8a) shows a curved basin with deep areas in the northern and southernmost part, separated by a bathymetric high. The reconstruction indicates a maximum depth exceeding 1000 m.

Large water depths prevailed in the early Pliocene, but in late **Pliocene** (Fig. 8b) the basin shallowed significantly, reaching a maximum of 600 m in the northernmost areas. The complete eastern flank was exposed, due to uplift of the southern Norway. This caused the basin to narrow considerably. Prograding systems entering the basin from the east are evident on both Sections 1 and 2 (Fig. 5). A westward shift of the basin occurred in the northernmost part.

During the **early Quaternary** (Fig. 8c), sediment influx from southern Norway filled the Pliocene deep, and thus reducing the basin relief. Glacial erosion along the coast of southern Norway created a N/NW–S/SE deepening trench (the Norwegian Channel), a prominent feature in the **present** sea floor bathymetry (Fig. 8d). The Norwegian Channel is shown clearly on Section 1 but is absent on Section 2 (Fig. 5), being located east of the seismic line.

Discussion

Restorations

As demonstrated by Kjennerud and Sylta (2001), the use of bathymetric data may significantly alter the results in hydrocarbon migration modelling. Although the quantitative consequences of improved palaeobathymetry have not been discussed in the present work, the high resolution quantitative maps presented here are ideal as an input in further basin modelling studies, e.g. secondary migration, hydrocarbon accumulation, pressure, maturation and tectonic modelling.

The uncertainty associated with depth estimates derived from micropalaeontological interpretations can be quite high (e.g. Kjennerud and Gillmore, 2003). The estimates are generally in the form of probable depth ranges, which may be in brackets of 100, 200, 500 m or more (e.g. Gradstein and Bäckström, 1996). A significant uncertainty factor may be connected with the position of the well where the sample is taken. Situated on a slope or down-flank from a steep slope, the sampled material may originate from a more shallow area. An example is the 30/10-6 well, where the geometrical restoration for late Turonian yields 900 m, whereas the micropalaeontological estimate is 300 m (Fig. 9). The well is situated in a slope environment, and it is thus likely that sediments were derived from the upper slope. Similar conclusions were drawn from the Palaeogene in the northern North Sea by Kjennerud and Gillmore (2003), where it was shown for down-slope of steep prograding systems, that micropalaeontological estimates commonly yield too low water depth, compared to what the clinoforms indicate. Further work is needed in order to construct an improved morphological and sediment environment driven approach that integrates micropalaeontology and geometric restorations of palaeobathymetry.

The Cretaceous restorations were integrated with micropalaeontological interpretations from Gillmore et al. (2001) (Fig. 9). The results derived from the pure geometrical restorations do generally show a good agreement, with a maximum deviation of 200 m for most cases. The 30/10-6 well shows a considerable discrepancy between 35–55 Ma. This is due to no intra-Eocene seismic sequence in the database. Integration was not performed for the Cenozoic due to the overall prograding setting.

The results from the restorations are generally in a good agreement with earlier works (e.g. Kjennerud, 2001; Kjennerud et al., 2001; Kjennerud and Sylta, 2001). The new maps do, however, show a drastic increase in spatial resolution (250 m × 250 m), revealing morphological features not evident from earlier works.

Geological implications

Results from the Cretaceous restorations show a typical post-rift basin, with decreasing influence

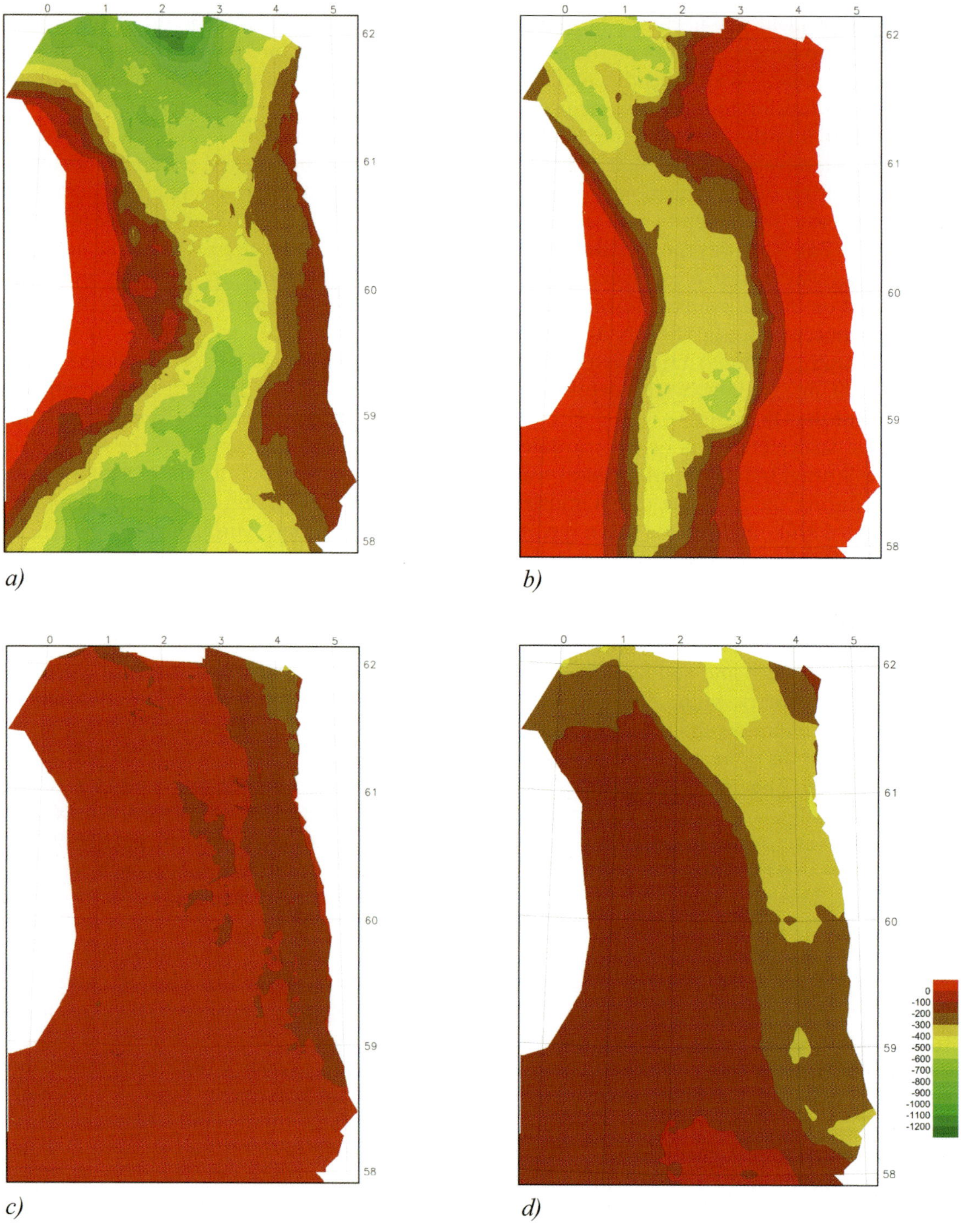

Fig. 8 Palaeo-water depth reconstructions of the (a) late Miocene, (b) late Pliocene, (c) early Quaternary and (d) present.

from the rift structures on the topography through time. This is in particular evident in the latest Campanian, where the depocenter was shifted westward of the Viking Graben. A similar transfer of basin axis was suggested by Gabrielsen et al. (2001).

A complex depocenter development through geologic time is revealed by the Cretaceous restorations. The late Ryazanian restoration does in particular show an intricate pattern of sub-basins along the rift axis. The earliest Cenomanian and the latest Turonian restorations show several more depocenters present along the rift axis, compared to earlier works by e.g. Gabrielsen et al. (2001) and Kjennerud et al. (2001).

On the latest Eocene restoration, a shallowing of the basin occurred in the north. This is probably due to the break-up of Norway and Greenland and

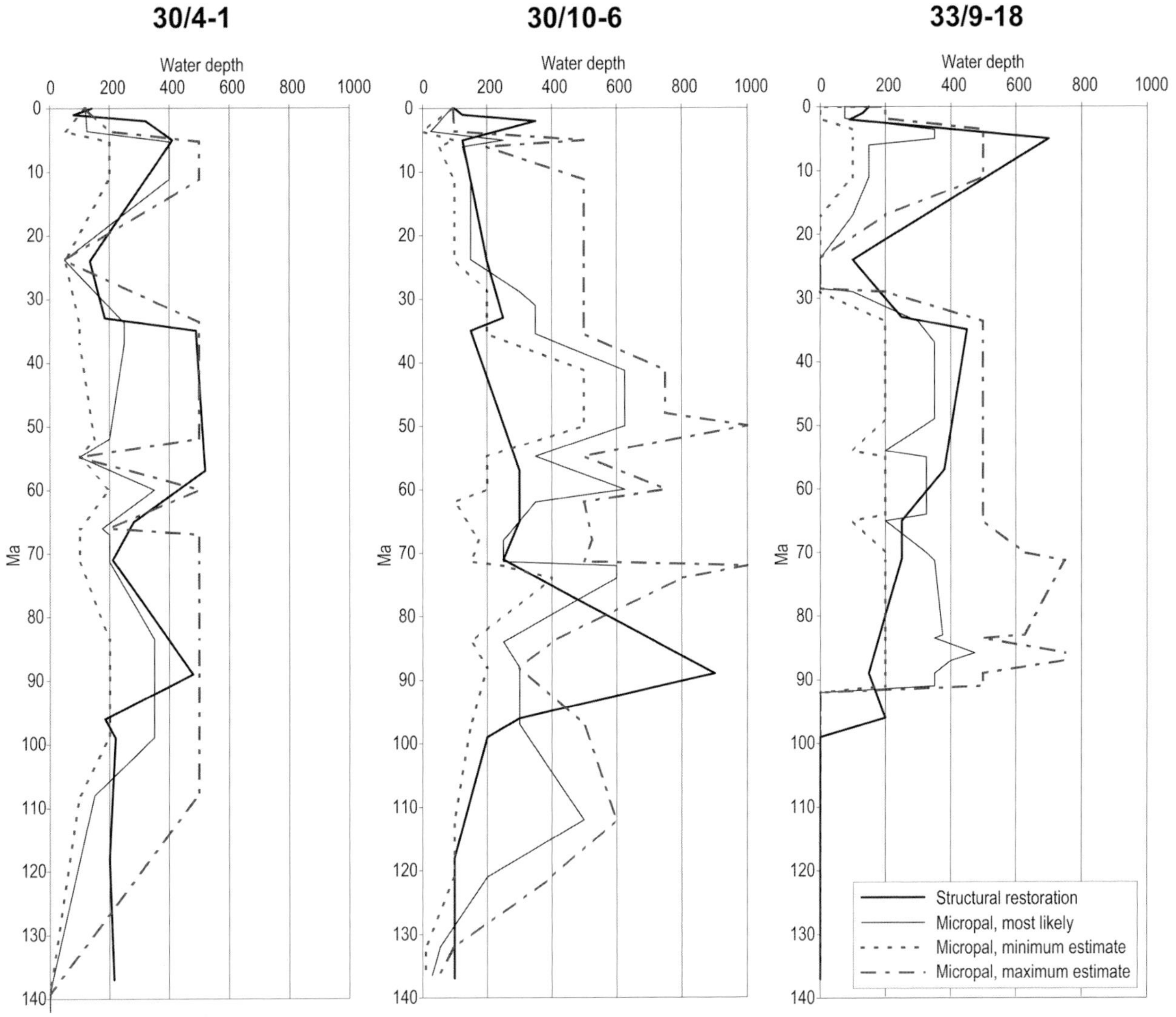

Fig. 9 Pseudowells from geometric bathymetry reconstruction plotted with micropalaeontological interpretations. Location of wells is showed in Fig. 1.

associated uplift and sediment influx from the north Atlantic margin (e.g. Brekke et al., 2001).

The Miocene bathymetry is poorly constrained from the restorations, due to no internal interpretations. In most wells, the Miocene is dominated by a large hiatus (e.g. Gradstein and Bäckström, 1996; Martinsen et al., 1999). It is an open question whether the basin stayed relatively deep during the Miocene and received very little sediment or if it was at, or near sea level. Martinsen et al. (1999) interpreted the Miocene hiatus to be due to non-deposition, although they suggested maximum water depth of 150–200 m for the Utsira Formation.

The late Pliocene restoration shows a sudden westward transfer of the eastern flank north of 61°N, implying increased sediment input in this area. Sediments may have been sourced from the Nordfjorden and Sognefjorden, whose outlets coincide with this area.

The restorations point out the significance of the episodic flank uplifts during Cenozoic. Three phases are recognised for southern Norway in late Paleocene, latest Eocene–early Oligocene, and late Pliocene. This is in agreement with Faleide et al. (2002). The present results do, however, give a better spatial description of the uplift. Earlier attempts to constrain these events have been sparse (e.g. Riis, 1996), and quantifying the effect of these events in basin modelling (e.g. pressure, hydrocarbon migration and accumulation, maturation) has not yet been resolved. Further emphasis should thus be placed on detailed basin modelling studies, concerning the vertical movements in the Cenozoic and its consequences. This is in particular evident for the late Neogene uplift, where large changes in the basin configuration and tilt occur rapidly in recent geological time.

Conclusions

An integrated 3D palaeobathymetric model has been constructed for the Cretaceous–Cenozoic for the northern North Sea (58–62°N). The results differ from earlier works in that they show high spatial resolution, and therefore more morphological details. The Neogene development, in particular, has been better resolved.

The Cretaceous post-rift development was characterised by gradual infill and widening of the late Jurassic syn-rift basin. Thermal equilibrium was reached in the latest Cretaceous (Gabrielsen et al., 2001).

The Cenozoic development was controlled by extra-basinal processes, such as the opening of the North Atlantic, which led to episodic uplift of the basin flanks and increased subsidence of the basin axis.

The present data set offers a crucial constraint for future 3D basin modelling studies in the northern North Sea. The influence of vertical movements can now be readily tested with respect to hydrocarbon migration and overpressure development.

Acknowledgements

The authors would like to thank Martin Hamborg, Are Tømmerås, Berit Fossum, Vegar Kleppe and Stephen Lippard. Felix Gradstein, Rudie van der Meer and Filippos Tsikalas are acknowledged for reviewing the manuscript.

References

Badley, M.E., Price, J.D., Rambech Dahl, C. and Agdestein, T., 1988. The structural evolution of the northern Viking Graben and its bearing upon extensional modes of basin formation. J. Geol. Soc., London, 145: 455–472.

Brekke, H., Sjulstad, H.I., Magnus, C. and Willams, R.W., 2001. Sedimentary environments offshore Norway — an overview. In: O.J. Martinsen and T. Dreyer (Editors), Sedimentary Environments Offshore Norway — Palaeozoic to Recent. Norwegian Petroleum Society (NPF) Special Publication 10, Elsevier, Amsterdam, pp. 7–37.

Christiansson, P., Faleide, J.I. and Berge, A.M., 2000. Crustal structure in the northern North Sea; an integrated geophysical study. In: A. Nøttvedt et al. (Editors), Dynamics of the Norwegian Margin. Geol. Soc., London. Spec. Publ., 167: 15–40.

Faleide, J.I., Fjeldskaar, W., Gabrielsen, R.H., Kjennerud, T., Kyrkjebø, R., Gillmore, G., Pascal, C., Sættem, J., Ter Voorde, M. and Clausen, J.A., 1999. Tectonic Impact on Sedimentary Processes in the Post-rift Stage — Improved models Vol. I and II. Report 23.2561.00/01/99.

Faleide, J.I., Kyrkjebø, R., Kjennerud, T., Gabrielsen, R., Jordt, J., Fanavoll, S. and Bjerke, M.D., 2002. Tectonic impact on sedimentary processes during the Cenozoic evolution of the northern North Sea and surrounding areas. In: A.G. Doré, J.A. Cartwright, M.S. Stoker, J.P. Turner and N.S. White (Editors), Exhumation of the North Atlantic Margin: Timing, Mechanisms and Implications for Petroleum Exploration. Geol. Soc., Spec. Publ., 196: 235–269.

Færseth, R.B., 1996. Interaction of Permo-Triassic and Jurassic extensional fault-blocks during the development of the northern North Sea. J. Geol. Soc., 153: 931–944.

Gabrielsen, R.H., Færseth, R.B., Steel, R.J. and Kløvjan, O.S., 1990. Architectural styles of basin fill in the northern Viking Graben. In: D. Blundell and A. Gibbs (Editors), Tectonic evolution of the North Sea Rifts. Oxford University Press, pp. 158–179.

Gabrielsen, R.H., Kyrkjebø, R., Faleide, J.I., Fjeldskaar, W. and Kjennerud, T., 2001. The Cretaceous post-rift basin configuration of the northern North Sea. Petrol. Geosci., 7: 137–154.

Gillmore, G.K., Kjennerud, T. and Kyrkjebø, R., 2001. The reconstruction and analysis of palaeo-water depth: a new approach and test of micropalaeontological approaches in the post-rift (Cretaceous to Quaternary) interval of the Northern North Sea. In: O.J. Martinsen and T. Dreyer (Editors), Sedimentary Environments Offshore Norway — Palaeozoic to Recent. Norwegian Petroleum Society (NPF), Special Publication 10, Elsevier, Amsterdam, pp. 365–379.

Gradstein, F. and Bäckström, S., 1996. Cenozoic biostratigraphy and palaeobathymetry, northern North Sea and Haltenbanken. Norsk Geol. Tidsskr., 76: 3–32.

Gradstein, F. and Ogg, J., 1996. A Phanerozoic time scale. International Union of Geological Science (IUGS), Ottawa, ON. Episodes 19 (1–2): 3–6.

Isaksen, D. and Tonstad, K., 1989. A revised Cretaceous and Tertiary lithostratigraphic nomenclature for the Norwegian North Sea, NPD Bull., 5: 59 pp.

Jordt, H., Faleide, J.I., Bjørlykke, K. and Ibraim, M.T., 1995. Cenozoic sequence stratigraphy of the central and northern North Sea Basin: tectonic development, sediment distribution and provenance areas. Mar. Petrol. Geol., 12 (8): 845–879.

Kjennerud, T., 2001. Palaeobathymetry and rift basin evolution — with particular reference to the northern North Sea Basin. PhD thesis, Norwegian University of Technology and Science.

Kjennerud, T., Faleide, J.I., Gabrielsen, R.H., Gillmore, G.K., Kyrkjebø, R., Lippard, S.J. and Løseth, H., 2001. Structural restoration of Cretaceous-Cenozoic (post-rift) palaeobathymetry in the northern North Sea. In: O.J. Martinsen and T. Dreyer (Editors), Sedimentary Environments Offshore Norway — Palaeozoic to Recent. Norwegian petroleum Society (NPF), Special Publication 10, Elsevier, Amsterdam, pp. 347–364.

Kjennerud, T. and Gillmore, G., 2003. Integrated Palaeogene palaeobathymetry of the northern North Sea. Petrol. Geosci., 9 (2): 125–132.

Kjennerud, T. and Sylta, Ø., 2001. Application of quantitative palaeobathymetry in basin modelling. Petrol. Geosci., 7 (4): 331–341.

Kyrkjebø, R., 1999. The Cretaceous — Tertiary if the northern North Sea: thermal and tectonic influences in a post-rift setting. PhD thesis, University of Bergen, Norway.

Kyrkjebø, R., Kjennerud, T., Gillmore, G.K., Faleide, J.I. and Gabrielsen, R.H., 2001. Cretaceous-Tertiary palaeobathymetry in the northern North Sea; integration of palaeowater depth estimates obtained by structural restoration and micropalaeontological analysis. In: O.J. Martinsen and T. Dreyer (Editors), Sedimentary Environments Offshore Norway — Palaeozoic to Recent. Norwegian petroleum Society (NPF), Special Publication 10, Elsevier, Amsterdam, pp. 321–345.

Martinsen, O.J., Bøen, F., Charnock, M.A., Mangerud, G. and Nøttvedt, A., 1999. Cenozoic development of the Norwegian margin 60–64°: sequences and sedimentary response to variable basin physiography and tectonic setting. In: A.J. Fleet and S.A.R. Boldy (Editors), Petroleum Geology of Northwest Europe: Proceedings of the 5th Conference. Geological Society, London, pp. 293–304.

Rattey, R.P. and Hayward, A.B., 1993. Sequence stratigraphy of a failed rift system: the Middle Jurassic to Early Cretaceous basin evolution of the Central and Northern North Sea. In: J.R. Parker (Editor), Petroleum Geology of Northwest Europe: Proceedings of the 4th Conference. Geological Society, London, pp. 215–249.

Riis, F., 1996. Quantification of Cenozoic vertical movements of Scandinavia by correlation of morphological surfaces with offshore data. In: A. Solheim, F. Riis, A. Elverhøi, J.I. Faleide, L.N. Jensen and S. Cloetingh (Editors), Impact of glaciations on basin evolution; data and models from the Norwegian margin and adjacent areas. Global Planet. Change, 12: 331–357.

Sclater, J.G. and Christie, P.A.F., 1980. Continental stretching; an explanation of the post-mid-Cretaceous subsidence of the central North Sea basin. J. Geophys. Res., 85: 3711–3739.

Steel, R.J. and Ryseth, A., 1990. The Triassic — early Jurassic succession in the northern North Sea: megasequence stratigraphy and intra-Triassic tectonics. In: R.P.F. Hardman and J. Brooks (Editors), Tectonic Events responsible for Britain's Oil and Gas Reserves. Geol. Soc., London, Spec. Publ., 55: 139–168.

Sylta, Ø., 1993. New techniques and their application in the analysis of secondary migration. In: A.G. Doré (Editor), Basin Modelling: Advances and Applications. Norwegian Petroleum Society (NPF), Special Publication 3, Elsevier, Amsterdam, pp. 385–398.

Underhill, J.R. and Partington, M.A., 1993. Jurassic thermal doming and deflation in the North Sea: implications of the sequence stratigraphic evidence. In: J.R. Parker (Editor), Petroleum Geology of Northwest Europe: Proceedings of the 4th Conference. Geological Society, London, pp. 337–346.

Øvrebø, L.K., Kjennerud, T., Lippard, S.J., Rivenæs, J.C. and Hamborg, M., 2001. Forward depositional modelling of the Cretaceous post-rift deposits in the northern North Sea. Norsk Geol. Tidskr. 81: 169–178.

Submarine slides on the Mid-Norwegian Continental Margin — A Challenge to the oil industry

Petter Bryn, Kjell Berg, Reidar Lien and Anders Solheim

Two large submarine slides, The Storegga and the Trænadjupet Slides, occurred on the Mid-Norwegian margin during the Holocene. The Ormen Lange gas field is located within the scar of the Storegga Slide. This gigantic submarine slide occurred about 8200 years ago, and caused large waves (tsunamis) that reached the coasts of Norway, Scotland, Shetland and the Faroe Islands. The objectives of this chapter are to present the challenges and the slide risk assessment related to the development the Ormen Lange gas field. The risk evaluation is based on a qualitative approach for large natural slides, and a quantitative approach for new small slides in the vicinity of the development area.

The work programme includes extensive, regional multi-disciplinary studies, carried out jointly by academia, industry and research institutions. The database includes an extensive grid of seismic data, detailed sea-floor morphology and sediment properties from a number of 'geoborings' (combined geological and geotechnical borings to sub-bottom depths of 200–400 m). Stability of the steepest slopes in the vicinity of the development area is calculated. Effects of excess pore pressures, earthquakes, reservoir compaction during depletion and underground gas blowouts into possible permeable layers have all been included in the stability calculations. To understand the recent slide history in the area and to find the frequency of the sliding, extensive sea-floor mapping and coring to date slide events are also included.

A geological model for the Plio–Pleistocene of the area explains the large-scale sliding as a response to climatic variability. Over long periods, marine deposition prevailed with focused deposition due to current effects in the locations of the Storegga and the Trænadjupet Slides. During short intervals of peak glacial conditions, till and glacial debris flow sediments were deposited at high rates directly on the continental slope. This created excess pore pressures in the thick marine deposits. The most likely triggering mechanism of the slides is a strong earthquake following the onshore uplift after the glaciation. This explains why the slides take place after a glacial period. Since all the soft unstable clays were removed from the Storegga Margin during the last slide, it is concluded that a new cycle with sedimentation of soft clays and deposition of glacial sediments in the upper slopes are needed, to create a new unstable situation in the Storegga area. At present, the slopes in the Ormen Lange area have high safety factors, and the likelihood of new slides, both local and regional, is considered very low.

Introduction

In 1996, the first Norwegian deep-water licenses were awarded in the Møre-Vøring deep-water area (15th round), in an area where two major Holocene submarine slides had occurred, the Storegga Slide and the Trænadjupet Slide (Bugge, 1983; Laberg et al., 2002) (Fig. 1). The deep-water operators decided to join efforts to cope with the new challenges in this area and formed The Norwegian Deep Water Project (NDP). One of the sub-projects was the Seabed Project, lead by Norsk Hydro.

The Storegga Slide was first studied in detail in the 1980s (Bugge, 1983). The slide was thought to have taken place as three main events in glacial to post-glacial times (Bugge et al., 1987). One of these events was subsequently linked to an anomalous sediment deposit found on the eastern coast of Scotland and interpreted as a tsunami deposit (Dawson et al., 1988). Detailed studies of the Norwegian coastline have since then identified similar and more extensive tsunami deposits, dated back to 7200 years BP (Bondevik et al., 1997).

The main challenge in the area was the presence of the two major submarine slides, both with regard to potential new slide risk, and the seafloor topography within the slides. The scope for the Seabed Project was to establish a regional seismic framework and drill some deep geotechnical boreholes in order to establish a geological model.

Onshore–Offshore Relationships on the North Atlantic Margin
edited by B. Wandas et al. NPF Special Publication 12, pp. 255–263,

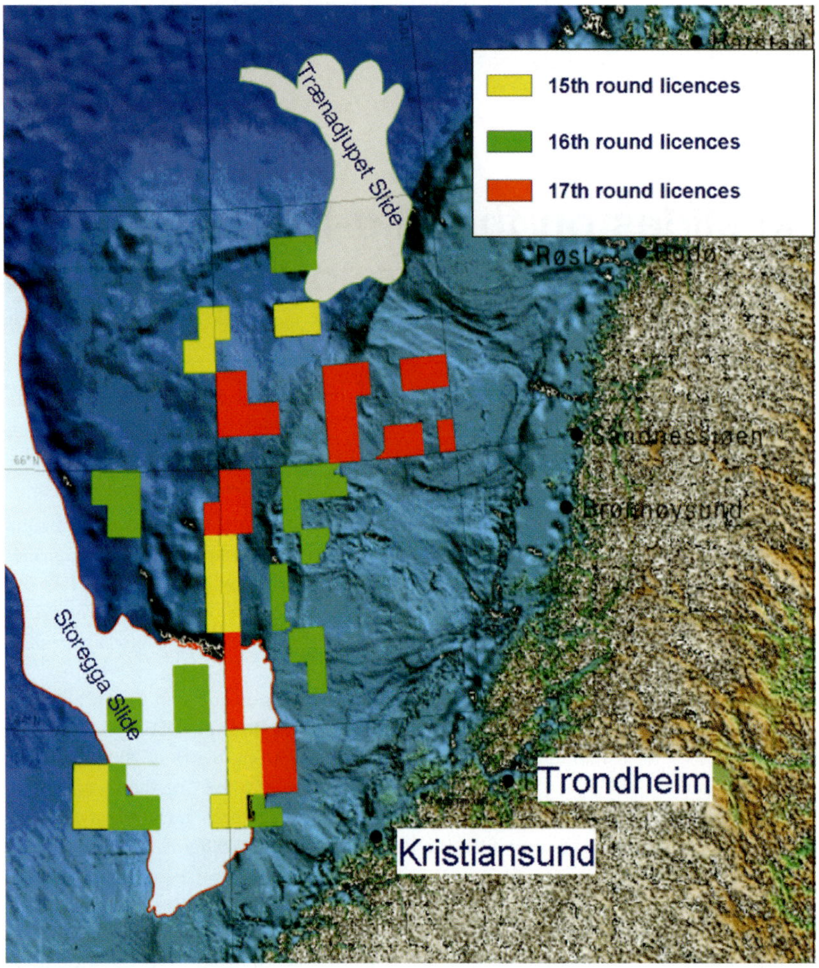

Fig. 1 Submarine slides and exploration licences in the Møre-Vøring deep-water area.

Also, seafloor mapping and dating of slide events were undertaken. The aim was to improve the understanding of geohazards in the area with special focus on the slide issue.

A close co-operation with the academia was established during 1996 and 1997, with focus on the on-going EU project ENAM (European North Atlantic Margin Project). The Universities of Tromsø, Bergen and Oslo, as well as GEOMAR in Kiel, were important partners in the co-operation. Since 1997, most of the activities related to geohazards, including all the fieldwork, have been planned and performed in close co-operation between these institutions and Norsk Hydro. Statoil has been responsible for the geotechnical fieldwork within the Seabed Project and assisted Norsk Hydro in the execution of the geotechnical fieldwork within the Ormen Lange Project.

In 1997, the major Ormen Lange gas field was discovered in the central slide scar close to the steep escarpments formed by the Storegga Slide. So far, this is the only commercial discovery in the area. New deep-water licences were awarded in the 16th round both inside and in the vicinity of the large slides and new licences were awarded in this area in the 17th round in 2002. Two of the 16th round blocks have been drilled (spring 2003) (Fig. 1).

The slide risk had to be clarified before the development of the Ormen Lange gas field, and the results are important for the other licences in this area because of the regional character and large size of the slides. The fact that a flood-wave (tsunami) hit the west coast of Norway (run up 10–12 m), Scotland (4–6 m), Shetland (20–30 m) and the Faeroes (>10 m) when the Storegga Slide occurred (Dawson et al., 1988, Bondevik et al., 2003) increases the focus on the slide risk issue in relation to hydrocarbon industry activity.

The main questions for the oil and gas industry was:

- Do we have access to this area?
- How can we explain a major slope failure when the slope angle was close to 1° prior to the sliding?
- Is the natural risk for new slides too high to operate in this area?
- Can our activity during field development influence the present-day slope stability?

- Is it safe to develop the Ormen Lange gas field close to the steep headwalls (30–40°) of the Storegga Slide?

Slide risk — method of approach

The slide risk evaluation covers the risk for new natural slides and risk of reduced slope stability related to the gas production. Effects of potential changes in the slope inclination due to the reservoir compaction and the increase in pore pressure due to a possible cross-flow of highly pressured gas from the reservoir to permeable shallow layers (<400 m) during an underground blowout are evaluated (Fig. 2).

The geological model describes the stratigraphy, sediment properties and the margin processes like sedimentation, fluid flow, diagenesis and tectonics. This model contains information of slide precondition and is combined with the event stratigraphy (climate changes and buried slides) to a slide explanation model and used in the qualitative risk evaluation and to constrain parameters in quantitative risk analysis.

The seabed topography was used to select critical profiles for dedicated slope-stability calculations. To understand the recent slide history in the area and to find the frequency of the sliding, the investigation programme has included extensive seafloor mapping and coring to date slide events. Evidence of tsunamis has been evaluated onshore and in fjords.

All known and relevant trigger mechanisms are included in the calculations of the present-day stability. Some potential trigger mechanisms were evaluated specifically for the Ormen Lange field:

- The compaction and subsidence during reservoir depletion may change the seafloor and cause potentially unfavourable slope inclination.
- A potential underground blowout from the reservoir into shallow layers will increase the pore pressure and may initiate slope failure.

The results show that the safety factor is very good, even for the worst-case scenarios regarding high pore pressures and large earthquakes (Kvalstad et al., 2002).

The Ormen Lange Project is, to our knowledge, the first large gas project to be developed within a relatively recent slide scar. The goal of the work is to perform a qualitative risk evaluation for largest natural slides, and to quantitatively evaluate the risk for new small slides in the vicinity of the development area. The work programme includes multi-disciplinary geosciences and a regional approach, and requires contributions from experts from universities, contractors and research institutions (Fig. 3).

Slide investigations

Since 1999 the Ormen Lange Project has intensified the slide studies based on the approach shown in Fig. 2. Extensive fieldwork, including geological drilling, seafloor mapping, seismic acquisition, pore pressure measurements, coring for dating of slide events, gas hydrate mapping and

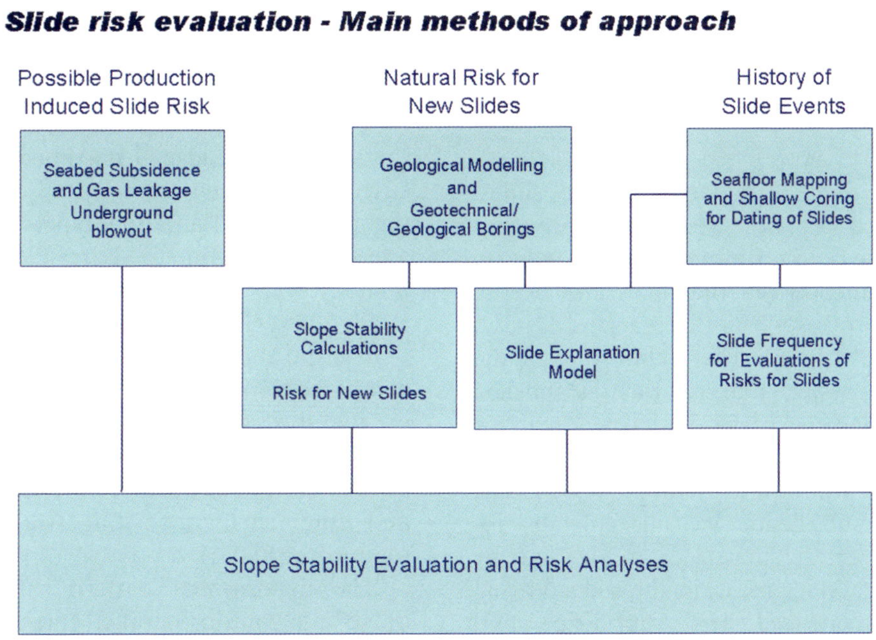

Fig. 2 Flow chart illustrating the main approaches used to assess the regional and local slide risk in the Møre-Vøring area.

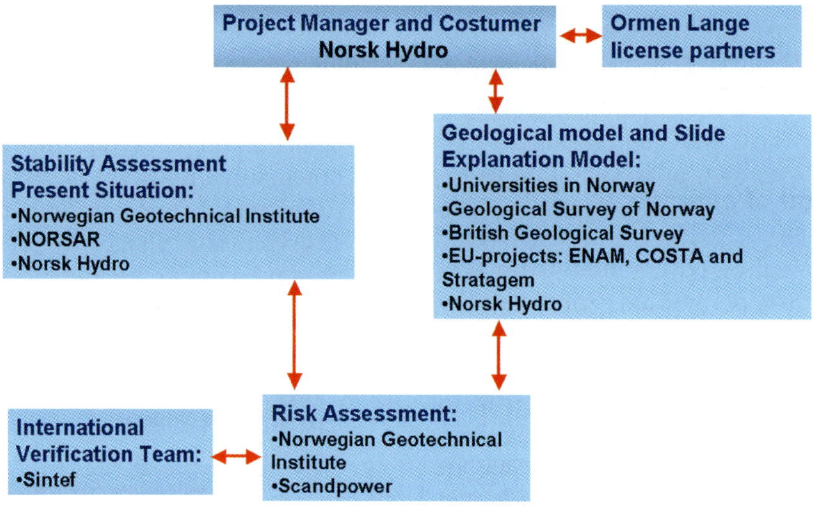

Fig. 3 Flow chart to illustrate the multidisciplinary approach to establish the database and technology required for the slope-stability risk assessment.

ROV and AUV surveys have been performed. Regional seismic interpretation, both of the shelf and of the deep-water margin, has been part of the programme to establish a regional geomodel to try to explain the sliding.

A geomodel, based on the regional work performed under the Seabed Project in the NDP programme, with an additional seismic interpretation, was established prior to the soil-sampling programme. The geomodel was used to select critical data points for geotechnical information to be used in slope stability analysis. Five sites covering the glide-planes and sediments involved in the Storegga Slide have been investigated (Fig. 8). Pore-pressure monitoring was established at four of these sites. The geomodel was calibrated with information from the geotechnical analysis and down-hole logs.

Regional geomodel and slide explanation model

The location of two major Holocene slides, Storegga Slide on the Møre Margin and Trænadjupet Slide on the Lofoten Margin, indicate a possible relation between the slides and deeper structures and processes on the margins (NGU, 2002; Evans et. al., 1996). The architecture of the MidNorwegian passive margin was established during earliest Eocene (c. 55 Ma) when seafloor spreading was initiated in the Norwegian–Greenland Sea (Talwani and Eldholm, 1977). The ensuing thermal subsidence was unevenly distributed with the greatest subsidence in the narrow Lofoten and Møre Margins, while the wider Vøring Margin has experienced less subsidence and represents a bathymetric high (NGU, 2002). Major tectonic lineaments define the boundaries between the three margin segments (Blystad et al., 1995). The Storegga and the Trænadjupet Slides are located southwest and northeast of these lineaments, respectively. The Neogene uplift of the Scandinavian mountains has the main elevation centres in Lofoten and Jotunheimen (Riis, 1996), and the Lofoten and Møre Margins are the areas where the contrast between onshore uplift and offshore subsidence is greatest.

Pliocene (5–2 million years ago) onshore uplift combined with glaciations produced large quantities of sediment that were transported from the mainland and built out onto the shelf (Riis, 1996). The main advance of the shelf took place in the Late Pliocene–Early Pleistocene (3.0–1.7 million years ago), as a response to onshore uplift/glaciation and brought the shelf edge close to the present location (NGU, 2002). From this time on, the main sediment transport was across the shelf during glacial maxima. Fast flowing ice-streams developed depocentres of glacial clays in the North Sea Fan, Buadjupet and Skjoldryggen areas during Mid to Late Pleistocene (0.7–0.2 million years ago) (Ottesen et al. 2000; Sejrup et al., 2000) (Fig. 4).

The durations of the peak glaciations are short compared to the periods with a limited ice cover on the shelf and the interglacial periods. During these long periods, normal marine and/or distal glacial marine deposition prevailed on the slope and outer shelf, depositing stratified fine-grained sediments (Fig. 5).

The depositional pattern shows an influence of strong contour-parallel currents with erosion/ transport mainly in the upper slope and deposition

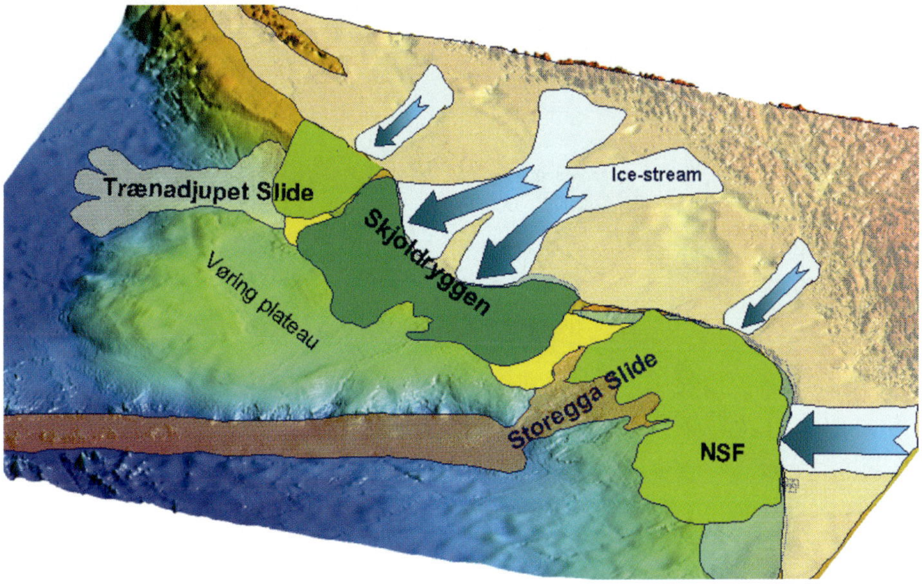

Fig. 4 Main depocentres of glacial clays. NSF = North Sea Fan.

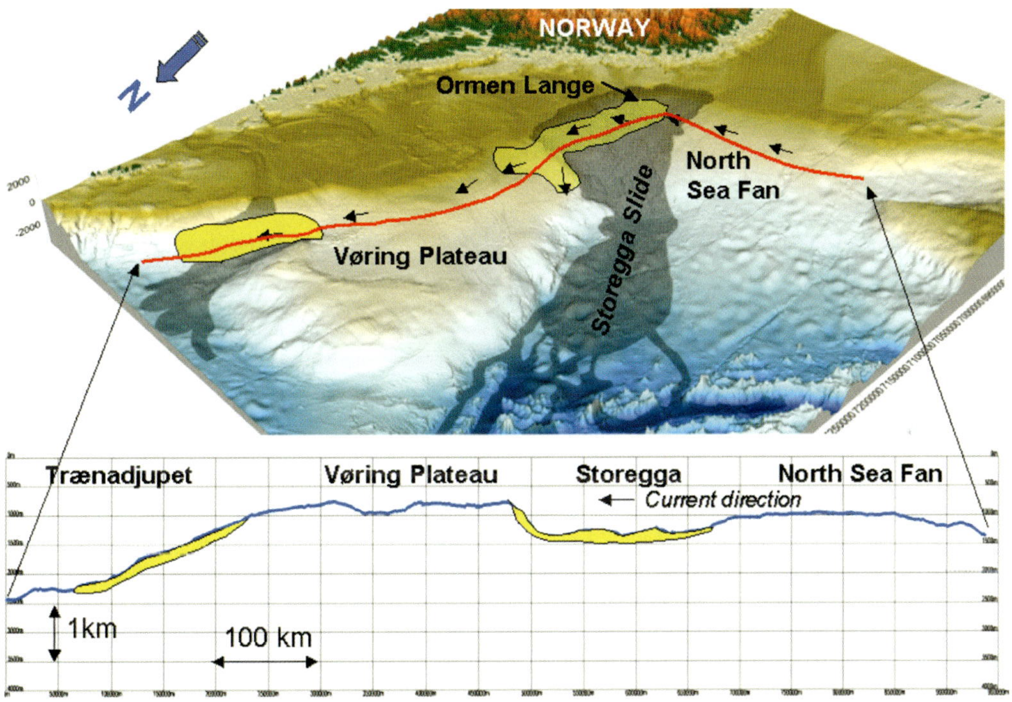

Fig. 5 Locations of the main contouritic drifts are controlled by the seabed topography, current direction and the thermocline.

in the lower/middle slope (Bryn et al., 2002). These contouritic deposits are commonly mounded, being up to 150 m thick where they infill older deep slide-scars. On the slope, the cyclic depositional environment produced interbeds of fine-grained marine clays and unsorted, coarse-grained and denser glacial clays.

In the area of the Storegga Slide, repeated sliding is documented during the last 1.7 million years (Evans, 1996, Bryn et al., 2003). Two of the older slides, (S and R slides) are linked to major glaciations and have a similar extent and character as the present Storegga Slide (Fig. 6). Glide-planes are formed within the laterally extensive units of marine clays, and the main driving forces are the weight of the overlying glacial clays. The seismic pattern indicates that the slide scars have an infill of contouritic drifts rather than glacial sediments and therefore the sliding seems to have occurred after a glacial maximum. Increased pore pressure in the marine clays due to the loading by rapidly deposited glacial clays is regarded as the main destabilisation force (Fig. 6). Melting of gas hydrate due to sealevel changes and increase in

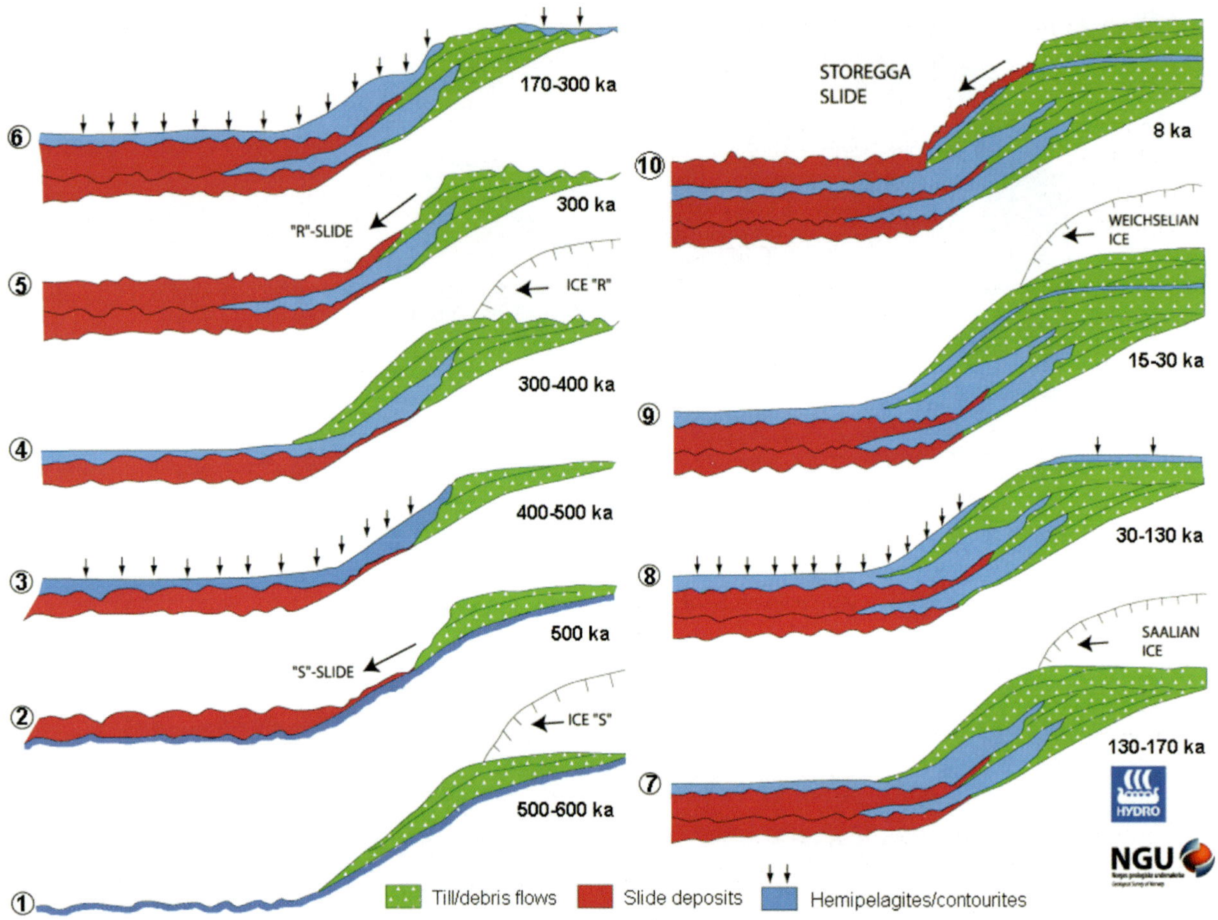

Fig. 6 Cyclic deposition and slide processes in the Storegga area. (Green = glacial clays. Red = slide deposits. Blue = marine clays.)

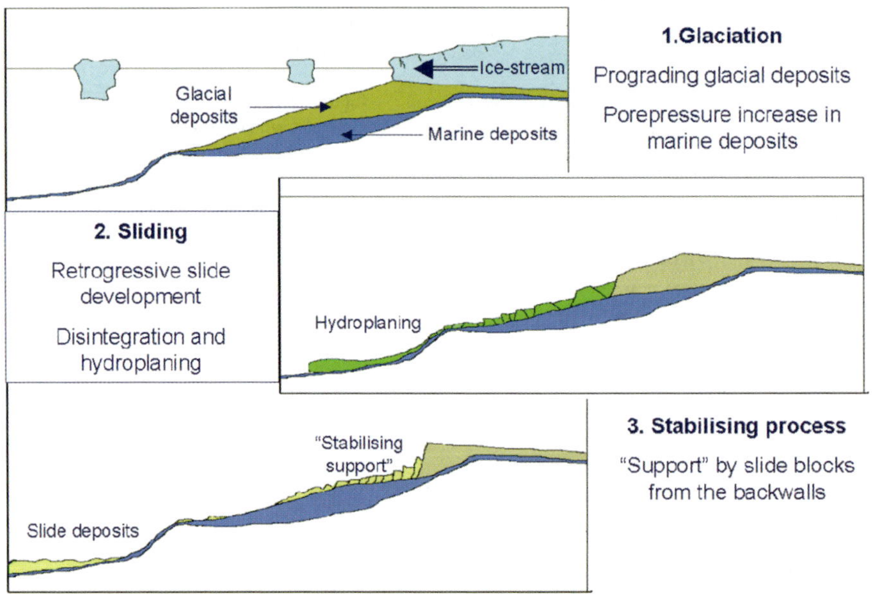

Fig. 7 Conceptual model of the Storegga Slide.

the bottom-water temperature are other potential destabilisation factors. The slide morphology indicates a retrogressive development with disintegration of the slide blocks and possible hydroplaning. The remaining slide blocks typically support the sidewalls and backwalls.

Glacio-isostatic rebounds cause high earthquake activity after deglaciation and a major earthquake at a fault zone in or close to the Storegga Slide is regarded to have most likely triggered off the slide (NORSAR, 2002).

In the upper slide scar, the soft and weak marine clays failed and were removed during the Storegga Slide event, 8200 years ago. The present day back-wall consists of over-consolidated glacial clays, which are very stable. A new cycle with infill of soft marine clays in the slide scar has started. However, a new glacial advance to the shelf edge and rapid deposition of glacial clays seem to be required to get a new unstable situation in the area.

Slope stability calculations and risk analysis

Slope stability calculations have been performed in the vicinity of the Ormen Lange gas field to clarify the present stability. A number of deep geoborings (Fig. 8) combined with seismic data and slope geometry, form the basis for the slope stability evaluation. The slope-stability calculations have included the present potential trigger mechanisms, of which earthquakes are the most significant. All potential triggers and destabilising factors prior to the Storegga Slide have been investigated to comprehend the historical repeated sliding and to be able to evaluate the significance of these triggers today. This includes the effects of high sedimentation rates during peak glaciations, gas hydrate melting, gas charging of shallow sediments, diapirism and earthquakes. A common factor for such processes is that they all increase the pore pressure of the sediments and decrease the effective soil strength. The result of the slope-stability calculation shows that the safety factor against sliding is relatively high even in combination with a large earthquake and high pore pressure (Kvalstad et al., 2002). The results of the calculation will be included in the probabilistic analyses before the final risk assessment is performed.

The risk of new slides related to the production is negligible because the maximum subsidence of the seafloor is conservatively estimated to less than 1.8 m and there are no permeable layers that becomes critical in an underground blowout scenario.

Dating of slides and tsunami sediments

A comprehensive radiocarbon (^{14}C), dating programme based on detailed slide morphology was prioritised in the Storegga Slide area, and extensive work has also been carried out in the Trænadjupet Slide area and in coastal areas exposed to possible tsunamis generated by the slides (Fig. 9).

The main objectives for the investigations in the different areas are summarised as follows:

- Date the Storegga Slide to conclude whether the slide occurred as one main event or in several separate stages, as proposed in the study by Bugge et al., 1987.
- Date the Trænadjupet Slide.
- Map and date flood sediments (turbidites) to evaluate their possible relation to the timing of submarine slides.

The results from these investigations enable a slide frequency evaluation to be used as input in risk analyses for future slides.

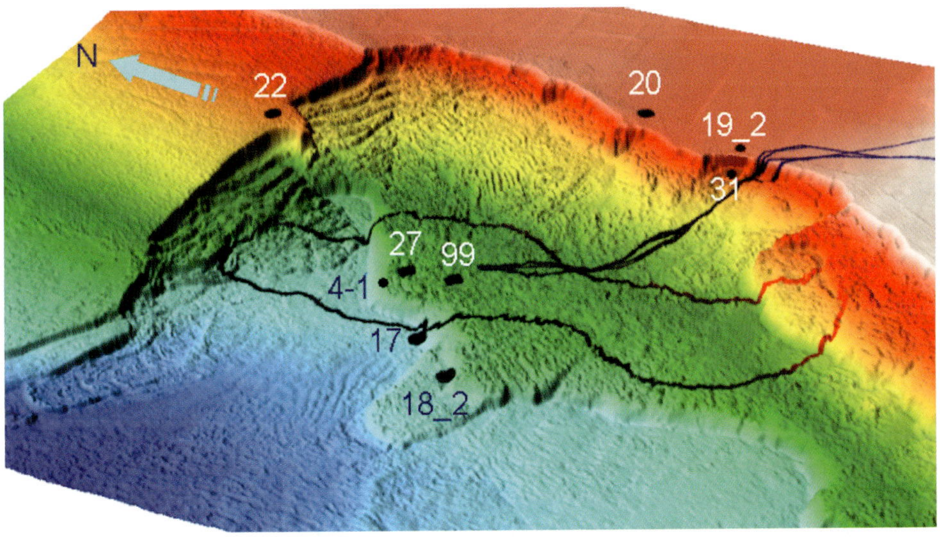

Fig. 8 3D bathymetry image of the geoboring positions in the Ormen Lange area.

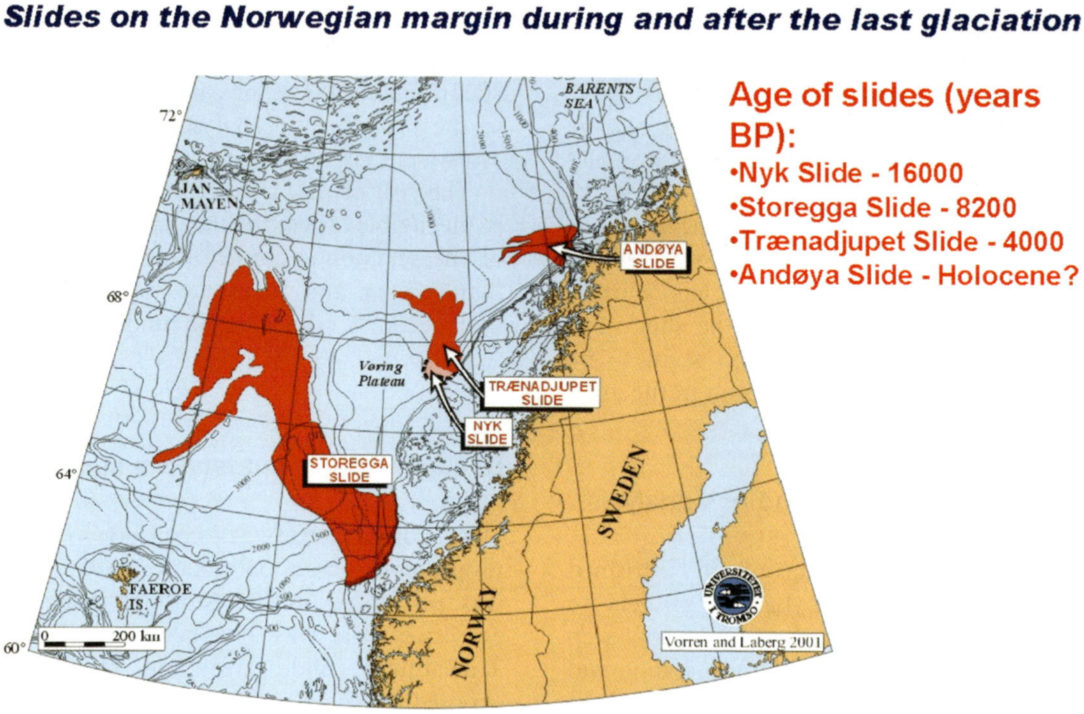

Fig. 9 The figure shows the major slides on the Norwegian continental margin.

The main conclusions of the ^{14}C datings are summarised below:

The Storegga Slide

- The Storegga Slide occurred as one main event approximately 8200 years BP.
- Some minor local slumping may have occurred from the northern flank area about 5000 and 2500 years BP (Haflidason et al., 2002).

The Trænadjupet Slide

- The Trænadjupet Slide occurred approximately 4000 years BP (Laberg et al., 2002).

Flood sediments and possible tsunamis

- Flood sediments most likely related to a tsunami generated from the Storegga Slide have been recorded in most of the investigated areas (coastal Norway, UK, Shetland and Faeroes). The investigation of the flood sediments indicates run-up heights up to 20–30 m.
- The search for flood sediments deposited by a possible tsunami generated from the Trænadjupet Slide, has been negative (Bondevik et al., 2003).
- Flood sediments presumably related to the Storegga Slide have been recorded on Shetland. In addition, two younger and less extensive flood sediment layers (approximately 4700 and 1500 years BP) with unknown cause are recorded on Shetland (Bondevik et al., 2003).
- Investigations in fjords (i.e. Voldafjorden, Sulafjorden, Syvdsfjorden and Ørstadfjorden) of western Norway reveal turbidites, which may be related to flood waves — three main time periods are identified, i.e. ~ 8200, 2800–3200 and 2000–2200 years BP. The oldest date most likely represents the Storegga Slide, while the two younger dates, based on geographical distribution and age differences, most likely relate to local sliding or climate changes and not a tsunami coming from offshore (Bøe et al., in press).
- Investigations in coastal lakes in western Norway give clear indications of a Storegga Slide tsunami, but no signs of younger tsunamis (Bondevik et al., 2003, Bøe et al., in press).

Conclusions

The depositional processes and the underlying structural architecture of the slope and margin are the governing factors for the location of the Storegga and the Trænadjupet Slides. In the Storegga Slide area, a glacial–interglacial cyclicity has caused repeated sliding through the glacial period. Loading by glacial clays (high density and low water content) on marine, mainly contouritic, clays (with low density and high water content) may lead to excess pore pressure and slope failure at low-slope angles. An earthquake is the most likely trigger of the unstable slope. Deposition of new soft sediments (on-going process) in the slide scar followed by new deposition of glacial sediments is required to start new major sliding in the Storegga area.

The stability calculation of the slopes in the vicinity of the Ormen Lange gas field, shows a high safety factor against sliding. This can be explained by the fact that the slide back wall has eroded into the shelf sediments where the marine clays are thin or absent and the glacial clays are predominantly overconsolidated. In addition the inclination of the potential glide planes decreases. Dating of the Storegga Slide and the tsunami event shows that the slide occurred as one major event 8200 years BP.

The final risk analyses are not yet complete (winter 2003), but all the results indicates that it is safe to develop the Ormen Lange field close to the main head wall of the Storegga Slide. Further work to clarify the local risk is most likely required for other licences in the slide areas.

References

Blystad, P, Færseth, R.B., Larsen, B.T., Skogseid, J. and Tørudbakken, B., 1995. Structural elements of the Norwegian continental margin between 62° and 69°N. NPD Bull. 8: 1–45.

Bondevik, S., Svendsen, J.I., Johnsen, G., Mangerud, J., and Kaland, P.E., 1997. The Storegga tsunami along the Norwegian coast, its age and runup. Boreas 26: 29–53.

Bondevik, S., Mangerud, J., Dawson, S., Dawson, A. and Lohne, Ø, 2003. Record-breaking height for 8000-year old tsunami in the North Atlantic. EOS, 84, pp. 289, 293.

Bryn, P., Berg, K., Lien, R., Solheim, A., Ottesen, D. and Rise, L., in press. The Storegga Geomodel and its use in slide risk evaluation. International Conference Offshore Site Investigations and Geotechnics, Sustainability through Diversity, London, 26–28 Nov. 2002.

Bryn P., Solheim A., Berg K., Lien R., Forsberg C.F., Haflidason H., Ottesen D. and Rise L., 2003. The Storegga Slide complex; Repeated large scale sliding in response to climatic cyclicity. In: J. Locat and J. Mienert (Editors), Submarine mass movements and their consequences, Kluwer Academic Publishers, Dordrecht, pp. 215–222.

Bugge, T., 1983. Submarine slides on the Norwegian continental margin, with special emphasis on the Storegga area. IKU, 110: 1–152.

Bugge, T., Belderson, R.H and Kenyon, N.H., 1988. The Storegga Slide, Philos. Trans. R. Soc. London, Ser. A, 325: 357–388.

Bøe, R., Longva, O., Lepland, A., Blikra, L.H., Sønstegaard, E., Haflidason, H., Bryn, P. and Lien, R., 2004: Postglacial mass movements and their causes in fjords and lakes in western Norway. Nor. J. Geol. (NGT), 84 (1): 35–55.

Dawson, A.G., Long, D. and Smith, D.E., 1988. The Storegga Slide; evidence from eastern Scotland for a possible tsunami. Mar. Geol., 82: 271–276.

Evans, D., King, E.L., Kenyon, N.H., Brett, C. and Wallis, D., 1996. Evidence for long-term instability in the Storegga Slide region off western Norway, Mar. Geol., 130: 281–292.

Haflidason, H., Sejrup, H.P., Bryn, P., Lien, R, Masson, D., Jacobs, C., Huehnerbach, V. and Berg, K., 2002. The architecture and slide mechanism of the Storegga Slide, Mid Norwegian margin. In: A. Hurst (Editor), Onshore–Offshore Relationships on the Nordic Atlantic Margin. NGF Abstracts and proceedings 2, 2002 of the Norwegian Petroleum Society (NPF) and Norwegian Geological Society (NGF) Conference, 7–9 Oct. Trondheim, pp. 80–81.

Kvalstad, T.J., Gauer,P., Kaynia, A.M., Nadim, F. and Bryn, P., 2002. Slope Stability at Ormen Lange. International Conference Offshore Site Investigation and Geotechnics, Sustainability through Diversity, 26–28 Nov. 2002.

Laberg, J.S., Vorren, T.O., Mienert, J., Bryn, P. and Lien, R., 2002. The Trænadjupet Slide: a large slope failure affecting the continental margin of Norway 4000 years ago. Geo-Mar. Lett., 22: 19–24.

Lundin, E.R. and Doré, A.G., 2002. Mid-Cenozoic post-breakup deformation in the 'passive' margins bordering the Norwegian–Greenland Sea. Mar. Petrol. Geol., 19: 79–93.

NGU, 2002. Large scale development of the mid Norwegian shelf and margin with emphasis on the last 3 million years. NGU Report 2002.015, 190 pp.

NORSAR, 2002. Site specific hazard for the Ormen Lange field and postglacial earthquakes in the larger Møre region. Report to Norsk Hydro.

Ottesen, D., Olsen, L. and Thorsnes, T., 2000. Ice sheet dynamics on the mid-Norwegian continental shelf based on regional and detailed bathymetric and seismic data. Norg. Geol. Unders. Rep. 2000.017, 52 pp.

Riis, F., 1996. Quantification of Cenozoic vertical movements of Scandinavia by correlation of morphological surfaces with offshore data. Global Planet. Change, 12: 331–357.

Sejrup, H.P., Larsen, E., Landvik, J., King, E.L., Haflidason, H. and Nesje, A., 2000. Quaternary glaciations in southern Fennoscandia: evidence from southwestern Norway and the northern North Sea region. Quaternary Sci. Rev., 19: 667–685.

Talwani, M. and Eldholm, O., 1977. Evolution of the Norwegian–Greenland Sea, Geol. Soc. Am. Bull., 88: 969–999.

Occurrence and implications of large Lophelia-reefs offshore Mid Norway

Martin Hovland, Dag Ottesen, Terje Thorsnes, Jan Helge Fosså and Petter Bryn

Large, up to 45 m high and at least 8600 year old coral reefs composed of the reef-building, stony coral *Lophelia pertusa* (L.) occur off Mid Norway. Their locations have been documented by side-scan sonar, multi-beam echosounder, and ROV (remotely operated vehicle) surveys carried out by the hydrocarbon industry and by authorities (Fosså and Mortensen, 1998; Hovland and Mortensen, 1999; Jung et al., 2001; Freiwald et al., 2002).

So far, the following large, continental shelf- and continental slope-based reefs have been found and mapped off Mid Norway: the Sula Ridge reef (including the Haltenpipe reef cluster), the Horse-Shoe Ridge reefs (Hesteskoen), the Træna Deep reefs, the Røst Bank reefs, and the Storegga escarpment reefs (Fig. 1). There are also numerous smaller coral structures of 1–3 m height, which include colonies of *Lophelia pertusa*. These are scattered on the general seafloor and tend to be located on bathymetric highs, such as moraine ridges, on glacial flutes and on iceberg plough marks.

Both the Sula Ridge reef and the Træna Deep reefs are located on top of sub-cropping Mesozoic sedimentary rocks that dip towards the west. The Horse-Shoe reefs are located on top of a moraine ridge, which is located over dipping Mesozoic strata. The Storegga escarpment reefs are located on top of Quaternary marine and glacimarine sediments. Thus, all these large reefs seem to have at least two conditions in common: They are located on top of a firm sea floor, and are on local heights.

Because of their importance as feeding and breeding grounds for some of the fish species, and also because of their obvious importance as biological resources, the large coral reefs, off Mid Norway must be carefully respected by the hydrocarbon industry. Large reefs are also known to occur along the coast and in some of the fjords of Mid Norway (Mortensen and Fosså, 2001). This means that the existence of reefs will have to be considered in all aspects of hydrocarbon exploration and exploitation, off Mid and Northern Norway, i.e., prior to and during exploration drilling, field development, and hydrocarbon transportation.

Description of some large reef occurrences

The Sula Reef

The second largest known Norwegian *Lophelia*-reef complex occurs on the Sula Ridge, about 75 km north of Kristiansund. Individual *Lophelia*-reefs here grow to heights of 30 m. The length of the main reef complex is about 13 km, and the width is up to 400 m. The depth-range for the *Lophelia*-reefs is from 260–330 m.

The Sula Reef Complex is constructed on top of the Sula Ridge (Figs. 1 and 2), which consists of Palaeocene sedimentary stratified rocks, with strong, enhanced acoustic reflections, indicating possible migration of hydrocarbon gases from the Jurassic source-rocks further west. The Sula Ridge is covered by a relatively thin layer (0–10 m thick) of sub-glacial till, which has been ploughed by icebergs during the deglaciation (Lien, 1983; Freiwald et al., 2002).

The Haltenpipe Reef Cluster

Reconnaissance mapping for the Haltenpipe gas trunk pipeline was performed between 1985 and 1990 (Mortensen et al., 1995; Mortensen et al., 2001). Within the 200 km long and 3 km wide corridor mapped, the total number of suspected coral reefs higher than 5 m was 57. Of these occurrences, a total of 14 suspected coral reefs were visually inspected by ROV, of which 12 turned out to be large, live *Lophelia*-reefs (Hovland et al., 1998). The highest reef was 31 m, and all the inspected reefs had base diameters of more than 50 m. Although they occur in local clusters with up to 9 reefs per km^2, the mean density of reefs

Onshore–Offshore Relationships on the North Atlantic Margin
edited by B. Wandas et al. NPF Special Publication 12, pp. 265–270,
Published by Elsevier B.V., Amsterdam

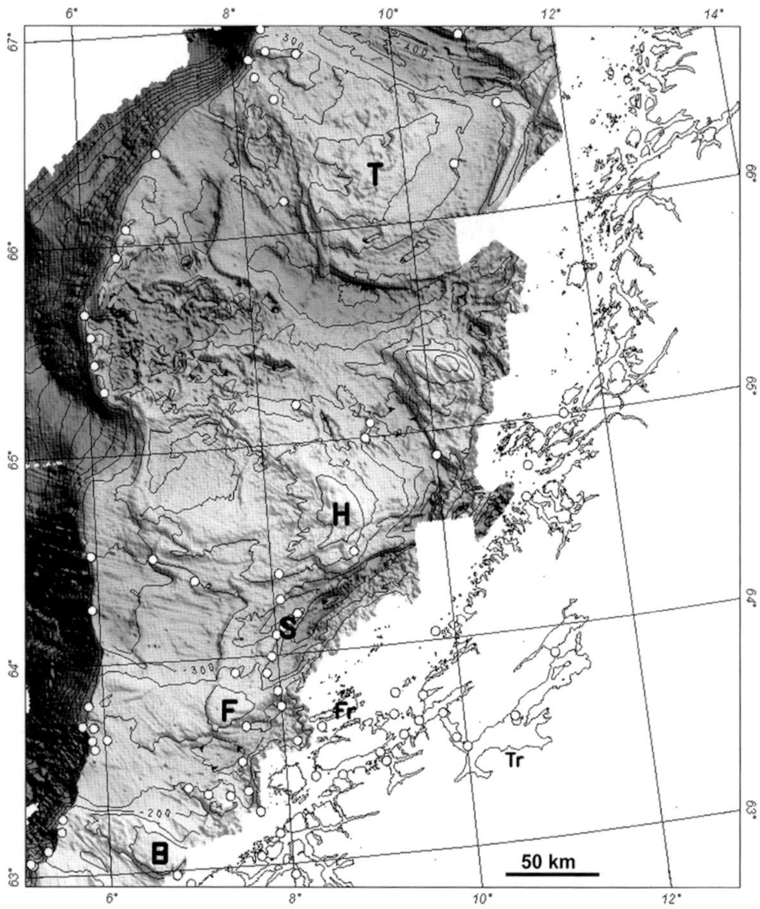

Fig. 1 Overview map of the Mid-Norwegian shelf with coral localities (white circles). The coral reefs mentioned in the text are the Røst Reef Complex which is off this map. Abbreviations: B = Buagrunnen; F = Frøyabanken; Fr = Frøya island; Tr = Trondheim; S = Sula Ridge; H = Haltenbanken; T = Trænabanken.

along the entire route transect is only 0.09 suspected reefs per km^2. However, there is a large regional density variation, with the highest regional density being 1.2 reefs per km^2 in an area of subcropping Palaeocene sedimentary rocks.

Since very little was known at that time, about the ecology and the robustness of deep-water coral reefs, a cluster of 9 large reefs, located near the Sula Reef Complex, was selected for further detailed studies. This cluster is now being used for reference studies and is known as the Haltenpipe Reef Cluster (HRC) (Figs. 3 and 4). Immediately to the SE of HRC, the less competent Cretaceous sedimentary rocks subcrop with numerous large pockmarks forming in the soft clays overlying them. These pockmarks occur within a distance of less than 500 m from the HRC.

In order to document baseline geochemical conditions of sea floor sediments, prior to pipeline construction, Statoil conducted a geochemical sampling and analysis investigation over the Palaeocene and Cretaceous subcropping rocks. Some of the samples targeted the HRC and also one of the large pockmark craters about 500 m to the east. These results prove that there is a natural input of light hydrocarbons (methane, C_1, ethane, C_2, propane, C_3, butane, C_4, and pentane, C_5) into the upper sediments, from the Jurassic source rocks.

It was found that light hydrocarbon concentrations increase well above background inside the pockmark and also near the two largest reefs of the HRC. Whereas the background value of the sum of adsorbed methane to pentane concentrations in the near-surface clay (~1.5 m below surface) is 100 ppbv (parts per billion, by volume), the value inside the large pockmark is about nine times higher, and at the reef bases, about six times higher than the background (Hovland et al., 1998). This geochemical investigation, therefore, confirms that both the large pockmark and the investigated reefs, are associated with the seepage of light hydrocarbons.

The Træna Deep Reefs

An area with equally large *Lophelia*-reefs was found during a pre-drilling site survey, in Norwegian concession block 6610/3 (Hovland and

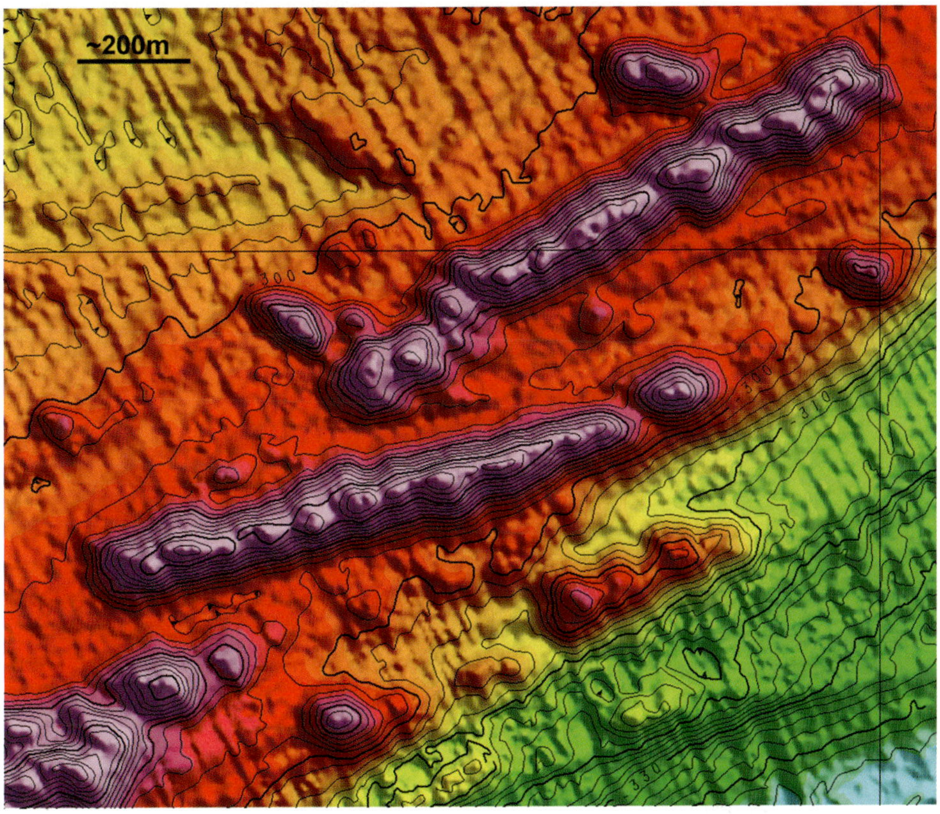

Fig. 2 Detail from the Sula Reef Complex. Colour shaded relief with metre contours. Strings of large *Lophelia*-reefs occur as violet ridges and mounds here.

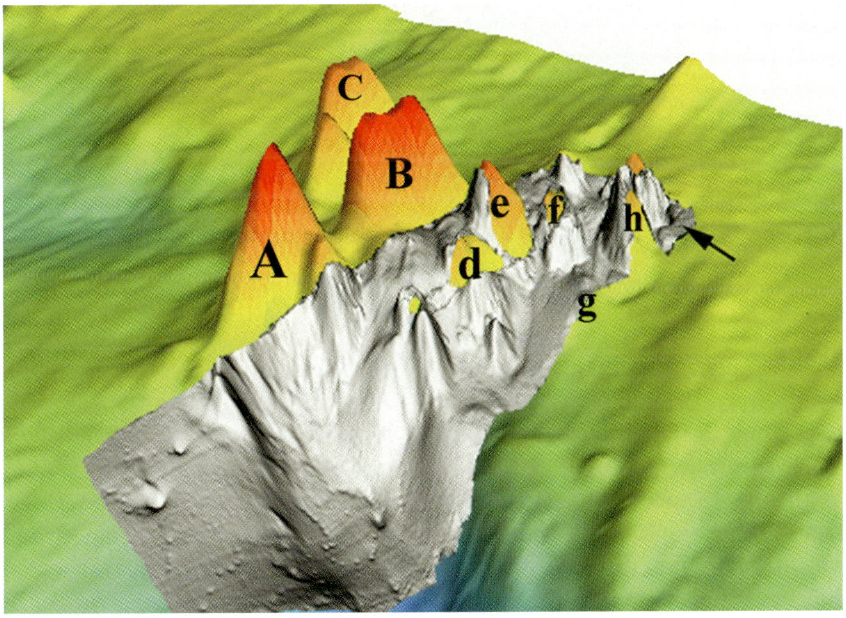

Fig. 3 A composite perspective view of the HRC, showing high-resolution swath bathymetry data (in grey), with small (unit) pockmarks at the base of at least two individual *Lophelia*-reefs (h and e). Direction of view is from northeast. Unit pockmarks (arrowed) are caused by the seepage of gas and/or porewater through seafloor sediments (Hovland and Judd, 1988). Apparent holes in the reefs are caused by data-gaps in the swath bathymetry data set. The background terrain model is based on vessel-hull mounted swath bathymetry.

Mortensen, 1999). These reefs are constructed on top of an Oligocene deltaic sandy fan deposit, with arcuate bedding, dipping towards the west. The water depth here is between 300 and 330 m. The largest reef within the 5×5 km^2 wide area that was mapped is 20 m high, 700 m long, and 150 m wide. There are 265 individual suspected reefs within this area, each wider than 15 m. The density of

Fig. 4 Detail from the HRC, showing immense biodiversity on reefs.

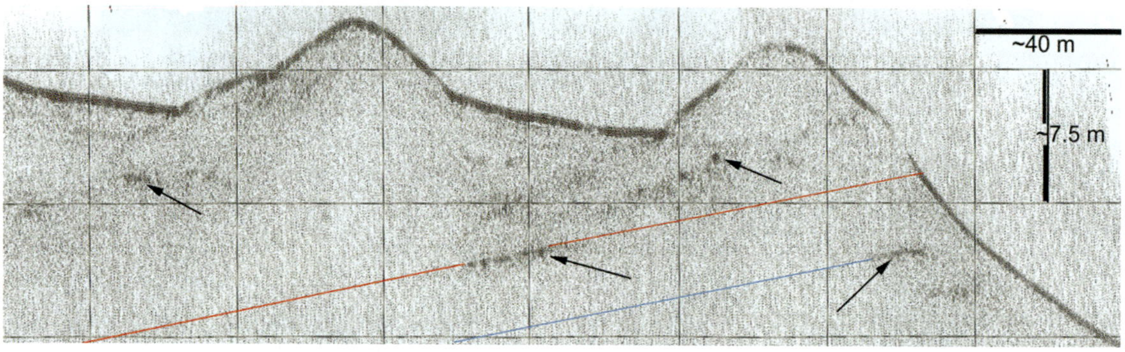

Fig. 5 High-resolution shallow seismic record across the HRC showing enhanced reflectors (arrowed). These indicate the existence of migrating hydrocarbons along permeable up-dipping layers. Two thin lines have been drawn to indicate dipping Palaeocene bedding planes (from other seismic data).

suspected reefs within the area is very high, at about 10 per km^2. On seismic records, there is a strong acoustic evidence that some of the layers below the reefs are gas-charged, and may represent hydrocarbon migration pathways to the surface (Fig. 5). The side-scan sonar and echosounder data collected during the site survey also contain clear evidence of acoustic plumes in the water column, probably representing free gas emitting directly to the water column from some of the dipping layers (Hovland and Mortensen, 1999).

The Horse-shoe reefs

The Horseshoe is located in the eastern part of the Trænabanken and is an arcuate hill rising about 100 m above the surrounding, rather flat c. 300 m deep sea floor (Fig. 6). The hill is c. 2.5 km across, and approximately 15 km long. Sættem (1990) reported till material from a vibrocore sample on top of the ridge. On top of the southern and the south-western part of the ridge, several mounds occur. During a video inspection by the Institute of Marine Research (IMR), these mounds proved to comprise of living *Lophelia* corals.

The Røst Reef Complex

The newly discovered Røst reef, was found by IMR Institute by employing a simrad multi-beam echosounder (on the research vessel 'Håkon Mosby') with automatic seabed classification ('Roxar'). The reef complex stretches for 35 km,

Fig. 6 A perspective view of the The Horse-shoe structure, based on multi-beam bathymetry, viewed from the NW. The Horse-shoe *Lophelia* reefs occur adjacent to this characteristic push-up moraine, and can be seen here as spikes to the south of the morainic ridge, which is 15 km long and 2.5 km wide.

along the upper Storegga continental slope shoulder and will be reported elsewhere.

Future research

Currently, there is a high activity level of research fieldwork partly attempting to verify or disprove the apparent association between *Lophelia*-reefs and seep indicators such as pockmarks, especially in the European GEOMOUND and ECOMOUND projects (van Weering et al., 2001; Henriet et al., 2001).

Further sampling and geochemical/microbiological analysis of sediments, porewater, and nearby seawater could yield some more information on the geochemical and bacterial/archaeal concentration gradients. Also, further stable isotope analyses, especially of C and N from organisms on the reefs may prove useful in acquiring a better understanding of why cold-water corals and carbonate mounds develop only on specific locations.

Because any mechanical bottom activity, such as anchoring, trawling, and construction near the *Lophelia*-reefs is likely to disturb the environment (Hall-Spencer et al., 2002; Fosså et al., 2002; Husebø et al., 2002), it is important for the government and industry to continue performing detailed surveys on the Norwegian coral reefs.

Acknowledgement

We thank our companies for the permission to publish these data.

References

Fosså, J.H. and Mortensen, P.B., 1998. Artsmangfoldet på Lophelia korallrev og metoder for kartlegging og overvåkning. Fisken og havet 17, 95 pp (in Norwegian).

Fosså, J.H., Mortensen, P.B. and Furevik, D.M., 2002. The deep-water coral Lophelia pertusa in Norwegian waters; distribution and fishery impacts. Hydrobiologia, 417: 1–12.

Freiwald, A., Hühnerbach, V., Lindberg, B., Wilson, J.B. and Campbell, J., 2002. The Sula Reef Complex, Norwegian Shelf. Facies, 47: 179–200.

Hall-Spencer, J., Allain, V. and Fosså, J.H., 2002. Trawling damage to Norteast Atlantic coral reefs. Proc. R. Acad. Sciences B., Online publications, DOI: 10.1098/rspb.2001.1910.

Henriet, J.-P., De Mol, B., Dullo, W.-C., Freiwald, A., Joergensen, B.B. and Parkes, J., Patching, J.W., 2001. Deciphering the message of carbonate mounds: the Porcupine Scientific Drilling Project. (abstr.) European Union of Geologists 11th Conference Proceedings (session RCM6).

Hovland, M. and Mortensen, P.B., 1999. Norske korallrev og prosesser i havbunnen (Norwegian coral reefs and seabed processes), John Grieg, Bergen, Norway, 167 pp. (in Norwegian with English summary).

Hovland, M., Mortensen, P.B., Brattegard, T., Strass, P. and Rokoengen, K., 1998. Ahermatypic coral banks off mid-Norway: Evidence for a link with seepage of light hydrocarbons. Palaios, 13: 189–200.

Husebø, Å., Nøttestad, L., Fosså, J.H., Furevik, D.M. and Jørgensen, S.B., 2002. Distribution and abundance of fish in deep-sea coral habitats. Hydrobiologia, 417, 91–99.

Jung, W.-Y., Vogt, P., Haflidason, H. and Parsons, B., 2001. US Navy submarine NR-1 dives in the upper Storegga slide area, Norwegian margin. (Abstract) European Union of Geologists 11th Conference Proceedings (session RCM4).

Lien, R., 1983. Iceberg scouring on the Norwegian continental shelf. Proceedings Offshore Technology Conference (OTC), Houston Texas 15, pp. 41–45.

Mortensen, P.B. and Fosså, J.H., 2001. Coral reefs and other bottom habitats on the Tautra Ridge in Trondheimsfjorden. Fisken og havet, 7.

Mortensen, P.B., Hovland, M., Brattegard, T. and Farestveit, R., 1995. Deep water bioherms of the scleractinian coral Lophelia pertusa (L.) at 64°N on the Norwegian shelf: structure and associated megafauna. Sarsia, 80: 145–158.

Mortensen, P.B., Hovland, M.T., Fosså, J.H. and Furevik, D.M., 2001. Distribution, abundance and size of Lophelia perusa coral reefs in mid-Norway in relation to seabed characteristics. J. Mar. Biol. Ass. U.K,. 81: 581–597.

Sættem, J. 1990: Glaciotectonic forms and structures on the Norwegian continental shelf: observations, processes and implications. Norsk Geol. Tidsskr. 70: 81–94.

van Weering, T.C.E., de Haas, H., Lykke-Andersen, H. and de Stigter, H., 2001. Giant carbonate mounds in the Rockall Trough, NE Atlantic Ocean. (abstr.) European Union of Geologists 11th Conference Proceedings (session RCM6).

Arctic Gas Hydrate Provinces along the Western Svalbard Continental Margin

Maarten Vanneste, Stéphanie Guidard and Jürgen Mienert

In 2001, a high-resolution seismic survey was conducted for the detailed study of the distribution, both spatially and vertically, of gas hydrate and free gas accumulations west of Svalbard, as part of the HYDRATECH and INGGAS projects.

High-resolution single-channel seismic reflection and the 4-component ocean-bottom seismometer (OBS) data illustrate the widespread nature of gas hydrates and free gas accumulations north of the Knipovich Ridge off Western Svalbard, by the presence of a nearly-continuous polarity-reversed bottom-simulating reflection (BSR) on down-slope seismic profiles. In the absence of a distinct and/or a continuous BSR, it is the sudden change in reflection amplitude and frequency content that marks the base of the hydrate zone. The BSR coincides with the top of the free gas zone. Compressional wave velocity analyses and modelling reveal increased velocities above the BSR attributed to a gradual increase of partial hydrate saturation (6–10% of pore volume). A sharp drop in compressional-wave velocity across the BSR is due to free gas accumulation. The sub-bottom depth of the BSR closely matches the calculated stability limit for methane hydrates.

To the east of the Knipovich Ridge, mud diapirism is observed in a deeper basin (~2250 m water depths). The domes rise from an extensive chaotic source zone buried under a 200–400 ms thick sediment drape, and are more pronounced in the south. At some places, there is evidence of stratigraphically-controlled shallow gas accumulations (bright spots) and short cross-cutting BSR-like features that might point towards the presence of hydrate and/or free gas. The diapiric movement is believed to be a recent and still ongoing process of mass mobilisation.

In both the cases, the nearby and tectonically-active slow-spreading, Knipovich Ridge is assumed to play an important role in the generation of elevated heat and methane fluxes as well as faulting and subsequent fluid migration. As a result, shallow subsurface hydrates (<10% of pore volume) may form and mud diapirs may develop.

Introduction

Gas hydrates are solid compounds formed by hydrogen-bonded water molecules enclosing single low-molecular-weight gas molecules. The stability of these structures requires specific conditions of pressure (high) and temperature (low), in the presence of sufficient water and gas molecules (Sloan, 1998). Gas hydrates are typically found in the pore spaces of the uppermost hundreds of metres of continental margin sediments in oceans and inland seas, where water depths exceed 300–500 m, and in the arctic permafrost areas (Kvenvolden and Barnard, 1983). Methane is by far the most abundant clathrated gas in natural environments, and can be of either biogenic or thermogenic origin (Kvenvolden, 1998). Small amounts of gases other than methane (e.g. carbon dioxide, ethane, propane, hydrogen sulphide, etc.) may be present as well, which would shift the hydrate stability conditions or even change its structure (Sloan, 1998). Evidence of hydrates is commonly inferred from the presence of a high-amplitude reversed-polarity cross-cutting reflection that parallels the sea floor, and is therefore called a bottom-simulating reflection (BSR) (Shipley et al., 1979). This reflection often coincides with the theoretical limits of methane hydrate stability, and can therefore be used to derive the geothermal field (Yamano et al., 1982; Bangs and Brown, 1995; Bouriak et al., 2000; and many others). The BSR is mainly the result of the acoustic impedance contrast of gas-bearing sediments overlain by partially hydrate-saturated sediments (Holbrook et al., 1996; Pecher et al., 1996). However, Hydrate-bearing sediments might also exist in areas lacking a BSR, as is the case on

Onshore–Offshore Relationships on the North Atlantic Margin
edited by B. Wandas et al. NPF Special Publication 12, pp. 271–284,
Published by Elsevier B.V., Amsterdam

Blake Ridge (e.g. ODP-164 site 994) (Holbrook et al., 1996) and Lake Baikal (Vanneste et al., 2001). The academic and industrial interests in gas hydrates are increasing steadily due to its potential as a future energy resource, a submarine geohazard, its role in the global carbon cycle, and a critical factor for global climate change (Henriet and Mienert, 1998; Mienert et al., 2001b; and many others).

The Norwegian–Barents–Svalbard continental margin is a highly dynamic area showing abundant evidence of fluid migration processes, submarine mass wasting, fan development, cold water reefs, faulting, hydrocarbon accumulation, and the inferred presence of gas hydrates (e.g. Vorren et al., 1998; Vogt et al., 1999a; Mienert and Weaver, 2003; Lindberg et al., in press). Along this passive margin, gas hydrates have so far only been retrieved from the Håkon Mosby mud volcano (Ginsburg et al., 1999), but the presence of BSRs at the Storegga Slide area (Mienert et al., 1998; Bouriak et al., 2000; Bünz et al., 2003), in the Barents Sea (Andreassen et al., 1990; Laberg and Andreassen, 1996) and along the continental slope of Western Svalbard (Eiken and Hinz, 1993; Posewang and Mienert, 1999), illustrates that gas hydrates and free gas accumulation are common features in this region. The previous high-resolution studies west of Svalbard indicate that the BSR marks a sharp transition from higher interval velocities above, attributed to the occurrence of gas hydrates, to low interval velocities below, attributed to free gas accumulations (Posewang and Mienert, 1999).

The aims of this chapter are: (1) to present new high-resolution single-channel seismic data sets and preliminary results of OBS records, (2) to illustrate the presence and the extent of gas hydrates and free gas accumulations, and the report first estimates of partial hydrate saturation, based on preliminary velocity modelling, (3) to discuss the evidence of subsurface mass mobilisation in the form of mud diapirism, and (4) to discuss a possible relationship between these features and the nearby active Knipovich Ridge system. The geophysical data were acquired using the R/V Jan Mayen (University of Tromsø), during the summer of 2001, as part of the HYDRATECH and INGGAS projects. The objective of the HYDRATECH Project (EU 5th framework) is to develop a technique for the quantification of gas hydrates in continental margin sediments, along the European margin, based on multi-component ocean-bottom seismometer (OBS) arrays and tomography. The INGGAS Project concentrates on an integrated geophysical characterisation and quantification of gas hydrates. Such methods should be viable, both in the presence and absence of a BSR. Hence, the Norwegian–Svalbard continental margin has one of the best potentials to meet these objectives.

Geological and Tectonic Setting

The study areas were selected, based on an earlier work by Posewang and Mienert (1999). The two target sites lie proximal to the Knipovich Ridge in water depths of 1250–1750 m in Area 1, and at about 2250 m in Area 2 (Fig. 1). The slow-spreading and segmented Knipovich Ridge is the northernmost extension of the active mid-Atlantic Ridge, and occupies an asymmetrical position in the Norway–Greenland Sea. The rift axis reaching depths of >3000 m, ends bathymetrically at 78.5°N, where it abuts the lower slope of the Svalbard Margin. From there, it continues as a buried feature in north-northwestern direction (Myhre and Thiede, 1995). A complex system of short spreading centres and transform faults (e.g. the Molloy Ridge and Transform, fig. 1) connect the Knipovich Ridge with the arctic Gakkel Ridge (Fig. 1) (Myhre and Thiede, 1995). Subsequent to post-Caledonian extensional episodes, the passive rifted and sheared Norwegian–Svalbard margin has been influenced by both tectonic (Myhre and Eldholm, 1988) and sedimentary processes (Vorren et al., 1998) during the Cenozoic. Breakup followed by sea floor spreading started in the early Eocene in the south of the Norwegian–Greenland Sea. A change in plate movements in the Oligocene forced rifting along the continental transform between the Barents Sea and Greenland, leading to the northwards stepwise propagation of and spreading along the Knipovich Ridge and culminated in the continental separation of Greenland and Svalbard (Lundin and Doré, 2002). As a result, the Fram Strait (Fig. 1) developed as the only deep-water passage between the North-Atlantic and the Arctic, playing a key role in large-scale oceanic circulation processes (Eiken and Hinz, 1993). The Norwegian–Svalbard margin has been further shaped by movement of the Fennoscandian and Barents Sea ice sheets. During Late Pliocene and Pleistocene, glaciers reached the Svalbard shelf break frequently (Vorren et al., 1998). Sedimentation rates in this region are relatively high, exceeding 30 cm/ka since the Miocene (Crane et al., 1988), resulting in sediment deposition between 1 and 6 km thick off Western Svalbard (Eiken and Hinz, 1993),

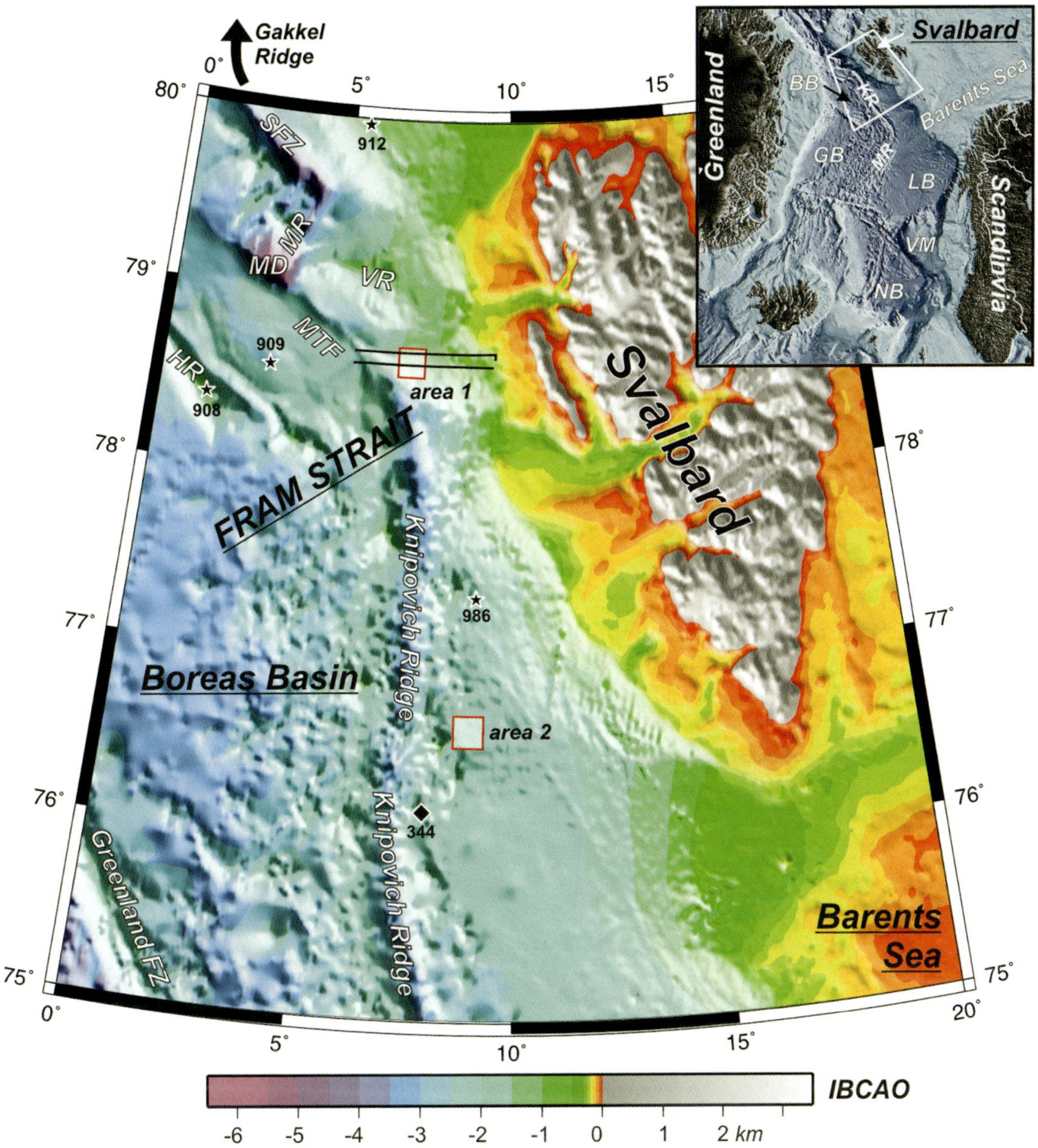

Fig. 1 Shaded relief bathymetry/topography (IBCAO, Jakobsson et al., 2000) map with the main structural features of the Fram Strait and the Western Svalbard Margin. The inset shows its location in the northern North-Atlantic Ocean. The two red boxes are the study areas presented in this chapter, in the immediate vicinity of the slowly-spreading Knipovich Ridge. Black stars and squares represent ODP respectively DSDP drill holes. The two black lines are the long slope profiles mentioned in the text. This map was created with GMT software (Wessel and Smith, 1998). (VR = Vestnesa Ridge, SFZ = Spitsbergen Fracture Zone, MD = Molloy Deep, MR = Molloy Ridge, HR = Hovgård Ridge; abbreviations in inset: BB = Boreas Basin, GB = Greenland Basin, LB = Lofoten Basin, VM = Vøring Margin, NB = Norway Basin, KR = Knipovich Ridge, MR = Mohns Ridge).

comprising Late Pliocene and Quaternary glacial fans, wedges, turbidites and contourites (Eiken, 1994; Faleide et al., 1996). Crustal heat flow is elevated (>75 mW m^{-2}) and reaches extreme values along the plate boundaries (Sundvor et al., 2000). The present-day surface water system is dominated by the relatively warm West-Spitsbergen Current and the cold East-Greenland Current. Deep-water formation in the polar North-Atlantic is the result of heat loss of northward flowing surface waters, and drives the ocean conveyor belt (Bauch et al., 2001).

Geophysical Data Acquisition

A total of ~400 km of seismic reflection profiles were obtained in each of the two study areas. A double-sleeve gun array (2 × 0.65 l volume) operated at 130–140 bar and towed at 4 m below the sea surface generated an acoustic pulse with frequency content between 30 and 450 Hz, centred around ~100 Hz. The data were recorded with a single-channel (SC) streamer towed at the surface at short offset (55 m) from the source, and simultaneously by a regular grid of 20, 4-component (1 hydrophone and 3 geophone components) OBS instruments, spaced about 400 m apart, and sampling at 1 kHz. The seismic experiment was designed to provide high-resolution P-wave and converted S-wave data suitable for tomography, pre-stack depth migration, full-waveform inversion, 2D and 3D ray trace modelling and anisotropy (Mienert et al., 2001a). Based on logistical constraints in combination with the results from forward modelling of travel times and amplitudes, the profiling was set up to form a set of line-oriented down-slope and a set oriented parallel to slope. Line spacing was about 200 m. Several diagonal lines across the survey area provided azimuthal variations. The seismic profiles are typically ~11 km long, with a shot spacing of 20–25 m. Additionally, we recorded 3.5 kHz echo-sounding data.

Data processing consisted of OBS and shot-point relocation, minimum phase bandpass frequency filtering, static header corrections, geometry loading, muting, and memory single Stolt migration. Additionally, semblance velocity analysis (Area 2) and ray tracing (Area 1) were used to estimate the interval velocity profile.

From the CTD casts, we know that the average water velocity is 1470 m s^{-1}, while the bottom water temperature is −0.90°C in Area 1 and −0.86°C in Area 2.

Observations from the NW Svalbard site

The first study area is situated just north of the termination of the Knipovich Ridge and east of the Molloy Transform fault on the Svalbard continental slope (Figs. 1 and 2).

SC seismic reflection data

The single-channel seismic profiles show a well-stratified wavy sequence of reflections down to the limit of penetration at ~500 ms sub-bottom depths

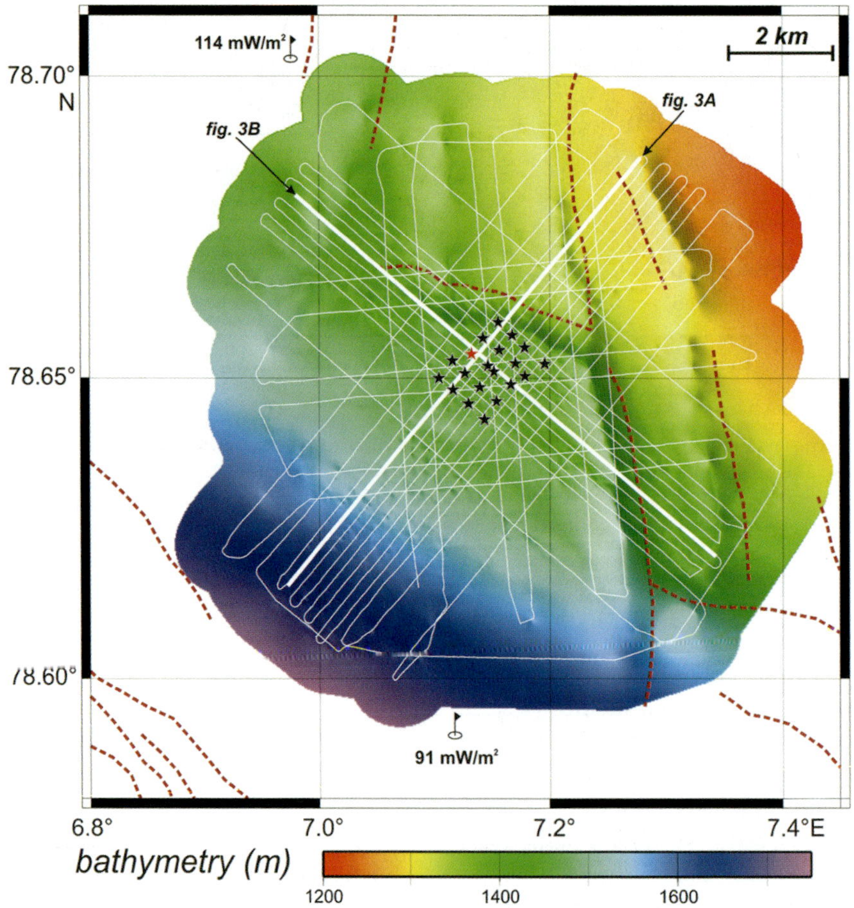

Fig. 2 Shaded relief bathymetry of the NW Svalbard study area, extracted from single-channel seismic data (white lines) (Area 1, for location see Fig. 1). The two bold white lines represent the profiles shown in Fig. 3. The dashed brown-red lines represent rift escarpments observed on SeaMARC II data and digitised from Crane et al. (2001), which coincide reasonably well with the fault structures observed on the single-channel data in this study. Also OBS locations (stars) and heat flow measurements are shown. The red star is the location of the OBS discussed in this manuscript (see Fig. 5). This map was created with GMT software (Wessel and Smith, 1998).

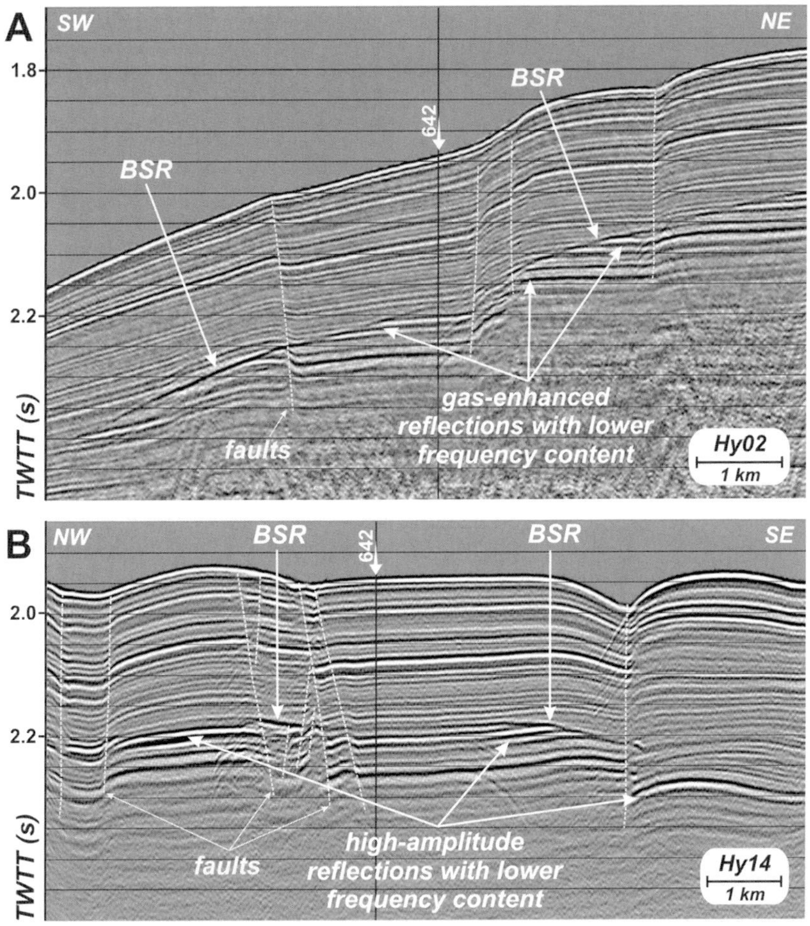

Fig. 3 Examples of a down-slope (A, Hy02) and along-slope (B, Hy14) seismic profile from the study area north of the Knipovich Ridge. The black line marks the intersection of these two lines. The location of the OBS station is marked as well.

(Fig. 3) that belongs entirely to the sediment sequence YP-3 and the upper part of YP-2 of Eiken and Hinz (1993). These sequences are interpreted as depositional from contour currents (Eiken and Hinz, 1993). Our data illustrate that several normal faults interrupt the strata. We also observe ample evidence of shallow gas hydrate accumulations (BSRs) and free gas zones.

On the down-slope profiles (Fig. 3A), a distinct BSR feature is observed at ~250 ms TWT sub-bottom depth. It is a single, polarity-reversed reflection that cuts obliquely across the stratigraphic units. The continuous BSR generally has a high reflection amplitude, although it varies laterally. It marks a clear zonation in terms of amplitude and frequency content. The reflection strength of lithologic boundaries underneath the BSR is enhanced. Spectral analyses, after horizon flattening relative to the BSR and averaged over 400 traces and 50 ms windows, revealed a sudden and drastic drop in peak frequencies with about 30 Hz across the BSR (Fig. 4), while the loss in frequency between the sea floor reflection and the window just above the BSR is much less (Fig. 4). This also becomes clear on instantaneous frequency displays. Such low-frequency shadows are more pronounced in places where a strong BSR feature or enhanced stratigraphic reflections occur, i.e. in places where we expect higher gas saturation. The enhanced reflections underneath the BSR also show reversed reflection polarity, indicating an inversed acoustic impedance contrast. Occasionally, lithologic reflections crossing the BSR change their polarity. Amplitude blanking, often thought indicative of hydrate presence (Lee et al., 1994), is not observed here.

The along-slope profiles (Fig. 3B) show less clear evidence of gas hydrates or free gas, mainly due to the sub-parallel stratigraphy. Nevertheless, shorter cross-cutting polarity-reversed BSR-like features appear on top of a series of enhanced reflections having lower frequency content. Although no continuous BSR is present, we can mark the transition from hydrate to free gas presence from the different amplitude and frequency character of the stratigraphic reflections.

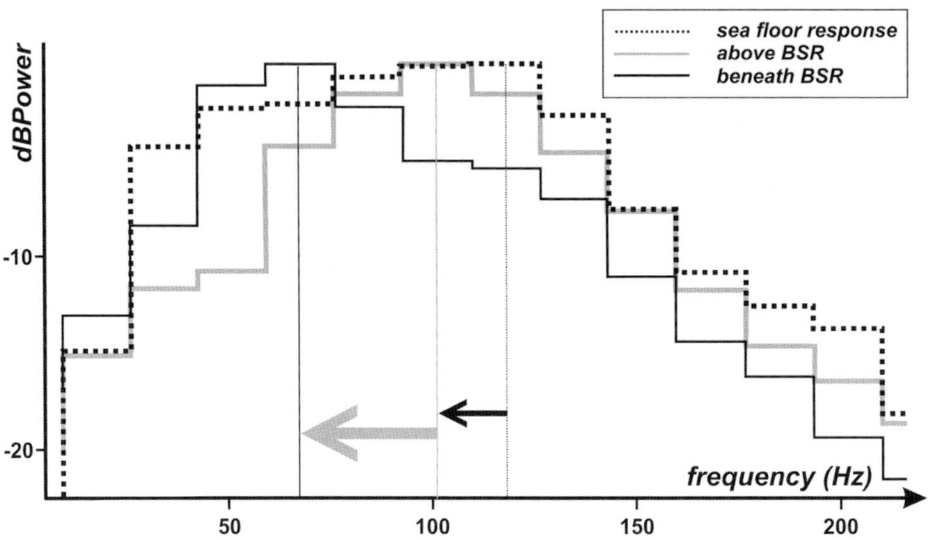

Fig. 4 Spectral analysis averaged over 400 traces in 50 ms windows just above (grey line) and beneath the BSR (black line) along line Hy02 (see Fig. 3) after horizon flattening with respect to the BSR. The dashed black line shows the sea floor response. While the frequency loss between the sea floor and the BSR is limited (black arrow), a sharp, significant drop in frequency occurs across the BSR interface (grey arrow).

Preliminary results from OBS data

Till date, we have only investigated the hydrophone and the vertical components of one of the OBS instruments (station 642). The instrument recorded clear seismic arrivals throughout the uppermost 500 miles of sediments. Preliminary P-wave interval velocity analyses from ray-tracing on the hydrophone component of OBS instrument 642 (Fig. 5) show a gradual increasing velocity trend that is abruptly inverted at the depths of the BSR, at ~205 mbsf (~245 miles TWT). A 100 m thick layer above the BSR has higher interval velocities compared to the generally expected velocity increase due to sediment compaction (Hamilton, 1980). Beneath the BSR, the velocity reduces in two steps over an interval of ~20 m to anomalously low values of ~1500 $m\,s^{-1}$. The thickness of this low-velocity gas-charged zone exceeds 50 m. These results are in agreement with previous investigations by Posewang and Mienert (1999), although we have no evidence of a P-wave velocity inversion within the hydrate stability zone.

In a first attempt to estimate partial hydrate saturation, we modelled the positive excursion of the OBS-derived velocity profile relative to the Hamilton regression curve (Hamilton, 1980), based on the methodology described by Tinivella (1999). As an outcome, we obtain slightly increasing hydrate occupations of pore volume with depths from 6.0% (103–120 mbsf), 7.5% (120–156 mbsf), 9.5% (156–182 mbsf) and 9.0% (182–202 mbsf), the latter just above the BSR.

Discussion: hydrates and free gas at the NW Svalbard site

As mentioned above, the SC seismic reflection data provide information on the spatial distribution of gas hydrates, while the OBS data give us an idea about the vertical extent of gas hydrates and free gas off Western Svalbard. Gassy sediments are known to affect the acoustic signatures in terms of attenuation, propagation and reflectivity (Anderson and Hampton, 1980). Even small amounts of pore space gas can significantly reduce acoustic velocity. Gas also attenuates or absorbs the higher frequencies of acoustic signals, and results in low frequency shadows, just underneath the gas-rich sediments. Therefore, we attribute the reversals of reflection polarity, the sudden low-frequency shadows, the enhanced reflectivity, and the reduced in situ P-wave velocities observed beneath the BSR to the presence of free gas pockets within the strata trapped under hydrate-rich sediments. As such, the BSR in our study area originates at the interface between partially hydrate-saturated sediments above, and gas-containing sediments underneath. The OBS data suggest that the gas-rich layer is at least 50 m thick. We cannot as yet predict the amount of gas saturation beneath the BSR. Such quantitative results await the further detailed analyses of the full 4-component OBS data (e.g. converted wave investigations and especially AVO analysis and modelling). As estimated, hydrate saturations increase gradually above the BSR and are less than 10% of pore volume, or less than 5%

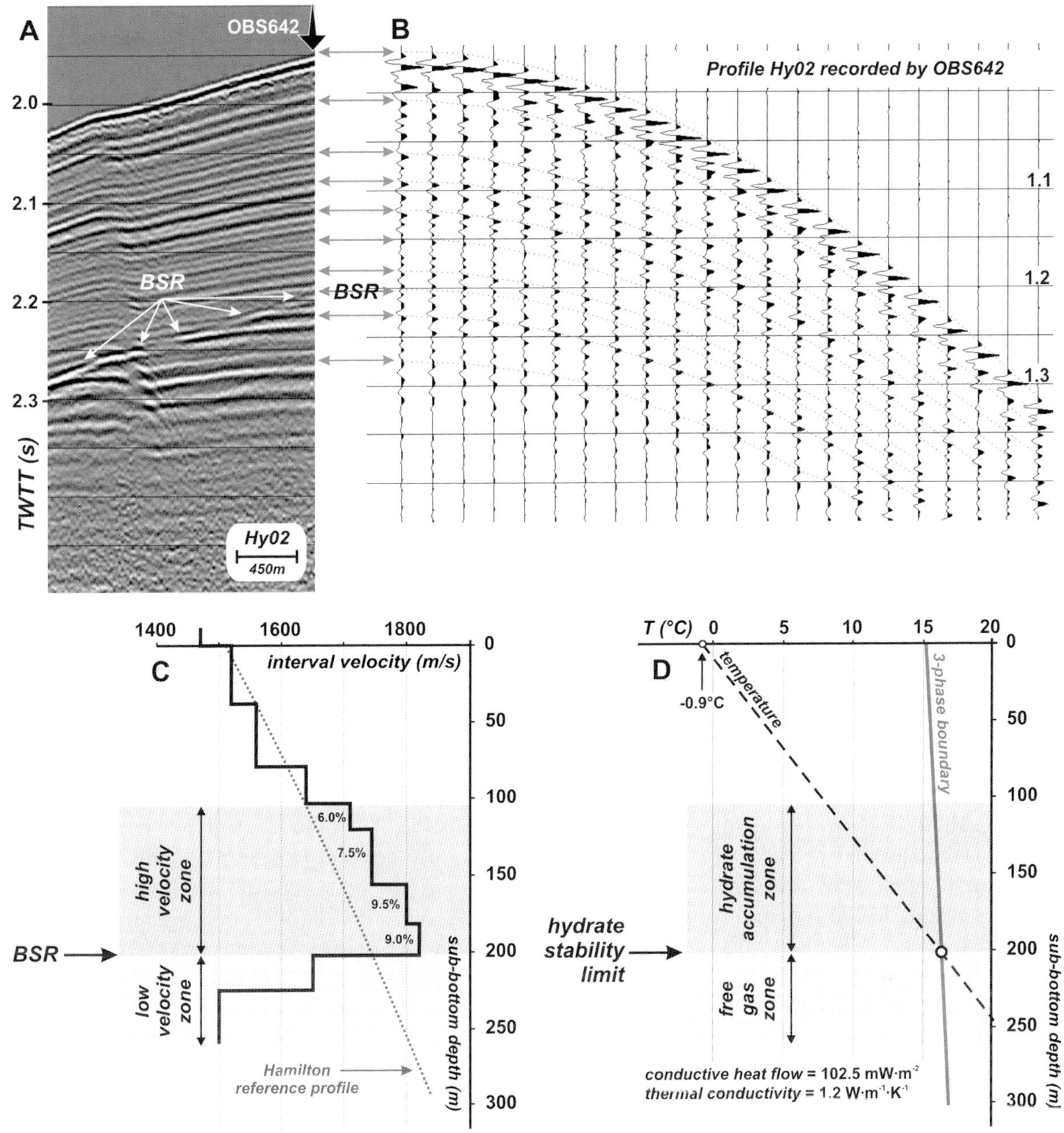

Fig. 5 Part of line Hy02 from surface streamer (A) and OBS 642 (B), and modelled interval velocity profile from ray tracing (C) across the BSR with estimated hydrate saturations. (D) Theoretical methane hydrate — seawater phase diagram for the study area. For location, see Fig. 2.

of bulk sediment volume. The calculated velocity profile also suggests that hydrates do not occur in the uppermost 100 m of the sediment.

The distribution of BSRs, hydrates and free gas accumulations off West Svalbard, is neither restricted solely to this OBS study area, nor the area discussed in Posewang and Mienert (1999). Two 90 km long down-slope profiles revealed the presence of continuous BSRs over a distance of some 50 km in water depths ranging from 800 m to 2300 m, i.e. from the middle part of the Western Svalbard Margin towards the depression of the Molloy Transform Fault (Fig. 1), where elevated heat flow (Sundvor et al., 2000) is responsible for shoaling of the BSR. Our new data confirm the observations by Eiken and Hinz (1993) from the conventional low-frequency seismic data. The surveyed area connects the northern extension of the Knipovich Ridge with the Vestnesa sediment ridge (Fig. 1), where pockmark or mud diapir belts were discovered earlier and described in silty-clay sediments (Vogt et al., 1994; Vogt et al., 1999b), and where the observed bright spots were suggestive of free gas accumulation in the deeper sediments as well (Eiken and Hinz, 1993).

Although several faults interrupt the overall well-stratified sedimentation pattern, and most of them extend from the sea floor to beneath the BSR, they do not significantly affect the position of the BSR, as evidenced on the down-slope profiles (Fig. 3A). However, the reflection amplitude becomes fainter in the vicinity of such faults. We carefully mapped the fault zones after extracting the bathymetry from our seismic data set, and found that several of these faults closely match the sea floor escarpments, observed on the SeaMARC II side-scan sonar images (Fig. 2). Because of their structural connection to the mid-oceanic ridge, Crane et al. (2001) interpret these as tectonic features related to propagation of the Knipovich Ridge in north-northwestern direction towards the Vestnesa sedimentary ridge, and thereby eventually deactivating the Molloy Ridge and Transform. Hence, gas hydrate forms and accumulates in an area of incipient slow rifting, with the rift escarpments linking the inferred hydrate zones with pockmark fields. The nearly-continuous character and the bottom-simulating behaviour of the hydrate/free gas interface illustrate that heat and fluid pulses through these rift escarpments are not intense yet, or that it is in a premature phase. Otherwise, BSR irregularities would appear at these places, and the reflection would lose its sea floor mimicking character, as reported from other locations (Vanneste et al., 2002; Wood et al., 2002). This incipient change in heat and fluid flow regime might in the future result in similar BSR anomalies off Western Svalbard.

Hydrates have not been sampled off Western Svalbard, so their geochemical and isotopic composition is neither known at present; nor do we have any ground-truthing (e.g. deep drilling) on their distribution and saturation. At present, the sub-bottom position of the BSR fits well with inversion from mean heat flow data from the area (102.5 $mW\,m^{-2}$) using a methane—average salt water—hydrate mixture, the measured (CTD) bottom water temperatures of $-0.90°C$, and a pure conductive sub-bottom temperature profile (Fig. 5D). This suggests that methane hydrates probably have a microbial origin. Slow-spreading mid-oceanic ridge segments may have elevated methane output (Bougault et al., 1993) that subsequently used the rift escarpments as fluid conduits into the hydrate stability zone. Gases could also be enriched in heavier hydrocarbons, which would change the in situ hydrate stability conditions. Such a mechanism may provide an additional gas source in the deeper part of the Svalbard slope.

Observations from the SW Svalbard site

The second area under investigation (Figs. 1 and 6) lies approximately 40 km east of a transition between two of the Knipovich Ridge segments and northeast of an off-axial seamount belt (Crane et al., 2001). It has a dominantly uniform sea bed morphology at 2260–2280 m water depth in the ocean basin.

SC seismic reflection data

The seismic data from the SW Svalbard area (Fig. 6) show a well-stratified basin-fill sequence of reflections (Fig. 7) over several hundreds of ms. At variable sub-bottom depths, the stratification suddenly changes to a rather extensive and chaotic interval lacking internal structures, but well above the rough acoustic basement typically found at 1 s sub-bottom depths. The boundary between the stratified sedimentary units and the incoherent zone undulates from south to north, and in places, is significantly deformed by dome-like features or diapirs. As a result, the strata are bent upwards, an effect that increases towards the south where the largest diapiric structures are found. There, the diapirs rise to ~150 ms below the sea bed, which is slightly uplifted (a few meters) as a result of the doming process (Figs. 6–9). From these data, it seems that the chaotic zone forms the source of these mud diapirs having an extent of at least 130 km^2. Several stratigraphically-controlled bright spots developed on top of these diapirs (typically >50 ms above the domes) at different levels. The polarity of these bright spots is reversed relative to the reflection of the sea bed (Fig. 8), indicative of a drop in acoustic impedance, and possibly, the presence of free gas. In some places, the sediments draping the diapirs are offset by normal faults (Fig. 8). Echo-sounding data show that some of these faults are active, slightly offsetting the seafloor (Fig. 8). Such faults may be caused by the diapiric deformation process in the underlying sub-surface. Only occasionally do we discern short BSR-like features as relatively faint reflections cross-cutting the bigger diapiric structure in the southern part of the study area (Fig. 7) at ~180–200 ms (or 145–170 m) below the sea floor. These reflections also display negative polarity and can only be traced over a distance of about 1 km. Noteworthy, these BSRs occur exactly in the area where the sea bed is slightly uplifted (Figs. 6 and 7), and hence, where the mass movement related to doming is most pronounced.

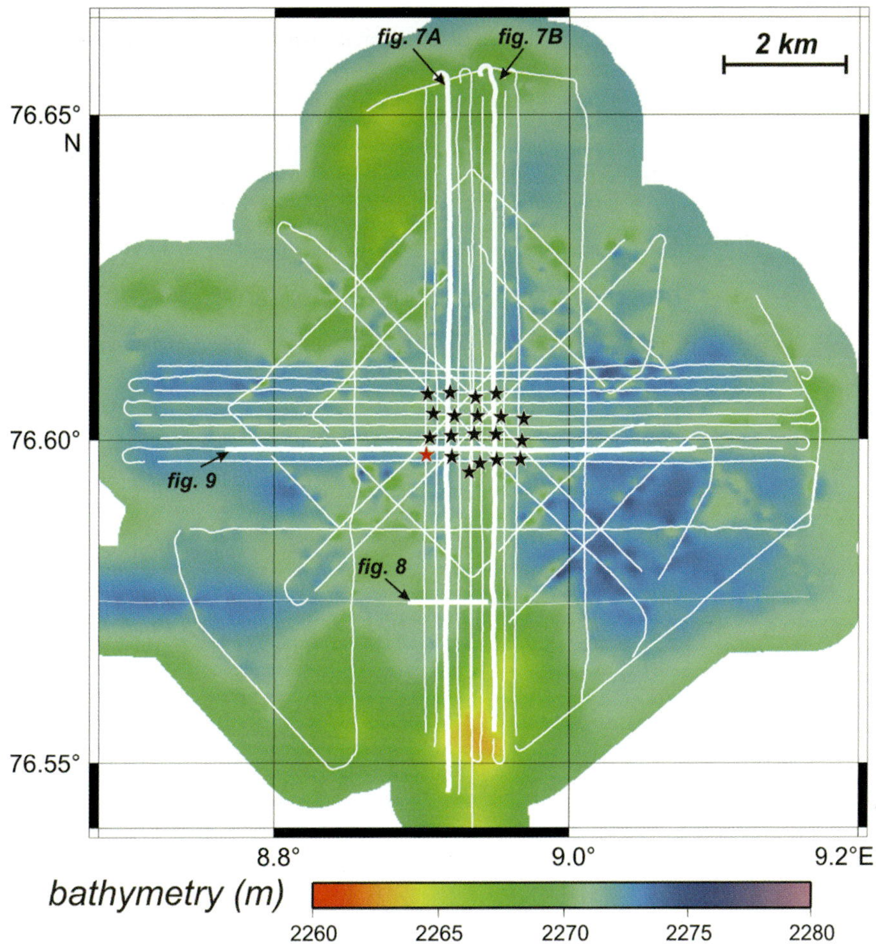

Fig. 6 Bathymetry of the SW Svalbard study area, extracted from single-channel seismic data (white lines) (Area 2, for location see Fig. 1), gridded with GMT software (Wessel and Smith, 1998). The small local sea floor uplift in the south is a result of the mud diapirism observed (see also Figs. 7 and 8). The bold white lines are shown in Figs. 7–9. OBS locations are shown as stars. The red star represents OBS station 716 discussed in this chapter (Fig. 9).

Preliminary results from OBS data

Preliminary results from semblance P-wave velocity analyses on the hydrophone component of one OBS station (716) for both, positive and negative offsets, are shown in fig. 9, together with the coinciding single-channel seismic profile Hy11s. With an exception of the uppermost ~100 ms, the P-wave interval velocities are high, compared to the Hamilton background velocity. In situ velocities also exceed those from the northern study area (see Fig. 5 for comparison). We find an interstitial layer of high interval velocities up to 1880 $m\,s^{-1}$ at shallow depths below the sea floor (100–125 m or 128–158 ms). The higher interval velocities may be related to either overconsolidation of the sediments or alternatively to partial hydrate saturation, since it falls within the zone of theoretical hydrate stability (see below). Beneath this layer, the P-wave velocity falls back to commonly expected values according to the Hamilton reference profile. The depth of the pronounced velocity inversion lies slightly shallower (~30 m) than the short cross-cutting BSR features (145–170 mbsf) observed at the southern end of the perpendicular lines (see Fig. 7). We also stress that this OBS instrument lies just off the bright spots (Fig. 9). Therefore, the lack of a velocity inversion at the depths of these negative-polarity reflection anomalies observed on the neighbouring seismic profiles, cannot be used as a criterion to exclude the presence of gas in the study area. Analyses of the full wavefield recorded in one of the other OBS stations on top of such a bright spot may resolve the ambiguity regarding whether or not hydrates and free gas are associated with these stratigraphically-controlled bright spots.

Discussion: hydrates and mud diapirism at the SW Svalbard site?

Whether the short BSR-like features in the southern part of the study area (Fig. 7) are related

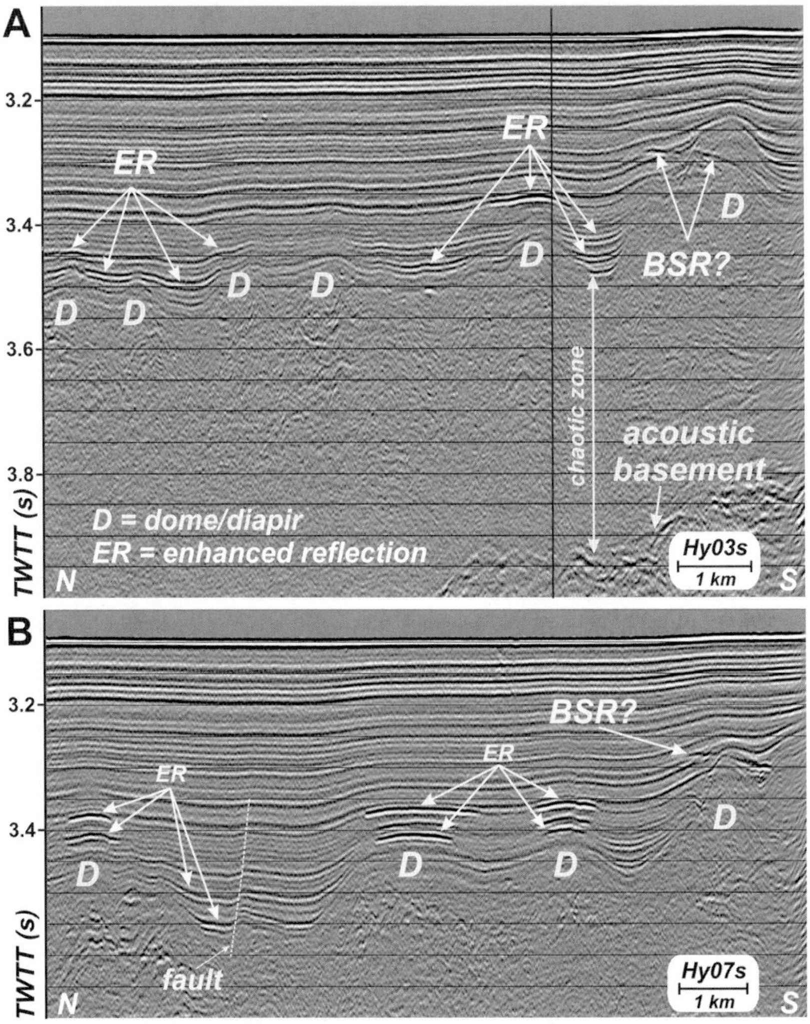

Fig. 7 Mud diapirs and bright spots or enhanced reflection on parallel seismic profiles from the SW Svalbard margin. Line Hy03s (A) lies 900 m to the W of line Hy07s (B). The black line marks the intersection with profile Hyconf1s shown in Fig. 8.

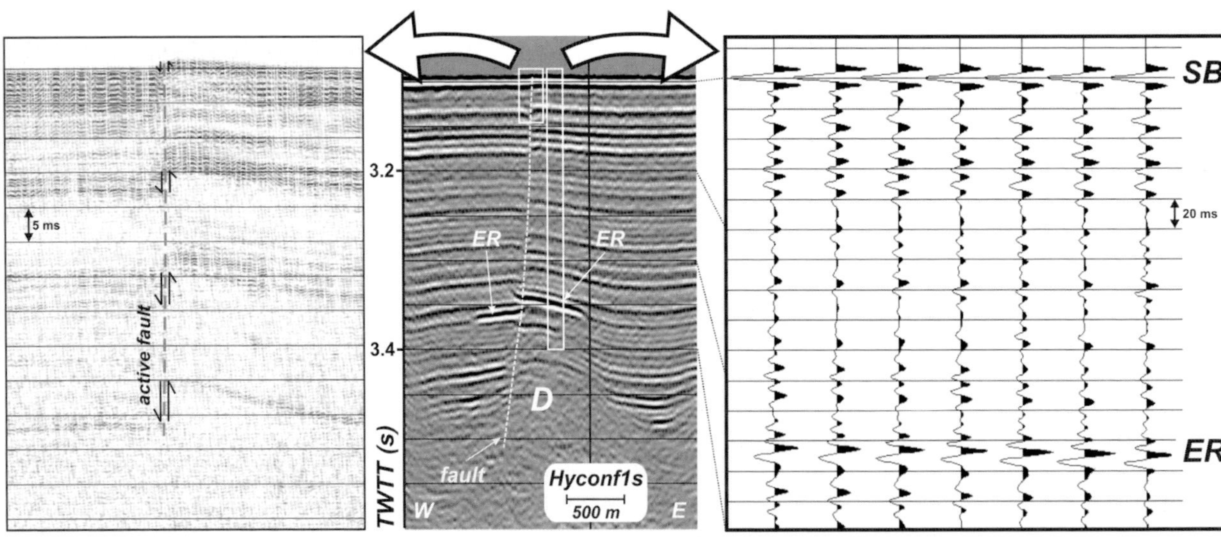

Fig. 8 Profile Hyconf1s shows a nice example of a bright spot or enhanced reflection (labelled 'ER') on top of a diapiric structure (labelled 'D'). The enhanced reflection has reversed reflection polarity relative to the reflection of the seabed (labelled 'SB') (right). Simultaneously recorded 3.5 kHz echo-sounder data also reveal that the normal fault cutting the bright spot is active, slightly offsetting the seafloor (left). The black line is the intersection of profile Hyconf1s with profile Hy03s (for location, see Fig. 6).

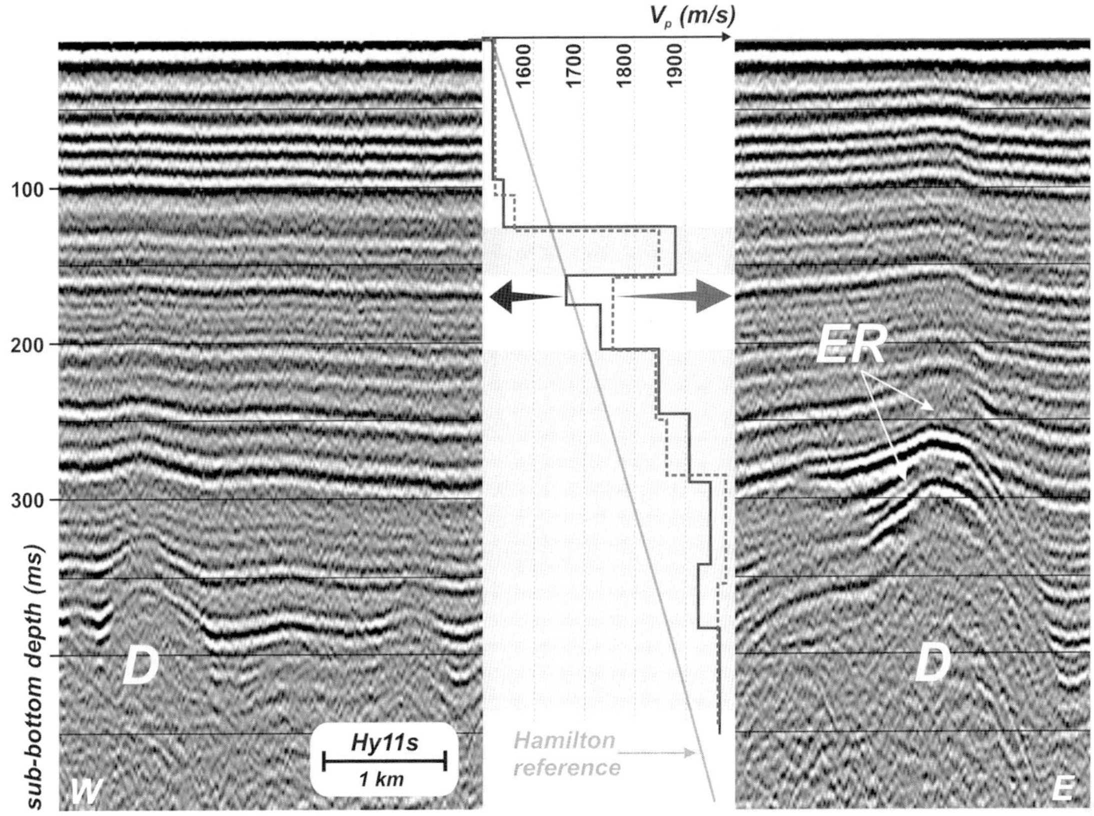

Fig. 9 Continuous profile Hy11s split at the location of OBS 716, where we show the sub-surface interval P-wave velocities from semblance velocity analysis on the hydrophone component, southern study area (for location, see Fig. 6). The grey shading represents zones with high interval velocity relative to the Hamilton reference (Hamilton, 1980). Note the velocity inversion at ~160 ms sub-bottom depth. (ER = enhanced reflection, D = diapir.)

to a gas hydrate/free gas transition is questionable. One way of eliminating the doubt would be to estimate the limit of hydrate stability in the area. Unfortunately, the geothermal data from this area compiled in the HEAT database are sparse (Planke, 1989). The heat flow east of the Knipovich Ridge falls between the 100 and 150 mW m^{-2} contours with an average thermal conductivity value of 1.1 W m^{-1} K^{-1} (Sundvor et al., 2000). The closest measurement of 118 mW m^{-2} is located approximately 35 km, off target. Combining these values with the measured CTD bottom-water temperature of −0.86°C and water-depths of ~2250 m, then methane hydrate could be stable in the uppermost 150–225 m of sediments. Thus, the short BSR falls within these theoretical limits. Assuming that this is truly a BSR, a heat flow of 130–155 mW m^{-2} would be inferred. Additionally, the reversed polarity may point towards the presence of free gas beneath this reflection. Unfortunately, since no OBS instrument was deployed directly on top of these BSR-like features, we cannot derive the internal velocity structure across this typical reflection, and thus the exact nature of these reflections remains ambiguous.

The limited extent of the seismic data presented here does not allow a full description and explanation of the observed diapirs. Mud diapirism is defined as the spontaneous rise of a muddy unit, driven by density differences (Hovland, 1990; Brown, 1990; Sumner and Westbrook, 2001). It forms a major component in subsurface mass movement (Kopf, 2002), and occurs here in an area where small amounts of hydrate cannot be excluded. Amongst the probable causes of mud diapirism in our study area are (1) high sedimentation rates, (2) the presence of permeability barriers that may help to generate an overpressured zone, (3) gas generation, (4) tectonic stresses, or (5) a combination of any of these factors. We do not have on-site sedimentation rates, but estimates based on results from deep drilling suggest that the domes pierce through relatively young sediments of about 2.5–4.6 Ma (DSDP-344, for location, see Fig. 1) to 0.7–1.5 Ma (ODP-986, for location, see Fig. 1). From ODP-986, it appears that clay, typically having low permeability, becomes the dominant size-fraction in the sediments deposited over the last 1.6 Ma, i.e. the top hundreds of m (Butt et al., 2000). Also the minimum conditions for in situ generation of biogenic methane are fulfilled in ODP-986 (~1% TOC) (Butt et al., 2000). Hence, the presence of gas within the chaotic muddy zone and its expansion during upward mobilisation is a possibility, as suggested by Hovland (1990) for diapirs on the mid-Norwegian margin. The onset of diapirism may in turn be facilitated by tectonic activity or instability of the nearby Knipovich Ridge segments (Fig. 1), with associated heat and fluid pulses affecting a wider area.

Most probably, this mud diapirism is a recent and still ongoing process, most pronounced in the southern part of this study area, where it slightly uplifts the seafloor (Fig. 6, Fig. 8). Diapirism and subsequent sediment deformation is believed to be responsible for changes in the subsurface fluid flow regime. Knowing that methane is frequently observed in association with mud volcanism and diapirism (Hovland and Judd, 1988; Milkov, 2000), we believe that the bright spots may also have originated from gas migration and accumulation subsequent to doming. In these cases, the gas is trapped under a stratigraphic seal in a local high, just on top of the diapirs (Fig. 7). Continued diapirism might also result in sedimentary faulting (e.g. Fig. 8) and differential compaction of the overlying sediment units. With time, this will lead to the formation of typical fluid or mass expulsion features at the seabed (pockmarks, mud domes, fluid vents, mud volcanoes, etc.). Additionally, such a complex scheme of stratigraphically-controlled and fault-controlled fluid migration may result in small hydrate accumulations upon entering the hydrate stability zone.

Conclusions

Our seismic data clearly illustrate the widespread nature of gas hydrates and free gas accumulations north of the Knipovich Ridge, off Western Svalbard, by the presence of a nearly-continuous polarity-reversed BSR on down-slope seismic profiles. At locations where distinct and continuous BSRs are not observed, a sudden change in reflection amplitude and frequency content defines the base of the hydrate zone and coincides with the top of the free gas zone. Velocity analyses reveal (1) high P-wave velocities above the BSR attributed to a gradual increase of partial hydrate saturation (6–10% of pore volume) and (2) a sharp, significant drop of acoustic velocity across the BSR due to free gas accumulation. The sub-bottom depth of the BSR closely matches the calculated stability limit for methane hydrates. The deep-water methane hydrate zone lies in an area characterised by mid-ocean ridge escarpments related to the northwards propagation of the Knipovich Ridge in its early stage. Tectonic activity related to incipient rifting and faulting may eventually result in changes to the heat and fluid flow regimes, gas composition and origin, and hydrate accumulation.

Mud diapirism occurs east of the Knipovich Ridge, rising from an extensive chaotic seismic source zone, buried under a 200–400 ms thick sediment drape. Several negative-polarity bright spots are present on top of the domes within the strata and are interpreted to be caused by trapped gas, resulting from sediment mobilisation and subsequent changes in fluid flow patterns. Short reflections having inversed polarity and obliquely crossing the strata might indicate the local formation and accumulation of gas hydrates. The origin of diapirism is unclear, but is most probably caused by a combination of overpressured gas, continuous loading of clay-rich sediments, and neotectonic activity of the Knipovich Ridge.

Acknowledgements

Special thanks are addressed to Stefan Bünz and Steinar Iversen for their valuable assistance and support during the geophysical data acquisition. We also thank the captain, crew, Science Party of the R/V Jan Mayen 2001 expedition. We are grateful to the Editors and Reviewers, W.P. Dillon, M. Hovland, W. Winters and E. D. Sloan for their comments and suggestions. We acknowledge the support of by the EU 5th framework project, HYDRATECH (EVK3-CT-2000–00043) and the German BMBF-project INGGAS. The University of Tromsø acknowledges the use of Landmark Graphics via the Landmark University Grant Program. The HEAT database was kindly made available to us by the University of Oslo, Norway. We also express our gratitude to Björn Lindberg for improving the text.

References

Anderson, A.L. and Hampton, L.D., 1980. Acoustics of gas-bearing sediments I. Background. J. Acoust. Soc. Am., 67 (6): 1865–1889.

Andreassen, K., Hogstad, K. and Berteussen, K.A., 1990. Gas hydrates in the southern Barents Sea, indicated by a shallow seismic anomaly. First Break, 8 (6): 235–245.

Bangs, N.L. and Brown, K.M., 1995. Regional heat flow in the vicinity of the Chile triple junction constrained by the depth of the bottom simulating reflection. In: S.D. Lewis, J.H. Behrmann, R.J. Musgrave and S.C. Cande (Editors), Proceedings of the Ocean Drilling Program, Scientific Results. Ocean Drilling Program, College Station, Texas, pp. 253–258.

Bauch, H.A., Erlenkeuser, H., Spielhagen, R.F., Struck, U., Matthiesen, J., Thiede, J. and Heinemeier, J., 2001. A multiproxy reconstruction of the evolution of deep and surface waters in the subarctic Nordic seas over the last 30,000 yr. Quaternary Sci. Rev., 20: 659–678.

Bougault, H., Charlou, J.-L., Fouquet, Y., Needham, H.D., Vaslet, N., Appriou, P., Baptiste, P.J., Rona, P.A., Dmitriev, L., Silantiev, S., 1993. Fast and Slow Spreading Ridges: Structure and Hydrothermal Activity, Ultramafic Topographic Highs, and CH4 Output. J. Geophys. Res., 98 (B6): 9643–9651.

Bouriak, S., Vanneste, M. and Saoutkine, A., 2000. Inferred gas hydrates and clay diapirs near the Storegga Slide on the southern edge of the Vøring Plateau, offshore Norway. Mar. Geol., 163: 125–148.

Brown, K.M., 1990. The nature and hydrogeologic significance of mud diapirs and diatremes for accretionary systems. J. Geophys. Res., 95 (B6): 8969–8982.

Bünz, S., Mienert, J. and Berndt, C., 2003. Geological controls on the Storegga gas-hydrate system of the mid-Norwegian margin. Earth Planet. Sci. Lett., 209 (3–4): 291–307.

Butt, F.A., Elverhøi, A., Solheim, A. and Forsberg, C.F., 2000. Deciphering Late Cenozoic development of the Western Svalbard Margin from ODP site 986 results. Mar. Geol., 169: 373–390.

Crane, K., Doss, H., Vogt, P.R., Sundvor, E., Cherkashov, G., Poroshina, I. and Devorah, J., 2001. The role of the Spitsbergen shear zone in determining morphology, segmentation and evolution of the Knipovich Ridge. Mar. Geophys. Res., 22: 153–205.

Crane, K., Sundvor, E., Foucher, J.-P., Hobart, M., Myhre, A.M., LeDouaran, S., 1988. Thermal evolution of the Western Svalbard Margin. Mar. Geophys. Res., 9: 165–194.

Eiken, O., 1994. Seismic Atlas of Western Svalbard. Norsk Polarinstitutt, Oslo, Meddelelser 130.

Eiken, O. and Hinz, K., 1993. Contourites in the Fram Strait. Sediment. Geol., 82: 15–32.

Faleide, J.I., Solheim, A., Fiedler, A., Hjelstuen, B.O., Andersen, E.S., Vanneste, K., 1996. Late Cenozoic evolution of the western Barents Sea — Svalbard continental margin. Global Planet. Change, 12 (1–4): 53–74.

Ginsburg, G.D., Milkov, A.V., Soloviev, V.A., Egorov, A.V., Cherkashev, G.A., Vogt, P.R., Crane, K., Lorenson, T.D. and Khutorskoy, M.D., 1999. Gas hydrate accumulation at the Håkon Mosby Mud Volcano. Geo-Mar. Lett., 19: 57–67.

Hamilton, E.L., 1980. Geo-acoustic modelling of the seafloor. J. Acoust. Soc. Am., 68 (5): 1313–1340.

Henriet, J.-P. and Mienert, J. (Editors), 1998. Gas Hydrates: Relevance to World Margin Stability and Climatic Change. Geol. Soc., London, Spec. Publ., 137: 338 pp.

Holbrook, S.W., Hoskins, H., Wood, W.T., Stephen, R.A., Lizzarralde, D. and ODP Leg 164 Scientific Party, 1996. Methane gas-hydrate and free gas on the Blake Ridge from vertical seismic profiling. Science, 273: 1840–1843.

Hovland, M., 1990. Suspected gas-associated clay diapirism on the seabed off Mid-Norway. Mar. Petrol. Geol., 7: 267–276.

Hovland, M. and Judd, A.G., 1988. Seabed pockmarks and seepages: impact on geology, biology and the marine environment. Graham and Trotman Ltd., London, 293 pp.

Jakobsson, M., Cherkis, N., Woodward, J., Coakley, B.J. and Macnab, R., 2000. A new grid of Arctic Bathymetry: a significant resource for scientists and mapmakers. EOS, Trans. Am. Geophys. U., 81 (9): 89, 93, 96.

Kopf, A.J., 2002. Significance of mud volcanism. Geophys. Rew., 40 (2): 1005 DOI:10.1029/2000RG000093.

Kvenvolden, K.A., 1998. A primer on the geological occurrence of gas hydrate. In: J.-P. Henriet and J. Mienert (Editors), Gas Hydrates: Relevance to World Margin Stability and Climatic Changes. Geological Society, London, Special Publication, 137, pp. 9–30.

Kvenvolden, K.A. and Barnard, L.A., 1983. Hydrates of natural gas in continental margins. In: J.S. Watkins and C.L. Drake (Editors), Studies in Continental Margin Geology. AAPG Mem., 34: 631–640.

Laberg, J.S. and Andreassen, K., 1996. Gas hydrate and free gas indications within the Cenozoic succession of the Bjørnøya Basin, western Barents Sea. Mar. Petrol. Geol., 13 (8): 921–940.

Lee, M.W., Hutchinson, D.R., Agena, W.F., Dillon, W.P., Miller, J.J. and Swift, B.A., 1994. Seismic character of gas hydrates on the Southeastern U.S. Continental Margin. Mar. Geophys. Res., 16: 163–184.

Lindberg, B., Berndt, C. and Mienert, J., in press. The Fugløy Reefs on the Norwegian-Barents Continental Margin: Cold-water corals at 70°N, their acoustic signature, geologic, geomorphologic and oceanographic setting. In: J.-P. Henriet and C. Dullo (Editors), Modern Carbonate Mound Systems — A window to Earth History. International Journal of Earth Sciences (Special Issue).

Lundin, E. and Doré, A.G., 2002. Mid-Cenozoic post-breakup deformation in the 'passive' margins bordering the Norwegian-Greenland Sea. Mar. Petrol. Geol., 19: 79–93.

Mienert, J., Bünz, S., De Roeck, Y.-H., Foucher, J.-P., Guidard, S., Harmegnies, F., Landure, J.-Y., Leythaeuser, T., Nouzé, H., Rossi, G., Schwenk, A., Vanneste, M. and Westbrook, G.K., 2001a. Western Svalbard Continental Margin — Gas Hydrates. R/V Jan Mayen 2001 HYDRATECH Cruise Report, Department of Geology, University of Tromsø, Norway, 124 pp.

Mienert, J., Posewang, J. and Baumann, M., 1998. Gas hydrates along the north-eastern Atlantic Margin: possible hydrate bound margin instabilities and possible release of methane. In: J.-P. Henriet and J. Mienert (Editors), Gas Hydrates: Relevance to World Margin Stability and Climatic Change. Geol. Soc., London, Spec. Publ., 137: 275–291.

Mienert, J., Posewang, J. and Lukas, D., 2001b. Changes in the Hydrate Stability Zone on the Norwegian Margin and their Consequences for Methane and Carbon Releases into the Oceanosphere. In: P. Schlaefer, W. Ritzrau, W. Schlueter and J. Thiede (Editors), The Northern North Atlantic: A Changing Environment. Springer, Berlin, pp. 259–280.

Mienert, J. and Weaver, P. 2003 (Editors). European Margin Sediment Dynamics: Side-Scan Sonar and Seismic images. Springer-Verlag, Berlin, Heidelberg, New York, 310 pp.

Milkov, A., 2000. Worldwide distribution of submarine mud volcanoes and associated gas hydrates. Mar. Geol., 167: 29–42.

Myhre, A.M. and Eldholm, O., 1988. The western Svalbard margin (74–80°N). Mar. Petrol. Geol., 5: 134–156.

Myhre, A.M. and Thiede, J., 1995. North-Atlantic-Arctic Gateways. In: A.M. Myhre, J. Thiede and J.V. Firth (Editors), Proceedings of the Ocean Drilling Program, Initial Reports. Ocean Drilling Program, College Station, pp. 5–26.

Pecher, I.A., Minshull, T.A., Singh, S. and Huene, R.v., 1996. Velocity structure of a bottom simulating reflector offshore Peru: Results from full waveform inversion. Earth Planet. Sci. Lett., 139: 459–469.

Planke, S., 1989. HEAT — a heat flow data base program. Marine and Applied Geophysics Research Group, Department of Geology, University of Oslo, Oslo.

Posewang, J. and Mienert, J., 1999. High-resolution seismic studies of gas hydrates west of Svalbard. Geo-Mar. Lett., 19: 150–156.

Sheriff, R.E. and Geldart, L.P., 1995. Exploration Seismology. Cambridge University Press, Cambridge, 592 pp.

Shipley, T.H., Houston, M.K., Fuffler, R.T., Shaub, F.J., McMillan, K.J., Ladd, J.W., Worzel, J.L., 1979. Seismic reflection evidence for the widespread occurrence of possible gas hydrate horizons on continental slopes and rises. AAPG, Bull., 63: 2201–2213.

Sloan, E.D. Jr., 1998. Clathrate Hydrates of Natural Gases. Marcel Dekker Inc., New York and Basel, 705 pp.

Sumner, R.H. and Westbrook, G.K., 2001. Mud diapirism in front of the Barbados accretionary wedge: the influence of fracture zones and North America–South America plate motions. Mar. Petrol. Geol., 18: 591–613.

Sundvor, E., Eldholm, O., Gladczenko, T.P. and Planke, S., 2000. Norwegian-Greenland Sea thermal field. In: A. Nøttvedt (Editor), Dynamics of the Norwegian Margin. Geol. Soc., London, Spec. Publ.: 397–410.

Tinivella, U., 1999. A method for estimating gas hydrate and free gas concentrations in marine sediments. Bolletino di Geofisico Teoretico et Applicata, 40 (1): 19–30.

Vanneste, M., De Batist, M., Golmshtok, A.Y., Kremlev, A. and Versteeg, W., 2001. Multi-frequency seismic study of gas hydrate-bearing sediments in Lake Baikal, Siberia. Mar. Geol., 172: 1–21.

Vanneste, M., Poort, J., De Batist, M. and Klerkx, J., 2002. Atypical heat flow near gas hydrate irregularities and cold seeps in the Baikal Rift Zone. Mar. Petrol. Geol., 19 (10): 1257–1274.

Vogt, P.R., Crane, K., Sundvor, E., Max, M.D. and Pfirman, S.L., 1994. Methane-generated (?) pockmarks on young, thickly sedimented oceanic crust in the Arctic: Vestnesa Ridge, Fram Strait. Geology, 22: 255–258.

Vogt, P.R., Gardner, J. and Crane, K., 1999a. The Norwegian-Barents-Svalbard (NBS) continental margin: Introducing a natural laboratory of mass wasting, hydrates, and ascent of sediment, pore water and methane. Geo-Mar. Lett., 19: 2–21.

Vogt, P.R., Gardner, J., Crane, K., Sundvor, E., Bowles, F. and Cherkashev, G., 1999b. Ground-truthing 11- to 12-kHz side-scan sonar imagery in the Norwegian-Greenland Sea: Part I: Pockmarks on the Vestnesa Ridge and Storegga Slide margin. Geo-Mar. Lett., 19: 97–110.

Vorren, T.O., Laberg, J.S., Blaume, F., Dowdeswell, J.A., Kenyon, N.H., Mienert, J., Rumohr, J. and Werner, F., 1998. The Norwegian-Greenland Sea continental margins: morphology and Late Quaternary sedimentary processes and environment. Quaternary Sci. Rev., 17 (1–3): 273–302.

Wessel, P. and Smith, W.H.F., 1998. Improved version of the Generic Mapping Tools released. Eos Trans. Am. Geophys. U., 79: 579.

Wood, W.T., Gettrust, J.F., Chapman, N.R., Spence, G.D. and Hyndman, R.D., 2002. Decreased stability of methane hydrates in marine sediments owing to phase-boundary roughness. Nature, 420: 656–660.

Yamano, M., Uyeda, S., Aoki, Y. and Shipley, T.H., 1982. Estimates of heat flow derived from gas hydrates. Geology, 10: 339–342.

Gas hydrate dissociation and sea-floor collapse in the wake of the Storegga Slide, Norway

Christian Berndt, Jürgen Mienert, Maarten Vanneste and Stefan Bünz

Two-dimensional seismic data from the Mid-Norwegian margin provide evidence for sediment liquefaction and fluid mobilisation within the sediments that were located at the base of the hydrate stability zone before the Storegga Slide occurred. The disturbed subsurface sediments are overlain by a prominent roll-over structure and sea-floor collapse. This indicates fluid escape from the formerly hydrated sediment and suggests that the landslide caused a pressure drop strong enough to dissociate the gas hydrates. We calculate that this fluid escape must have taken place within less than 250 years after the slide, as the effect of pressure decrease on hydrate stability was later compensated by a temperature decrease, related to the slumping process. The volume of expelled fluids from the collapse structure exceeds the volume of the gas hydrate dissociation products, implying that gas hydrate dissociation significantly affected the surrounding sediments.

Introduction

Natural gas hydrates are clathrates of light hydrocarbons, such as the greenhouse gas methane, which are captured in water-ice crystals. They occur under pressure/temperature conditions frequently encountered in ocean sediments at water depths greater than 500 m. Kvenvolden (1993) estimated that gas hydrates bound more than half of the Earth's carbon that could potentially influence climate. Therefore, it is necessary to assess the mobility of this reservoir. So far, evidence for natural gas hydrate dissociation is sparse and most of the reported examples are from settings in which gas hydrates dissociate slowly, as, for example, in areas of rapid sedimentation (Dillon et al., 1998; Milkov, 2000), tectonic uplift (von Huene and Pecher, 1999) or ocean warming (Mienert et al., 2001). Here, we present geophysical evidence from the hydrated sediments of the Norwegian Margin (Fig. 1) that gas hydrates have decomposed and released fluids adjacent to the side wall rapidly after the Storegga Slide event. Fast fluid escape is a prerequisite for rapid impact of gas hydrate dissociation on climate, and may support the 'clathrate gun hypothesis' (Kennett et al., 2003).

Geological background

The Mid-Norwegian margin is a passive continental margin that developed during the continental break-up between Fennoscandia and Laurentia, 54 Ma ago (Saunders et al., 1997). The two top-most sedimentary formations relevant for this study include the Miocene/earliest Pliocene Kai Formation and the Plio-/Pleistocene Naust Formation (Dalland et al., 1988; Rokoengen et al., 1995). The Kai Formation is generally characterised by fine-grained hemipelagic oozes. The overlying Naust formation is characterised by pronounced changes in lithology. It encompasses sediments of the Plio-/Pleistocene glacial–interglacial cycles consisting of debris-flow deposits and hemipelagic sediments, respectively. The main depocentres of the glacigenic deposits are located along the shelf break in front of glacial trough-mouth fans (Dahlgren et al., 2002; Henriksen and Vorren, 1996; Vorren et al., 1998). Sediments deposited at the shelf break were remobilised and transported mostly downslope, as the debris flows. Contour currents sustained high sedimentation rates in the basin during deglaciation and interglacials (Laberg et al., 2001). The resulting hemipelagic contourite drift deposits

Onshore–Offshore Relationships on the North Atlantic Margin
edited by B. Wandas et al. NPF Special Publication 12, pp. 285–292,

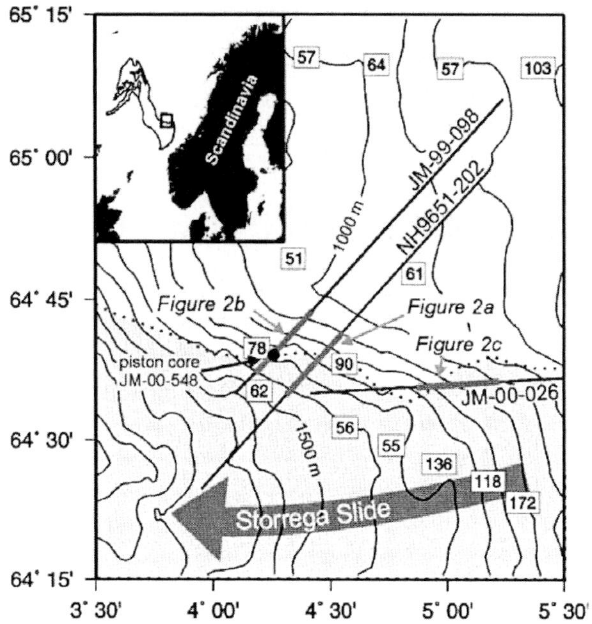

Fig. 1 Study area at the northern rim of the Storegga Slide. Gray lines indicate seismic profiles shown in Fig. 2. Numbers in boxes indicate heat flow measurements in mW/m^2 after Sundvor et al., 2000. Dotted line, boundary of the Storegga Slide; VB, Vøring Basin, MB, Møre Basin.

frequently interlayer the glacigenic debris flow deposits (Bryn et al., 2003).

One of the largest known submarine slides on the continental margins, the Storegga Slide (Fig. 1), cuts deep into the sediments of the Naust Formation of the Møre Basin (Bugge et al., 1987). This submarine slide had a maximum run-out from the shelf break to the abyssal plain of 800 km. The eastern headwall reaches up to 300 m in height and extends for about 300 km from north to south, along the shelf break. The northern sidewall is up to 100 m high and runs roughly in an east-west direction along the border between the Vøring and the Møre Basins (Fig. 1). The submarine landslide transported approximately 3400 km^3 of sediments (Bryn et al., 2003), and occurred in up to nine phases, which took place within a very short period of time at ~8.2 calendar ka B.P. (Bryn et al., 2003). The triggering mechanism is possibly a combination of various effects, e.g. earthquakes and gas hydrate dissociation. (Bugge et al., 1987) and (Mienert et al., 2002) suggested that gas hydrate dissociation contributed to slope instability as there is a bottom-simulating reflector (BSR) in an area of ~4000 km^2 in water depth of 550–1300 m (Bünz et al., 2003). The BSR is shallowest with 860 ms two-way travel time (twt) beneath sea surface (180 ms twt sub-bottom depth) in the vicinity of the headwall, and deepest with 2150 ms twt (385 ms twt sub-bottom depth) at the northern sidewall. On the continental slope, the lithology of glacigenic debris flow deposits and pre-glacial basin deposits of the Kai Formation prevent gas hydrate formation, because of reduced pore size, reduced water content and fine-grained sediment composition (Bünz et al., 2003). Towards the continental shelf, the shoaling and the pinch out of the gas hydrate stability zone terminates the area of gas hydrate growth.

Observations

In this study, we use two seismic reflection data sets. The 96-channel data (Fig. 2a) are provided by Norsk Hydro ASA, whereas we acquired the single-channel data (Fig. 2b and c) in 1999 and 2000, during the R/V Jan Mayen cruises. Both the data sets image the top 500 m of the sediment with high resolution, as the sleeve gun sources generated seismic signals with frequency bands from 10 to 130 and from 50 to 250 Hz, respectively. Both surveys have been processed including Stolt migration.

The high-resolution seismic reflection data show a distinct roll-over structure that is most pronounced towards a normal fault that coincides with the 110 m-high side wall of the submarine Storegga Slide (Fig. 2a and b). The roll-over involves the upper 150 ms twt of the seismic section, whereas the underlying strata are undisturbed. The throw of the fault is of the order of 60 ms twt or approximately 50 m. The seismic character of the strata directly above the Top Kai reflector (Fig. 2) at approximately 2.15 s twt is chaotic, and this chaotic seismic facies continues between 3 and 10 km downslope of the slide scarp. Beyond this zone the seismic reflectors are undisturbed. The depth of the base of this chaotic facies and the top of the undisturbed reflections coincide with the depth of a BSR observed outside the slide area, which has been attributed to gas accumulation beneath gas hydrates (Mienert et al., 1998). Whereas the chaotic facies coincides with the laterally projected BSR on some lines (Fig. 2a), it is located above the BSR on others (Fig. 2b). However, if it is above the projected BSR, the top of the Kai Formation bounds it at its base. Locally, a BSR is visible within the slide area (Fig. 2c; Bouriak et al., 2000).

The area of the collapse structure varies along the slide side wall between 7475 and 183000 m^2 (Fig. 3). The area of collapse along individual transects is given in Table 1. However, the values for lines SG9801-305, SG9801-114 and JM00-010 must be considered as not representative,

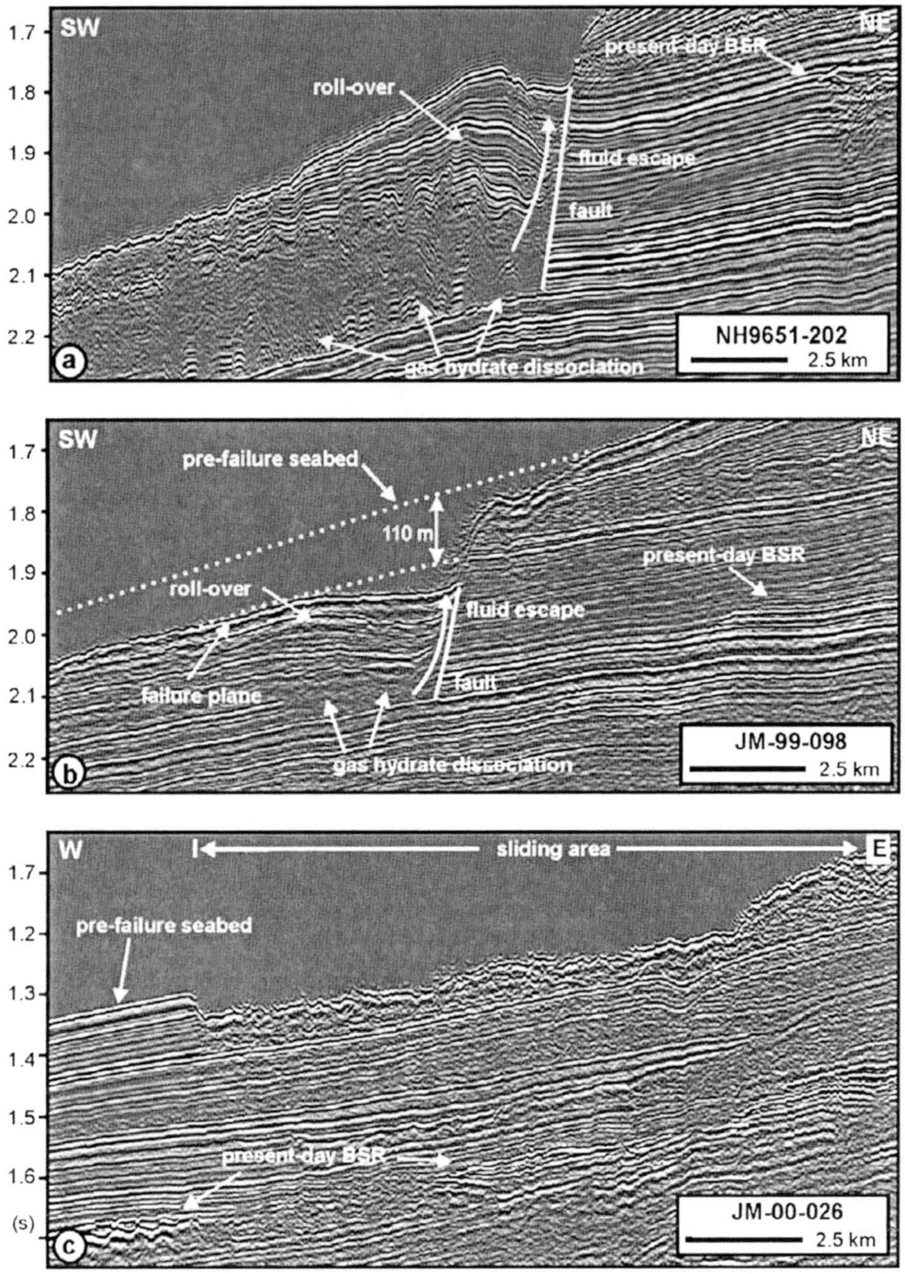

Fig. 2 Migrated high-resolution seismic data. (a) multi-channel data. (b and c) single-channel data. Note, that the bottom simulating reflector (BSR) within the slide area (c) mimicks the sea floor indicating that it has adapted to the new pressure/temperature conditions.

because the lines strike parallel to the collapse feature. Therefore the typical area of collapse in any given seismic cross-section across the collapse structure is between 7 475 and 56 818 m^2.

Discussion

Evidence for subsurface mass movement

The fact that the strata underlying the roll-over structure remain undisturbed (Fig. 2a and b) implies that mass has been transported away from the base of the roll-over structure. Subsurface transport of mass in a direction perpendicular to the seismic lines must have been minor, because all parallel lines show a lack of material at the base of the fault. Therefore, we conclude that most of the missing mass must have escaped to the surface. The most likely conduit for this transport is along the fault as there is no seismic evidence for other transport mechanisms, such as mud diapirs. This does not exclude, however, that the intensity of fluid flow might have varied along the collapse structure. We interpret the chaotic seismic facies between the base of the roll-over structure and the

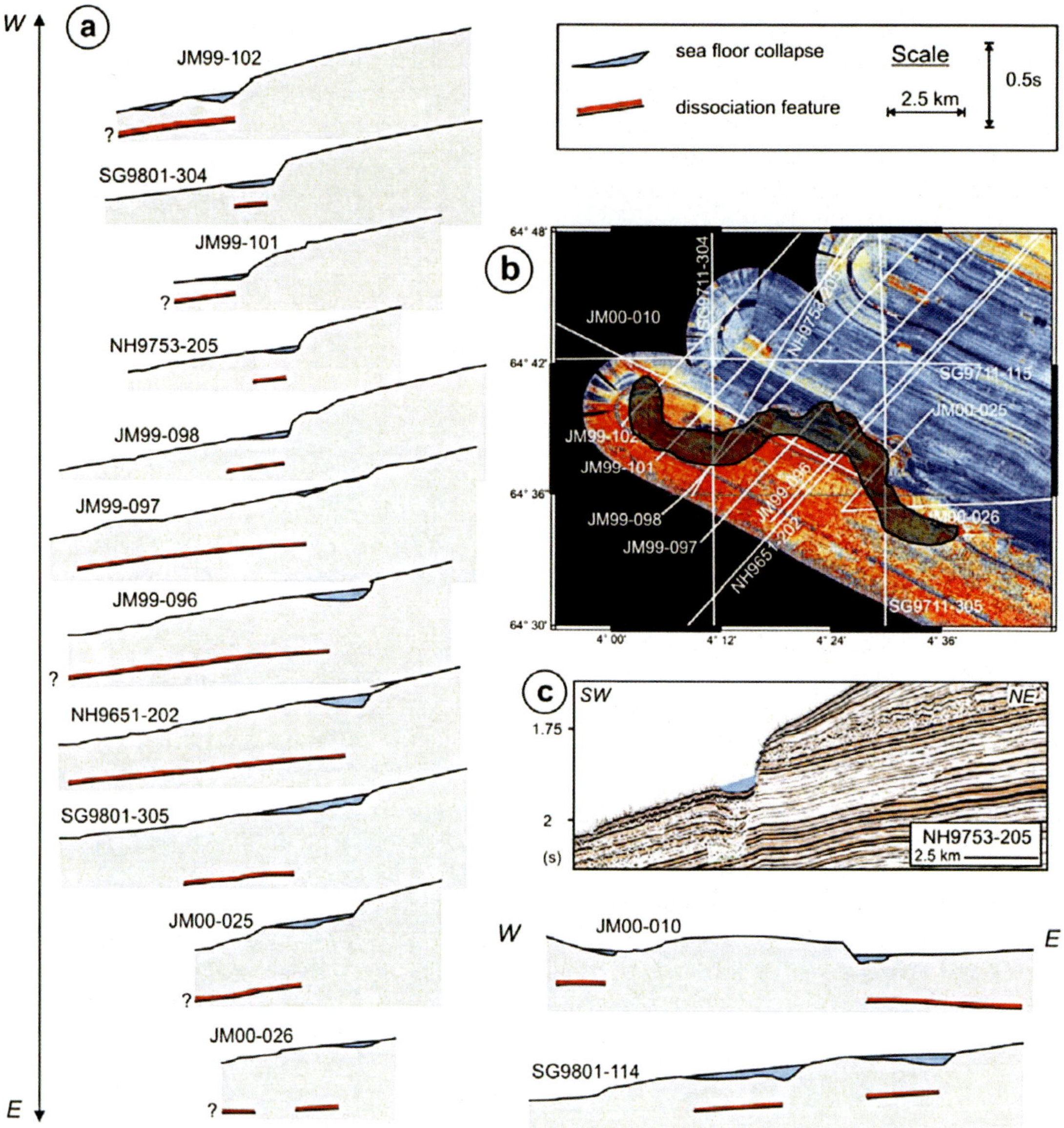

Fig. 3 (a) Geometry and size of sediment disturbance and the associated sea floor collapse structure for all seismic transects. Numerical estimates for the areas, lengths, and computed volumes are given in Table 1. (b) Side-scan sonar imagery of the slide area with sea floor collapse in dark grey, red colors — high backscatter, blue colors — little backscatter. (c) seismic type example.

Table 1 Areas of sea floor collapse and gas hydrate dissociation. Dissociation length, length of observed chaotic zone in the seismic data; Dissociation area, product of dissociation length and 8 m.

Seismic Line	Collapse area (m^2)	Dissociation length (m)	Dissociation area (assuming 8 m) (m^2)	Ratio (%)
JM99-102	35 100 + 40 000 = 75 100	min. 4700	min. 37 600	max. 199
SG9801-304	31 800	1200	9 600	331
JM99-101	7 475	min. 2560	min. 20 480	max. 36
NH9753-205	14 150	1200	9 600	147
JM99-098	19 550	2670	21 360	91
JM99-097	8 200 + 5000 = 13200	7740	61 920	21
JM99-096	43 680	min. 9800	min. 78 400	max. 58
NH9651-202	45 500	min. 6900	min. 55 200	max. 82
SG9801-305	11 4000	560 + 2600 = 3160	25 300	450 (striking the collapse structure)
JM00-025	56 818	min. 5400	min. 43 200	max. 131
JM00-026	21 400	1900 + min. 1900 = min. 3800	min. 30 800	max. 70
JM00-010	20 400 + 11100 = 31 500	2250 + 6750 = 9000	72 000	43
SG9801-114	91 000 + 92000 = 18 3000	2800 + 3600 = 6400	51 200	357 (striking the collapse structure)
Sum	**65 7173**		**516660**	**127**

top of the undisturbed sediments to be the result of liquefaction of the sediments and mobilisation of fluids. It appears that both fault and roll-over structure, developed as the result of sediment and fluid removal from the base of the roll-over structure, as the seismic data show no indications for a pre-existing fault at this location.

Mobilisation due to gas hydrate dissociation

The extent of the disturbed sediments is confined to the vicinity of the fault and our data do not show such features farther inside the slide area. This localised disturbance would not be anticipated, if the disturbance was caused by regional, seismicity-related liquefaction. Therefore, it is more likely that a different mechanism must have caused these structures. The disturbed sediments lie at the projected depth of the pre-slide gas hydrate stability zone. This is evident from their location just above the BSR observed outside the slide area (Fig. 2a). The presence of gas hydrate-bearing sediments is commonly inferred from the presence of this characteristic reflector, which is caused by the impedance contrast between hydrated sediments and free gas that is trapped underneath (Pecher et al., 1996; Mienert et al., 2001). The coincidence of the depth of disturbed sediments and the base of the gas hydrate stability zone strongly suggests that the sediment disturbance is related to gas hydrates. On lines on which the base of the Naust Formation lies above the projected base of the gas hydrate stability zone, the chaotic zone starts not directly above the gas hydrate stability zone, but higher up at the base of the Naust Formation. We attribute this to the fact that gas hydrates only exist within the Naust Formation and not in the Kai Formation (Bünz et al., 2003). This supports the interpretation that gas hydrate dissociation causes the sediment mobilisation.

The lithostatic pressure decrease due to the Storegga Slide event ~8.2 calendar ka ago, is a possible explanation for a sudden shoaling of the gas hydrate stability zone. Increased seismic amplitudes at the base of the BSR (Fig. 2) indicate that gas is present underneath the gas hydrates. This reservoir is most likely under close to critical pressure (Flemings et al., 2003; Hornbach et al., 2004) and it is reasonable to assume that the pre-landslide conditions at the base of the gas hydrate stability zone were similar. The pore pressure decrease due to the landslide depends on the amount of this overpressure. For our argument we assume that the pore-pressure drop approximately equals the difference in lithostatic pressures before and after the landslide, as close-to-critical pore pressures seem to be normal for the free gas zones under gas hydrates (Hornbach et al., 2004). A stepwise decrease in pore pressure might be an oversimplification as the pore pressure might be as low as hydrostatic pressure during the depressurisation event. However, as a result of the pressure decrease, the lowermost hydrated sediments were no longer within the stability zone. Provided that the surrounding sediments supplied enough thermal energy, the hydrates started to dissociate releasing fluids that could propagate upwards through the fault. Subsequently the geothermal field adjusted to the new conditions leveling out the temperature difference between the cold bottom-water and the newly exposed, i.e. warmer, sediments at the sea floor. The sediments progressively cooled downwards until the conductivity-controlled geothermal gradient was re-established. Gas hydrate dissociation must have stopped when the thermal signal had reached the pre-landslide depth of the gas hydrate stability zone and re-stabilized the gas hydrates. This is evidenced by the observation of a re-adjusted BSR within the slide area (Fig. 2c).

Quantification of the thermal evolution

In order to estimate the maximum time of gas hydrate dissociation, we calculate the thermal evolution of the sedimentary column since the sliding event. Because the landslide removed the overburden and, consequently, exposed the warmer, deeper material at the seabed, we assume that an instantaneous sea-floor temperature change has imposed on a homogeneous conductive medium (Vanneste, 2000). This does not take into account the possible heat transport due to fluid flow. However, such heat transport will shorten the time available for dissociation because it will re-establish the thermal equilibrium faster than conductive heat transport alone. For the calculation we also assume lithostatic pressures. This might overestimate the effect of the pressure drop leading to a larger thickness of dissociating hydrates, and hence, our calculations give an upper limit. In case the pressure distribution is purely hydrostatic, removal of part of the overburden would not result in *in situ* pressure drop and hydrate would not dissociate. Given these caveats, the temperature T as a function of time t, and the sub-bottom depth z can be described by the initial conditions and a complementary error function, Erfc (Carslaw and Jaeger, 1959):

$$T(z,t) = T_0 + Gz + \Delta T Erfc\left(\frac{z}{2\sqrt{\kappa t}}\right)$$

with T_0: the pre-landslide temperature at the slide plane, G: the geothermal gradient, ΔT the temperature difference at the new sea floor exposed by the slide, and κ the average thermal diffusivity of the sediments involved. The physical sediment properties used in the calculation are based on the logging results of a piston core taken at the slide's side wall (Fig. 1) and from the seismic constraints (Table 2). The calculations indicate that the landslide initially moved the lowermost 8 m of sediment out of the pre-slide gas hydrate stability zone initiating gas hydrate dissociation within this layer (Fig. 4). After 60 years, the thermal signal reached the pre-landslide depth of the gas hydrate stability zone beginning to re-stabilise the gas hydrates at the top of the 8 m layer and 180 years after the landslide, the new base of the gas hydrate stability zone had reached pre-landslide depth. At that time, gas hydrate dissociation must have terminated. Our model predicts that re-adjustment of the geotherm still continues. Today, the base of the gas hydrate stability zone is ~80 m deeper than the pre-slide base. The error of these calculations depends mainly on the thermal diffusivity, and is on the order of 40%. The errors of the remaining parameters such as the parameters deduced from seismic interpretation and the effects of additional physical processes, such as latent heat generation, are small compared to this. This implies that fluid escape must have happened within a maximum of 180 ± 70 years after the landslide and perhaps even faster.

Table 2 Modelling Parameters.

Property	Value	Reference
Bulk density	1850 kg/m^3	Piston core JM-00-548
Bottom water temperature	0°C	Mienert et al., 1998
Thermal conductivity	1.2 W/mK	Sundvor et al., 2000
Bulk specific heat capacity	800–1300 J/kg K	Buntebarth, 1984
Geothermal gradient Outside the slide area	62 K/km	Hydrate stability conditions
Thickness of removed sediments	110 m	Seismic line JM-99-098

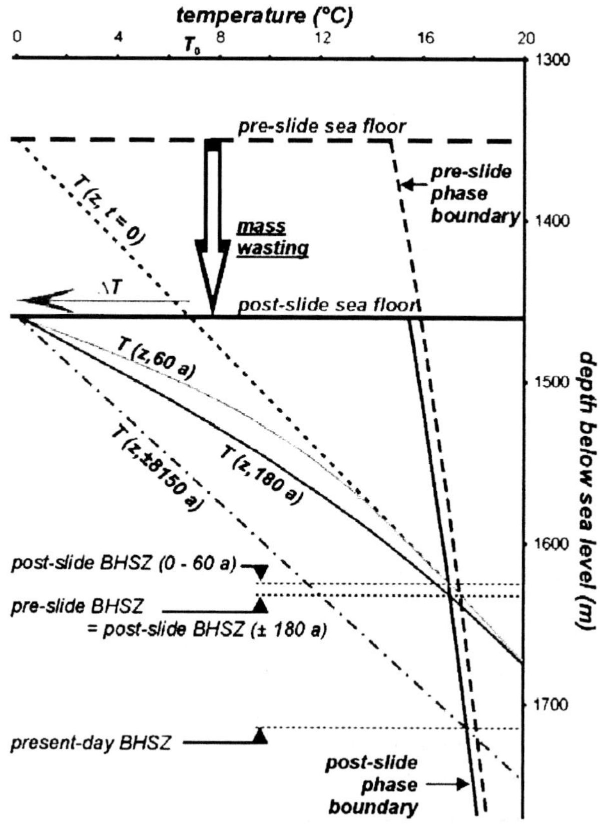

Fig. 4 Thermal evolution at the foot of the Storegga Slide side wall. Note the initial shoaling of the hydrate stability zone immediately after the occurrence of the landslide and subsequent deepening. Hydrate dissociation and fluid escape must occur within 180 +/− 70 years after the sliding event. BHSZ, base of the hydrate stability zone; T, temperature, T0, temperature difference at the sea floor due to the slides.

Implications for fluid expulsion

Because the upper limit of gas hydrate dissociation is not imaged in the seismic data due to disturbance of the sediments, it is necessary to use an indirect approach to quantify the area of dissociation, along individual seismic transects. Therefore, we mapped the length of the zone where chaotic reflectors and dissociation occurred (Fig. 3) and multiplied the length with the theoretically expected height of 8 m, assuming in the absence of a detailed information that this height is representative for the entire collapse structure. The area varies between 9600 and 78 400 m^2 for each transect (Table 1).

Comparing the areas of gas hydrate dissociation and the sea-floor collapse as a proxy of how much mass was mobilised due to gas hydrate dissociation, yields percentages between 21 and 331% (excluding the two strike lines). On an average, the collapse area is slightly bigger (127%) than the sediment layer in which dissociation has occurred (Table 1). This implies that generally fluid volumes were expelled from the sediments that were much larger than the total volume of sediments in which hydrate dissociation has happened on the basis of theoretical considerations, because hydrates constitute only a small fraction of the total sediment volume in this area (Andreassen et al., 2003). As the uncertainty of the theoretical considerations is up to 40%, this implies that fluids, in addition to those released by gas hydrate dissociation, were expelled from the subsurface. There is no evidence for sediment distortion underneath the dissociation feature (Figs. 2 and 3c), but the sediments 50–100 m above the 8 m-thick basal layer of the gas hydrate stability zone are disturbed (Figs. 2 and 3c). Having this in mind, we conclude that the gas hydrate dissociation caused liquefaction of the overlying sediments within the gas hydrate stability zone, and that not only the hydrate dissociation products were expelled but also fluids and sediments from the surrounding sediments contribute to the amount of expelled fluid. As there is no obvious seismic or side-scan sonar evidence of expelled sediments at the present seabed we conclude that most of the transported material was pore water and that only little sediment made its way to the surface.

Conclusions

Sea-floor collapse within the Storegga Slide region and above dissolved gas hydrates shows that the gas hydrate reservoir is highly dynamic. Our calculations show that it can release fluids from the base of the hydrate stability zone to the surface within decades. This makes the release of methane due to submarine landslides a process that potentially can influence climate. However, ice core data from Greenland show that it takes very large methane input, i.e. of the order of 4000 Tg, to increase the global mean temperatures by 0.3 to 1 K (Thorpe et al., 1998), and it is not clear whether submarine slope failures, such as the Storegga Slide events did release enough methane from the gas hydrate reservoir to the atmosphere to influence climate (Raynaud et al., 1998).

The Storegga Slide head wall extends for 300 km (Bugge et al., 1987) and has developed simultaneously (Haflidason et al., 2001). So far, it is not clear what caused the slope failure at such large regional scale. The observation of induced sea floor collapse in the wake of landslides suggests a possible propagation mechanism: an initial local slope failure, possibly induced by an earthquake, may not only have started the gas hydrate dissociation and fluid expulsion under the initial slide area but also in the adjacent parts of the slope, because the pressure drop will also effect the vicinity of the slide. A subsequent sea-floor collapse and the rise of fluids perhaps, decreased the shear strength of the sedimentary overburden leading to more regional slope failure.

Acknowledgements

We thank the captains of R/V Jan Mayen and their crew for their help during data acquisition. Special thanks go to Steinar Iversen for skillfully handling and maintaining the seismic equipment. The manuscript improved through constructive input from the editor Johan Petter Nystuen and the reviewers Deborah Hutchinson, Dendy Sloan, Jr. and Martin Hovland. GMT (Wessel and Smith, 1991) was extensively used for figure preparation. This work is a contribution to the COSTA Project funded under FP5 of the European Commission EVK-CT-1999–00006 and the DFG Project MI306/10-1. Additional funding was provided by the Ormen Lange Licensing Group, contract NHT-B44-VK0768-00. Norsk Hydro ASA kindly provided seismic data presented in Fig. 2a through the COSTA-Seabed Project co-operation. Furthermore, we acknowledge support by the Landmark Graphics University Programme.

References

Andreassen, K., Berteussen, K.A., Sognnes, H., Henneberg, K., Langhammer, J. and Mienert, J., 2003. Multi-component ocean bottom cable data in gas-hydrate investigations. J. Geophys. Res., 108: 2399.

Bouriak, S., Vanneste, M. and Saoutkine, A., 2000. Inferred gas hydrates and clay diapirs near the Storegga Slide on the southern edge of the Vøring Plateau, offshore Norway. Mar. Geol., 163: 125–148.

Bryn, P., Solheim, A., Berg, K., Lien, R., Forsberg, C.F., Haflidason, H., Ottesen, D. and Rise, L., 2003. The Storegga Slide complex: Repeated large scale sliding in response to climatic cyclicity. In: J. Locat and J. Mienert (Editors), Submarine Mass Movements and their Consequences. Kluwer Academic Publishers, Kluwer, pp. 215–222.

Bugge, T., Befring, S., Belderson, R.H., Eidvin, T., Jansen, E., Kenyon, N.H., Holtedahl, H. and Sejrup, H.P., 1987. A giant three-stage submarine slide off Norway: Geo-Mar., Lett., 7: 191–198.

Bünz, S., Mienert, J. and Berndt, C., 2003. Geological controls on the Storegga gas-hydrate system of the mid-Norwegian continental margin. Earth Planet. Sci. Lett., 209: 291–307.

Buntebarth, G.,1984. Geothermics; an introduction, Springer Verlag, Berlin.

Carslaw, H.S. and Jaeger, J.C., 1959. Conduction of Heat in Solids, 2nd ed., Clarendon Press, Oxford, UK.

Dahlgren, K.I.T., Vorren, T.O. and Laberg, J.S., 2002. Late Quaternary glacial development of the mid-Norwegian margin — 65 to 68°N. Mar. Petrol. Geol., 19: 1089–1113.

Dalland, A., Worsley, D. and Ofstad, K., 1988: A lithostratigraphic scheme for the Mesozoic and Cenozoic succession offshore mid- and northern Norway. NPD Bull., 4: 65 pp.

Dillon, W.P., Danforth, W.W., Hutchinson, D.R., Drury, R.M., Taylor, M.H. and Booth, J.S., 1998. Evidence for faulting related to dissociation of gas hydrate and release of methane off the southeastern United States. In: J.-P. Henriet and J. Mienert (Editors), Gas Hydrates: Relevance to World Margin Stability and Climate Change, Geol. Soc., London, Spec. Publ., 137: 275–291.

Flemings, P.B., Liu, X. and Winters, W.J., 2003. Critical pressure and multiphase flow in Blake Ridge gas hydrates. Geology, 31: 1057–1060.

Haflidason, H., Sejrup, H.P., Bryn, P. and Lien, R., 2001. The Storegga Slide; chronology and flow mechanism, paper presented at the XI European Union of Geoscientists meeting, 8–12th Apr., Strasbourg, France.

Henriksen, T. and Vorren, T.O., 1996. Late Cenozoic sedimentation and uplift history on the mid-Norwegian continental shelf. Global Planet. Change, 12: 171–199.

Hornbach, M.J., Saffer, D.M. and Holbrook, W.S., 2004. Critically pressured free-gas reservoirs below gas-hydrate provinces. Nature, 427: 142–144.

Kennett, J.P., Cannariato, K.G., Hendy, I.L. and Behl, R.J., 2003. Methane Hydrates in Quaternary Climate Change: The Clathrate Gun Hypothesis. Am. Geophys. U., Spec. Publ. 54.

Kvenvolden, K.A., 1993. Gas hydrates as a potential energy resource; a review of their methane content. In: D.G. Howell, K. Wiese, M. Fanelli, L.L. Zink and F. Cole (Editors), The future of energy gases, U. S. Geological Survey Reston, VA, Professional Paper, pp. 555–561.

Laberg, J.S., Dahlgren, K.I.T., Vorren, T.O., Haflidason, H. and Bryn, P., 2001. Seismic analyses of Cenozoic contourite drift development in the Northern North Sea. Mar. Geophys. Res., 22: 401–416.

Mienert, J., Posewang, J. and Baumann, M., 1998. Gas hydrates along the northeastern Atlantic margin: possible hydrate-bound margin instabilities and possible release of methane. In: J.-P. Henriet and J. Mienert (Editors), Gas Hydrates: Relevance to World Margin Stability and Climate Change, Geol. Soc., London, Spec. Publ. 137: 275–291.

Mienert, J., Posewang, J., and Lukas, D., 2001. Changes in the hydrate stability zone on the Norwegian Margin and their consequence for methane and carbon releases into the ocean. In: P. Schäfer, W. Ritzrau, M. Schlüter and J. Thiede (Editors), The Northern North Atlantic: A Environment, Springer, Berlin, pp. 259–280.

Mienert, J., Berndt, C., Laberg, J.S. and Vorren, T.O., 2002. Submarine Landslides on Continental Margins. In: G. Wefer, D.S.M. Billet, D. Hebbeln, B.B. Jørgensen, M. Schlüter and T. van Veering (Editors), Ocean Margin Systems. Springer, Berlin, pp. 179–193.

Milkov, A.V., 2000. Worldwide distribution of submarine mud volcanoes and associated gas hydrates. Mar. Geol. 167: 29–42.

Pecher, I., Minshull, T.A., Singh, S.C. and von Huene, R., 1996. Velocity structure of a Bottom Simulating Reflector offshore Peru. Earth Planet. Sci. Lett., 139: 459–469.

Raynaud, D., Chappellaz, J. and Blünier, T., 1998. Ice-core data of atmospheric methane changes: relevance to climatic changes and possible gas hydrate sources. In: J.-P. Henriet and J. Mienert (Editors), Gas Hydrates: Relevance to World Margin Stability and Climate Change, Geol. Soc., London, Spec. Publ., 137: 327–331.

Rokoengen, K., Rise, L., Bryn, P., Frengstad, B., Gustavsen, B., Nygaard, E. and Sættem, J., 1995. Upper Cenozoic stratigraphy on the Mid-Norwegian continental shelf. Norsk Geol. Tidsskr., 75: 88–104.

Saunders, A.D., Fitton, J.G., Kerr, A.C., Norry, M.J. and Kent, R.W., 1997. The North Atlantic Igneous Province. In: J.J. Mahoney and M.F. Coffin (Editors), Large Igneous Provinces: Continental, Oceanic, and Planetary Flood Basalt Volcanism. Am. Geophys. U., Geophys. Monograph., pp. 45–93.

Sundvor, E., Eldholm, O., Gladczenko, T. and Planke, S., 2000. Norwegian-Greenland Sea thermal field. In: A. Nøttvedt (Editor), Dynamics of the Norwegian Margin, Geol. Soc., London, Spec. Publ., 176: 397–410.

Thorpe, R.B., Pyle, J.A. and Nisbet, E.G., 1998. What does the ice-core record imply concerning the maximum climatic impact of possible gas hydrate release at Termination 1A? In: J.-P. Henriet and J. Mienert, (Editors), Gas Hydrates: Relevance to World Margin Stability and Climate Change, Geol. Soc., London, Spec. Publ., 137: 319–326.

Vanneste, M., 2000. Gas hydrate stability and destabilisation processes in lacustrine and marine environments: Results from theoretical analyses and multi-frequency seismic investigations, Ph.D. thesis, University of Ghent, Ghent, Belgium.

Vorren, T.O., Laberg, J.S., Blaume, F., Dowdeswell, J.A., Kenyon, N.H., Mienert, J., Rumohr, J. and Werner, F., 1998. The Norwegian-Greenland Sea continental margins: Morphology and late Quarternary sedimentary processes and environment. Quaternary Sci. Rev., 17: 273–302.

von Huene, R. and Pecher, I., 1999. Vertical tectonics and the origins of BSRs along the Peru margin. Earth Planet. Sci. Lett., 166: 47–55.

Wessel, P. and Smith, W.H.F., 1991. Free software helps map and display data. EOS, 72: 441, 445–446.

Reference Index

Subject Index